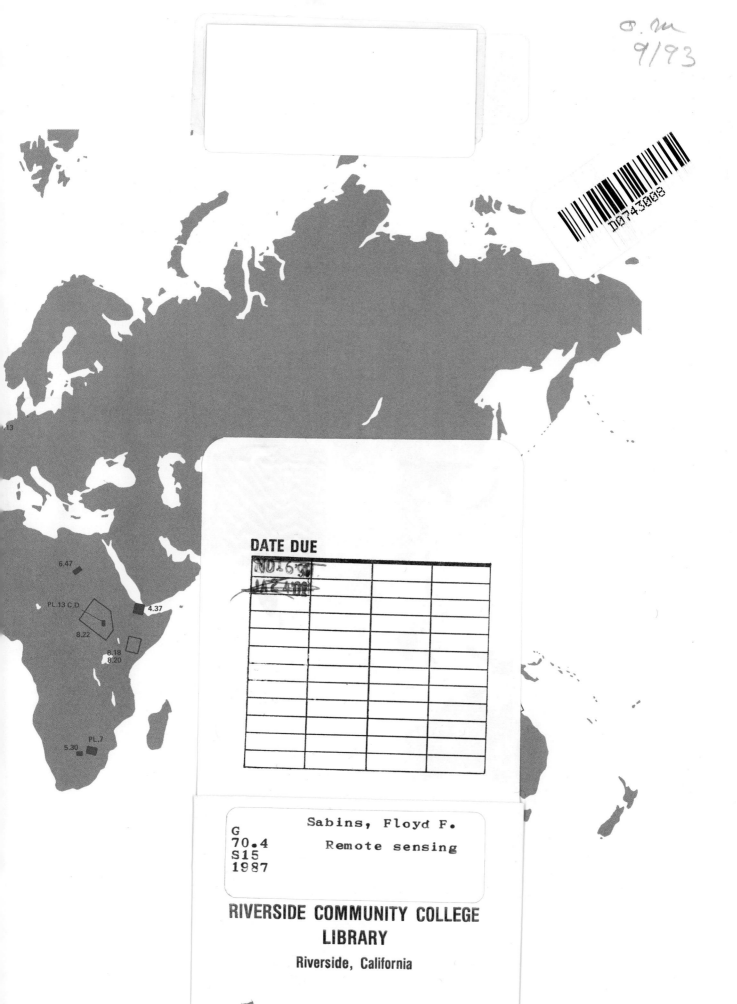

REMOTE SENSING

REMOTE SENSING

PRINCIPLES AND INTERPRETATION

Second Edition

FLOYD F. SABINS, JR.

Chevron Oil Field Research Company
and
University of California, Los Angeles

W. H. FREEMAN AND COMPANY
NEW YORK

COVER IMAGE: Salton Sea and Imperial Valley, southern
California and Mexico. This infrared color image was acquired
December 12, 1982, by the unmanned NASA/NOAA Landsat
thematic mapper from an altitude of 705 km. The image, at a scale
of 1:500,000, was digitally processed at Chevron Oil Field Research
Company. The Salton Sea, appearing in the center of the image, is
a shallow, salty lake. Irrigated agriculture causes the bright red
tones in the Coachella Valley north of the Salton Sea and in the
Imperial Valley to the south. The border between the United States
and Mexico trends east-west across the southern part of the
Imperial Valley, where it is marked by the different patterns of land
ownership in the two countries. The city of Mexicali is located on
the Mexican side of the border, whereas El Centro is the largest
city in the U.S. portion of the Imperial Valley. The Imperial Valley
is bordered by desert terrain with white tones caused by windblown
sand. The mountain ranges consist largely of igneous and
metamorphic bedrock that has eroded to form extensive alluvial
fans of sand and gravel.

Library of Congress Cataloging in Publication Data

Sabins, Floyd F.
  Remote sensing.

  Bibliography: p.
  Includes index.
  1. Remote sensing.   I. Title.
G70.4.S15   1986   621.36′78   86-2144
ISBN 0-7167-1793-X

Printed in the United States of America

  2 3 4 5 6 7 8 9 0   KP   5 4 3 2 1 0 8 9 8 7

To Janice, Barbara, and Edward

Contents

Preface

When the first edition of this book went to press in late 1977, the basic framework of the science of remote sensing had been established: images were acquired by aircraft in all regions of the electromagnetic spectrum; the Landsat Multispectral Scanner System acquired satellite images of moderate spatial resolution in the visible and reflected IR spectral regions; and digital processing of image data was available. In the nine years since 1977 many new airborne and spaceborne systems have been deployed:

Return Beam Vidicon System (Landsat 3)

Thematic Mapper (Landsat 4 and 5)

Shuttle Imaging Radar (Space Shuttle)

Large Format Camera (Space Shuttle)

Seasat Radar

Heat Capacity Mapping Mission

Advanced Very High Resolution Radiometer

Coastal Zone Color Scanner

Airborne Imaging Spectrometer

Thermal Infrared Multispectral Scanner

These new systems provide images with improved spectral and spatial resolution over the 1977-vintage systems. The deployment of thermal IR, radar, and high-resolution cameras on satellites has increased coverage to encompass the entire world. The use of digital image processing systems is becoming commonplace, and many new programs are available. A major goal of this second edition is to communicate these developments to the reader.

I have incorporated all these new advances into the original, successful organization used in the first edition. Yet the revised text remains sufficiently concise to be covered in a single semester or quarter at the upper-division or graduate level. The book should also be useful for short courses and as a reference for workers in the remote sensing field. As was true of the first edition, no previous training in remote sensing is required. Courses in introductory physics, physical geography, and physical geology would provide useful, but not essential, background for users of this book.

Another goal of this revision has been to broaden the scope from the admitted geologic bias of the first edition. To gain insight into nongeologic aspects, I taught a remote sensing course in the Geography Department at UCLA. As a result of this and other experience, the second edition is more versatile than its predecessor. A chapter on land use and land cover analysis has been added, and examples incorporated into all chapters. The chapter on environmental applications was expanded. To provide background information for nongeological readers, the text now contains an illustrated appendix "Geology for Remote Sensing."

OTHER FEATURES OF THE SECOND EDITION

1. The text contains well over 600 images and diagrams, many of which are new. Most of the images are accompanied by an interpretation map at a matching scale. Index maps inside the front and back covers show locations of the images.

2. The number of color images has been expanded from 12 to 42. All but four of these are new to this edition.

3. A set of questions has been added to each chapter.

4. A glossary of technical terms appears at the end of the text.

Both the first and second editions were strongly influenced by my continued teaching of a remote sensing course in the Earth and Space Sciences Department of UCLA and short courses for the American Association of Petroleum Geologists and the Geological Society of America. I am convinced that remote sensing cannot be taught as a lecture-only topic. Students should have "hands-on" training at interpreting actual images. To provide materials for this training I have prepared a "Remote Sensing Laboratory Manual" that is available from:

Remote Sensing Enterprises, Inc.
P.O. Box 2893
La Habra, CA 90631

Sets of 35-mm slides to accompany each chapter of the second edition are also availabe from the above address.

OUTLINE OF THE TEXT

The first chapter of the book summarizes the fundamental characteristics of electromagnetic radiation and the interactions of radiation with matter that are the basis of remote sensing. The vital concepts of spatial resolution and detection are explained using the eye as an example of a remote sensing system. Each of the next five chapters describes one of the following remote sensing systems: aerial photographs, manned satellite images, Landsat, thermal infrared images, and radar images. For each system the following topics are covered:

1. Physical properties and electromagnetic interactions of the materials that control the imaging process

2. Design and operation of the imaging system

3. Characteristics of the images, including defects and geometric distortion that may distract or confuse the interpreter

4. Guidelines and examples for interpreting images

There is a growing trend toward quantitative interpretation of images and toward the use of mathematical models to understand the interaction between electromagnetic radiation and materials. The text describes and illustrates these techniques and gives practical examples.

A chapter on digital image processing describes computer techniques for restoring and enhancing images and for extracting information. This rapidly expanding technology should be included in any remote sensing curriculum. The remaining chapters describe practical applications of remote sensing to the following fields: resource exploration, environmental applications, land use and cover analysis, and natural hazards. Chapter 12 compares a variety of images covering two test sites and demonstrates the advantages and disadvantages of various types of image for different applications.

ACKNOWLEDGMENTS

A number of images were provided by colleagues in government, universities, and industry, and I am grateful for this help. Sources of images and information are acknowledged in the appropriate figure captions and text.

The manuscript for all chapters was reviewed by Professor Ronald J. Wasowski of the University of Notre Dame and by Marcus Borengasser of Mackay School of Mines, University of Nevada. Chapters 7 and 8 were also reviewed by William S. Kowalik of Chevron Oil Field Research. Most of the digitally processed images were prepared at Chevron Oil Field Research Company by my co-workers William S. Kowalik and Todd F. Battey. Susan Middleton copy edited the text. Jerry Lyons and Susan Moran of W. H. Freeman and Company guided the book through the production process.

I am grateful for the support provided by my employer for the past 30 years, the Chevron Oil Field Research Company. My fellow employees in the word processing, drafting, photographic, and reproduction departments contributed greatly to preparing the manuscript and illustrations.

Floyd F. Sabins, Jr.
La Habra, California
March 1986

REMOTE SENSING

CHAPTER

1

Fundamental Considerations

Remote sensing is broadly defined as collecting and interpreting information about a target without being in physical contact with the object. Aircraft and satellites are the common platforms for remote sensing observations. The term *remote sensing* is commonly restricted to methods that employ electromagnetic energy (such as light, heat, and radio waves) as the means of detecting and measuring target characteristics. This definition of remote sensing excludes electrical, magnetic, and gravity surveys that measure force fields rather than electromagnetic radiation. Magnetic and radioactivity surveys are frequently made from aircraft but are considered airborne geophysical surveys rather than remote sensing.

Aerial photography is the original form of remote sensing and remains the most widely used method. Aerial photography analyses have played major roles in the discovery of many oil and mineral deposits around the world. These successes, using the visible portion of the electromagnetic spectrum, suggested that it might be possible to obtain comparable results by using other wavelength regions. In the 1960s, technologic developments enabled the acquisition of images at other wavelengths, including thermal infrared (IR) and microwave. (A later section, "Electromagnetic Spectrum," describes the wavelength bands in greater detail.) The development and deployment of manned and unmanned earth satellites began in the 1960s and provided an orbital vantage point for acquiring images of the earth. Descriptions of all these methods and interpretations of the images form the subject of this book. For a review of the history of remote sensing, see Fischer and others (1975).

UNITS OF MEASURE

This text employs the *International System of units* (SI), a modernized metric system adopted in 1960 that uses the following standard units and abbreviations:

meter	m
second	sec
kilogram	kg
gram	g
radian	rad
milliradian	mrad
hertz	Hz
watt	W

(The SI abbreviation for second is actually *s*, but this text uses *sec* to avoid confusion with other abbreviations.)

TABLE 1.1 Metric nomenclature for distance

Unit	Symbol	Equivalent	Comment
Kilometer	km	$1000 \text{ m} = 10^3 \text{ m}$	
Meter	m	$1.0 \text{ m} = 10^0 \text{ m}$	Basic unit
Centimeter	cm	$0.01 \text{ m} = 10^{-2} \text{ m}$	
Millimeter	mm	$0.001 \text{ m} = 10^{-3} \text{ m}$	
Micrometer	μm	$0.000001 \text{ m} = 10^{-6} \text{ m}$	Formerly called *micron* (μ)
Nanometer	nm	10^{-9} m	

Distance is expressed in the multiples and fractions of meters shown in Table 1.1. Where appropriate for clarity, English units for distance will be used, with metric equivalents shown in parentheses.

Frequency (ν) is the number of wave crests passing a given point in a specified period of time. Frequency was formerly expressed as "cycles per second," but today we use *hertz,* the unit for a frequency of one cycle per second. The terms for designating frequencies are shown in Table 1.2.

Temperature is given in degrees Celsius (°C) or in degrees Kelvin (°K), which is also known as the absolute temperature scale. A temperature of -273°C is equivalent to 0°K. (The formal SI system omits the degree symbol for Kelvin temperatures, but the letter K alone may be confused with other constants that employ this letter; therefore "°K" is used in this text.) A few temperatures commonly given in degrees Fahrenheit (°F) will remain in that scale where conversion to degrees Celsius would be inconvenient.

In fractional statements, units in the denominator are identified by a negative superscript. For example, the property called *thermal inertia* (P) for a particular rock type is expressed as

$$P = 0.53 \text{ cal} \cdot \text{cm}^{-2} \cdot \text{sec}^{-1/2} \cdot °\text{C}^{-1}$$

This expression means that for the particular rock type, thermal inertia (P) equals 0.53 calories per square centimeter per second to the square root per degree Celsius.

TABLE 1.2 Terms used to designate frequencies

Unit	Symbol	Frequency, cycles · sec
Hertz	Hz	1
Kilohertz	kHz	10^3
Megahertz	MHz	10^6
Gigahertz	GHz	10^9

ELECTROMAGNETIC ENERGY

Electromagnetic energy refers to all energy that moves with the velocity of light in a harmonic wave pattern. A harmonic pattern consists of waves that occur at equal intervals in time. The wave concept explains how electromagnetic energy propagates (moves), but this energy can only be detected as it interacts with matter. In this interaction, electromagnetic energy behaves as though it consists of many individual bodies called *photons* that have such particlelike properties as energy and momentum. When light bends (refracts) as it propagates through media of different optical densities, it is behaving like waves. When a light meter measures the intensity of light, however, the interaction of photons with the light-sensitive photodetector produces an electric signal that varies in strength proportional to the number of photons. Suits (1983) describes the characteristics of electromagnetic energy that are significant for remote sensing.

Properties of Electromagnetic Waves

Electromagnetic waves can be described in terms of their velocity, wavelength, and frequency. All electromagnetic waves travel at the same speed (c). This velocity is commonly referred to as the *speed of light,* which is one form of electromagnetic energy. For electromagnetic waves moving through a vacuum, $c = 299{,}793$ km · sec^{-1} or, for practical purposes, $c = 3 \times 10^8$ m · sec^{-1}.

The *wavelength* (λ) of electromagnetic waves is the distance from any point on one cycle or wave to the same position on the next cycle or wave. The *micrometer* (μm) is a convenient unit for designating wavelength of both visible and IR radiation. Optical scientists commonly employ *nanometers* (nm) for measurements of visible light to avoid using decimal numbers.

Unlike velocity and wavelength, which change as electromagnetic energy is propagated through media of different densities, frequency remains constant and is therefore a more fundamental property. Electronic engineers use frequency nomenclature for designating radio and radar energy regions. This book will use wavelength rather than frequency to simplify comparisons between all portions of the electromagnetic spectrum. Velocity (c), wavelength (λ), and frequency (ν) are related by

$$c = \lambda\nu \qquad (1.1)$$

Interaction Mechanisms

Electromagnetic energy that encounters matter, whether solid, liquid, or gas, is called *incident* radiation. Interactions with matter can change the following properties of the incident radiation: intensity, direction, wavelength, polarization, and phase. The science of remote sensing detects and records these changes. Scientists then interpret the resulting images and data to identify the characteristics of the matter that produced those changes.

During interactions between electromagnetic radiation and matter, mass and energy are conserved according to basic physical principles. Figure 1.1 illustrates the five common results of those interactions. The incident radiation may be

1. transmitted, that is, passed through the substance. Transmission through media of different densities, such as from air into water, causes a change in the velocity of electromagnetic radiation. The ratio of the two velocities is called the *index of refraction* (n) and is expressed as

$$n = \frac{c_a}{c_s} \qquad (1.2)$$

where c_a is the velocity in a vacuum and c_s is the velocity in the substance.

2. absorbed, giving up its energy largely to heating the matter.

3. emitted by the substance, usually at longer wavelengths, as a function of its structure and temperature.

4. scattered, that is, deflected in all directions. Surfaces with dimensions of *relief,* or roughness, comparable to the wavelength of the incident energy produce scattering. Light waves are scattered by molecules and particles in the atmosphere that have sizes similar to the dimensions of wavelengths of light.

5. reflected, that is, returned from the surface of a material with the angle of reflection equal and opposite to the angle of incidence. Reflection is caused by

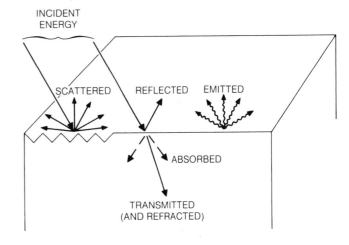

FIGURE 1.1 Interaction mechanisms between electromagnetic energy and matter.

surfaces that are smooth relative to the wavelength of incident energy. *Polarization*, or direction of vibration, of the reflected waves may differ from that of the incident wave.

Emission, scattering, and reflection are called *surface phenomena* because these interactions are determined primarily by properties of the surface, such as color and roughness. Transmission and absorption are called *volume phenomena* because these interactions are determined by the internal characteristics of matter, such as density and conductivity. The particular combination of surface and volume interactions with any particular material depend both on the wavelength of the electromagnetic radiation and the specific properties of that material. These interactions between matter and energy are recorded on remote sensing images, from which one may interpret the characteristics of matter.

ELECTROMAGNETIC SPECTRUM

The *electromagnetic spectrum* is the continuum of energy that ranges from meters to nanometers in wavelength, travels at the speed of light, and propagates through a vacuum such as outer space. All matter radiates a range of electromagnetic energy, with the peak intensity shifting toward progressively shorter wavelengths with increasing temperature of the matter.

Wavelength Regions

Figure 1.2 shows the electromagnetic spectrum, which is divided on the basis of wavelength into regions that are described in Table 1.3. The electromagnetic spec-

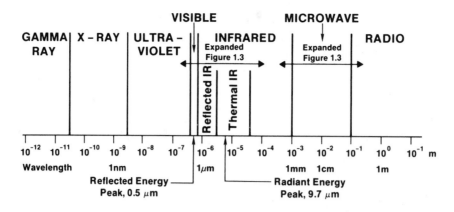

FIGURE 1.2 Electromagnetic spectrum. Expanded versions of the visible, infrared regions, and microwave regions are shown in Figure 1.3.

TABLE 1.3 Electromagnetic spectral regions

Region	Wavelength	Remarks
Gamma ray	< 0.03 nm	Incoming radiation is completely absorbed by the upper atmosphere and is not available for remote sensing.
X-ray	0.03 to 3.0 nm	Completely absorbed by atmosphere. Not employed in remote sensing.
Ultraviolet	0.03 to 0.4 μm	Incoming wavelengths less than 0.3 μm are completely absorbed by ozone in the upper atmosphere.
Photographic UV band	0.3 to 0.4 μm	Transmitted through atmosphere. Detectable with film and photodetectors, but atmospheric scattering is severe.
Visible	0.4 to 0.7 μm	Imaged with film and photodetectors. Includes reflected energy peak of earth at 0.5 μm.
Infrared	0.7 to 100 μm	Interaction with matter varies with wavelength. Atmospheric transmission windows are separated by absorption bands.
Reflected IR band	0.7 to 3.0 μm	Reflected solar radiation that contains no information about thermal properties of materials. The band from 0.7 to 0.9 μm is detectable with film and is called the *photographic IR band*.
Thermal IR band	3 to 5 μm, 8 to 14 μm	Principal atmospheric windows in the thermal region. Images at these wavelengths are acquired by optical-mechanical scanners and special vidicon systems but not by film.
Microwave	0.1 to 30 cm	Longer wavelengths can penetrate clouds, fog, and rain. Images may be acquired in the active or passive mode.
Radar	0.1 to 30 cm	Active form of microwave remote sensing. Radar images are acquired at various wavelength bands.
Radio	> 30 cm	Longest wavelength portion of electromagnetic spectrum. Some classified radars with very long wavelength operate in this region.

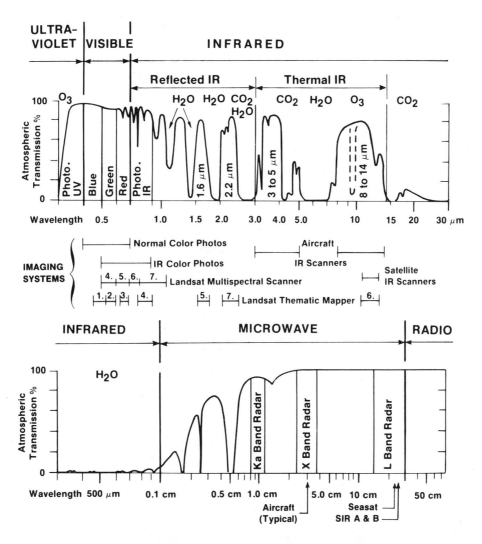

FIGURE 1.3 Expanded diagrams of the visible and infrared regions (upper) and the microwave region (lower) showing atmospheric windows. Wavelength bands of commonly used remote sensing systems are indicated. Gases responsible for atmospheric absorption are shown.

trum ranges from the very short wavelengths of the gamma-ray region (measured in fractions of nanometers) to the long wavelengths of the radio region (measured in meters). The horizontal scale in Figure 1.2 is logarithmic in order to portray the shorter wavelengths. Notice that the visible region (0.4-to-0.7-μm wavelengths) occupies only a small portion of the spectrum. Energy reflected from the earth during daytime may be recorded as a function of wavelength. The maximum amount of energy is reflected at 0.5 μm wavelength, which corresponds to the green band of the visible region, and is called the *reflected energy peak* (Figure 1.2). The earth also radiates energy both day and night, with the maximum energy radiating at 9.7 μm wavelength. This *ra-diant energy peak* occurs in the thermal band of the IR region (Figure 1.2).

The earth's atmosphere absorbs energy in the gamma-ray, X-ray, and most of the ultraviolet (UV) region; therefore, these regions are not used for remote sensing. Remote sensing records energy in the microwave, infrared, and visible regions, as well as the long-wavelength portion of the UV region. Details of these regions are shown in Figure 1.3. The horizontal axes again show wavelength on a logarithmic scale; the vertical axes show percent atmospheric transmission of electromagnetic energy. Wavelength regions with high transmission are called *atmospheric windows* and are used to acquire remote sensing images. The major remote sensing regions (visible, infrared, and microwave) are further subdivided into *bands,* such as the blue, green, and red bands of the visible region. The upper curve of Figure 1.3 spans the UV through thermal IR regions. Bars below the curve show wavelength bands recorded by major imaging systems such as photography and scanners. For the Landsat systems the numbers identify specific bands recorded by these systems. Characteristics of the remote sensing regions are summarized in Table 1.3.

Passive remote sensing systems record the energy that naturally radiates or reflects from an object. An *active* system supplies its own source of energy, directing it at the object in order to measure the returned energy. Flash photography is an example of active remote sensing, in contrast to available-light photography, which is passive. The other common form of active remote sensing is radar (Table 1.3), which provides its own source of electromagnetic energy at microwave wavelengths.

Atmospheric Effects

Our eyes inform us that the atmosphere is essentially transparent to light, and we tend to assume that this condition exists for all electromagnetic energy. In fact, the gases of the atmosphere absorb electromagnetic energy at specific wavelength intervals called *absorption bands.* Figure 1.3 shows these absorption bands, together with the gases responsible for the atmospheric absorption.

Wavelengths shorter than 0.3 μm are completely absorbed by the ozone (O_3) layer in the upper atmosphere (Figure 1.3). This absorption is essential to allow life on earth, because prolonged exposure to the intense energy of these wavelengths destroys living tissue. For example, sunburn occurs more readily at high mountain elevations than at sea level. Sunburn is caused by UV energy, much of which is absorbed by the atmosphere at sea level. At higher elevations however, there is less atmosphere to absorb the UV energy.

Clouds consist of aerosol-sized particles of liquid water that absorb and scatter electromagnetic radiation at wavelengths less than about 0.3 cm. Only radiation of microwave and longer wavelengths is capable of penetrating clouds without being scattered, reflected, or absorbed.

IMAGE CHARACTERISTICS

In general usage, an *image* is any pictorial representation, irrespective of the wavelength or imaging device used to produce it. A *photograph* is an image that records wavelengths of 0.3 to 0.9 μm that have interacted with light-sensitive chemicals in photographic film. Images can be described in terms of certain fundamental properties regardless of the wavelength at which the image is recorded. These common fundamental properties are scale, brightness, contrast, and resolution. Tone and texture of images are functions of the fundamental properties.

Scale

Scale is the ratio of the distance between two points on an image to the corresponding distance on the ground. A common scale on U.S. Geological Survey topographic maps is 1:24,000, which means that one unit on the map equals 24,000 units on the ground. Thus 1 cm on the map represents 24,000 cm (240 m) on the ground, or 1 in. represents 24,000 in. (2000 ft). The maps and images of this book show scales graphically as bars. Image scale is determined by

1. the effective focal length of the remote sensing device

2. the altitude from which the image is acquired

3. the magnification factor employed in reproducing the image

The deployment of imaging systems on satellites has changed the concepts of image scale. In this book, scales of images are designated as follows:

Small scale (greater than 1:500,000)	1 cm = 5 km or more (1 in. = 8 mi or more)
Intermediate scale (1:50,000 to 1:500,000)	1 cm = 0.5 to 5 km (1 in. = 0.8 to 8 mi)
Large scale (less than 1:50,000)	1 cm = 0.5 km or less (1 in. = 0.8 mi or less)

These designations differ from the traditional scale concepts of aerial photography. Twenty years ago, 1:62,500 was the minimum scale of commercially available photographs and was considered small-scale. Today sensing systems on high-altitude aircraft and satellites can acquire photographs and images of excellent quality at much smaller scales. Optimum image scale is determined by how the images are to be interpreted. With the advent of satellite images, many investigators have been surprised at the amount and types of information that can be interpreted from very small scale images.

FIGURE 1.4 Gray scale.

Brightness and Tone

Remote sensing systems detect the intensity of electromagnetic radiation that an object reflects, emits, or scatters at particular wavelength bands. Variations in intensity of electromagnetic radiation from the terrain are commonly displayed as variations in brightness on images. On positive images, such as those in this book, the brightness of objects is directly proportional to the intensity of electromagnetic radiation that is detected from that object.

Brightness is the magnitude of the response produced in the eye by light; it is a subjective sensation that can be determined only approximately. *Luminance* is a quantitative measure of the intensity of light from a source and is measured with a device called a photometer, or light meter. People who interpret images rarely, if ever, make quantitative measurements of brightness variations on an image. Variations in brightness may be calibrated with a gray scale such as the one in Figure 1.4. Each distinguishable shade from black to white is a separate *tone*. In practice, most interpreters do not use an actual gray scale the way one would use a centimeter scale; they characterize areas on an image as light, intermediate, or dark in tone, using their own mental concept of a gray scale.

On aerial photographs the tone of an object is primarily determined by the ability of the object to reflect incident sunlight, although atmospheric effects and the spectral sensitivity of the film are also factors. On images acquired in other wavelength regions, tone is determined by other physical properties of objects. On a thermal IR image, the tone of an object is proportional to the heat radiating from the object. On a radar image the tone of an object is determined by the intensity at which the transmitted beam of radar energy is scattered back to the receiving antenna.

Contrast Ratio

Contrast ratio (*CR*) is the ratio between the brightest and darkest parts of the image and is defined as

$$CR = \frac{B_{max}}{B_{min}} \qquad (1.3)$$

where B_{max} is the maximum brightness of the scene and B_{min} is the minimum brightness. Figure 1.5 shows images of high, medium, and low contrast together with profiles of brightness variations across each image. On a brightness scale of 0 to 10, the images in Figure 1.5 have the following contrast ratios:

A. High contrast $\qquad CR = \dfrac{9}{2} = 4.5$

B. Medium contrast $\qquad CR = \dfrac{5}{2} = 2.5$

C. Low contrast $\qquad CR = \dfrac{3}{2} = 1.5$

Note that when $B_{min} = 0$, *CR* is infinity; when $B_{min} = B_{max}$, *CR* is unity. This discussion was summarized from Slater's (1983) extensive review, in which he describes other terms for contrast. In addition to describing an entire scene, contrast ratio is also used to describe the ratio between the brightness of an object on an image and the brightness of the adjacent background. Contrast ratio is a vital factor in determining the ability to resolve and detect objects.

Images with a low contrast ratio are commonly referred to as "washed out," with monotonous, nearly uniform tones of gray. Low contrast may result from the following causes:

1. The objects and background of the scene may have a nearly uniformly electromagnetic response at the particular wavelength band that the remote sensing system records. In other words, the scene has an inherently low contrast ratio.

2. Scattering of electromagnetic energy by the atmosphere can reduce the contrast of a scene. This effect is most pronounced in the shorter wavelength portions of the photographic remote sensing band as described in Chapter 2.

3. The remote sensing system may lack sufficient sensitivity to detect and record the contrast of the terrain. Incorrect recording techniques can also result

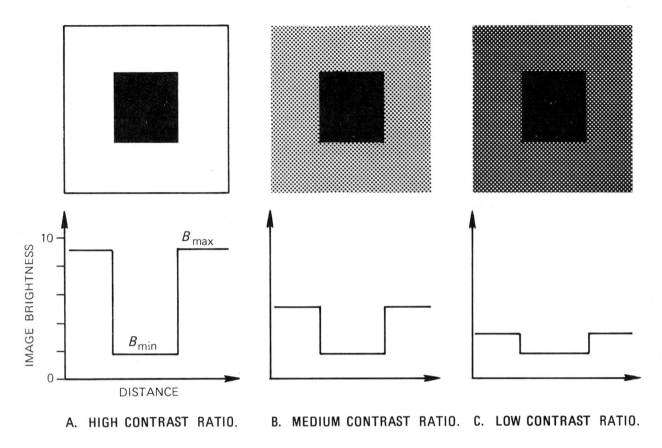

A. HIGH CONTRAST RATIO.   B. MEDIUM CONTRAST RATIO.   C. LOW CONTRAST RATIO.

FIGURE 1.5   Images of different contrast ratios with corresponding brightness profiles.

in low-contrast images even though the scene has a high contrast ratio when recorded by other means.

A low contrast ratio, regardless of the cause, can be improved by digital methods, as described in Chapter 7.

Spatial Resolution and Resolving Power

This text defines *spatial resolution* as the ability to distinguish between two closely spaced objects on an image. More specifically, it is the minimum distance between two objects at which the images of the objects appear distinct and separate. Objects spaced together more closely than the resolution limit will appear as a single object on the image. Forshaw and others (1983) discuss alternate definitions of spatial resolution.

Resolving power and spatial resolution are two closely related concepts. The term *resolving power* applies to an imaging system or a component of the system, whereas spatial resolution applies to the image produced by the system. For example, the lens and film of a camera system each have a characteristic resolving power that, together with other factors, determines the resolution of the photographs.

Spatial resolution of a photographic system is customarily determined by photographing a standard resolution target, such as the one shown in Figure 1.6A, under specified conditions of illumination and magnification. The resolution targets, or *bar charts,* consist of alternating black and white bars of equal width that are called *line-pairs.* Spacing of resolution targets is expressed in line-pairs per centimeter. For the target with 5 line-pairs $\cdot$ cm^{-1}, each black bar is 0.1 cm wide and separated by a white bar of the same width. The photograph is viewed under magnification, and the observer determines the most closely spaced set of line-pairs for which the bars and spaces are discernable. Spatial resolution of the photographic system is stated as the number of line-pairs per millimeter of the resolved target. Human judgment and visual characteristics are critical components in this analysis, which therefore is not completely objective and reproducible. Spatial resolution is different for objects of different shape, size, arrangement, and contrast ratio.

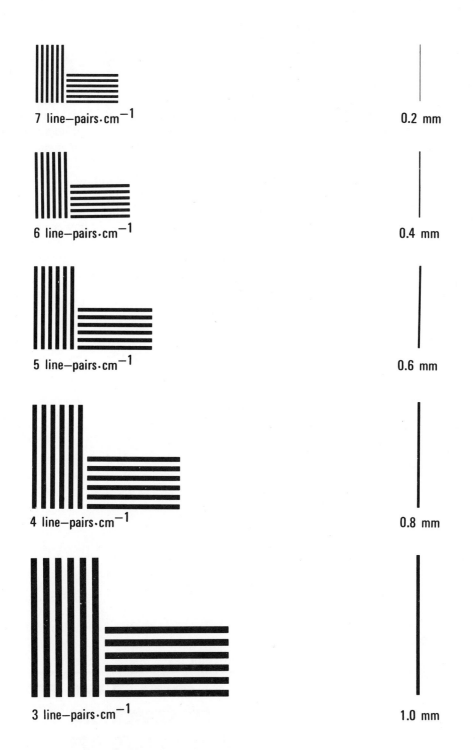

7 line–pairs·cm^{-1}

6 line–pairs·cm^{-1}

5 line–pairs·cm^{-1}

4 line–pairs·cm^{-1}

3 line–pairs·cm^{-1}

0.2 mm

0.4 mm

0.6 mm

0.8 mm

1.0 mm

A.   RESOLUTION TARGETS.   B.   DETECTION TARGETS.

FIGURE 1.6   Resolution and detection targets with high contrast ratio. View this chart from a distance of 5 m (16.5 ft). For A, determine the most closely spaced set of bars you can resolve. For B, determine the narrowest bar you can detect.

An alternate method of describing resolution is the *modulation transfer function* (MTF), which employs a bar chart with progressively closer spacing of the bars (McKinney, 1980). *Angular resolving power* is defined as the angle subtended by imaginary lines passing from the imaging system and two targets spaced at the minimum resolvable distance. Angular resolving power is commonly measured in radians. As shown in Figure 1.7, a *radian* (rad) is the angle subtended by an arc BC of a circle having a length equal to the radius AB of the circle. Because the circumference of a circle has a length equal to 2π times the radius, there are 2π, or 6.28, rad in a circle. A radian corresponds to 57.3°, or 3438 min, and a milliradian (mrad) is 10^{-3} rad. In the radian system of angular measurement,

$$\text{Angle} = \frac{L}{r} \text{ rad} \qquad (1.4)$$

where L is the length of the subtended arc and r is the radius of the circle. A convenient relationship is that at a distance r of 1000 units, 1 mrad subtends an arc L of 1 unit. Figure 1.8 illustrates the angular resolving power of a remote sensing system (the eye) that can resolve the center bar chart of Figure 1.6 at a distance of 5 m. This chart has 5 line-pairs · cm^{-1}, and the bars are separated by 1 mm. For these targets with a high contrast ratio, the angular resolving power is 0.2 mrad.

Resolving power and spatial resolution will be discussed for each remote sensing system described in this book, but you should remember the following points:

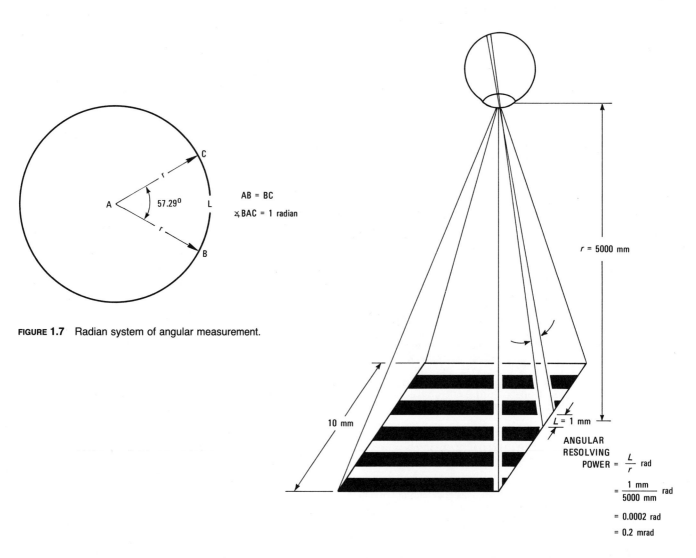

FIGURE 1.7 Radian system of angular measurement.

FIGURE 1.8 Angular resolving power (in milliradians) for a remote sensing system that can resolve 5 line-pairs · cm^{-1} at a distance of 5 m.

1. Theoretical resolving power of a system is rarely achieved in actual operation.

2. Resolution alone does not adequately determine whether an image is suitable for a particular application.

3. Resolution is the minimum separation between two objects for which the images appear distinct and separate; it is *not* the size of the smallest object that can be seen. By knowing the resolution and scale of an image, however, one can estimate the size of smallest detectable object.

Other Characteristics of Images

Detectability is the ability of an imaging system to record the presence or absence of an object, although the identity of the object may be unknown. An object may be detected even though it is smaller than the theoretical resolving power of the imaging system.

Recognizability is the ability to identify an object on an image. Objects may be detected and resolved and yet not be recognizable. For example, roads on an image appear as narrow lines that could also be railroads or canals. Unlike resolution, there are no quantitative measures for recognizability and detectability. It is important for the interpreter to understand the significance and correct use of these terms. Rosenberg (1971) summarizes the distinctions between them.

A *signature* is the expression of an object on an image that enables the object to be recognized. Characteristics of an object that control its interaction with electromagnetic energy determine its signature. For example, the spectral signature of an object is its brightness measured at a specific wavelength of energy.

Texture is the frequency of change and arrangement of tones on an image. *Fine, medium,* and *coarse* are some terms used to describe texture.

An *interpretation key* is a characteristic or combination of characteristics that enables an object to be identified on an image. Typical keys are size, shape, tone, and color. The associations of different characteristics are valuable keys. On images of cities, one may recognize single-family residential areas by the association of a dense street network, lawns, and small buildings. The associations of certain landforms and vegetation species are keys for identifying different types of rocks.

VISION

Of our five senses, two detect electromagnetic radiation. Some of the nerve endings in our skin detect thermal IR radiation as heat but do not form images. Vision is the most important sense and accounts for most of the information input to our brain. Vision is not only an important remote sensing system in its own right, but it is also the means by which we interpret the images produced by other remote sensing systems. The following section analyzes the human eye as a remote sensing system. Much of the information is summarized from Gregory (1966).

Structure of the Eye

For such a complex structure, the human eye in cross section, shown on Figure 1.9, appears deceptively simple. Light enters through the clear *cornea*, which is separated from the lens by fluid called the *aqueous humor*. The *iris* is the pigmented part of the eye that controls the variable aperture called the *pupil*. It is commonly thought that variations in pupil size allow the eye to function over a wide range of light intensities. However, the pupil varies in area over a ratio of only 16:1 (maximum area 16 times the minimum area), whereas the eye functions over a brightness range of about 100,000:1. The pupil contracts to limit the light rays to the central and optically best part of the lens, except when the full opening is needed in dim light. The pupil also contracts for near vision, increasing the depth of field for near objects.

A common misconception is that the lens refracts (bends) the incoming rays of light to form the image. The amount that light bends when passing through two adjacent media is determined by the difference in the refractive indices (n) of the two media; the greater the difference, the greater the bending. For the eye the maximum difference is between air ($n = 1.0$) and the cornea ($n = 1.3$), and this interface is where the maximum light refraction occurs. Although the lens is relatively unimportant for forming the image, it is important in *accommodating,* or focusing, for near and far vision. In

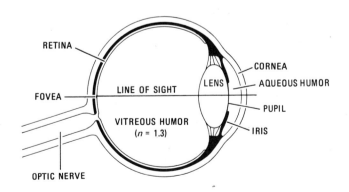

FIGURE 1.9 Structure of the human eye.

cameras this accommodation is done by changing the position of the lens relative to the film. In the human eye, the shape rather than the position of the lens is changed by muscles that vary the tension on the lens. For near vision the muscles release tension, allowing the lens to become thicker in the center and assume a more convex cross section. With age the cells of the lens harden and the lens becomes too rigid to accommodate for different distances; as a result, bifocal glasses may become necessary to provide for near and far vision.

An inverted image is focused on the *retina,* a thin sheet of interconnected nerve cells that includes the light receptor cells, called rods and cones, that convert light into electrical impulses. Rods and cones receive their names from their longitudinal shapes when viewed microscopically. The cones function in daylight conditions to give color, or *photopic,* vision. The rods function under low illumination and only give vision in tones of gray, called *scotopic* vision. Rods and cones are not uniformly distributed throughout the retinal surface. The maximum concentration and organization of receptor cells is in the *fovea* (Figure 1.9), a small region at the center of the retina that provides maximum visual acuity. You can demonstrate the existence and importance of the fovea by concentrating on a single letter on this page. The rest of the page and even the nearby words and letters appear indistinct because they are outside the field of view of the fovea. The eye is in continual motion to bring the fovea to bear on all parts of the page or scene. Close to the fovea is the blind spot, where the optic nerve joins the eye and there are no receptor cells. The electrical impulses from the receptor cells are transmitted to the brain, which interprets them as the visual perception.

Resolving Power of the Eye

The diameter of the largest receptor cells in the fovea determines the resolving power of the eye. Multiplying this maximum diameter (3 μm) by the refractive index of the vitreous humor ($n = 1.3$) determines an effective diameter (4 μm) for the receptor cells. The *image distance,* or distance from the retina to the lens, is about 20 mm, or 20,000 μm. The effective width of the receptors is 4/20,000, or 1/5000, of the image distance. Image distance is proportional to *object distance,* which is the distance from the eye to the object. An object forms an image that fills the width of a receptor if the object width is 1/5000 the object distance. Therefore adjacent objects must be separated by 1/5000 the object distance for their images to fall on alternate receptors and be resolved by the eye.

You may demonstrate the resolving power of the eye by viewing the resolution targets of Figure 1.6A at a distance of 5 m (16.5 ft) and determining the most closely spaced set of line-pairs that can be resolved. Also determine the narrowest of the bars on Figure 1.6B that you can detect. Make these determinations now, before reading further, because the following text may influence the your perception of the targets.

For the high-contrast resolution targets of Figure 1.6A at a distance of 5 m, the normal eye should be able to resolve the middle set that has 5 line-pairs $\cdot$ cm^{-1}. The black and white bars are 1 mm wide. The *instantaneous field of view* (IFOV) of any detector is the solid angle through which a detector is sensitive to radiation. Equation 1.4 is used to calculate the IFOV of the eye, where the radius (r) is 5000 mm and the length of the subtended arc (L) is 1 mm:

$$\text{IFOV} = \frac{L}{r} \text{ rad}$$

$$= \frac{1 \text{ mm}}{5000 \text{ mm}} \text{ rad}$$

$$= 0.2 \times 10^{-3} \text{ rad}$$

$$= 0.2 \text{ mrad}$$

Figure 1.8 shows the relationships of the resolution targets to the IFOV of the eye. The 0.2-mrad IFOV of the eye means that at a distance of 1000 units, the eye can resolve high-contrast targets that are spaced no closer than 0.2 units.

Detection Capability of the Eye

When the detection targets of Figure 1.6B are viewed from a distance of 5 m, most readers can detect the narrowest bar, which is 0.2 mm wide. Recall, however, that at this distance the minimum separation at which bar targets can be resolved is 1.0 mm. This test illustrates the difference between resolution and detection. Detection is influenced not only by the size of objects but also by their shape and orientation. For example, if dots are used in place of lines in Figure 1.6B, the diameter of the smallest detectable dot would be considerably larger than 0.2 mm.

Effect of Contrast Ratio on Resolution and Detection

The resolution and detection targets in Figure 1.10 have the same spacing as those in Figure 1.6, but the contrast ratio has been reduced by the addition of a gray background. To evaluate the effect of the lower contrast ratio, view Figure 1.10 from a distance of 5 m and determine which targets can be resolved and detected.

7 line–pairs·cm^{-1}

0.2 mm

6 line–pairs·cm^{-1}

0.4 mm

5 line–pairs·cm^{-1}

0.6 mm

4 line–pairs·cm^{-1}

0.8 mm

3 line–pairs·cm^{-1}

1.0 mm

A.   RESOLUTION TARGETS.   B.   DETECTION TARGETS.

FIGURE 1.10   Resolution and detection targets with low contrast ratio. View this chart from a distance of 5 m (16.5 ft). For A, determine the most closely spaced set of bars you can resolve. For B, determine the narrowest bar you can detect. Compare these values with those determined from Figure 1.6.

Using this figure, most readers only resolve 3 line-pairs · cm^{-1} and the smallest detectable target is the line 0.6 mm wide. These dimensions are larger than the 5 line-pairs · cm^{-1} and the 0.2-mm line of the high-contrast target and demonstrate the effect of a lower contrast ratio on resolution and detection.

REMOTE SENSING SYSTEMS

The eye is a familiar example of a remote sensing system. The inorganic remote sensing systems described in this text belong to two major categories: framing systems and scanning systems.

Framing Systems

Framing systems instantaneously acquire an image of an area, or frame, on the terrain. Cameras and vidicons are common examples of such systems (Figure 1.11). The human eye can also be considered a framing system. A *camera* employs a lens to form an image of the scene at the *focal plane,* which is the plane at which the image

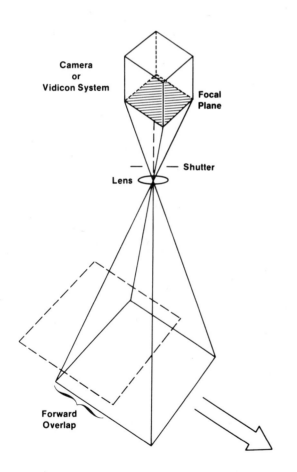

FIGURE 1.11 Framing system for acquiring remote sensing images.

is sharply defined. A shutter opens at selected intervals to allow light to enter the camera, where the image is recorded on photographic film. A *vidicon* is a type of television camera that records the image on a photosensitive electronically charged surface. An electron beam then sweeps the surface to detect the pattern of charge differences that constitutes the image. The electron beam produces a signal that may be transmitted and recorded on magnetic tape for eventual display on film.

Successive frames of camera and vidicon images may be acquired with *forward overlap,* as shown in Figure 1.11. The overlapping portion may be viewed with a stereoscope to produce a three-dimensional view, as described in Chapter 2. Film is sensitive only to portions of the UV, visible, and reflected IR regions (0.3 to 0.9 μm). The sensitivity range of special vidicons extends into the thermal band of the IR region. A framing system can instantaneously image a large area because the system has a dense array of detectors located at the focal plane. The eye has a network of rods and cones. The emulsion of camera film contains tiny grains of silver halide. A vidicon surface is coated with sensitive phosphors.

Scanning Systems

A *scanning system* employs a single detector with a narrow field of view which sweeps across the terrain to produce an image. When photons of electromagnetic energy radiated or reflected from the terrain encounter the detector, an electrical signal is produced that varies in proportion to the number of photons. The electrical signal is amplified, recorded on magnetic tape, and played back later to produce an image. All scanning systems sweep the detector's field of view across the terrain in a series of parallel scan lines. There are four common scanning modes: cross-track scanning, circular scanning, along-track scanning, and side scanning, which are diagrammed in Figure 1.12.

Cross-Track Scanning System The widely used *cross-track scanning systems* employ a faceted mirror that is rotated by an electric motor, with a horizontal axis of rotation aligned parallel with the flight direction (Figure 1.12A). The mirror sweeps across the terrain in a pattern of parallel scan lines oriented *normal* (perpendicularly) to the flight direction. Energy radiated or reflected from the ground is focused onto the detector by secondary mirrors (not shown).

The angular resolving power of a detector, measured in milliradians, determines the IFOV of the detector. As shown in Figure 1.12A, the IFOV subtends an area on the terrain called a *ground resolution cell*. Dimensions of a ground resolution cell are determined by the de-

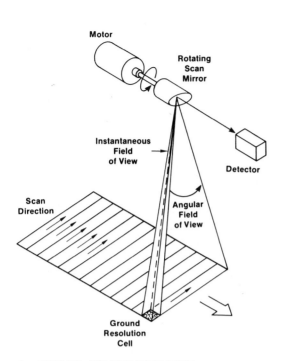

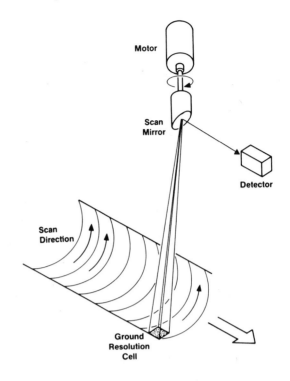

A. CROSS–TRACK SCANNER.

B. CIRCULAR SCANNER.

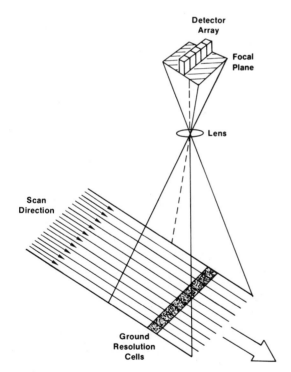

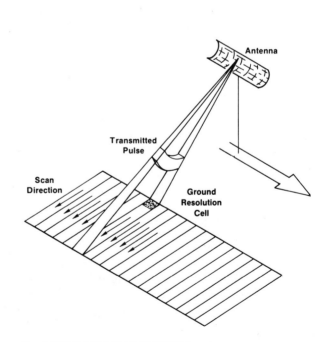

C. ALONG–TRACK SCANNER.

D. SIDE SCANNING SYSTEM.

FIGURE 1.12 Scanning systems for acquiring remote sensing images.

tector's IFOV and the altitude of the scanning system. A detector with an IFOV of 1 mrad at an altitude of 10 km has a ground resolution cell of 10 by 10 m.

The *angular field of view* (Figure 1.12A) is that portion of the mirror sweep, measured in degrees, that is recorded as a scan line. The angular field of view and the altitude of the system determine the *ground swath,* which is the width of the terrain strip represented by the image. Ground swath is calculated as

$$\text{Ground swath} = \tan\left(\frac{\text{angular field of view}}{2}\right) \times \text{altitude} \quad (1.5)$$

The distance between the scanner and terrain is greater at the margins of the ground swath than at its center. As a result, ground resolution cells are larger toward the margins than at the center, which results in a geometric distortion characteristic of cross-track scanner images. Examples of this distortion are illustrated in Chapter 5. At the high altitude of satellites, a narrow angular field of view is sufficient to cover a broad swath of terrain. For this reason, the rotating mirror is replaced by a flat mirror that oscillates back and forth through an angle of approximately 15°. An example is the multispectral scanner of Landsat described in Chapter 4.

The strength of the signal generated by a detector is a function of the following factors:

Energy flux The amount of energy reflected or radiated from terrain is the *energy flux.* For visible detectors, this flux is lower on a dark day than on a sunny day.

Altitude For a given ground resolution cell, the amount of energy reaching the detector is inversely proportional to the square of the distance. At greater altitudes the signal strength is weaker.

Spectral bandwidth of the detector The signal is stronger for detectors that respond to a broader-wavelength range of energy. For example, a detector that is sensitive to the entire visible range will receive more energy than a detector that is sensitive to a narrow band, such as visible red.

Instantaneous field of view Both the physical size of the sensitive element of the detector and the effective focal length of the scanner optics determine the IFOV. A small IFOV is required for high spatial resolution but also restricts the *signal strength* (amount of energy received by the detector).

Dwell time The time required for the detector IFOV to sweep across a ground resolution cell is the *dwell time.* A longer dwell time allows more energy to impinge on the detector, which creates a stronger signal.

For a cross-track scanner, the dwell time is determined by the detector IFOV and by the velocity at which the scan mirror sweeps the IFOV across the terrain. As shown in Figure 1.13A, a typical airborne scanner with a detector IFOV of 1 mrad, a 90° angular field of view, operating at 2×10^{-2} sec per scan line, at an altitude of 10 km, has a dwell time of 1×10^{-5} sec per ground resolution cell. It is instructive to compare the dwell time with the ground speed of the aircraft. At a typical ground speed of 720 km · h^{-1}, or 200 m · sec^{-1}, the aircraft crosses the 10 m dimension of a ground resolution in 5×10^{-2} sec. The cross-track scanner time of 1×10^{-5} is 5×10^3 times faster than the ground velocity of the aircraft. The high scanner speed relative to ground speed is required to prevent gaps between adjacent scan lines.

The short dwell time of cross-track scanners imposes constraints on the other factors that determine signal strength. For example, the IFOV and spectral bandwidth must be large enough to produce a signal of sufficient strength to overcome the inherent electronic noise of the system. The signal-to-noise ratio must be sufficiently high for the signal to be recognizable.

Circular Scanning System In a *circular scanning system,* the scan motor and mirror are mounted with a vertical axis of rotation that sweeps a circular path on the terrain (Figure 1.12B). Only the forward portion of the sweep is recorded to produce images. An advantage of this system is that the distance between scanner and terrain is constant and all the ground resolution cells have the same dimensions. The major disadvantage is that most image processing and display systems are designed for linear scan data; therefore the circular scan data must be extensively reformatted prior to processing. Circular scanners have short dwell times comparable to those of cross-track scanners.

Circular scanners are used for reconnaissance purposes in helicopters and low-flying aircraft. The axis of rotation is tilted to point forward and acquire images of the terrain well in advance of the aircraft position. The images are displayed in real time on a screen in the cockpit to guide the pilot.

Along-Track Scanning System For scanner systems to achieve finer spatial and spectral resolution, the dwell time for each ground resolution cell must be increased. One method is to eliminate the scanning mirror and provide an individual detector for each ground resolution cell across the ground swath (Figure 1.12C). The detectors are placed in a linear array in the focal plane of the image formed by a lens system.

The long axis of the linear array is oriented normal to the flight path, and the IFOV of each detector sweeps a ground resolution cell along the terrain parallel with the flight track direction (Figure 1.12C). *Along-track scanning* refers to this movement of the ground resolution cells. These systems are also called pushbroom

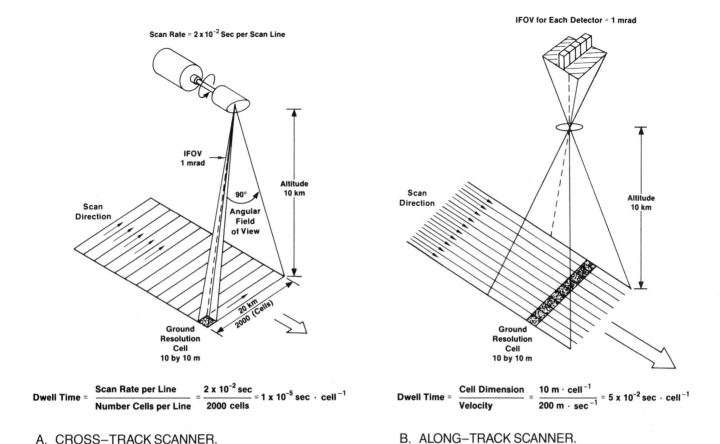

Scan Rate = 2 x 10⁻² Sec per Scan Line

IFOV for Each Detector = 1 mrad

A. CROSS–TRACK SCANNER.

$$\text{Dwell Time} = \frac{\text{Scan Rate per Line}}{\text{Number Cells per Line}} = \frac{2 \times 10^{-2} \text{ sec}}{2000 \text{ cells}} = 1 \times 10^{-5} \text{ sec} \cdot \text{cell}^{-1}$$

B. ALONG–TRACK SCANNER.

$$\text{Dwell Time} = \frac{\text{Cell Dimension}}{\text{Velocity}} = \frac{10 \text{ m} \cdot \text{cell}^{-1}}{200 \text{ m} \cdot \text{sec}^{-1}} = 5 \times 10^{-2} \text{ sec} \cdot \text{cell}^{-1}$$

FIGURE 1.13 Dwell time calculated for cross-track and along-track scanners.

scanners because the detectors are analogous to the bristles of a broom pushed along the floor.

For along-track scanners, the dwell time of a ground resolution cell is determined by the ground velocity, as Figure 1.13B illustrates. For a jet aircraft flying at 720 km · h⁻¹ or 200 m · sec⁻¹, the along-track dwell time for a 10-m cell is 5×10^{-2} sec, which is 5×10^3 times greater than the dwell time for a comparable cross-track scanner. The increased dwell time allows two improvements: (1) detectors can have smaller IFOVs, which provide finer spatial resolution, and (2) detectors can have a narrower spectral bandwidth, which provides higher spectral resolution. The airborne imaging spectrometer, described in Chapter 2, is an along-track scanner with a spectral bandwidth of 0.01 μm. Typical cross-track scanners have bandwidths of 0.10 μm, which is a spectral resolution coarser by one order of magnitude.

Side Scanning System   The three types of scanners just described are passive systems, since they detect and record energy naturally reflected or radiated from the terrain. Active systems, which provide their own

energy sources, operate primarily in the side-scanning mode. The example in Figure 1.12D is a radar system that transmits pulses of microwave energy to one side of the flight path (range direction) and records the energy scattered from the terrain back to the antenna, as described in Chapter 6. Another system is side-scanning sonar, which transmits pulses of sonic energy in the ocean to map bathymetric features (Chapter 9).

Multispectral Systems

The framing and scanning systems described above record a single image that represents a single spectral band. For many remote sensing applications, it is essential to record a scene with *multispectral images,* multiple images acquired at different spectral bands. Multispectral images may be acquired by several means. Multiple cameras or vidicons may be mounted together and aligned to photograph the same area. The shutters are linked together and triggered simultaneously. Filters that transmit selected bands of energy are used to acquire black-and-white photographs. A typical cluster of four multispectral cameras records three visible bands (blue at 0.4 to 0.5 μm, green at 0.5 to 0.6 μm; and red at 0.6 to 0.7 μm) and one reflected IR band (0.7 to 0.8 μm). Multispectral photographs acquired by the Skylab S-190A experiment are illustrated in Chapter 3. Some multispectral cameras employ a single camera body and focal-plane shutter with four lenses and filters that produce four photographs of the same area on a single piece of film. Any three of the multispectral black-and-white positive films may be registered and projected in the additive primary colors (blue, green, and red) to produce a color composite photograph.

Framing systems for multispectral imaging have two major disadvantages:

1. Each image is acquired with a separate lens system, and it may be difficult to register the multiple images.

2. The spectral range of camera systems is restricted to the visible region and part of the reflected IR region.

Today virtually all multispectral image data are acquired by scanner systems. Cross-track scanners employ a spectrometer to disperse the incoming energy into a spectrum (Figure 1.14A). Detectors are positioned to record specific wavelength bands of energy (denoted λ_1, λ_2, λ_3 and λ_4 in the figure). The Landsat thematic mapper (Chapter 4) is a cross-track multispectral scanner that records seven bands of data: three visible bands, three reflected IR bands, and one thermal IR band. A typical aircraft cross-track scanner, described in Chapter 2, records 11 bands of data. As described earlier, the short dwell time of cross-track systems requires that the detector bandwidth be fairly broad, on the order of 0.05 to 0.10 μm.

Along-track multispectral scanners employ multiple arrays of linear detectors with each array recording a separate band of energy (Figure 1.14B). Because of the extended dwell time, the detector bandwidth may be narrow and still produce an adequate signal. The SPOT satellite system (Chapter 4) uses a multispectral along-track scanner. The airborne imaging spectrometer is described in Chapter 2. More advanced along-track systems employ a spectrometer to disperse the incoming energy onto the detector arrays. Slater (1985) provides a survey of multispectral systems.

SOURCES OF REMOTE SENSING INFORMATION

Several scientific journals devoted to remote sensing are published on a regular schedule by the technical societies listed in Table 1.4. Membership in these societies

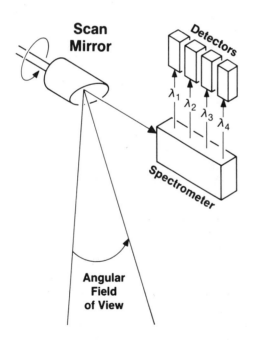

A. CROSS–TRACK
MULTISPECTRAL SCANNER.

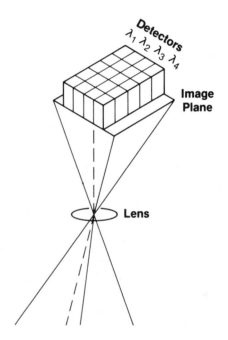

B. ALONG–TRACK
MULTISPECTRAL SCANNER.

FIGURE 1.14 Multispectral scanner systems.

TABLE 1.4 Remote sensing journals and societies

Journal	Publishing Society
Canadian Journal of Remote Sensing	Canadian Aeronautics and Space Institute Saxe Building 60–75 Sparks Street Ottawa, Canada K1P 5A5
IEEE Transactions on Geoscience and Remote Sensing	IEEE Remote Sensing and Geoscience Society Institute of Electrical and Electronics Engineers 445 Hoes Lane Piscataway, NJ 08854
International Journal of Remote Sensing	Remote Sensing Society c/o Taylor and Francis, Ltd. Rankine Road Basingstoke, Hants RG24 OPR United Kingdom
Photogrammetric Engineering and Remote Sensing	American Society for Photogrammetry and Remote Sensing 210 Little Falls Street Falls Church, VA 22046
Remote Sensing of the Environment	Elsevier Publishing Co. 52 Vanderbilt Avenue New York, NY 10017
Geo Abstracts—G. Remote Sensing, Photogrammetry, and Cartography	Geo Abstracts Regency House 34 Duke Street Norwich NR3 3AP United Kingdom
Remote Sensing Reviews	Harwood Academic Publishers 50 West 23rd Street New York, NY 10010

is open to users of remote sensing data. There are also several journals published in languages other than English. In addition to these journals, many articles on remote sensing are published in journals devoted to other disciplines such as geology, geography, and oceanography. Several of the societies listed in Table 1.4 conduct remote sensing conferences on a regular basis.

In addition to the technical societies, there are a number of academic and research organizations (listed in Table 1.5) that conduct conferences and workshops on various aspects of remote sensing. These organizations can provide information on forthcoming activities.

Under the editorship of R. N. Colwell (1983), the American Society for Photogrammetry and Remote Sensing published the second edition of the *Manual of remote sensing,* which is a useful reference for students of the subject.

COMMENTS

Remote sensing is broadly defined as collecting information about a target without being in physical contact with the target. This book emphasizes remote sensing processes that (1) record information by detecting, or sensing, the interaction between the target and electromagnetic radiation, and (2) produce an image of the target.

The electromagnetic spectrum is divided into wavelength regions. The regions employed for remote sens-

TABLE 1.5 Remote sensing organizations and conferences

Organization	Address	Conference
Environmental Research Institute of Michigan	P.O. Box 618 Ann Arbor, MI 48107	"Remote Sensing of Environment"; "Thematic Conferences on Remote Sensing"
EROS Data Center of U.S. Geological Survey	Sioux Falls, SD 57198	"Pecora Symposium on Remote Sensing" (annual conference)
Geosat Committee	153 Kearny Street San Francisco, CA 94108	Annual workshop on exploration applications of remote sensing
Jet Propulsion Laboratory	4800 Oak Grove Drive Pasadena, CA 91103	Publishes reports; conducts conferences and workshops
Laboratory for Applications of Remote Sensing, Purdue University	1220 Potter Drive East Lafayette, IN 47906	"Machine Processing of Remote Sensed Data" (annual symposium)
IEEE Geoscience and Remote Sensing Society	345 East 47th Street New York, NY 10017	"IEEE International Geoscience and Remote Sensing Symposium" (annual symposium)

ing range from short-wavelength UV energy to long-wavelength microwave and radio energy. The electromagnetic regions are further subdivided into narrow wavelength bands. Electromagnetic energy interacts with matter by being scattered, reflected, transmitted, absorbed, or emitted. Subsequent chapters of this book describe these interactions between matter and electromagnetic radiation of the different wavelength bands together with the technology employed in sensing the radiation. Remote sensing systems operate in the framing mode or the scanning mode. Multispectral systems acquire several images of a scene at different wavelengths of energy. Three of the image bands may be combined in red, green, and blue to produce a color composite image.

The interpretation of an image depends on its scale, texture, contrast ratio, and spatial resolution. In this text, spatial resolution refers to the minimum distance between two objects at which they can be distinguished on an image.

QUESTIONS

1. Use Equation 1.1 to calculate the wavelength in centimeters of radar energy at a frequency of 10 GHz. What is the frequency in gigaherz of radar energy at a wavelength of 25 cm?

2. What is temperature of boiling water at sea level in degrees Kelvin?

3. Distinguish between the earth's radiant energy peak and the reflected energy peak.

4. The atmosphere is essential for life on earth, but it causes problems for remote sensing. Describe these problems.

5. Use Equation 1.3 to calculate the contrast ratio between a target with a brightness of 17 and a background with brightness of 8.

6. On images acquired by the multispectral scanner system of Landsat (910-km altitude), targets on the ground separated by 80 m can be resolved. Use Equation 1.4 to calculate the angular resolving power (in mrad) of the scanning system.

7. Assume that your eyes have normal resolving power (0.2 mrad) and that you are an airline passenger at an altitude of 9 km. For targets on the ground with high contrast ratios, what is the minimum separation (in meters) at which you can resolve these targets?

8. An airborne cross-track scanner has the following characteristics: IFOV = 1.5 mrad; angular field of view = 45°; and the scan mirror rotates at 4000 revolutions per minute. The aircraft altitude is 10 km. Calculate the following: size of ground resolution cell = _____ by _____ m; ground swath = _____ km; dwell time for a ground resolution cell = _____ sec.

9. An along-track scanner has detectors with a 2 mrad IFOV. The scanner is carried in an aircraft at an altitude of 15 km and a ground speed of 600 km · h^{-1}. Calculate the following: ground resolution cell = _____ by _____ m; dwell time for a ground resolution cell = _____ sec.

10. Compare the advantages and disadvantages of acquiring images by multispectral cameras and multispectral scanners.

REFERENCES

Colwell, R. N., ed., 1983, Manual of remote sensing, second edition: American Society for Photogrammetry and Remote Sensing, Falls Church, Va.

Fischer, W. A., and others, 1975, History of remote sensing in Reeves, R. G., ed., Manual of remote sensing: ch. 2, p. 27–50, American Society for Photogrammetry and Remote Sensing, Falls Church, Va.

Forshaw, M. R., A. Haskell, P. F. Miller, D. J. Stanley, and J. R. G. Townshend, 1983, Spatial resolution of remotely sensed imagery—a review paper: International Journal Remote Sensing, v. 4, p. 497–520.

Gregory, R. L., 1966, Eye and brain, the psychology of seeing: World University Library, McGraw-Hill Book Co., New York.

McKinney, R. G., 1980, Photographic materials and processing in Slama, C. C., ed., Manual of photogrammetry, fourth edition: ch. 6, p. 305–366, American Society for Photogrammetry and Remote Sensing, Falls Church, Va.

Rosenberg, P., 1971, Resolution, detectability, and recognizability: Photogrammetric Engineering and Remote Sensing, v. 37, p. 1244–1258.

Slater, P. N., 1983, Photographic systems for remote sensing in Colwell, R. N., ed., Manual of remote sensing, second edition: ch. 6, p. 231–291, American Society for Photogrammetry and Remote Sensing, Falls Church, Va.

Slater, P. N., 1985, Survey of multispectral imaging systems for earth observations: Remote Sensing of Environment, v. 17, p. 85–102.

Suits, G. H., 1983, The nature of electromagnetic radiation in Colwell, R. N., ed., Manual of remote sensing, second edition: ch. 2, p. 37–60, American Society for Photogrammetry and Remote Sensing, Falls Church, Va.

ADDITIONAL READING

Colwell, R. N., and others, 1963, Basic matter and energy relationships involved in remote reconnaissance: Photogrammetric Engineering and Remote Sensing, v. 29, p. 761–799.

Southworth, C. S., 1985, Characteristics and availability of data from earth-imaging satellites: Bulletin 1631, U. S. Geological Survey.

Teleki, P., and C. Weber, eds., 1984, Remote sensing for geological mapping: Bureau de Recherchés Géologiques et Minières, Orleans, France.

Watson, K., 1985, Remote sensing—a geophysical perspective: Geophysics, v. 50, p. 2595–2610.

Watson, K., and R. D. Regan, 1983, Remote sensing: Geophysics reprint series no. 3, Society of Exploration Geophysicists, Tulsa, Oklahoma.

Aerial Photographs and Multispectral Images

Aircraft are used as platforms for a wide range of remote sensing systems; this chapter describes airborne cameras and multispectral scanners that operate in the UV, visible, and reflected IR spectral regions (Figure 1.2). Aerial photographs were the first form of remote sensing imagery, and they remain the most widely used images today. Knowing the techniques for interpreting aerial photographs is essential background for understanding other remote sensing images. Indeed, aerial photographs are used throughout this text to aid in explaining thermal IR, radar, and other kinds of images. In the enthusiasm for satellite images and new forms of airborne remote sensing, one should not overlook the advantages of aerial photography. Topographic maps are made from aerial photographs, and many engineering projects employ aerial photographs. Soil conservation studies, agricultural crop inventories, and city planning all use aerial photographs. Geologic mapping and exploration commonly begins with an analysis of photographs. As recently as the early 1970s interpretation of aerial photographs led to the discovery of several valuable oil fields in Irian Jaya, Indonesia.

Multispectral scanners, the second airborne system described in this chapter, employ optical-mechanical scanners to acquire data that are recorded on magnetic tape and converted into images. In addition to recording the same spectral range as does film (0.3 to 0.9 μm), the

scanners cover the full range of reflected IR energy to wavelengths of 3.0 μm. These longer wavelengths provide valuable information about vegetation, soil, and rocks.

INTERACTIONS BETWEEN LIGHT AND MATTER

As with other forms of electromagnetic energy, light may be reflected, absorbed, or transmitted by matter. Aerial photographs record the light reflected by a surface, which is determined by the property called *albedo*. Albedo is the ratio of the energy reflected from a surface to the energy incident on the surface. Dark surfaces have a low albedo and bright surfaces have a high albedo. Light that is not reflected is transmitted or absorbed by the material. During its transmission through the atmosphere, light interacts with the gases and particulate matter in a process called *scattering*, which has a strong effect on aerial photographs.

Atmospheric Scattering

Atmospheric scattering results from multiple interactions between light rays and the gases and particles of the atmosphere, as shown in Figure 2.1. The two major processes, selective scattering and nonselective

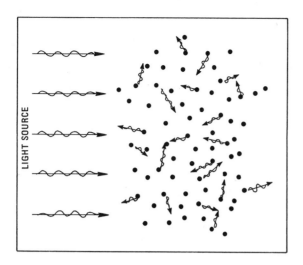

FIGURE 2.1 Scattering of light waves by particles in the atmosphere.

scattering, are related to the size of particles in the atmosphere. In *selective scattering* the shorter wavelengths of UV energy and blue light are scattered more severely than the longer wavelengths of red light and IR energy. Selective scattering is caused by smoke, fumes, and by gases such as nitrogen, oxygen, and carbon dioxide. The selective scattering of blue light causes the blue color of the sky.

In *nonselective scattering* all wavelengths of light are equally scattered. Nonselective scattering is caused by dust, clouds, and fog in which the particles are much larger than the wavelengths of light. Clouds and fog are white because they scatter all wavelengths equally. The curves in Figure 2.2 show relative scattering as a function of wavelength. Nonselective scattering is shown by the horizontal line.

Scattering in the atmosphere results from a combination of selective and nonselective processes. The range of atmospheric scattering is shown by the shaded area in Figure 2.2. The lower curve of the shaded area represents scattering in a clear atmosphere, and the upper curve scattering in a hazy atmosphere. Typical atmospheres have scattering characteristics that are intermediate between these extremes. The important point for aerial photography is that the earth's atmosphere scatters UV and blue wavelengths at least twice as strongly as it does red light.

Light scattered by the atmosphere illuminates shadows, which are never completely dark but are bluish in color. This scattered illumination is referred to as *skylight* to distinguish it from direct sunlight. A striking characteristic of photographs taken by Apollo astronauts on the surface of the moon is the black appearance

of the shadows. The lack of atmosphere on the moon precludes any scattering of light into the shadowed areas.

Effects on Aerial Photographs

Scattered light that enters the camera is a source of illumination but contains no information about the terrain. This extra illumination reduces the contrast ratio of the scene, thereby reducing the spatial resolution and detectability of the photograph. Figure 2.3 diagrams the effect of scattered light on the contrast ratio of a scene in which a dark area (brightness of 2) is surrounded by a brighter background (brightness of 5). For the original scene with no scattered light (Figure 2.3A,B), the contrast ratio is determined from Equation 1.3 as

$$CR = \frac{B_{max}}{B_{min}}$$

$$= \frac{5}{2}$$

$$= 2.5$$

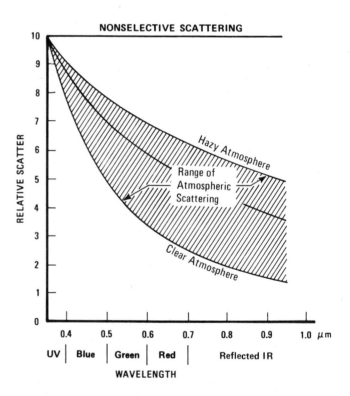

FIGURE 2.2 Atmospheric scattering as a function of wavelength. The shaded region indicates the range of scattering caused by typical atmospheres. From Slater (1983, Figure 6-15).

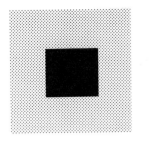

A. ORIGINAL SCENE.

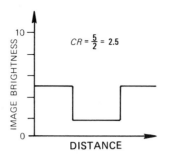

B. BRIGHTNESS PROFILE AND CONTRAST RATIO OF IMAGE WITH NO SCATTERED LIGHT.

C. IMAGE OF SCENE WITH FIVE BRIGHTNESS UNITS ADDED BY SCATTERED LIGHT.

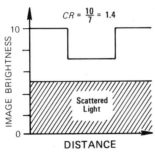

D. BRIGHTNESS PROFILE AND CONTRAST RATIO OF IMAGE WITH SCATTERED LIGHT.

FIGURE 2.3 Effect of scattered light on contrast ratio of an image.

Figure 2.3C shows the appearance of the scene in conditions of heavy haze, where the atmosphere contributes five brightness units of scattered light. As the brightness profile of Figure 2.3D shows, scattered light adds uniformly to all parts of the scene and results in a contrast ratio of

$$CR = \frac{10}{7}$$

$$= 1.4$$

Thus atmospheric scattering has reduced the contrast ratio of the scene from 2.5 to 1.4, which lowers the spatial resolution on a photograph of that scene. Chapter 1 demonstrated this relationship between contrast ratio and resolving power. The effect of atmospheric scattering on aerial images is illustrated later in this chapter.

Filtering out the selectively scattered shorter wavelengths before they reach the film reduces the effects of atmospheric scattering. Superposed on the scattering diagram of Figure 2.4 are the spectral transmittance curves

of typical filters used in aerial photography showing the wavelengths absorbed and transmitted. There is a trade-off with filters: although they reduce haze, they also remove the spectral information contained in the wavelengths that are absorbed.

FILM TECHNOLOGY

Photographic film consists of a flexible transparent base coated with a layer of light-sensitive emulsion approximately 100 μm in thickness (Figure 2.5A). The *emulsion* is initially a suspension in solidified gelatin of grains of silver halide salts a few micrometers or less in diameter. The grains have been precipitated from solution rather rapidly to make them irregular, with numerous *points of imperfection* on the surface. After the emulsion is deposited on the film base, further processing of the grains increases their sensitivity to light. *Photographic exposure* is the photochemical reaction between the silver halide grains in the emulsion and photons of light incident on them. When a photon strikes one of the grains, an electron in the silver halide crystal is given enough energy to move freely about until it becomes trapped at a point of imperfection in the grain. By combining with a silver ion lacking one electron (due to liberation by another photon), the electron then converts the ion into a silver atom (Jones, 1968, p. 116). This atom cannot remain an atom for long by itself, but if two electrons in the grain are liberated within about a second, a stable combination of silver atoms will form at the point of imperfection (Figure 2.5B).

The success of the photographic method depends on the requirement that a silver halide grain receive more than one photon within a short time. If only one photon were needed to expose each grain, random photons caused by normal ionic vibration would soon convert all the grains to silver. The stable combinations of silver atoms are large enough to trigger the conversion of the entire silver halide grain to metallic silver when the film is later chemically developed. Before chemical development, however, exposed silver halide grains have the same appearance as unexposed grains, so at this stage the film is said to contain a *latent image*.

Developing is the chemical process of changing the latent image into a real image by converting the exposed silver halide grains into opaque grains of silver. The film is immersed in the developer, which is a water solution containing a reducing agent that is unable to interact with unexposed silver halide grains. For exposed grains with silver atoms at a point of imperfection, however, the agent starts reducing the silver halide to silver. As Figure 2.5C shows, the entire grain converts to metallic

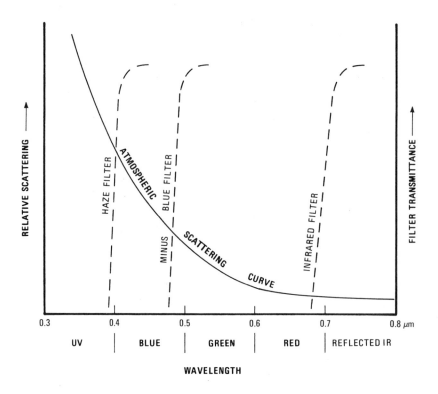

FIGURE 2.4 Atmospheric scattering diagram and transmission curves of filters used in aerial photography. Shorter wavelengths to the left of each filter curve are absorbed. From Slater (1983).

silver once the reduction process begins. The next step is the *fixing process*, which removes unexposed grains, leaving clear areas in the emulsion. The resulting film is called a *negative film* because bright targets form dark images on the film. When the film is printed onto photographic paper, the dark negative images are reversed and bright targets appear bright on the print.

One advantage of photographic remote sensing is the enormous amplification that occurs in the development process. A few photons absorbed in a grain of silver halide with a volume of 1 μm^3 will produce more than 10^{10} atoms of developed silver, which is an amplification of more than a billion. Other advantages of photographic remote sensing are high resolving power, low cost, versatility, and ease of operation. Yet another advantage is the film's capacity to store large amounts of information. On the developed film, each grain, whether exposed or unexposed, records information about the scene. There are more than 150 million (1.5×10^8) such grains on a 6.5-cm^2 (1-in.^{2}) piece of film. For comparison, the same area of a 1600 bit · in. magnetic computer tape typically contains only 2.9×10^4 magnetic signals that record information. James (1966) is a standard reference on the photographic process.

The three major disadvantages of photographic remote sensing are the following:

1. It is restricted to the spectral region from 0.3 to 0.9 μm.

2. It is restricted by weather, lighting conditions, and atmospheric effects.

3. Information is recorded in a nondigital format that is not readily available for computer processing.

CHARACTERISTICS OF AERIAL PHOTOGRAPHS

Aerial photographs are acquired with a variety of cameras, films, and filters. Characteristics such as resolution, scale, and relief displacement are common to all aerial photographs. The sections that follow discuss these characteristics in some detail.

Spatial Resolution of Photographs

Spatial resolution, or resolving power, of aerial photographs is influenced by several factors:

1. Atmospheric scattering, which was discussed earlier

2. Vibration and motion of the aircraft, which are minimized by vibration-free camera mounts and motion compensation devices

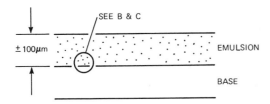

A. CROSS SECTION OF FILM.

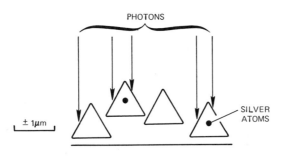

B. EXPOSURE OF SILVER HALIDE GRAINS.

C. DEVELOPED FILM NEGATIVE.

FIGURE 2.5 Film technology.

3. Resolving power of lenses

4. Resolving power of films

All of these factors combine to determine the spatial resolution of a photograph.

Resolving Power of Lenses The resolving power of a lens is determined by its optical quality and size. If a lens is used to photograph a resolution target, such as the one shown in Figure 1.6, there is an upper limit to the number of line-pairs within the space of a millimeter that can be resolved on the resulting photograph. This maximum number of resolvable line-pairs per millimeter is a measure of the *resolving power of the lens*.

Resolving Power of Film The *resolving power of film* is determined by several factors, the most important of which is the film's *granularity*. The two factors that largely determine granularity are the size distribution of silver halide grains in the emulsion and the nature of the development process. Films with high granularity have lower resolving power than those with low granularity.

There is a trade-off between granularity and the *speed* of film: films with high granularity are *faster*, meaning they are more sensitive to light.

One method for expressing the resolving power of film is to photograph a resolution target and determine the maximum number of line-pairs per millimeter one can distinguish on the developed film. As illustrated earlier in Figures 1.6 and 1.10, targets with high contrast ratios produce better resolution than those with low contrast ratios; terrain features typically have low contrast ratios. Film resolving power is commonly stated for both targets with high contrast ratios and those with low contrast ratios. A widely used black-and-white film, Kodak Panatomic X aerial film, has a resolving power of 300 line-pairs $\cdot$ mm^{-1} for high-contrast targets and 80 line-pairs $\cdot$ mm^{-1} for low-contrast targets. *System resolution* (R_s) of a camera and film combination results from the resolving powers of the lens and film and typically ranges from about 25 to 100 line-pairs $\cdot$ mm^{-1}.

Ground Resolution *Ground resolution* expresses the ability to resolve ground features on aerial photographs.

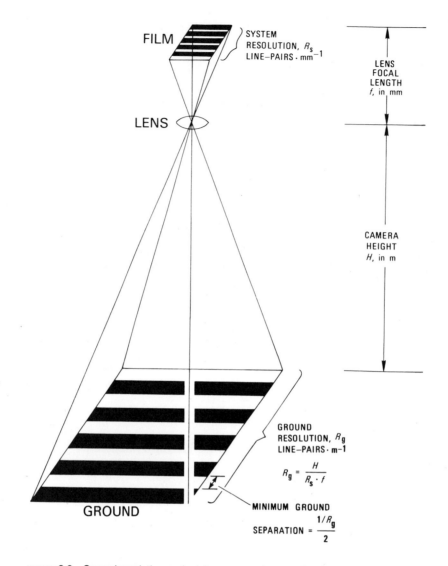

FIGURE 2.6 Ground resolution and minimum ground separation on aerial photographs.

System resolution is converted into ground resolution by the formula

$$R_g = \frac{H}{R_s f} \qquad (2.1)$$

where

R_g = ground resolution in line-pairs per millimeter

H = camera height above ground in meters. (Do not confuse this with aircraft altitude above mean sea level.)

R_s = system resolution in line-pairs per millimeter

f = camera focal length in millimeters

Figure 2.6 shows the geometric basis for this relationship. For a camera lens with focal length of 152 mm producing photographs with a system resolution of 20 line-pairs · mm^{-1} acquired at a camera height of 6100 m, the ground resolution, using equation 2.1, is

$$R_g = \frac{H}{R_s f}$$

$$= \frac{6100 \text{ m}}{20 \text{ line-pairs} \cdot \text{mm}^{-1} \times 152 \text{ mm}}$$

$$= 2.0 \text{ line-pairs} \cdot \text{m}^{-1}$$

Under the conditions specified in this example, the most closely spaced resolution target on the ground that can be resolved on the photograph consists of 2.0 line-pairs $\cdot$ m^{-1}. The width of an individual line-pair in meters is determined by the reciprocal:

$$\frac{1.0 \text{ line-pairs}}{R_g}$$

and is 0.5 m in this example. *Minimum ground separation* is the minimum distance between two objects on the ground at which they can be resolved on the photograph. As Figure 2.6 shows, it is the separation between lines or bars in the resolution target and is determined by

Minimum ground separation

$$= \frac{1.0 \text{ line-pairs}/R_g}{2} \qquad (2.2)$$

$$= \frac{1.0 \text{ line-pairs}/2.0 \text{ line-pairs} \cdot \text{m}^{-1}}{2}$$

$$= 0.25 \text{ m}$$

Table 2.1 lists minimum ground separation values for typical aerial photographs acquired with camera systems of medium and high system resolutions. The camera heights and lens focal length of Table 2.1 correspond to the medium-resolution aerial photographs in Figure 2.7. Inspection of the photographs with a magnifier indicates that these values for minimum ground separation are appropriate, although the photographs lack a ground-resolution target, which is necessary for precise measurement. Note that prints and enlargements may have poorer resolution than the original negative from which they are made.

Table 2.2 lists features that may be identified on photographs with different ground-separation values. These are only guidelines to illustrate the general relationship between ground resolution and recognition.

Photographic Scale

Scale of aerial photographs is determined by the relationship:

$$\text{Scale} = \frac{1}{H/f} \qquad (2.3)$$

Both H and f must be given in the same units, typically meters. For example, the scale of a photograph acquired at a camera height of 3050 m with a 152-mm lens (Figure 2.7B) is

$$\frac{1}{3050 \text{ m}/0.152 \text{ m}} = \frac{1}{20,000}, \text{ or } 1:20,000$$

A scale of 1:20,000 means that 1 cm on the photograph represents 20,000 cm (or 200 m) on the ground (1 in. = 20,000 in. = 1667 ft). Figure 2.7 illustrates the different scales that result from photographing the same area at different altitudes with the same camera.

Relief Displacement

Figure 2.8 illustrates the geometric distortion, called *relief displacement*, that is present on all vertical aerial photographs that are acquired with the camera aimed directly down. The tops of objects such as buildings appear to "lean" away from the *principal point*, or optical center, of the photograph. The amount of displacement increases at greater radial distances from the center and reaches a maximum at the corners of the photograph. Figure 2.9A shows the geometry of image displacement, where light rays are traced from the terrain through the camera lens and onto the film. Prints made from the film appear as though they were in the position shown by the plane of photographic print in Figure 2.9A. The vertical arrows on the terrain represent objects of various heights located at various distances

TABLE 2.1 Minimum ground separation on typical aerial photographs acquired at different heights (focal length of camera lens is 152 mm)

Camera height (H), m	Scale of photographs	Minimum ground separation for system resolution R_s of	
		40 line-pairs $\cdot$ mm^{-1}, m	100 line-pairs $\cdot$ mm^{-1}, m
1525	1:10,000	0.12	0.05
3050	1:20,000	0.25	0.10
4575	1:30,000	0.37	0.15
6100	1:40,000	0.50	0.20

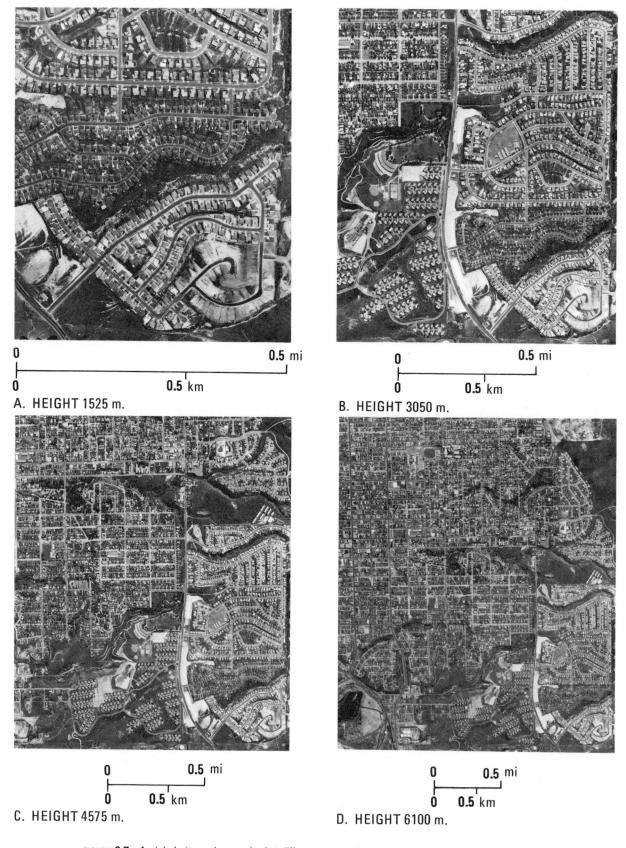

0 _____ 0.5 mi

0 _____ 0.5 km

A. HEIGHT 1525 m.

0 _____ 0.5 mi

0 _____ 0.5 km

B. HEIGHT 3050 m.

0 _____ 0.5 mi

0 _____ 0.5 km

C. HEIGHT 4575 m.

0 _____ 0.5 mi

0 _____ 0.5 km

D. HEIGHT 6100 m.

FIGURE 2.7 Aerial photographs acquired at different camera heights with a 152-mm focal-length lens. Minimum-ground-separation values for this medium-resolution system are given in Table 2.1. The southeast corner is common to all four photographs. Palos Verde Peninsula, southern California.

TABLE 2.2 Features recognizable on aerial photographs at different minimum-ground-separation values

Minimum ground separation, m	Recognizable features
15.0	Identify geographic features such as shorelines, rivers, mountains, and water
4.50	Differentiate settled areas from undeveloped land
1.50	Identify roadways
0.15	Distinguish front from the rear of automobiles
0.05	Count people, particularly if there are shadows and if the individuals are not in crowds

Source: After Rosenblum (1968, Table 2).

from the principal point. The light ray reflected from the base of object A intersects the plane of the photographic print at position A, and the ray from the top (or point of the arrow) intersects the print at A′. The distance A–A′ is the relief displacement (*d*) shown in the plan view (Figure 2.9B).

The amount of relief displacement (*d*) on an aerial photograph is

1. directly proportional to the height (*h*) of the object. For objects A and C (Figure 2.9A) at equal distances from the principal point, *d* is greater for A, which is the taller object.

2. directly proportional to the radial distance (*r*) from the principal point to the point on the displaced image corresponding to the top of the object (Figure 2.9B). For objects A and B, which are of equal height, *d* is greater for A because it is located farther from the principal point.

3. inversely proportional to the height (*H*) of the camera above the terrain.

These relationships are expressed mathematically as

$$d = \frac{hr}{H}$$

which may be transposed to

$$h = \frac{Hd}{r} \qquad (2.4)$$

This equation may be used to determine the height of an object from its relief displacement on an aerial photograph. For the building in the lower right corner of Figure 2.8, *d* and *r* are measured using the scale of the

photograph, and the height is calculated from Equation 2.4 as

$$h = \frac{212 \text{ m} \times 40 \text{ m}}{260 \text{ m}}$$

$$= 32.6 \text{ m}$$

PHOTOMOSAICS

Aerial photographs are typically acquired at scales of 1:80,000 or larger and therefore cover a relatively small area. Taking photographs on a series of parallel flight lines provides broader coverage. Along a flight line, successive photographs are acquired with 60 percent forward overlap (see Chapter 1). Flight lines are spaced to provide 30 percent *sidelap*, which is the overlap between adjacent strips of photographs. A *photomosaic* is a composite of these individual photographs that covers an extended area. Figure 2.10A is a photomosaic of the northern Coachella Valley in southern California. Flight lines are oriented north-south. One can tell that this is a ''homemade'' mosaic because the borders of individual photographs are visible. Professionals who make mosaics process the individual photographs in the dark room to a uniform tone and contrast; after these are assembled, the contacts between adjacent photographs are almost impossible to recognize.

The Coachella Valley photomosaic (Figure 2.10) includes a variety of natural and man-made features. Windblown sand, which covers much of the area, has a bright tone. Outcrops of sedimentary bedrock at Garnet Hill and Edom Hill are eroded into ridges and canyons. Vegetation has a dark signature as seen along the White Water River and the golf course on the north edge of Palm Springs. The San Andreas fault strikes northwest through the area and is marked by a pronounced linear feature in the vicinity of Palm Drive, where the fault forms the boundary between windblown sand on the south and vegetated terrain on the north. The fault is a barrier to the southward movement of groundwater in the subsurface. The water table is shallower on the north side of the fault and thus supports the growth of native vegetation. This expression of geologic features by vegetation patterns is called a *vegetation anomaly*. The transportation network, urban areas, and other cultural features are clearly visible.

STEREO PAIRS OF AERIAL PHOTOGRAPHS

A pair of successive overlapping photographs along a flight line constitutes a *stereo pair*, which may be viewed

PRINCIPAL
POINT

FIGURE 2.8 Vertical aerial photograph of Long Beach, California, showing relief displacement. Courtesy J. Van Eden.

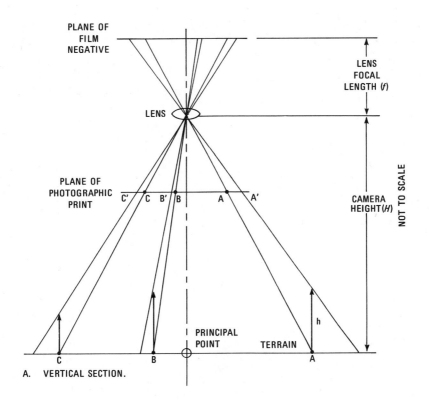

A. VERTICAL SECTION.

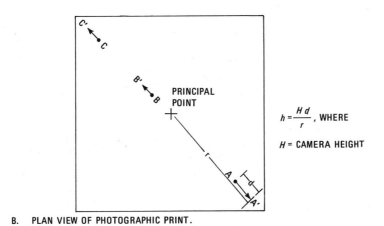

$$h = \frac{H\,d}{r}, \text{ WHERE}$$

H = CAMERA HEIGHT

B. PLAN VIEW OF PHOTOGRAPHIC PRINT.

FIGURE 2.9 Geometry of relief displacement on a vertical aerial photograph.

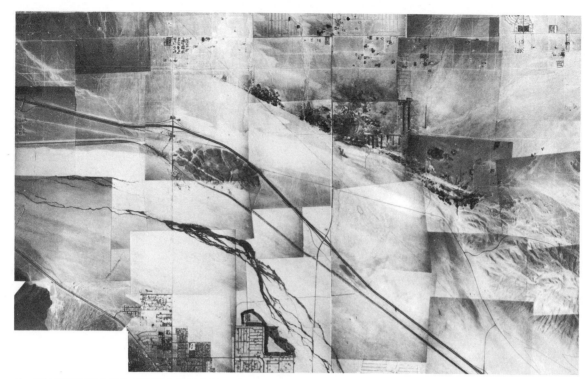

A. PHOTOMOSAIC. FROM SABINS (1973a, FIGURE 6).

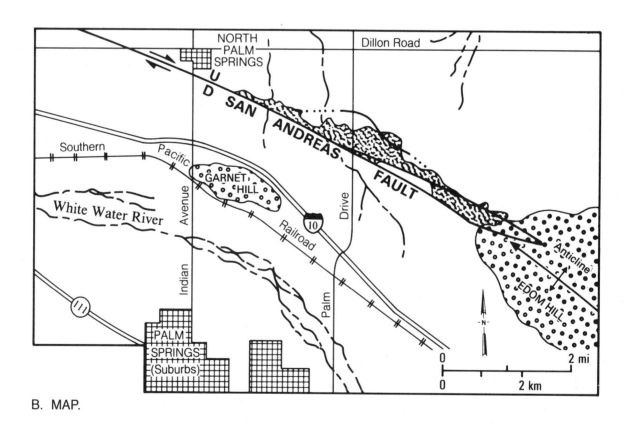

B. MAP.

FIGURE 2.10 Photomosaic and map of northern Coachella Valley, California.

with a stereoscope to produce a three-dimensional image called a *stereo model*. Figure 2.11 is a stereo pair of the Alkali anticline in the eastern part of the Bighorn Basin, Wyoming. The first impression on viewing Figure 2.11 with a stereoscope is the image's extreme vertical relief. This is caused by a characteristic of stereo models called vertical exaggeration. Figure 2.12 includes a topographic and geologic map of the Alkali anticline.

Vertical Exaggeration

Vertical exaggeration (*VE*) results because the perceived vertical scale is larger than the horizontal scale in a stereo model. The amount of exaggeration may be approximately calculated by

$$VE = \left(\frac{AB}{H}\right)\left(\frac{AVD}{EB}\right) \qquad (2.5)$$

where

AB = *air base*, or ground distance, which is the distance on the ground between the centers of successive overlapping photographs.

H = height of the camera above terrain.

AVD = the *apparent stereoscopic viewing distance*. This is *not* the distance from the stereoscopic lenses to the photographs but from the lenses to the plane in space where the image appears to be. The stereo model always appears to be somewhere below the table top on which the stereoscope sits. Several interpreters using a variety of stereoscopes have estimated an average value for *AVD* of 45 cm (Wolf, 1974).

EB = *eye base*, which is the distance between the interpreter's eyes. For the average adult, *EB* is 6.4 cm.

For the stereo pair of Figure 2.11, with an air base of 1700 m and height of 3000 m, vertical exaggeration may be estimated from Equation 2.5 as

$$VE = \left(\frac{1700 \text{ m}}{3000 \text{ m}}\right)\left(\frac{45 \text{ cm}}{6.4 \text{ cm}}\right)$$
$$= 4.0 \times$$

On viewing Figure 2.11 with a stereoscope, the average interpreter should perceive vertical distances to be exaggerated four times the equivalent horizontal distances.

In Equation 2.5 the terms *AVD* and *EB* are constant; therefore the amount of vertical exaggeration is determined by the ratio *AB/H*, which is called the *base–height ratio*. Figure 2.13 shows the vertical exaggeration for stereo pairs with various base–height ratios.

Figures 2.12A and 2.14 illustrate the effect of vertical exaggeration. Figure 2.14 shows two topographic profiles constructed along line AB in the contour map of Figure 2.12A. The profile in Figure 2.14A has the same vertical and horizontal scales (1×), which is equivalent to no vertical exaggeration. The lower profile (Figure 2.14B) was constructed with the vertical scale four times that of the horizontal scale, which is the same as that of the stereo model. Comparing the topographic profiles demonstrates the effect of vertical exaggeration.

Although vertical relief is exaggerated by a factor of 4 on a stereo model, the angles of topographic slope and structural dip do not increase four times. Figure 2.15 illustrates the *slope exaggeration* of a hill with a height (BC) of 270 m and a width (AB) of 1000 m. The tangent of the true-slope angle (CAB) is BC/AB, which is 0.27, or 15°. On the stereo model, the exaggerated height DB is 1080 m and the tangent of the exaggerated-slope angle (DAB) is BD/AB, which is 1.08, or 47°. The effect of the 4× vertical exaggeration on topographic slopes is illustrated by the lower cross section in Figure 2.14B. La Prade (1972, 1973) describes the geometry of stereoscopic photographs and reviews various hypotheses for explaining vertical exaggeration.

Geologic Interpretation of Stereo Pairs

Geologic maps may be prepared by interpreting stereo pairs. For those of you unfamiliar with geology, the Appendix is a condensed description of geologic concepts employed in interpreting remote sensing images. The first step in interpreting an area underlain by sedimentary rocks, such as the Alkali anticline, is to define the geologic units, or formations. Figure 2.12 lists the formations and their map symbols; Figure 2.11 shows contacts between formations, which are mapped in Figure 2.12B. The formations have the following topographic and tonal characteristics in the photograph:

Cody Shale A thick unit of shale with medium gray tone that is readily eroded and forms broad valleys cut by numerous streams. Cody Shale is the youngest formation exposed at the Alkali anticline.

Frontier Formation Medium gray sandstone with shale interbeds. The sandstone is resistant to erosion and forms ridges. The nonresistant shale forms narrow valleys. The prominent dip slopes of the sandstone are incised by numerous closely spaced minor stream channels, producing a distinctive serrated appearance.

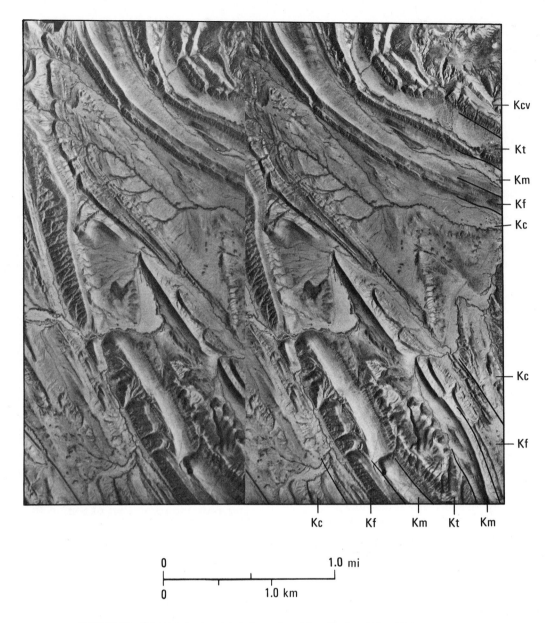

Kcv
Kt
Km
Kf
Kc

Kc

Kf

Kc   Kf   Km   Kt   Km

0   1.0 mi
0   1.0 km

FIGURE 2.11 Stereo pair of aerial photographs of the Alkali anticline, Bighorn Basin, Wyoming, with formation contacts indicated. Formation symbols are explained in Figure 2.12.

Mowry Shale Siliceous shale that weathers to a slope with a very light gray tone.

Thermopolis Shale Very dark gray shale with thin, light-toned interbeds that cause a banded appearance.

Cloverly Formation Alternating beds of resistant sandstone and nonresistant shale that weather to ridges and valleys. The Cloverly is the oldest formation exposed in the area.

These characteristics are used to map the contacts between formations.

At several localities, the contacts are interrupted and offset by *normal faults* (Appendix), which are mapped with heavy lines and symbols to show the displacement along the faults (Figure 2.12B). *Attitudes* of the beds are readily seen in the stereo model and are recorded with symbols showing *strike* and *dip*. The amount of dip is vertically exaggerated in the same fashion as topographic slopes shown in Figure 2.15. Techniques for converting exaggerated dips to true dips are described by Miller (1961). Outcrop patterns and the attitude of beds are used to locate the axial trace of the Alkali anticline and the adjacent *syncline*.

As mentioned earlier, the topographic map in Figure

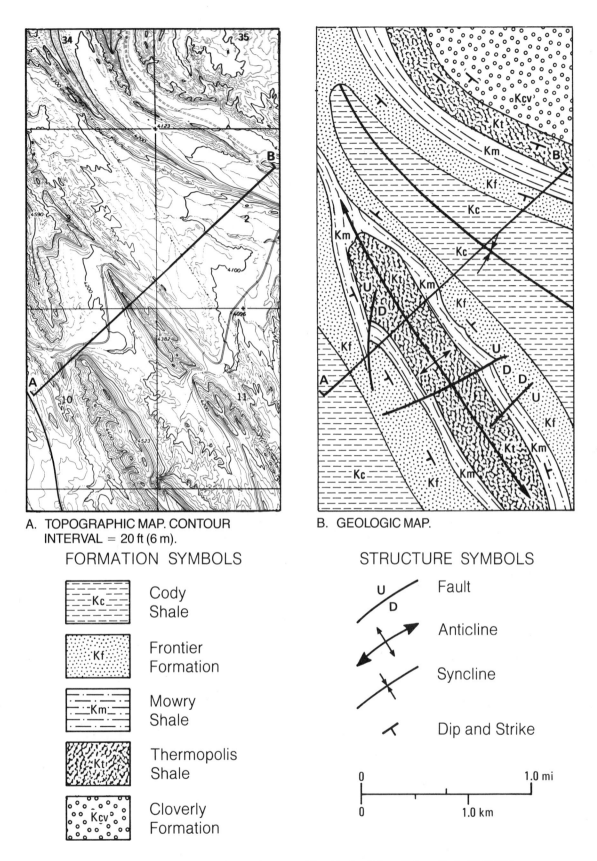

A. TOPOGRAPHIC MAP. CONTOUR
INTERVAL = 20 ft (6 m).

B. GEOLOGIC MAP.

FORMATION SYMBOLS

Kc	Cody Shale	
Kf	Frontier Formation	
Km	Mowry Shale	
Kt	Thermopolis Shale	
Kcv	Cloverly Formation	

STRUCTURE SYMBOLS

U / D — Fault

Anticline

Syncline

Dip and Strike

0 ———————— 1.0 mi
0 ———————— 1.0 km

FIGURE 2.12 Topographic and geologic maps of the Alkali anticline, Bighorn Basin, Wyoming.

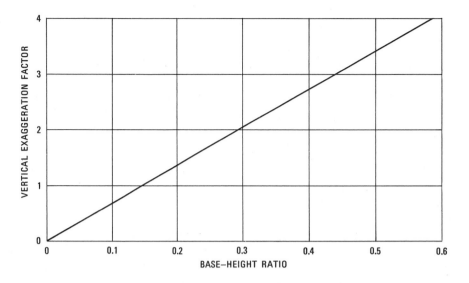

FIGURE 2.13 Relationship between the base–height ratio and the vertical exaggeration factor for stereo models. From Thurrell (1953, Figure 5).

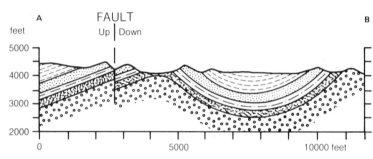

A. 1X VERTICAL EXAGGERATION.

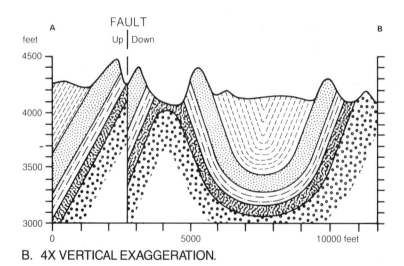

B. 4X VERTICAL EXAGGERATION.

FIGURE 2.14 Topographic profiles and geologic cross sections of the Alkali anticline, showing effects of vertical exaggeration.

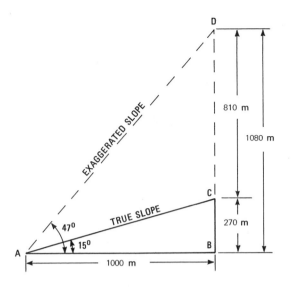

FIGURE 2.15 Slope exaggeration on a stereo model with 4× vertical exaggeration.

2.12A was used to draw the topographic profile for the cross section with no vertical exaggeration (Figure 2.14A). The geologic cross section was filled in by adding the faults, formation contacts, and strike-and-dip information from the geologic map (Figure 2.12B). Simple geometric techniques were used to produce the cross section with 4× vertical exaggeration (Figure 2.14B). This photogeologic interpretation should not be considered complete until the map has been checked in the field.

LOW-SUN-ANGLE PHOTOGRAPHS

Aerial photographs are normally acquired between 10:00 a.m. and 3:00 p.m., when the sun is at a high angle above the horizon and shadows have a minimum extent. These photographs are desirable for topographic mapping, which requires unobscured terrain, but for geologic interpretation, photographs acquired with lower sun angles are often valuable. The photograph in Figure 2.16A was acquired shortly after sunrise, when the sun was only 15° above the horizon. A photograph of the same area acquired at midday (Figure 2.16B) is shown for comparison. The area is a gravel-covered slope on the east flank of the Carson Range south of Reno, Nevada. The low-sun-angle photograph includes several north-trending linear features with prominent bright or dark signatures. As shown in the map and cross section (Figure 2.17), these features are low topographic *scarps*, which are caused by active normal faults that cut the surface. The east-facing scarps are strongly illuminated by the

early morning sun and have bright signatures, or highlights. The west-facing scarps are shadowed and have dark signatures. Orientation of the highlights and shadows provides information on the sense of displacement along the faults, shown in the cross section (Figure 2.17B). Note, however, that if the scarps trended east-west, essentially parallel with the sun azimuth, the shadows and highlights would not be present in the photograph.

In the midday photograph, highlights and shadows are minimal and the fault scarps are inconspicuous. There are some subtle linear gray traces along the scarps that are not due to illumination, but are local concentration of sage brush. Runoff from rainfall is concentrated at the foot of the scarps and supports a higher density of vegetation. This pattern is one of several types of vegetation anomaly that aid in recognizing geologic features.

Acquiring good low-sun-angle photographs in the summer is complicated by the limited number of hours in the morning and evening, when the desired illumination occurs. Illumination values are low and change rapidly during these times; therefore, proper camera exposures may be difficult to achieve. At middle and high latitudes, the sun is at relatively low elevations throughout the day in the winter, which may be an optimum season for acquiring these photographs. Low-sun-angle photographs are widely used to recognize subtle topographic features associated with active faults.

BLACK-AND-WHITE PHOTOGRAPHS

Many types of black-and-white films are available for acquiring aerial photographs at wavelengths ranging from UV, through visible, and into the photographic portion of the IR region (Figure 1.2).

Panchromatic Black-and-White Photographs

Black-and-white photographs exposed by visible light are called *panchromatic photographs*. As shown in Figure 2.4, panchromatic aerial photographs are normally acquired with a Kodak Wratten 12 or equivalent filter over the lens, which eliminates the UV and blue wavelengths selectively scattered by the atmosphere. Examples of these *minus-blue* photographs are shown in Figures 2.7, 2.8, and 2.11.

These photographs are a widely used and readily available remote sensing product. Stereo coverage of most of the United States is available at modest prices from the agencies listed in the section "Sources of Aerial Photographs" later in this chapter. These photographs are used to compile topographic maps, geologic surveys, engineering studies, and crop inventories. Color photographs are superior for many applications, but

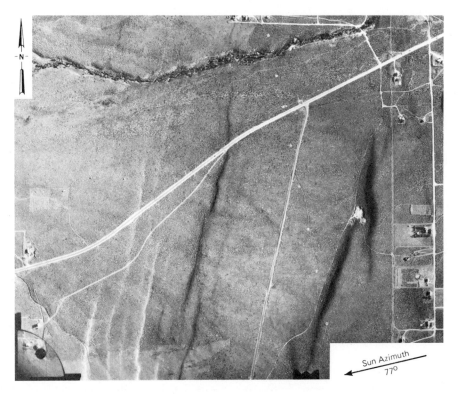

Sun Azimuth 77°

A. LOW–SUN–ANGLE PHOTOGRAPH ACQUIRED JUNE 23, 1972, AT 5:30 A.M. LOCAL SUN TIME
WITH A SUN ELEVATION OF 15°. FROM WALKER AND TREXLER (1977, FIGURE 3).

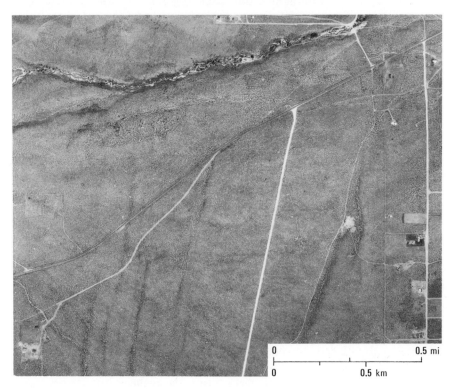

B. HIGH–SUN–ANGLE PHOTOGRAPH ACQUIRED MAY 21, 1966, AT MIDDAY BY THE
U.S. GEOLOGICAL SURVEY.

FIGURE 2.16 Low-sun-angle photograph and high-sun-angle photograph of eastern
flank of the Carson Range, Nevada.

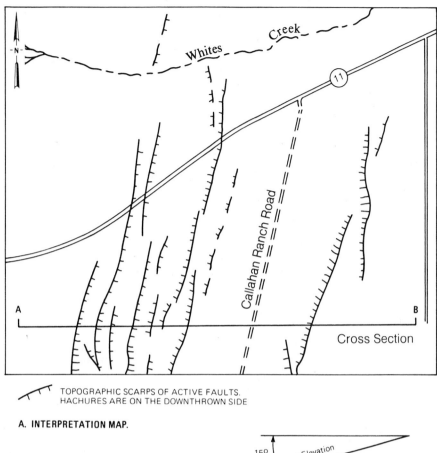

TOPOGRAPHIC SCARPS OF ACTIVE FAULTS. HACHURES ARE ON THE DOWNTHROWN SIDE

A. INTERPRETATION MAP.

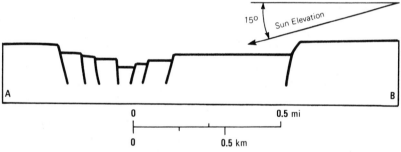

B. VERTICALLY EXAGGERATED CROSS SECTION ALONG LINE A–B.

FIGURE 2.17 Interpretation map and cross section of a low-sun-angle photograph.

panchromatic photographs are still a major source of remote sensing information.

IR Black-and-White Photographs

By using IR-sensitive film and a filter such as the Kodak Wratten 89B, which transmits only reflected IR energy (Figure 2.18), one can obtain photographs in the portion of the IR spectral region at wavelengths from 0.7 to 0.9 μm. This reflected solar radiation is called *photographic IR energy* and should not be confused with thermal IR energy, which occurs at wavelengths from 3 to 14 μm. Figure 2.19 illustrates simultaneously acquired panchromatic and IR black-and-white photographs of the Massachusetts coast. This example demonstrates the following advantages of IR photographs:

1. Haze penetration improves because the filter eliminates the severe atmospheric scattering that occurs in the visible and UV regions. Eliminating most scattered light results in a higher contrast ratio and there-

fore higher spatial resolution on the IR photograph, as discussed earlier.

2. Maximum reflectance from vegetation occurs in the photographic IR region, as shown by the bright tones in the IR photograph. In addition, maximum spectral differences between vegetation types, such as hardwoods and conifers, show up in the photographic IR region, which is advantageous for mapping plant communities.

3. IR energy is almost totally absorbed by water, which causes water to have a dark tone on IR photographs. For this reason, boundaries between land and water show up more clearly on IR photographs than on panchromatic photographs. Note the tidal-flat area in the upper part of each photograph in Figure 2.19. The shoreline and individual tidal channels are clearly distinguishable on the IR photograph (Figure 2.19B). Such distinctions are not possible on the panchromatic photograph (Figure 2.19A) because light penetrates the shallow water and does not differentiate submerged areas from the land.

IR color film, described in a later section, combines these properties of IR black-and-white film with the advantages of color.

UV Photographs

The UV spectral region extends from 3 nm to 0.4 μm; however, the atmosphere only transmits UV wavelengths from 0.3 to 0.4 μm, which is known as the *photographic UV region*. Photographs may be acquired in the photographic UV region with film-and-filter combinations such as Kodak Plus-X Aerographic film 2402 and the Kodak Wratten 18A filter (Figure 2.20). The Kodak Wratten 39 filter transmits both UV and blue energy and for most applications is almost as useful as the Wratten 18A filter. Most camera lenses absorb UV energy of wavelengths less than about 0.35 μm, but special quartz lenses can transmit shorter wavelengths.

UV photographs have low contrast ratios and poor spatial resolution because of severe atmospheric scattering (Figure 2.2). As a result, the UV spectral region is rarely employed in remote sensing, except for special applications such as monitoring oil films on water (illustrated in Chapter 9). UV photographs have largely been replaced by images acquired with multispectral scanners, as described later in this chapter.

COLOR SCIENCE

The average human eye can discriminate many more shades of color than it can tones of gray. This greatly

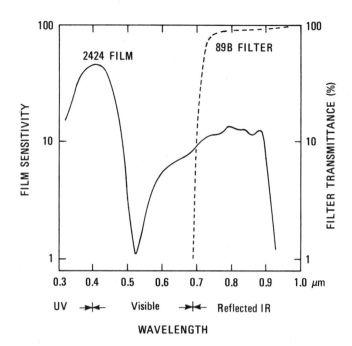

FIGURE 2.18 Combination of film and filters for IR black-and-white photographs. From Vizy (1974, Figure 7).

increased information content is a major factor favoring the use of color photography. Color aerial photographs of much of the United States are now available from government agencies listed in a later section.

Visible Spectrum

Figure 2.21 (upper part) shows the six gradational colors into which white light can be divided, with arrows indicating the center of each wavelength range. The eye responds as though it were a three-receptor system, and many colors can be synthesized by adding different proportions of blue, green, and red. These are called the three *additive primary colors* because synthetic white light forms when equal amounts of blue, green, and red light are added by superposition.

Additive Primary Colors

The range of each additive primary color spans one-third of the visible spectrum. Figure 2.21 (lower part) shows the characteristics of three filters, each of which transmits one additive primary color and absorbs the other two. A color image can be produced by acquiring three separate black-and-white pictures of a subject, each exposure using one of the three primary filters. Positive films of the three pictures can then be *registered* (superposed to align precisely) on a screen with three pro-

A. PANCHROMATIC PHOTOGRAPH.

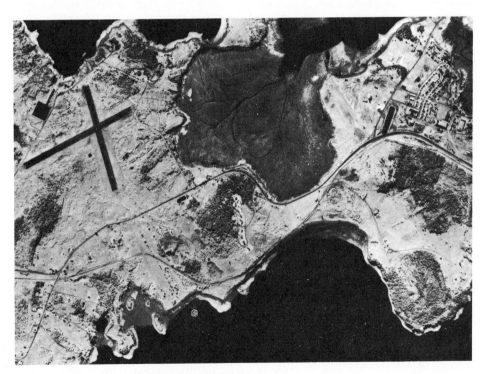

B. IR BLACK–AND–WHITE PHOTOGRAPH.

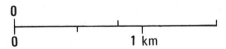

FIGURE **2.19** Panchromatic and IR black-and-white aerial photographs of the Massachusetts coast.

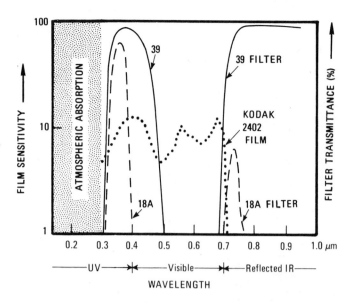

FIGURE 2.20 Combinations of film and filters for UV photographs. From Vizy (1974, Figure 6).

jectors, each projector using the primary color filter appropriate for that black-and-white photograph. The result is a color image.

For most purposes it is impractical to use three projectors to produce color photographs, and additive primaries cannot be mixed by directly superposing the films. As Figure 2.21 shows, because each additive primary filter absorbs two-thirds of the spectrum, no light is transmitted where any two filters are superposed. Color television employs additive primaries not by superposing but by juxtaposing blue, green, and red specks small enough to blend together when observed by the eye.

Subtractive Primary Colors

In order to mix colors by superposition of films, the three *subtractive primary colors* yellow, magenta, and cyan must be used. As the lower part of Figure 2.21 shows, each theoretical subtractive primary filter absorbs one-third of the visible spectrum and transmits the remaining two-thirds.

The yellow subtractive primary filter absorbs blue light. The magenta primary is a bluish-red filter that absorbs green. Cyan is a bluish-green filter that absorbs red. Figure 2.22 relates the additive primaries at the corners of the triangle to the subtractive primaries along the sides. Each subtractive primary absorbs the additive primary at the opposite corner and transmits the additive primaries at the adjacent corners. When any two different subtractive primary filters are superposed and illuminated with white light, the color transmitted will

be the additive color located at their common corner on the triangle. For example, white light projected through overlapping yellow and cyan filters appears green. Superposition of all three subtractive filters absorbs all light and is perceived as black. *Complementary colors* are pairs of colors that produce white light when added together, such as magenta and green. On the color triangle (Figure 2.22), complementary colors are located opposite each other. A color film system using a single projector is possible with subtractive primary colors, as the following section describes.

Color Film Technology

Color prints are color photographs with an opaque base. *Color film* is a transparent medium that may be either positive or negative. On conventional negative film, the color present is complementary to the color of the subject photographed, and the density on the film is the inverse of the brightness of the subject. For example, a bright green subject is portrayed by a dark magenta image on negative film. Negative film is normally employed in making color prints. On positive color film (except IR color film), the subject is represented by an image in its true color. Kodachrome and Ektachrome are positive color films manufactured by Kodak that are familiar to amateur photographers.

Color negative film consists of a transparent base coated with three emulsion layers (Figure 2.23). The emulsion layers are similar to those on black-and-white film but with the following differences:

1. Each layer is sensitive to one of the additive primary colors—blue, green, or red.

2. During developing, each emulsion layer forms a color dye that is complementary to the primary color that exposed the layer: the blue-sensitive emulsion layer forms a yellow negative image; the green-sensitive layer forms a magenta negative image; and the red-sensitive layer forms a cyan negative image.

Figure 2.23 illustrates a color subject and the manner in which its image forms on negative and positive film. The silver halide salts of the green-sensitive and red-sensitive layers are also sensitive to blue light. A yellow filter layer beneath the upper emulsion layer prevents blue light from exposing the green-sensitive and red-sensitive layers. The yellow filter layer is dissolved and removed during film processing.

On negative film the red-sensitive, bottom layer produces a complementary cyan image of a red subject. Green and blue subjects produce magenta and yellow images respectively. The white subject exposes all three

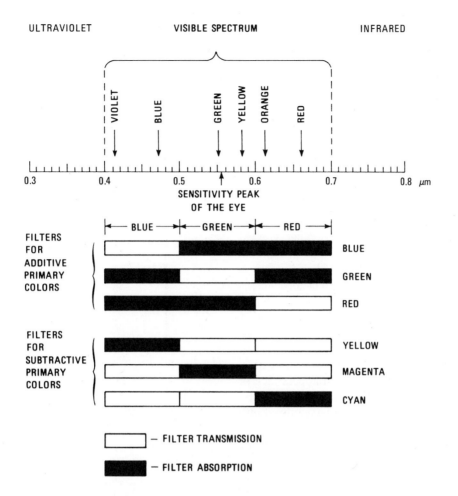

FIGURE 2.21 Visible spectrum with transmission and absorption characteristics of filters for additive and subtractive primary colors.

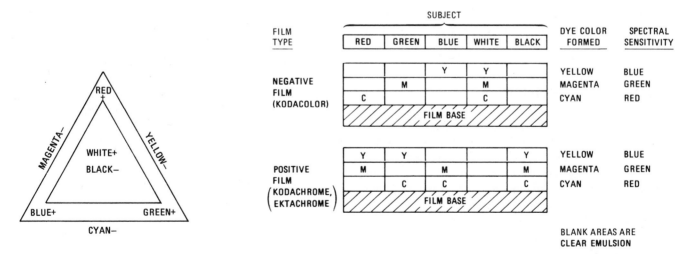

FIGURE 2.22 Color triangle showing the relationship among additive (+) and subtractive (−) primary colors.

FIGURE 2.23 Cross sections of positive and negative color films showing how images are formed on the three emulsion layers.

layers, resulting in an image that transmits no light; the black subject results in a clear image because none of the layers is exposed. As Figure 2.23 shows, the film records the color of a subject as its complementary color on the negative. The image on a negative color film is projected onto photographic paper coated with sensitive emulsions that is developed to produce a color print.

A widely used product is positive color film, which records a subject in its true color, not in its complementary color. As shown in the diagram of positive film in Figure 2.23, a red subject forms a clear image on the red-sensitive cyan-colored layer. The red subject also forms a magenta and a yellow image respectively on the green-sensitive and blue-sensitive layers. When viewed with transmitted white light, the yellow and magenta images absorb blue and green respectively and allow a red image to be projected. A white subject forms clear images on all three layers.

A haze filter is normally employed in aerial color photography to absorb UV radiation that is strongly scattered by the atmosphere (Figure 2.4). Blue light is not removed by the haze filter because this would destroy the color balance of the film. The bluish appearance of most high-altitude aerial color photographs is caused by the additional blue light that is selectively scattered into the lens.

With positive color film, the original film transparency may be viewed on a light table, which provides maximum resolution. However, the film rolls require special handling and viewing equipment and are not suitable for use in the field. Black-and-white prints, color prints, and color transparencies can be made from negative color film. Paper prints, despite their slightly lower resolution, are more versatile and easily used in the field than transparencies. The photographs described above are called *normal color photographs* to distinguish them from photographs in which the emulsion layers are sensitive to wavelengths other than the three primary colors. Plate 1A is a print of a normal color aerial photograph.

IR COLOR PHOTOGRAPHS

Spectral sensitivity of the three layers of color emulsion may be changed to respond to energy of other wavelengths, including photographic IR (0.7 to 0.9 μm). This *IR color film* is sold as Kodak Aerochrome Infrared film, type 2443, which is available only as positive film. Plate 1B is an IR color photograph of the area covered by the normal color photograph in Plate 1A. IR color film was originally designed for military reconnaissance and was called *camouflage detection films. False color film* is occasionally used, but *IR color film* is the preferred name.

Film Characteristics

IR color film is best described by comparing it with normal color positive film. Figure 2.24A,B shows the spectral sensitivity of the three emulsion layers that produce blue, green, and red images. In normal color film (Figure 2.24A) each layer is exposed by the corresponding wavelength band of light. The blue-imaging layer is exposed by blue light, the green-imaging layer by green light, and the red-imaging layer by red light. In IR color film (Figure 2.24B) the photochemistry of

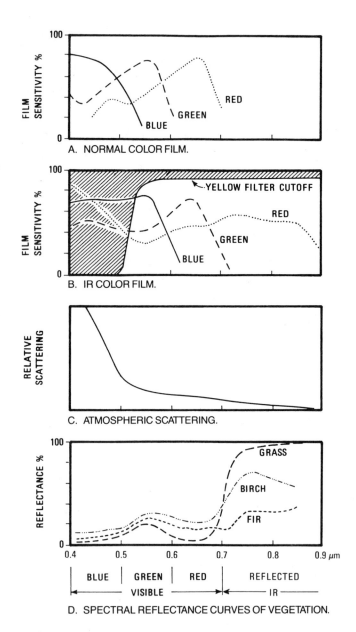

FIGURE 2.24 Spectral sensitivity of normal color film and IR color film, together with an atmospheric scattering diagram and vegetation reflectance spectra. From Sabins (1973b, Figure 2).

the layers is changed so they are sensitive to different wavelengths of light: the blue-imaging layer is exposed by green light; the green-imaging layer is exposed by red light; and the red-imaging layer is exposed by reflected IR energy. All three layers are also sensitive to blue light, which is eliminated by placing a yellow (minus-blue) filter over the camera lens. The diagonal line pattern in Figure 2.24B shows the blue wavelengths removed by the Wratten 12 yellow filter. Removing these strongly scattered wavelengths (Figure 2.24C) improves the contrast ratio and spatial resolution of IR color film.

Because the term *infrared* suggests heat, some users mistakenly assume that the red tones on IR color film record variations in temperature. A few moments' thought will show that this is *not* the case. If the IR-sensitive layer were sensitive to ambient heat, it would be exposed by the warmth of the camera body itself. As Chapter 1 pointed out, thermal radiation occurs at wavelengths longer than 3 μm, which is beyond the sensitivity range of IR film. To repeat, the red-imaging layer of IR color film is exposed by reflected IR energy at wavelengths somewhat longer than those in the visible red band.

Comparison of IR Color and Normal Color Photographs of the UCLA Campus

Plate 1 illustrates an IR color photograph and a normal color photograph that were simultaneously acquired of the UCLA campus in the western part of Los Angeles. Figure 2.25 is a location map. Comparing these photographs is a useful way to understand their different characteristics. Table 2.3 compares the color signatures of common subjects on the two types of photographs.

The most striking difference is the red color of healthy vegetation in the IR color photograph, which is explained by the spectral reflectance curves of vegetation shown in Figure 2.24D. *Spectral reflectance curves* show the percentage of incident energy reflected by a material as a function of wavelength. Blue and red light are absorbed by chlorophyll in the process of photosynthesis. Up to 20 percent of the incident green light is reflected causing the familiar green color of leaves on normal color photographs. The spectral reflectance of vegetation increases abruptly in the photographic IR region, which includes the wavelengths that expose the red-imaging layer in IR color film. These relationships explain the red color of vegetation on IR color photographs. Figure 2.24D shows that different types of vegetation have a wider range of reflectance values in the photographic IR region than in the green region, which makes it easier to discriminate vegetation types on IR color than on normal color film. For example, the grass of the golf course has a bright red signature, which distinguishes it from the red-blue signature of native vegetation in the Santa Monica Mountains. In the normal color photograph, however, there is only a subtle difference between the green hues of these different vegetation types.

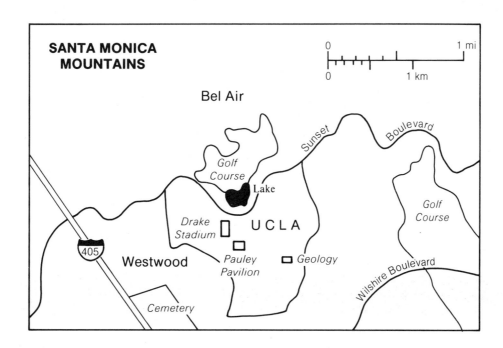

FIGURE 2.25 Location map of the UCLA area, Los Angeles, California.

TABLE 2.3 Terrain signatures on normal color film and IR color film

Subject	Normal color film	IR color film
Healthy vegetation:		
Broadleaf type	Green	Red to magenta
Needle-leaf type	Green	Reddish brown to purple
Stressed vegetation:		
Previsual stage	Green	Pink to blue
Visual stage	Yellowish green	Cyan
Autumn leaves	Red to yellow	Yellow to white
Clear water	Blue-green	Dark blue to black
Silty water	Light green	Light blue
Damp ground	Slightly darker	Distinct dark tones
Shadows	Blue with details visible	Black with few details visible
Water penetration	Good	Green and red bands: same; IR band: poor
Contacts between land and water	Poor to fair discrimination	Excellent discrimination
Red bed outcrops	Red	Yellow

The high IR reflectance of leaves is not caused by chlorophyll but by the internal leaf tissue, or mesophyll, which consists of water-filled cells and numerous air spaces. The boundaries between cell walls and air spaces strongly reflect IR energy. When vegetation is stressed because of drought, disease, insect infestation, or other factors that deprive the leaves of water (moisture stress), the internal cell structure begins to collapse and the IR reflectance decreases. This decreased reflectance diminishes the red color on IR color photographs. The loss of IR reflectance is called a *previsual symptom* of plant stress because it often occurs days or even weeks before the visible green color begins to change. Advanced plant stress produces a cyan color on the film. The previsual effect may be used for early detection of disease and insect damage in crops and forests. Evidence of plant stress is seen in the intramural playing field east of Drake Stadium (Figure 2.25 and Plate 1), which is watered by a sprinkler system. In the normal color photograph the field is entirely green, but in the IR color photograph the red signature is interrupted by blue strips that indicate inadequately watered turf.

When leaves of deciduous trees turn red and brown in the fall, they appear white or yellow on IR color film. The green chlorophyll has decayed, unmasking the red and yellow pigments of the leaf and allowing additional ones to form. The resulting reflectance values in the green, red, and photographic IR bands are nearly equal, resulting in the white to yellow signature on IR color photographs. Gausman (1985) gives details of optical properties of plant leaves in the visible and reflected IR regions.

The small lake north of UCLA has a dark green signature in the normal color photograph and in the IR color photograph a dark blue signature that contrasts with the red signature of the surrounding vegetation. This ability to enhance the difference between vegetation and water is especially valuable for mapping drainage patterns in heavily forested terrain. Silty water has a light blue signature in IR color photographs. One can recognize damp ground on IR color photographs by its relatively darker signature, caused by absorption of IR energy. Shadows are darker on IR color photographs than on normal color photographs because the yellow filter eliminates blue light.

The IR color photograph in Plate 1 has a better contrast ratio than the normal color photograph for two reasons:

1. The yellow filter eliminates blue light, which is preferentially scattered by the atmosphere, as shown by the curve in Figure 2.24C. Eliminating much of the scattering improves the contrast ratio.

2. For vegetation, soils, and rocks, reflectance differences are commonly greater in the photographic IR region than in the visible region.

The higher contrast ratio of the IR color photograph results in improved spatial resolution, which is evident when one compares finer details of the two photographs. On the slopes of the Santa Monica Mountains (upper left corner), for example, closely spaced shrubs may be separated more readily in the IR color photograph. In the urban areas, individual buildings are more distinctly separate in the IR color example.

Experimental Use

In addition to large sizes for aerial cameras, IR color film is available in the 35-mm size for use in ordinary cameras. Cost of the 20-exposure casettes and processing is comparable to that of normal color films. The characteristics and applications of IR color film can be evaluated at minimal expense with this format. It is instructive to acquire normal color photographs simultaneously with IR color photographs and compare them. (IR color film may deteriorate with time and excessive heat; if you keep the film for more than a few weeks, store it in a freezer. You should allow the frozen film to reach room temperature before opening the sealed container. This will prevent moisture from condensing on the cold film.)

A yellow (minus-blue) filter, such as the Kodak Wratten 12, is used with IR color film. This film and filter combination has an approximate speed of ASA 100. Some experimentation will be necessary to determine the optimum exposure because conventional light meters do not measure the same spectral region to which the film is sensitive. Some cameras have an IR setting on the focusing ring that is intended for IR black-and-white film. Do not use this setting for IR color film because two of the three emulsion layers are sensitive to visible wavelengths.

HIGH-ALTITUDE AERIAL PHOTOGRAPHS

Aerial photographs have traditionally been acquired at altitudes of 6000 m or less, resulting in scales of 1:40,000 or larger (Table 2.1). In the 1970s, improvements in cameras and film enabled acquisition of photographs at higher altitudes, which provide adequate resolution for many applications. The advantage of the high-altitude, smaller-scale photographs is that fewer photographs are required to cover an area.

NASA High-Altitude Photographs

For a number of years, NASA has been acquiring photographs of the United States from U-2 and RB-57 reconnaissance aircraft at altitudes of 18 km above terrain with standard aerial cameras (152-mm focal length) on film with a 23-by-23-cm format. The resulting photographs cover 839 km^2 at a scale of 1:120,000. Black-and-white, normal color, or IR color film is used; many missions employ two cameras to acquire photographs with two different film types.

Coverage of NASA photographs is concentrated over numerous large regional test sites for which repeated coverage over several years may be available. Many areas lack this coverage.

National High Altitude Photography Program

The National High Altitude Photography (NHAP) program, coordinated by the U.S. Geological Survey, began in 1978 to acquire coverage of the United States with a uniform scale and format. From aircraft at an altitude of 12 km, two cameras (23-by-23-cm format) acquire black-and-white photographs and IR color photographs. The black-and-white photographs are acquired with a camera of 152-mm focal length to produce photographs at a scale of 1:80,000, which cover 338 km^2. Figure 2.26 shows an NHAP photograph of Washington, D.C. A stereo pair of these photographs (not illustrated) covers the area of a standard 7.5-minute topographic quadrangle and can be used to produce new maps or update existing maps.

The IR color photographs are acquired with a camera of 210-cm focal length to produce photographs at a scale of 1:58,000, which cover 178 km^2. Plate 2A illustrates an NHAP IR color photograph of Coeur d'Alene, Idaho, which includes a wide range of land-use and land-cover categories. In addition to the urban and suburban complex, various types of agriculture and forest land are present. Two sawmills are located on the lake at either side of the town. Rafts of logs form the blue patches in the lake.

LAND-USE AND LAND-COVER ANALYSIS

Aerial photographs of different types and scales are widely used, together with other remote sensing images, to analyze land use and land cover, as Chapter 10 describes and illustrates. Chapter 10 also presents a three-level system for recognizing and classifying categories of land use and land cover. The levels are listed in the order of increasing complexity and increasing scale of images employed. Level I has only 9 categories and is interpreted from Landsat images (described in Chapter 4) at scales of 1:250,000 or smaller. Level II has about 40 categories and is interpreted from high-altitude photographs at scales between 1:120,000 and 1:80,000. Level III has about 70 categories and is based on medium-altitude photographs at scales between 1:80,000 and 1:20,000.

0   4 mi.

0   4 km.

FIGURE 2.26   National High Altitude Photography program photograph of
Washington, D.C., acquired from an altitude of 12 km.

REFLECTANCE SPECTRA

Spectral reflectance curves, or *spectra*, are measured by instruments called *spectrometers* that record the energy, usually sunlight, reflected from materials as a function of wavelength. Figure 2.24D illustrates spectra of various types of vegetation. Spectra of materials such as soils, rocks, and minerals were originally measured in the laboratory using small, pulverized, unweathered samples. The spectra were valuable for understanding the relationship between mineralogy and reflectance (Hunt, 1980). In the real world, however, rocks weather to different degrees, contain various amounts of moisture, and are partially covered by vegetation and soils. Portable spectrometers have been developed to record, in the field, spectra of the actual surfaces that are imaged by remote sensing devices. Figure 2.27 illustrates a combination field-and-laboratory spectrometer.

Figure 2.28 shows spectra of typical volcanic and sedimentary rocks that were measured by a portable spectrometer designed and operated by Jet Propulsion Laboratory. Such curves provide a comparison standard for identifying spectra of unknown materials. The curves also indicate optimum wavelengths for differentiating various rocks. For example, the spectra of basalt, andesite, shale, and arkose have very similar reflectance values in the blue and green region (0.4 to 0.6 μm). In the red region (0.6 to 0.7 μm) and reflected IR region (0.7 to 0.9 μm), however, these rocks are readily differentiated. Figure 2.29 shows how soil types may be distinguished by spectral differences; it also illustrates the fact that moisture reduces reflectance, particularly in the reflected IR region.

MULTISPECTRAL PHOTOGRAPHS

Multispectral photographs are an alternative to normal color and IR color photographs. One multispectral system employs four cameras aimed at the same centerpoint on the ground with the shutters linked to trigger simultaneously. Each camera uses a combination of black-and-white film and filter that records a particular spectral band, typically blue, green, red, and photographic IR. One can compare the photographs by inspecting them individually or by preparing color composites. Some satellite photographs have been acquired in this fashion; Chapter 3 gives such an example from the Apollo 9 mission. The term *multiband* is commonly used for multispectral photographs and cameras.

Another multispectral system uses a single camera body and shutter with four lenses and two film supplies, one for each pair of lenses. The resulting negatives are made into black-and-white positive transparencies that are registered and projected with colored lights in a special viewer to produce normal color or IR color pictures. Various color combinations will enhance specific features. Some systems have nine lenses and three types of film. These systems are described by Slater (1983).

Gilbertson, Longshaw, and Viljoen (1976) compared multispectral photographs with normal color and IR color aerial photographs for mineral exploration in South Africa. They concluded that multispectral photographs have no significant advantage over conventional photographs in this area. The color photographs composited from multispectral black-and-white photographs of South Af-

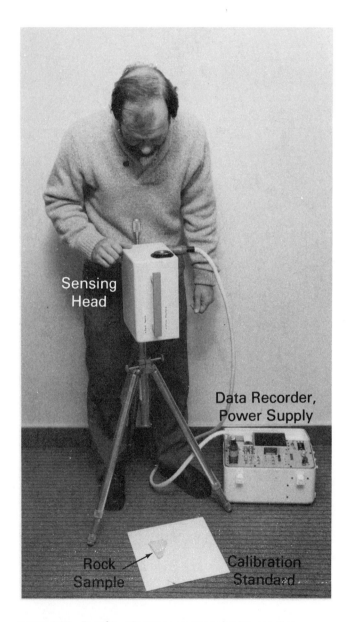

FIGURE 2.27 Portable reflectance spectrometer for recording spectra in the visible and reflected IR regions. Manufactured by Geophysical Environmental Research, Incorporated.

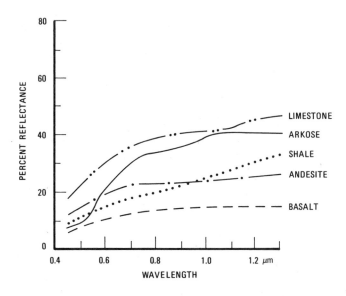

FIGURE 2.28 Reflectance spectra of typical volcanic and sedimentary rocks, as measured in the field. From Goetz (1976, Figure 1).

rica lack the ground resolution of normal color photographs and are 10 times as expensive to produce. The major advantage of multispectral photography is that the individual images provide more options for displaying the data. Multispectral scanners (which are discussed later in the chapter) have now largely replaced multispectral cameras.

SOURCES OF AERIAL PHOTOGRAPHS

The distribution center for aerial photographs acquired by the U.S. Geological Survey and NASA is

U.S. Geological Survey
EROS Data Center
Sioux Falls, SD 57198

An inquiry in the EROS Data Center (EDC) should specify the latitude and longitude boundaries of the desired area and the type of photography required. The major categories are

1. *aerial mapping photography*: typically at 1:40,000 scale or larger

2. *National High Altitude Photography*: 1:80,000 black-and-white and 1:58,000 color IR

3. *NASA aircraft photography:* 1:120,000 scale black-and-white, normal color, and IR color

EDC will provide computer listings of available photographs, price lists, and instructions for selecting and ordering the desired coverage.

The Agricultural Stabilization and Conservation Service has also photographed much of the United States. One can obtain a set of state index maps and ordering instructions from

Western Aerial Photography Laboratory
ASCS–USDA
Post Office Box 30010
Salt Lake City, UT 84130

Black-and-white photographs of U.S. national forests are available from regional offices of the U.S. Forestry Service. Many local aerial photography contractors have negatives and can furnish prints of areas over which they have flown. If necessary, a contractor can be hired to fly and take the needed photographs.

MULTISPECTRAL SCANNER IMAGES

The images described to this point have been detected and recorded using cameras with various combinations of films and filters that are sensitive in the UV, visible, and reflected IR spectral bands. Multiband cameras simultaneously image the same scene with a range of film-and-filter combinations to acquire photographs at various narrow spectral bands of energy. Multiband cameras have largely been replaced by *multispectral scanners*, which record several images at different wavelength bands. The data are recorded in digital form and then

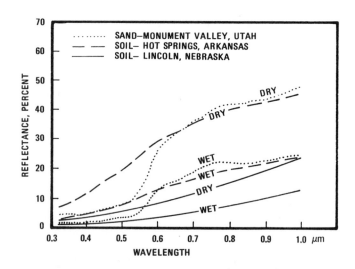

FIGURE 2.29 Reflectance spectra of wet and dry sandy soils. From Condit (1970, p. 955).

processed to produce multispectral images. Chapter 1 described multispectral versions of cross-track and along-track scanners. The following sections illustrate and interpret images acquired by these systems.

Cross-Track Multispectral Scanner Images

Cross-track multispectral scanners have been deployed on aircraft since the 1960s. Figure 2.30 shows images that a NASA U-2 aircraft acquired using a cross-track multispectral scanner. Table 2.4 summarizes the characteristics of the scanner and the 10 multispectral images.

Plate 2B is a normal color composite image that was prepared by projecting bands 2, 4, and 7 in blue, green, and red respectively. The IR color composite image of Plate 2C was prepared by projecting bands 4, 7, and 8 in blue, green, and red. These color composites of scanner images are analogous to the normal color and IR color photographs in Plate 1. Figure 2.31 shows the categories of land use and land cover in the eastern margin of San Pablo Bay, which is the northern extension of San Francisco Bay. The black-and-white images of the individual spectral bands (Figure 2.30) demonstrate the relationships between wavelength, atmospheric scattering (Figure 2.2), contrast ratio (Figure 2.3), and spatial

TABLE 2.4 Characteristics of the Daedalus aircraft multispectral scanner and images

Aircraft altitude	19.5 km
Scanner IFOV	1.25 mrad
Ground resolution cell	24 by 24 m
Scan angle	42°
Image-swath width	14.7 km

Band*	Wavelength, μm	Spectral region
1	0.38 to 0.42	UV and blue
2	0.42 to 0.45	Blue
3	0.45 to 0.50	Blue
4	0.50 to 0.55	Green
5	0.55 to 0.60	Green
6	0.60 to 0.65	Red
7	0.65 to 0.70	Red
8	0.70 to 0.80	Reflected IR
9	0.80 to 0.90	Reflected IR
10	0.90 to 1.10	Reflected IR

*A composite of bands 2, 4, 7 (in blue, green, and red) produce a normal color image. A composite of bands 4, 7, and 8 (in blue, green, and red) produce an IR color image.

resolution. Band 1 in the UV and blue region records the shortest wavelengths of all the bands and has the maximum atmospheric scattering, resulting in a low contrast ratio and poor spatial resolution. The spectral band of this image is similar to that of the Wratten 18A filter used in UV photography (Figure 2.20). The network of streets in the city of Vallejo is a useful resolution target; as the wavelength of the images increases, the ability to discern the streets improves and reaches a maximum in the reflected IR region (bands 8, 9, and 10). Bands 4 through 7 in the green and red spectral regions are analogous to the minus-blue photographs illustrated earlier in the chapter (Figures 2.7, 2.8, and 2.11). Bands 8 through 10 in the reflected IR region are analogous to the IR black-and-white photograph in Figure 2.19B.

Water, vegetation, and urban area are the major types of land cover and land use in the San Pablo Bay area (Figure 2.31). It is instructive to compare the signatures of vegetation in the various multispectral bands with the spectral reflectance curves of Figure 2.24D. The native vegetation in the Diablo Range has a somewhat higher reflectance in the green images (bands 4 and 5) than in the blue (bands 2 and 3) and red images (bands 6 and 7), where chlorophyll absorbs energy. Vegetation has a very bright signature in the reflected IR images (bands 8, 9, and 10). The signature of water is also different in the various spectral bands. In San Pablo Bay, patterns of suspended silt are obvious in the visible bands; in the IR bands, however, water has a uniform dark signature because these wavelengths are completely absorbed. In the IR color composite image (Plate 2C), the green and red images cause the water patterns. Vegetated areas border the tidal flats along the Napa River on the south; the tidal flats are not distinguishable from vegetation in the visible bands. In the IR bands, however, the vegetation has a bright signature and wet mudflats are dark. Some of the evaporating ponds for producing salt are red and pink because of microorganisms and have bright signatures in bands 6 and 7. These ponds are dark because water absorbs energy in the IR bands.

Along-Track Multispectral Scanner Images

The *airborne imaging spectrometer* (AIS) is an along-track multispectral scanner developed at Jet Propulsion Laboratory (Goetz, 1984). Figure 2.32 illustrates 32 images acquired by AIS in the spectral region from 1.20 to 1.51 μm. Each image records a separate spectral band 0.01 μm wide; thus the image at the left margin of the strip covers the spectral region from 1.20 to 1.21 μm. For the 32 spectral bands there is a linear array of 32 detectors, each with an IFOV of 1.9 mrad, that forms a ground resolution cell of 10 by 10 m at the flight altitude

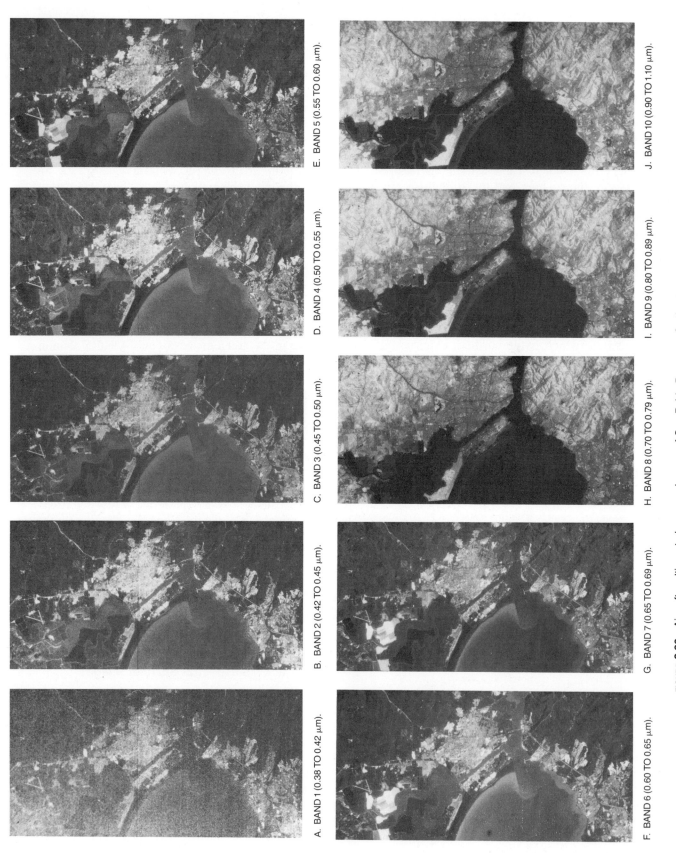

A. BAND 1 (0.38 TO 0.42 μm).

B. BAND 2 (0.42 TO 0.45 μm).

C. BAND 3 (0.45 TO 0.50 μm).

D. BAND 4 (0.50 TO 0.55 μm).

E. BAND 5 (0.55 TO 0.60 μm).

F. BAND 6 (0.60 TO 0.65 μm).

G. BAND 7 (0.65 TO 0.69 μm).

H. BAND 8 (0.70 TO 0.79 μm).

I. BAND 9 (0.80 TO 0.89 μm).

J. BAND 10 (0.90 TO 1.10 μm).

FIGURE 2.30 Aircraft multispectral scanner images of San Pablo Bay area, California, acquired March 28, 1980. Each image covers an area of 15 by 30 km. Courtesy NASA Ames Research Center.

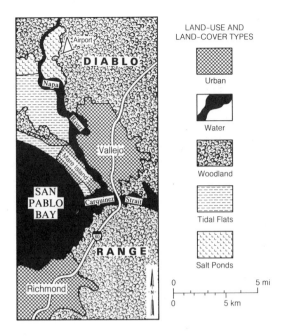

LAND-USE AND
LAND-COVER TYPES

Urban

Water

Woodland

Tidal Flats

Salt Ponds

0 ——————— 5 mi
0 ——————— 5 km

FIGURE 2.31 Land-use and land-cover types of the San Pablo Bay area, California, interpreted from aircraft multispectral scanner images.

Advantages of Multispectral Scanner Images

Multispectral scanners have the following advantages over multispectral cameras:

1. The scanner images are acquired with a single optical system and are perfectly registered to each other; this is not the case in many multiband photographs.

2. Scanner detectors can record wavelengths beyond the sensitivity range of film.

3. Advanced scanner systems can acquire hundreds of multispectral images at very narrow spectral bands, which are essential for detailed spectral analysis and mapping.

4. The digital data from scanners are calibrated and suitable for computer processing (see Chapter 7).

Some disadvantages of scanners are that they are complex, expensive, and require computer processing to produce images. In contrast, cameras are relatively inexpensive and simple to operate, and the film requires only chemical processing. The technology of airborne scanners is directly applicable to spacecraft, as Chapter 4 describes.

COMMENTS

Aerial photographs are a versatile and useful form of remote sensing for the following reasons:

1. The film provides excellent spatial resolution and has a high information content.

2. Photographs cost relatively little.

3. Different films provide a sensitivity range from the UV spectral region through the visible and into the reflected IR region.

4. Low-sun-angle photographs enhance subtle topographic features that are suitably oriented with respect to the sun's azimuth.

5. Stereo photographs are valuable aids for many types of interpretation.

The principal drawbacks of aerial photographs are that

1. Daylight and good weather are necessary to acquire them.

2. In the shorter wavelength regions, atmospheric scattering reduces their contrast ratio and resolving power.

of this survey. The ground swath of each image is 320 m. As the spectral diagram of Figure 1.2 shows, the spectral region (1.20 to 1.51 μm) of the AIS images is located in the reflected IR portion of the electromagnetic spectrum and includes an absorption band caused by water vapor in the atmosphere. This absorption effect causes the blurred appearance of the AIS images (Figure 2.32B) in the vicinity of 1.40 μm.

The AIS images (Figure 2.32B) cover the 320-m ground swath outlined in the aerial photograph of a school and its surroundings in the city of Van Nuys in the Los Angeles area (Figure 2.32A). In the photograph and images of Figure 2.32, area a is a well-watered lawn and area b is a dry field. In the AIS images, reflectance of the lawn (a) decreases at the longer wavelengths as the progressively darker signature shows. Reflectance of the dry field (b) increases at longer wavelengths. The responses are consistent with spectral reflectance curves measured for these types of terrain and degrees of moisture.

Jet Propulsion Laboratory is developing an advanced airborne sensor called *airborne visible and infrared imaging spectrometer* (AVIRIS) that will acquire 224 images, each with a spectral bandwidth of 9.6 nm in the region from 0.4 to 2.4 μm. When flown in a NASA U-2 aircraft, AVIRIS will cover a swath 11 km wide with 20-m ground resolution cells.

| 320 m |

A. AERIAL PHOTOGRAPH.

Water vapor
absorption band

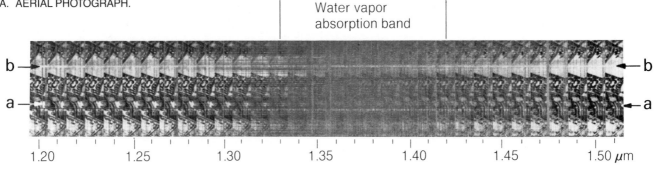

b →   ← b
a →   ← a

| 1.20 | 1.25 | 1.30 | 1.35 | 1.40 | 1.45 | 1.50 μm |

B. THIRTY-TWO SPECTRAL BANDS OF AIS IMAGES.

FIGURE 2.32 Images acquired by the airborne imaging spectrometer over Van
Nuys, California. Locality a is a well-watered lawn and locality b is a dry field. From
Goetz (1984). Courtesy A. F. H. Goetz, Jet Propulsion Laboratory.

3. Variations in reflectance are recorded in uncalibrated fashion, which precludes quantitative interpretation.

The advantages often outweigh the disadvantages, and one should evaluate aerial photographs as a possible data source for any remote sensing investigation.

Multispectral scanner images are a valuable alternative source of information.

QUESTIONS

1. Normal color photographs taken of subjects in shaded areas have a bluish cast. Explain this bluish cast.

2. Calculate the contrast ratio for a scene in which the brightest and darkest areas have brightness values of 6 and 2 respectively.

3. Suppose the scene in question 2 is covered by an atmosphere that contributes four brightness values of scattered light. What is the resulting contrast ratio? Panchromatic aerial photographs will be acquired of this scene. How can their contrast ratio be improved?

4. What is the ground resolution for aerial photographs acquired at a height of 5000 m with a camera having a system resolution of 30 line-pairs · mm^{-1} and a focal length of 304 mm?

5. What is the minimum ground separation in the photographs of question 4?

6. What is the scale of the photographs of question 4?

7. For Figure 2.8, calculate the height of the highest portion of the building in the extreme lower left corner.

8. The air base for two overlapping photographs is 1500 m. The photographs were acquired from a height of 3000 m. What is the base–height ratio of this stereo pair? What is the vertical exaggeration of the stereo model?

9. For the multispectral images of San Pablo Bay (Figure 2.30), select the single optimum band for each of the following uses:

   a. Distinguishing types of native vegetation in the Diablo Range
   b. Distinguishing between different salt evaporating ponds
   c. Identifying bridges
   d. Mapping urban land-use categories
   e. Mapping variations in the silt content of water

REFERENCES

Condit, H. R., 1970, The spectral reflectance of American soils: Photogrammetric Engineering and Remote Sensing, v. 36, p. 955–966.

Gausman, H. W., 1985, Plant leaf optical properties in visible and near-infrared light: Texas Tech University Graduate Studies, No. 29, Lubbock, Texas.

Gilbertson, B., T. C. Longshaw, and R. P. Viljoen, 1976, Multispectral aerial photography as exploration tool—an application in the Khomas Trough region, Southwest Africa; and cost effectiveness analysis and conclusions: Remote Sensing of Environment, v. 5, p. 93–107.

Goetz, A. F. H., 1976, Remote sensing geology—Landsat and beyond: Caltech/JPL Conference on Image Processing Technology, Data Sources, and Software for Commercial and Scientific Applications, Jet Propulsion Laboratory SP 43-30, p. 8-1 to 8-8.

Goetz, A. F. H., 1984, High spectral resolution remote sensing of the land: Society Photo-Optical Instrumentation Engineers, Proceedings, v. 475, p. 56–68.

Hunt, G. R., 1980, Electromagnetic radiation—the communication link in remote sensing in Siegal, B. S., and A. R. Gillespie, eds., Remote sensing in geology: ch. 2, p. 5–45, J. Wiley & Sons, New York.

James, T. H., 1966, The theory of the photographic process, third edition, Macmillan Co., New York.

Jones, R. C., 1968, How images are detected: Scientific American, v. 219, p. 111–117.

Kodak, 1970, Kodak filters for scientific and technical uses: Eastman Kodak Technical Publication B-3, Rochester, N.Y.

La Prade, G. L., 1972, Stereoscopy—a more general theory:
Photogrammetric Engineering and Remote Sensing, v. 38, p. 1177–1187.

La Prade, G. L., 1973, Stereoscopy—will data or dogma prevail: Photogrammetric Engineering and Remote Sensing, v. 39, p. 1271–1275.

Miller, C. V., 1961, Photogeology: McGraw-Hill Book Co., New York.

Rosenblum, L., 1968, Image quality in aerial photography: Optical Spectra, v. 2, p. 71–73.

Sabins, F. F., 1973a, Aerial camera mount for 70-mm stereo: Photogrammetric Engineering and Remote Sensing, v. 39, p. 579–582.

Sabins, F. F., 1973b, Engineering geology applications of remote sensing in Moran, D. E., ed., Geology, seismicity, and environmental impact: Association of Engineering Geologists, Special Publication, p. 141–155, Los Angeles, Calif.

Slater, P. N., 1983, Photographic systems for remote sensing in Colwell, R. N., ed., Manual of remote sensing, second edition: ch. 6, p. 231–291, American Society for Photogrammetry and Remote Sensing, Falls Church, Va.

Thurrell, R. F., 1953, Vertical exaggeration in stereoscopic models: Photogrammetric Engineering and Remote Sensing, v. 19, p. 579–588.

Vizy, K. N., 1974, Detecting and monitoring oil slicks with aerial photos: Photogrammetric Engineering and Remote Sensing, v. 40, p. 697–708.

Walker, P. M., and D. T. Trexler, 1977, Low sun-angle photography: Photogrammetric Engineering and Remote Sensing, v. 43, p. 493–505.

Wolf, P. R., 1974, Elements of photogrammetry: McGraw-Hill Book Co., New York.

ADDITIONAL READING

Avery, T. E., 1962, Interpretation of aerial photographs, third edition: Burgess Publishing Co., Minneapolis, Minn.

Colwell, R. N., ed., 1960, Manual of photographic interpretation: American Society for Photogrammetry and Remote Sensing, Falls Church, Va.

Cravat, H. R., and R. Glaser, 1971, Color aerial stereograms of selected coastal areas in the United States: U.S. Department of Commerce, National Oceanic and Atmospheric Administration, Washington, D.C.

Kodak, 1962, Color as seen and photographed, second edition: Eastman Kodak Co., Rochester, N.Y.

Lattman, L. H., and R. G. Ray, 1965, Aerial photographs in field geology: Holt, Reinhart & Winston, New York.

Mollard, J. D., and J. R. Jones, 1984, Airphoto interpretation and the Canadian landscape: Canadian Government Publishing Center no. M52-60/1984E, Hull, Canada.

Paine, D. P., 1981, Aerial photography: John Wiley & Sons, New York.

Ray, R. G., 1960, Aerial photographs in geologic mapping and interpretation: U.S. Geological Survey Professional Paper 373.

Smith, J. T., and A. Anson, eds., 1968, Manual of color aerial photography: American Society for Photogrammetry and Remote Sensing, Falls Church, Va.

Manned Satellite Images

The first photographs of the earth from very high altitudes (160 to 320 km) were acquired after World War II by small automatic cameras on unmanned sounding rockets launched from White Sands Missile Range, New Mexico. Despite the mediocre quality of these photographs, as judged by current standards, Merifield (1964) was able to interpret a number of geologic features. During the testing phase for the Mercury project, NASA placed an unmanned satellite in earth orbit with a camera that acquired several hundred high-angle oblique normal color photographs in a 70-mm format. These photographs were used to study the western Sahara, the first regional application of orbital photographs. The greatest value to remote sensing of these early unmanned photography experiments was the interest they created for imaging the earth from space.

Table 3.1 lists the U.S. manned satellite programs. The Mercury and Gemini missions were tests of people and equipment in preparation for the Apollo missions that landed humans on the moon. All three of these programs, however, acquired photographs of the earth. Skylab included several remote sensing experiments. The Space Shuttle has provided the widest spectral range of images of all the manned missions.

GEMINI AND APOLLO MISSIONS

The Gemini two-man missions orbited the earth in 1965 and 1966 primarily to test people, equipment, and procedures for the later Apollo lunar missions; terrain photography was a subordinate objective.

Gemini orbit paths ranged between latitudes 35°N and 35°S at altitudes from 160 to 320 km. The astronauts acquired photographs with hand-held 70-mm cameras through the spacecraft windows. All photography was on normal color film, except for one magazine of IR color film, which produced unsatisfactory results. NASA (1967) published a collection of color photographs from Geminis 3, 4, and 5. Approximately 1100 photographs were usable for geology, geography, or oceanography. Figure 3.1 is a black-and-white copy of a typical Gemini-4 oblique color photograph in western Texas with readily identifiable geologic and topographic features. The linear expression of the Hillside fault that trends northwest from Van Horn toward Sierra Blanca is especially pronounced. Rocks south of the fault near Van Horn are nearly vertical, north-striking metamorphosed strata of Precambrian age that form aligned ridges, which are dragged to the west along the fault. Gently dipping

TABLE 3.1 NASA manned satellite programs

Program	Dates	Crew	Remarks
Mercury	1962–1963	1	A few hand-held photographs were taken.
Gemini	1964–1965	2	Some hand-held photographs were taken.
Apollo	1968–1972	3	First multispectral photographs acquired from orbit.
Skylab	1973–1974	3	Skylab was occupied by three different crews that acquired photographs and multispectral scanner images of earth.
Space Shuttle	1981–present	3 to 7	LFC, SIR, hand-held cameras, and MOMS have been deployed on the Shuttle.*
Space Station	1995		Main station will be in a low-altitude orbit at 28.5° inclination; an associated manned polar platform will be in a near-polar, sun-synchronous orbit.

*A later section of this chapter ("Space Shuttle") describes MOMS, SIR, and LFC.

Cretaceous strata are exposed north of the Hillside fault.

Lowman and Tiedemann (1971) summarized geologic applications of the Gemini terrain photography and pointed out the following accomplishments:

1. A previously unmapped Quaternary volcanic field was discovered in northern Mexico. The field includes over 30 volcanoes with associated basalt flows.

2. The Gemini mission acquired photographs of the northern part of Baja California that showed the extent and geologic relationships of the Agua Blanca fault zone, which is a major structural feature of the region.

3. West of the area shown on Figure 3.1, the Texas lineament in southwest New Mexico and southeast Arizona was shown to be a broad zone of folds and faults rather than a major wrench fault. Chapter 4 summarizes the significance of lineaments.

4. A very large area in North Africa was found to have been eroded primarily by deflation and wind erosion, suggesting that wind erosion is more important in the Sahara than in North America and other deserts.

5. The unexpected views of regional geologic structure stimulated speculative thinking about regional tectonics. Abdel-Gawad (1969) recognized two major shear zones in the crystalline rocks on opposite sides of the Red Sea; the offset of these distinctive features is evidence that Arabia has moved northward 150 km

relative to Africa. (See Chapter 4 for a discussion of plate tectonics.)

The three-man Apollo missions carried the first astronauts to the moon, but during the testing phases they orbited the earth many times. During these earth orbits, the astronauts acquired several hundred hand-held 70-mm photographs.

The Apollo 9 earth-orbiting test mission included the SO-65 experiment, which acquired the first orbital multispectral photographs. Four 70-mm cameras were mounted in the window of the command module and triggered simultaneously. Three cameras were equipped with film-and-filter combinations to acquire black-and-white photographs in the green, red, and photographic IR spectral regions. The fourth camera used IR color film with a yellow filter. Ninety relatively cloud-free sets of photographs covered portions of northwest and southern Mexico, the Caribbean-Atlantic area, and the United States south of latitude 35°N. The resulting photographs were similar to those acquired during the later Skylab missions, which are illustrated in the following section. Lowman (1969) interpreted Apollo 9 photographs and demonstrated the value of multispectral satellite images for interpreting both arid and forested terrain.

SKYLAB

Skylab was the first manned mission dedicated to earth observations. The vehicle consisted of the third stage of a Saturn-V launch vehicle that had been converted

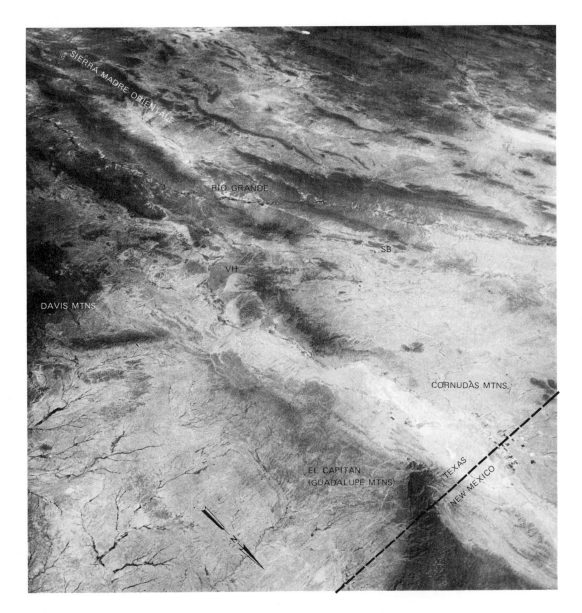

FIGURE 3.1 Gemini-4 photograph looking southwest across western Texas. Albritton and Smith (1956) proposed the Hillside fault between Van Horn (VH) and Sierra Blanca (SB) as the type locality for the Texas lineament. Courtesy P. D. Lowman, NASA Goddard Space Flight Center.

to living and working quarters for the astronauts. Skylab was launched unmanned and was occupied sequentially by three crews, each consisting of three men, who used Apollo vehicles to dock with Skylab and to return to earth. Observations of earth were made from May 25, 1973, through February 8, 1974. The 50° inclination of the orbit path provided coverage of the earth between latitudes 50°N and 50°S at an altitude of 435 km (Figure 3.2). The orbit paths were repeated every five days, but image coverage of the earth was incomplete because of clouds and astronauts' schedules. The Earth Resources Experiment Package (EREP) acquired remote sensing images of the earth (NASA 1974). Magnetic tapes and exposed camera film returned to earth with the crews. A circular scanner acquired multispectral images, but few of these data were analyzed. The most important systems were the multispectral camera and the earth-terrain camera, which are described in the following sections.

Multispectral Camera (S-90A Experiment)

The multispectral camera consisted of six cameras that were aimed at the same point on the earth and

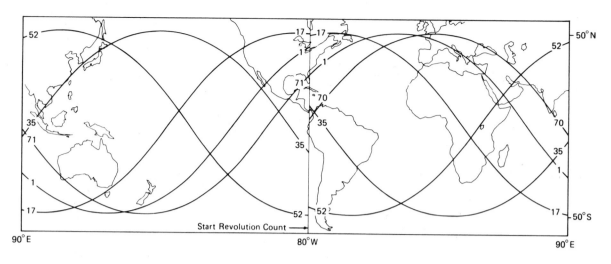

FIGURE 3.2 Typical Skylab orbit paths and numbers. From NASA (1974, Figure 4).

triggered simultaneously. The photographs were recorded on 70-mm film. The 21.2° angular field of view of the 152-mm focal-length lenses provided ground coverage of 163 by 163 km. The original film has a scale of 1:2,850,000 and can be enlarged more than 10 times with little loss of detail. Table 3.2 gives the spectral band, film type, and ground resolution for each camera.

Figure 3.3 shows three of the black-and-white multispectral photographs of the Salton Sea and the Imperial Valley, California. The green band provides the maximum information on turbidity distribution within the Salton Sea. The red band has maximum tonal contrast, which emphasizes differences in agriculture between the U.S. and Mexican parts of the Imperial Valley. On the IR photograph, the strong reflectance of vigorous vegetation produces the light tones in the Imperial Valley that are similar to the tone of the desert on either side of the valley. Vegetation is readily distinguished from desert, however, on the green and red bands. Very dark

rectangles in the Imperial Valley are reservoirs and fields that are flooded with water and can be distinguished only on the photographic IR band. This capability to distinguish targets in different spectral bands is a major advantage of multispectral photographs and images. On all three bands, the south-southeast-trending Elsinore fault is prominent, especially in Mexico. Several smaller, south-trending faults cross the international boundary west of the Elsinore fault zone and are marked by prominent linear tonal anomalies. The black-and-white transparencies from cameras 5, 6, and 1 or 2 can be composited with green, blue, and red light respectively to produce an IR color photograph. The optimum color balance can be obtained during the compositing process.

Earth-Terrain Camera (S-190B Experiment)

The *earth-terrain camera* (ETC) had a 45.7-cm focal-length lens and acquired photographs with a ground coverage of 109 by 109 km. Normal color, IR color, and black-and-white photographs were acquired at a scale of 1:950,000 on 11.4-cm square film. Ground resolution of second-generation images is 15 m for the black-and-white photographs and 30 m for the normal color and IR color photographs. According to Welch (1976), the black-and-white photographs can be enlarged to scales of 1:50,000 (20×) and the color photographs to 1:100,000 (10×) before blurring occurs.

Figure 3.4 is a typical ETC photograph that covers the Kaiparowits Plateau and Lake Powell in south-central Utah. Major features are shown in the location map of Figure 3.5. The photograph clearly shows the gentle dips of the Mesozoic strata in this part of the Colorado Plateau. The steeply east-dipping East Kaibab monocline separates the Kaibab Plateau on the west from the

TABLE 3.2 Skylab multispectral camera (S-190A) experiment

Camera	Spectral band, μm	Film type	Ground resolution, from Welch (1974), m
1*	0.7 to 0.8	IR black-and-white	145
2	0.8 to 0.9	IR black-and-white	145
3	0.5 to 0.9	IR color	145
4	0.4 to 0.7	Normal color	85
5*	0.6 to 0.7	Black-and-white	60
6*	0.5 to 0.6	Black-and-white	60

*Illustrated in Figure 3.3.

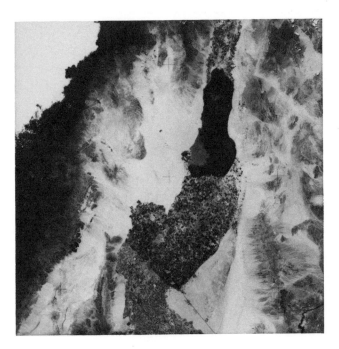

A. GREEN (0.5 TO 0.6 μm), CAMERA 6.

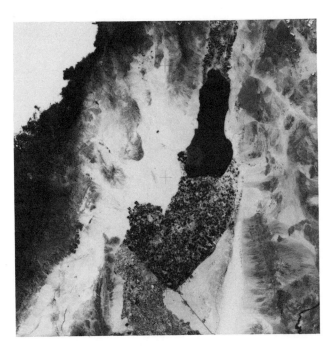

B. RED (0.6 TO 0.7 μm), CAMERA 5.

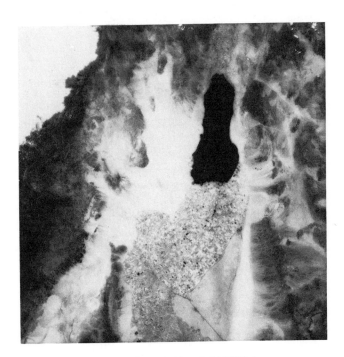

C. PHOTOGRAPHIC IR (0.7 TO 0.8 μm), CAMERA 1.

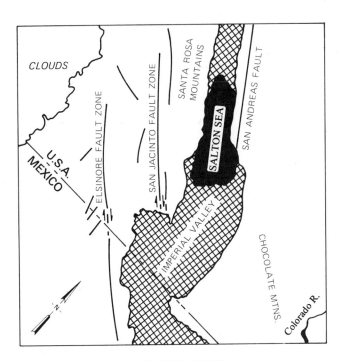

D. LOCATION MAP SHOWING MAJOR FAULTS.

FIGURE 3.3 Skylab multispectral camera (S-190A) photographs of southern California and Mexico. Acquired June 2, 1973.

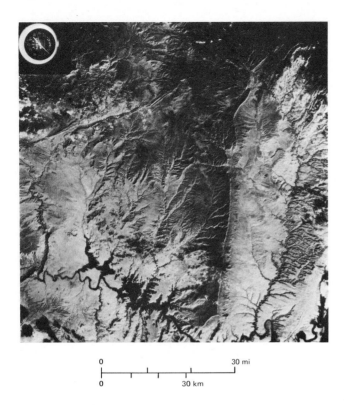

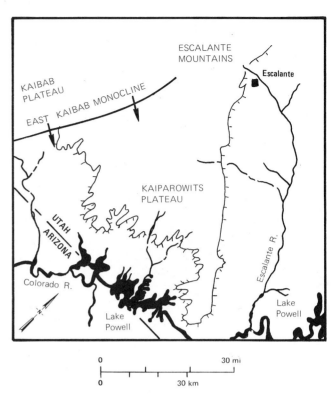

FIGURE 3.4 Skylab-3 earth-terrain camera (S-190B) photograph of southern Utah. The low sun elevation enhances topographic and structural features.

FIGURE 3.5 Location map for Skylab S-190B photograph of south-central Utah.

Kaiparowits Plateau. The high spatial resolution of the ETC photographs is shown by the detailed expression of badlands topography along the Escalante River in the eastern part of the photograph. Many of the ETC photographs were acquired with 60 percent overlap and may be viewed stereoscopically.

SPACE SHUTTLE

The preceding manned missions were all experimental and not designed for maximum operational efficiency or economy. With the exception of Skylab, the vehicles were small with no capacity for sophisticated remote sensing systems. Of necessity, the astronauts were primarily pilots rather than scientists trained to observe the earth and acquire data from space. All of the launch rockets were jettisoned into the ocean, and only the small command modules were retrieved at the end of a mission. The Space Shuttle program was designed to overcome these deficiencies.

The Space Shuttle vehicle (Figure 3.6) is as large as a medium-size commercial jet airliner and accommodates a crew of up to seven. Three crew members are primarily concerned with piloting the craft; the remaining members are mission specialists, scientist-astronauts who conduct various experiments during the missions, which last up to 9 days. Experiments include materials fabrication, chemical synthesis, and biological and medical testing. A number of communications satellites have been placed in earth orbit from the Shuttle, although there have been several failures not related to the Shuttle. The vehicle can dock with malfunctioning satellites, which the crew may then repair and refuel in orbit or return to earth in the cargo bay.

Figure 3.7 shows a Shuttle mission in condensed form. At launch, the Shuttle is attached to a large liquid-fuel tank feeding three engines plus two smaller solid-propellant rockets. Shortly after launch, the solid-propellant rockets are expended and return on parachutes to the ocean, where they are retrieved for future use. Later the liquid fuel is expended, and the external tank is jettisoned and disintegrates on reentering the atmosphere. Once the Shuttle is in orbit, it can maneuver with two small *orbital maneuvering system* (OMS) engines. Doors to the cargo bay open, and the Shuttle inverts to aim remote sensing systems at the earth (Figure 3.6). Image data are recorded on film and magnetic

FIGURE 3.6 The Space Shuttle in orbit with cargo bay doors open.

tape or telemetered to earth receiving stations. After completion of a mission, the OMS engines fire in retrograde fashion to cause reentry and an unpowered landing (Figure 3.7). From 1981 to 1986, Shuttle missions were launched from Kennedy Space Center at Cape Canaveral on the east coast of Florida into low, *oblique orbits* (which do not cover areas at high latitudes) similar to those of Skylab (Figure 3.2). Beginning in 1987 or 1988, Shuttle missions will be launched from Vanden-

berg Air Force Base, on the California coast, into near polar orbits that provide almost complete global coverage.

A number of remote sensing systems have been operated from the Shuttle, including the *Shuttle imaging radar* (SIR–A&B) experiments in 1981 and 1984 that are described in Chapter 6. Two Shuttle missions have carried the *modular optoelectronic multispectral scanner* (MOMS) of the German Aerospace Research Establishment (DFVLR). MOMS is an along-track scanner that records two spectral bands of data (0.575 to 0.625 μm and 0.825 to 0.975 μm) with ground resolution cells of 20 by 20 m. MOMS has acquired relatively few images, which are available only to investigators selected by DFVLR.

Of prime interest in this chapter are the photographs acquired by hand-held cameras and by the large-format camera. These are described in the following sections.

Hand-Held Photographs

The Shuttle has several windows that enable astronaut-observers to acquire photographs with hand-held cameras at various angles including vertical, that is, pointed directly downward. Most photographs are taken on normal color film in 70-mm and 140-mm formats. Figure 3.8 is a black-and-white reproduction of a color photograph of the Denham Sound on the northwest coast of Australia. Details of land features and shallow bathymetric features are visible. A few scattered clouds with shadows occur in the southeast portion of the photograph.

FIGURE 3.7 Profile of a Space Shuttle mission.

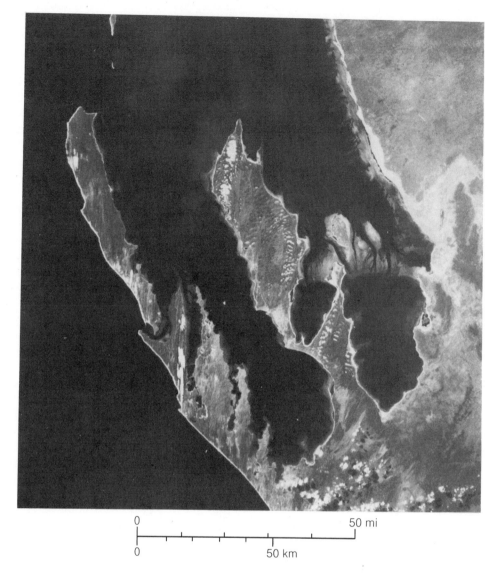

0   50 mi

0   50 km

FIGURE 3.8 Hand-held vertical photograph acquired from the Space Shuttle, June 1983. Denham Sound, northwest Australia.

Large-Format Camera

Commercial cameras modified for use in spacecraft acquired the satellite photographs described thus far. NASA contracted with the ITEK division of Litton Industries to design and build a *large-format camera* (LFC) specifically for the Shuttle. The large format refers to the film size of 23 by 46 cm, with the longer dimension in the direction of flight. Focal length of the lens is 30.5 cm, and spatial resolution is 80 line-pairs · mm^{-1}. LFC first orbited the earth on the October 1984 mission of the Shuttle, when it was mounted in the rear of the cargo bay, as Figure 3.9 illustrates. The camera was controlled

from Johnson Space Center and performed flawlessly. It acquired many exceptional scenes, some examples of which are illustrated in this chapter. LFC acquired 1520 black-and-white, 320 normal color, and 320 IR color photographs. Scale of the photographs ranged from 1:1,213,000 to 1:783,600 depending on the altitude of the Shuttle, which ranged from 370 to 239 km. Clouds were a problem, but 60 percent of the photographs show less than 36 percent cloud cover.

Plate 3 illustrates portions of two IR color photographs reproduced at reduced scale in order to accommodate the large size of the original photographs. Plate 3A covers part of the coast of Queensland, Australia,

LFC   SIR

FIGURE 3.9 Cargo bay of the Space Shuttle showing the large-format camera (LFC) and the Shuttle imaging radar (SIR) antenna.

and has a few scattered clouds over the land and the Coral Sea. The large white patches in the sea are shallow shoals of coral sand along the Great Barrier Reef. Closer to shore, a few vegetated islands form small red spots. North is to the left in this photograph. The city of Cairns, famed for marlin fishing, is located directly north of Cape Grafton, which is the prominent peninsula in the center of the photograph. The coastal area is relatively humid, supporting forest cover and agriculture, but inland the Atherton tableland is drier and lacks the red tones.

Plate 3B covers part of the eastern foothills of the Andes Mountains in Argentina. North-trending eroded, faulted, and folded sedimentary rocks form the prominent ridges in the western (right) portion of the photograph. The dark elliptical areas in the eastern portion are large eroded uplifts of bedrock. Vegetation is scarce except for a few patches at the town of San Jose de Jachal (left center) and the outskirts of the city of San Juan in the lower right part of the photograph.

Figure 3.10 is a contact print of a portion of a black-and-white negative at the original scale of 1:769,000. The photograph covers parts of Massachusetts, Rhode

Island, and the Atlantic Ocean. Figure 3.11 is a 15× enlargement of Boston that shows the high spatial resolution of LFC photographs and their potential for interpreting land use and land cover.

A number of LFC photographs were acquired with forward overlap ranging from 20 to 80 percent, as Figure 3.12 shows. The chart shows base–height ratio, at a 239-km altitude, for the different degrees of overlap. Vertical exaggeration of the resulting stereo models is determined from the chart in Figure 2.12. Figure 3.13 illustrates a stereo pair compiled from LFC contact prints with 60 percent overlap. The 4× vertical exaggeration is obvious when this pair is viewed with a stereoscope. The area is in the Mojave Desert of southeast California and adjacent Nevada and is part of the Basin Range province. The mountains are fault blocks surrounded by valleys filled with detritus eroded from the mountains. Geologic features are particularly well expressed in this stereo model. The Spring Mountains in the east portion (upper part) of the photographs consist of generally west-dipping sedimentary rocks. The light-toned rocks that form prominent cliffs along the east front of the Spring Mountains are sandstone of Jurassic age. The sandstone is overlain on the west by dark-toned carbonate rocks of Paleozoic age that have been thrust eastward for many kilometers over the sandstone. Strike-slip faults of the Death Valley and Garlock fault systems are clearly seen at the north flank of the Avawatz Mountains. The cluster of dark circles at the south margin of the photograph near the California-Nevada border are fields watered by centerpoint irrigation systems.

The high spatial resolution, regional coverage, and stereo capability make LFC photographs a very valuable form of remote sensing. It is hoped that future Shuttle missions will routinely carry this camera.

SOURCES OF PHOTOGRAPHS

The EROS Data Center (EDC), in Sioux Falls, SD 57198, is a repository for photographs acquired by the manned satellite missions. For Skylab photographs, the *Skylab Earth Resources Data Catalog* (NASA, 1974) provides a complete listing. Index maps showing worldwide Skylab coverage are available. Gemini, Apollo, Skylab, and Space Shuttle photographs may also be purchased from

Technology Application Center
University of New Mexico
P.O. Box 181
Albuquerque, NM

The Technology Application Center also provides catalogs of the available photographs.

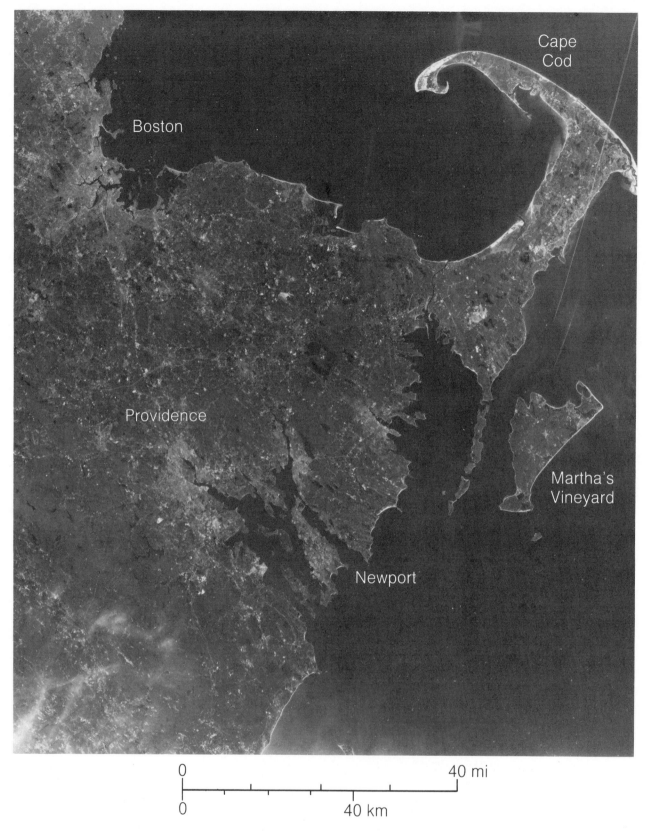

FIGURE 3.10 Portion of a LFC photograph of Massachusetts and Rhode Island at original scale (1:769,000). Acquired October 1984.

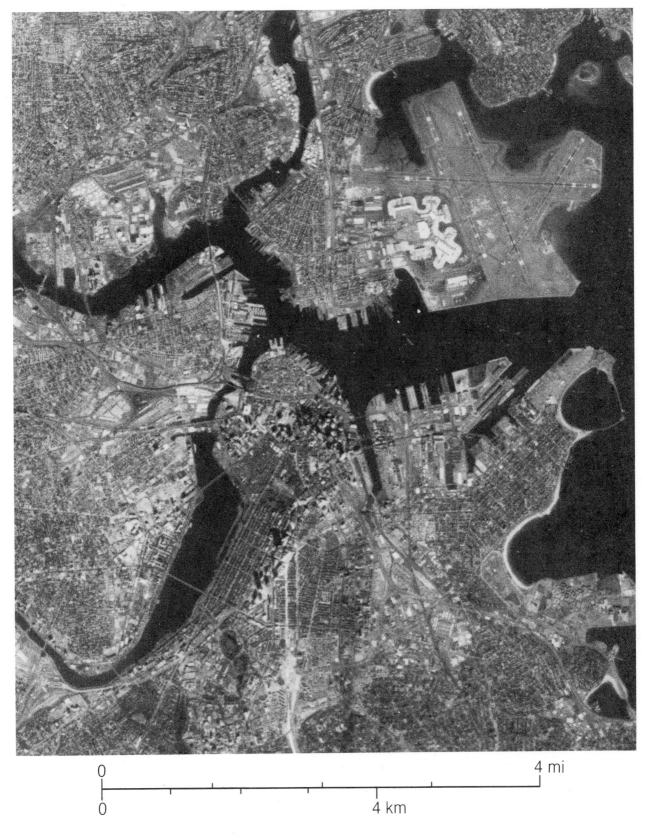

0   4 mi

0   4 km

FIGURE **3.11** LFC photograph of Boston, Massachusetts, enlarged 15×. Acquired October 1984.

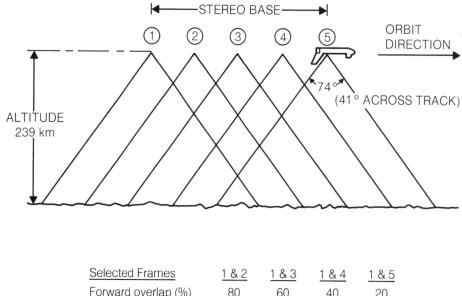

Selected Frames	1 & 2	1 & 3	1 & 4	1 & 5
Forward overlap (%)	80	60	40	20
Stereo base (km)	69	138	207	276
Base/height ratio	0.29	0.58	0.87	1.15
Vertical exaggeration	2.0	4.0	6.0	7.8

FIGURE 3.12 Geometry of overlapping photographs acquired by LFC on the Space Shuttle.

COMMENTS

The most important contribution to remote sensing made by the manned spacecraft missions, Mercury through Apollo, was to demonstrate the potential value and advantages of acquiring images from orbital altitudes. Despite the limited coverage, those images provided justification for Skylab and, more significantly, for the unmanned Landsat program, which is described in Chapter 4. The photographs acquired from Skylab demonstrated the utility of high-resolution photographs acquired from space.

The Space Shuttle is a manned platform which acquires LFC photographs, radar images, and other remote sensing data. LFC photographs with high spatial resolution, regional coverage, and stereo capability are valuable for many applications. The Space Shuttle will also provide transport for constructing a permanent, manned Space Station in the 1990s. The Space Station will be in a low-altitude equatorial orbit and will serve as the base for launching smaller unmanned satellites into polar orbits, from which remote sensing information can be relayed to earth. Astronauts from the Space Station will be able to rendezvous with the polar satellites to repair and refuel them. McElroy and Schneider (1984) describe the Space Station and polar orbiter concepts.

Later chapters describe unmanned satellite systems (Landsat and Seasat) for acquiring remote sensing images from the visible through radar spectral regions. These systems can function for up to several years while telemetering high-quality data to earth receiving stations. Unmanned systems have provided by far the bulk of satellite remote sensing data.

QUESTIONS

1. Summarize the advantages of manned satellite systems for gathering remote sensing data.

2. Summarize the disadvantages of manned satellite systems for acquiring remote sensing data.

3. What were the main contributions to remote sensing of manned missions before the Space Shuttle?

4. Summarize the qualification requirements for a remote sensing scientist-astronaut.

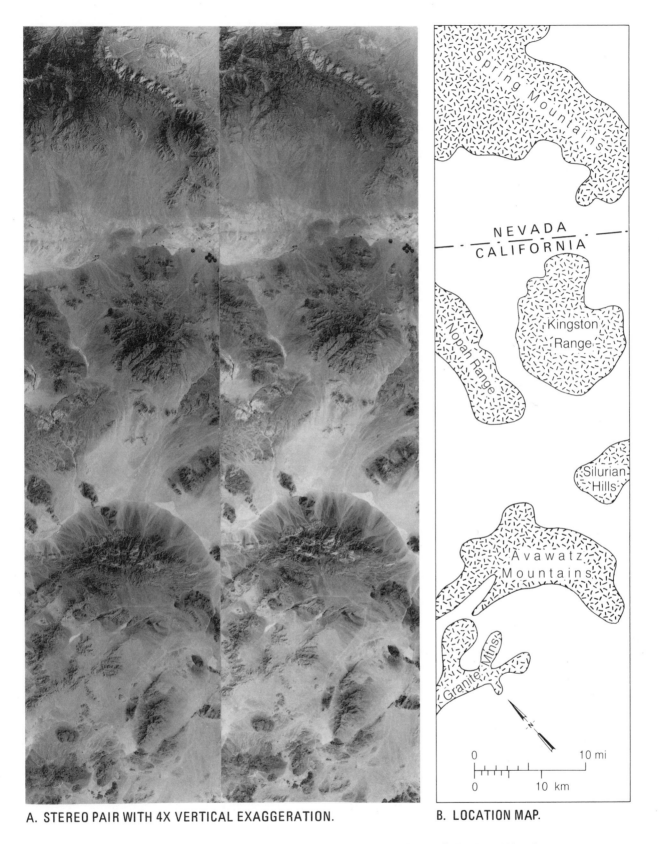

A. STEREO PAIR WITH 4X VERTICAL EXAGGERATION.

B. LOCATION MAP.

FIGURE 3.13   LFC photographs with 60 percent overlap. Mojave Desert, California and Nevada.

REFERENCES

Abdel-Gawad, M., 1969, New evidence of transcurrent movements in Red Sea area and petroleum implications: American Association Petroleum Geologists Bulletin, v. 53, p. 1466–1479.

Albritton, C. H., and J. F. Smith, 1956, The Texas Lineament *in* Tomo de Relaciones entre la tectonicas y la sedimentacion: 20th International Geological Congress, sec. 5, p. 501–518, Mexico, D.F.

Lowman, P. D., 1969, Geologic orbital photography—experience from the Gemini program: Photogrammetria, v. 24, p. 77–106.

Lowman, P. D., and H. A. Tiedemann, 1971, Terrain photography from Gemini spacecraft—final geologic report: NASA Goddard Space Flight Center Report X-644-71-15, Greenbelt, Md.

McElroy, J. H., and S. R. Schneider, 1984, Utilization of the polar platform of NASA's Space Station Program for operational earth observations: National Oceanic and Atmospheric Administration Technical Report NESDIS 12, Washington, D.C.

Merifield, P. M., 1964, Photo interpretation of White Sands rocket photography: Lockheed California Company Report LR17666, Burbank, Calif.

NASA, 1967, Earth photographs from Gemini 3, 4, and 5: NASA SP-129, Washington, D.C.

NASA, 1974, Skylab earth resources data catalog: NASA, JSC 09016, U.S. Government Printing Office, stock no. 3300-00586, Washington, D.C.

Welch, R., 1974, Skylab-2 photo evaluation: Photogrammetric Engineering and Remote Sensing, v. 40, p. 1221–1224.

Welch, R., 1976, Skylab S-190B ETC photo quality: Photogrammetric Engineering and Remote Sensing, v. 42, p. 1057–1060.

ADDITIONAL READING

Lowman, P. D., 1966, The earth from orbit: National Geographic, v. 129, p. 644–671.

Lowman, P. D., 1969, Apollo 9 multispectral photography—geologic analysis: NASA Goddard Space Flight Center, Report X-644-69-423, Greenbelt, Md.

NASA, 1977, Skylab explores the earth: NASA SP-380, Washington, D.C.

Nicks, O. W., ed., 1970, This island earth: NASA SP-250, Washington, D.C.

Landsat Images

Landsat is an unmanned system that prior to 1974 was called ERTS (Earth Resources Technology Satellite). Initially NASA operated Landsat, but in 1983 responsibility for operating the system transferred to the National Oceanic and Atmospheric Administration (NOAA). In 1985 operation of Landsat transferred to the EOSAT Company, a private corporation.

The system operates in the international public domain, which means that

1. an "open skies" policy allows images to be acquired of the entire earth without the United States first requesting permission from any government.

2. the EROS Data Center (EDC) archives all the images.

3. users anywhere in the world may purchase all images at uniform prices and priorities.

The science of remote sensing and the acquisition of images from satellites predated Landsat by many years, but the Landsat program is largely responsible for the growth and acceptance of remote sensing as a scientific discipline. For the first time, a repetitive world-wide database is available as a consistent set of images. The data are also available in a digital format suitable for computer processing. Present and future generations of

remote sensing specialists are indebted to the late William T. Pecora and William Fischer of the U.S. Geological Survey, who did so much to make Landsat a reality.

SATELLITE PLATFORMS AND ORBIT PATTERNS

Landsat satellites have been placed in orbit using Delta rockets launched from Vandenberg Air Force Base on the California coast between Los Angeles and San Francisco. The five Landsats belong to two generations of technology with different platforms and orbital characteristics (Table 4.1). Three different imaging systems have been orbited. Despite these differences, NASA and NOAA have maintained continuity in the flow of image data, and the image formats are compatible. Freden and Gordon (1983) give details of the Landsat platforms, orbits, and imaging systems.

Landsats 1, 2, and 3

The three satellites in this series were launched July 23, 1972, January 21, 1975, and March 5, 1978; all have ceased operation, but they produced hundreds of thousands of valuable images.

TABLE 4.1 Platforms and orbits of first and second generations of Landsat

	Landsats 1, 2, and 3	Landsats 4 and 5
Altitude	918 km	705 km
Orbits per day	14	14.5
Number of orbits (paths)	251	233
Image sidelap at equator	14 percent	7.6 percent
Crosses 40°N latitude at (local sun time, approximate)	9:30 a.m.	10:30 a.m.
Operational from	1972 to 1984	1982 to future
On-board data storage	Yes	No
Imaging systems:		
Multispectral scanner	Yes	Yes
Return-beam vidicon, panchromatic	Yes (Landsat 3)	No
Thematic mapper	No	Yes

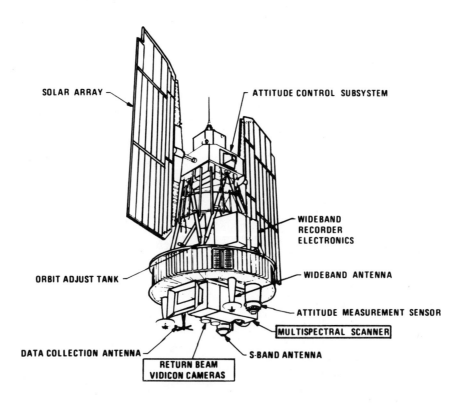

FIGURE 4.1 Satellite platform of Landsats 1, 2, and 3 showing location of multispectral scanner (MSS) and return-beam vidicon (RBV) systems.

Platform The platform (Figure 4.1) was modified from earlier Nimbus meteorological satellites. The solar arrays generated electrical power for operating the system. Fuel in the orbit adjust tank periodically operated small rockets in the attitude control subsystem to maintain the correct orbit and attitude. An array of microwave antennas received commands and transmitted image data to earth receiving stations. The *multispectral scanner* (MSS) and *return-beam vidicon* (RBV) were the imaging systems in this first generation of Landsat. Both systems are described in later sections.

Collecting the data produced by the imaging systems of all the Landsats has been a major concern. There are a number of Landsat receiving stations whose ranges are determined primarily by the curvature of the earth and the altitude of the satellite. When the satellite is within receiving range, it transmits image data to the station, where they are recorded on magnetic tape. There are many gaps that are not covered by the receiving stations, and a number of stations were not established until relatively late in the Landsat program. Image data for areas within the gaps were recorded on one of the two magnetic tape systems on-board each satellite (Figure 4.1). Reliability of tape recorders was a major problem for the first generation of Landsats. Several tape recorders failed to operate. The operational recorders performed well but had a design lifetime of 500 hours, after which the ferromagnetic tape coating began to deteriorate. The lifetime of the satellites and their imaging systems exceeded that of the on-board recorders, which

limited the coverage of areas in the gaps. The second generation of Landsats employs a different method of data collection.

The first-generation Landsats carried a nonimaging *data collection system* (DCS) that relayed information from sensors on earth to the receiving stations. Each sensor had a small antenna and power supply for transmitting data to the satellite, which sent it to the receiving stations. These sensors were installed in areas inaccessible to normal data transmission links. Flood gauges, seismometers, and tiltmeters on active volcanoes were typical examples.

Orbit Paths

Figure 4.2 shows the 14 south-bound daytime orbit segments covered by Landsats 1, 2, and 3 during a single day. The north-bound orbit segments cover the dark side of the earth. Rotation of the earth shifts the orbit paths westward each day so that at the end of 18 days, or 252 orbits, the entire earth is covered and the cycle starts over. Polar areas above latitude 81° are the only regions that Landsat orbits do not cover. As Figure 4.3 shows, at latitude 40° there is a 62-km, or 34 percent, sidelap of the 185-km-wide image swaths generated on successive days. The sidelap decreases to 14 percent at the equator and increases to 70 percent at polar latitudes. The sidelapping portions of adjacent images may be viewed stereoscopically, as shown later.

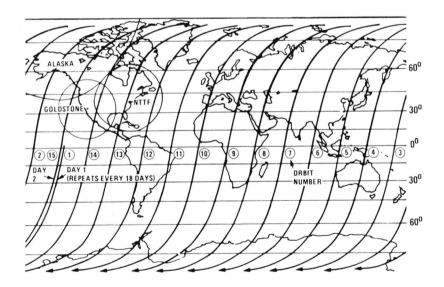

FIGURE 4.2 Daytime portion of orbits for Landsats 1, 2, and 3 for a single day. Each day the orbits shift westward 160 km at the equator to cover the earth in 18 days. Receiving ranges are shown for MSS ground stations in the United States. From NASA (1976, Figure 2.7).

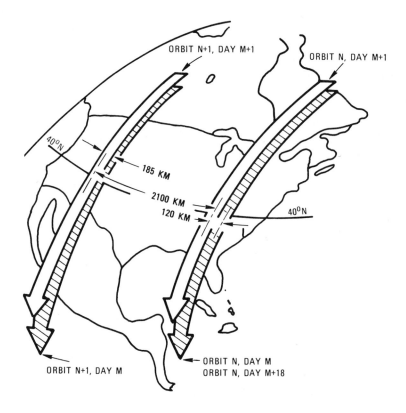

FIGURE 4.3 Landsat orbits over the United States on successive days. Note the 62-km sidelap of successive image swaths at the 40°N latitude.

NASA chose a *sun-synchronous orbit* pattern to acquire images at mid-morning hours. For example, on each south-bound pass, Landsat crosses the latitude of Los Angeles at approximately 10:00 a.m. local sun time. Each 18-day repetitive orbit occurs at the same local sun time, facilitating comparison of the images. The intermediate to low sun elevation in the morning causes highlights and shadows that enhance subtle linear features, many of which represent faults and fracture zones. Images acquired at very low sun elevations in mid-winter are particularly useful for recognizing linear features. These images are interpreted in the same manner as the low-sun-angle photographs in Chapter 2. Sun azimuth is from the southeast for images in the Northern Hemisphere and from the northeast in the Southern Hemisphere.

Landsats 4 and 5

The second generation of Landsat consists of two satellites launched July 16, 1982, and March 1, 1984. At the time of this writing, Landsat 4 is in orbit but not functioning because of problems with the power supply and data transmission systems. Landsat 5 is performing as planned.

Platform Landsats 4 and 5 (Figure 4.4) are larger and more complex than their predecessors. Their most conspicuous features are a single large solar array and a microwave antenna mounted on a mast for communication with other satellites. The major components, shown in Figure 4.5, include an MSS imaging system and a new system, the *thematic mapper* (TM).

Only a few of the existing receiving stations have recording equipment adequate to handle the very high volumes of image data transmitted from TM. For this reason, and to avoid the problems of on-board tape recorders, TM data are handled as shown in Figure 4.6. Tracking and Data Relay Satellites (TDRSs) are placed in geostationary orbits; when all the planned TDRSs are deployed, Landsat will always be in communication with a ground station via a TDRS. Image data are transmitted to TDRS, which relays them to the ground station at White Sands, New Mexico, which then relays the data via Domestic Communication Satellite (DOMSAT) to the Goddard Space Flight Center (GSFC), at Greenbelt, Maryland. GSFC converts the TM data into master film negatives and computer compatible tapes that it forwards to EDC for distribution to users. There have been problems in deploying TDRS platforms, which has hampered the acquisition of TM data. When all TDRS plat-

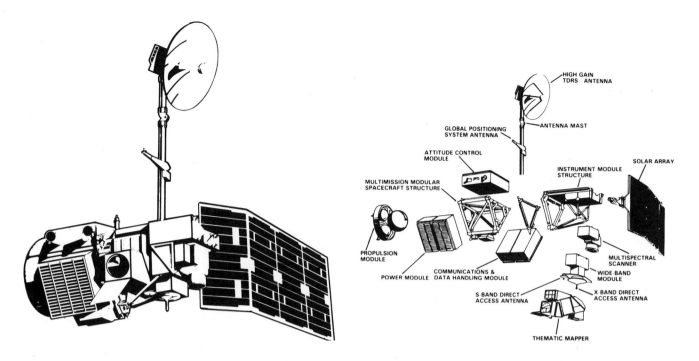

FIGURE 4.4 Satellite platform of Landsats 4 and 5.

FIGURE 4.5 Components of the platform of Landsats 4 and 5.

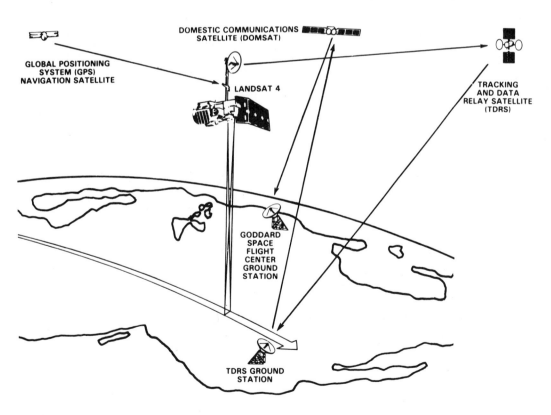

FIGURE 4.6 Communications network for Landsats 4 and 5.

forms are fully operational, TM images can be acquired on a worldwide basis. Landsats 4 and 5 also have *direct-access antennas* (Figure 4.5) that can transmit TM data to suitably equipped ground receiving stations when the satellite is within receiving range. All MSS data are transmitted directly to ground receiving stations.

Orbit Paths Table 4.1 lists the orbital characteristics of the two generations of Landsat. Because of the lower altitude of Landsats 4 and 5, only 233 orbits and 16 days are required to cover the earth. The smaller number of orbits and their wider spacing reduces the sidelap between adjacent image swaths to only 7.6 percent at the equator in contrast to 14 percent for Landsats 1, 2, and 3. The orbit paths for Landsats 4 and 5 are parallel, but not coincident, with those of their predecessors shown in Figure 4.2.

MULTISPECTRAL SCANNER SYSTEM

The three imaging systems deployed on various Landsats, listed in the order of improving spatial resolutions, are MSS, RBV, and TM. Table 4.2 lists their significant characteristics. All five Landsats carried MSS, which has produced the vast majority of images in this program.

System Characteristics

MSS is a cross-track scanning system with an oscillating mirror that scans a swath 185 km wide normal to the orbit path (Figure 4.7). For Landsats 1, 2, and 3 the scan angle was 11.56°; at the 705-km altitude of Landsats 4 and 5 the scan angle is increased to 14.9° to sweep a 185-km swath. Image data are recorded only during the east-bound mirror sweep. At the 918-km altitude, the 0.087-mrad IFOV produces a ground resolution cell that measures 79 by 79 m. Continuous strips of imagery are acquired along the orbit path and transmitted to a ground receiving station then recorded on magnetic tape. The image strips are subdivided into *scenes,* such as those shown in Figure 4.8, that cover a 185-by-185-km area on the ground. Additional details on MSS are given in Chapter 7 and in the *Landsat Data Users Handbook* (NASA, 1976).

Sunlight reflected from the terrain is separated by a spectrometer into four wavelengths, or spectral bands (Figure 4.7). There are six detectors for each spectral

TABLE 4.2   Characteristics of Landsat imaging systems

	Multispectral scanner (MSS)	Return-beam vidicon (RBV)	Thematic mapper (TM)
Spectral region			
Visible and reflected IR	0.5 to 1.1 μm	0.50 to 0.75 μm	0.45 to 2.35 μm
Thermal IR (TM band 6)	—	—	10.5 to 12.5 μm
Spectral bands	4	1	7
Terrain coverage			
East-west direction	185 km	99 km	185 km
North-south direction	185 km	99 km	170 km
Instantaneous field of view			
Visible and reflected IR	0.087 mrad	0.043 mrad	0.043 mrad
Thermal IR (TM band 6)	—	—	0.17 mrad
Ground resolution cell			
Visible and reflected IR	79 by 79 m	40 by 40 m	30 by 30 m
Thermal IR (TM band 6)	—	—	120 by 120 m
Number of picture elements			
Single band	7.6×10^6	6.1×10^6	39×10^6
All bands	30.4×10^6	6.1×10^6	273×10^6

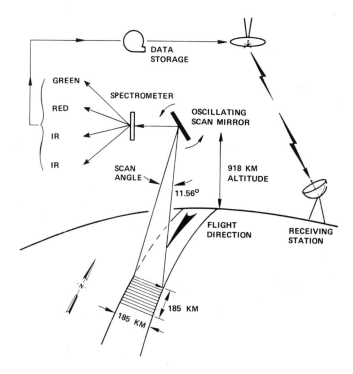

FIGURE 4.7 MSS on Landsats 1, 2, and 3. On Landsats 4 and 5, the altitude is 705 km and the scan angle is 14.9°. For each terrain scene, four images are transmitted to a receiving station. From NASA (1976, Figure 2.4).

band; thus for each sweep of the mirror, six scan lines are simultaneously generated for each of the four spectral bands. The detectors sense the energy and convert it into electrical signals for recording and transmission as image data. A blue image is not acquired; with the signal-to-noise characteristics of MSS, atmospheric scattering would severely degrade a blue image. Table 4.3 lists the wavelength and color recorded by each of the MSS bands.

TABLE 4.3 Landsat multispectral scanner bands

MSS Band*	Wavelength, μm	Color	Projection color for IR color composite image
4	0.5 to 0.6	Green	Blue
5	0.6 to 0.7	Red	Green
6	0.7 to 0.8	Reflected IR	
7	0.8 to 1.1	Reflected IR	Red

*These band designations apply to Landsats 1, 2, and 3 MSS images, while bands 1, 2, and 3 designate images from an RBV color system used only during the early days of Landsats 1 and 2. For Landsats 4 and 5, the four MSS bands are designated 1, 2, 3, and 4.

Typical Multispectral Scanner Images

Figure 4.8 shows the four spectral bands for an MSS scene of the Los Angeles region. During playback of the data tape to produce an image, successive scan lines are offset to the west to compensate for earth's rotation during the approximately 25 sec required to scan the terrain. This offset accounts for the slanted parallelogram outline of the images. The four bands on Figure 4.8 were digitally processed at Jet Propulsion Laboratory to produce optimum contrast in the land areas.

Image Annotation Each Landsat image is annotated with useful information. Following is an explanation from left to right of the annotation for an image of the Los Angeles region:

21 Oct 72	Date image was acquired
C N34-33/W118-24	Geographic center (C) of the image, in degrees and minutes of latitude and longitude
N N34-31/W118-19	Nadir (N) of the spacecraft
MSS	Multispectral scanner image
4, 5, 6, or 7	MSS spectral band
SUN EL39	Sun elevation, in degrees above horizon
AZ148	Sun azimuth, in degrees clockwise from north
190	Spacecraft heading, in degrees
1255	Orbit revolution number
G, A, or N	Ground recording station: G = Goldstone, California; A = Alaska; N = Network Test and Training Facility at GSFC
1090-18012	Unique frame identification number composed as follows:
1	Landsat 1
090	Days since launch; this was October 21, 1972
18	Hour at time of observation, Greenwich mean time (GMT)
01	Minutes
2	Tens of seconds

There are several versions of this annotation format with minor differences. EDC provides a detailed explanation.

Infrared Color Composite Images

The individual MSS bands may be combined in color, using the projection colors listed in Table 4.3 to produce

A. BAND 4: GREEN (0.5 TO 0.6 μm).

B. BAND 5: RED (0.6 TO 0.7 μm).

C. BAND 6: PHOTOGRAPHIC IR (0.7 TO 0.8 μm).

D. BAND 7: PHOTOGRAPHIC IR (0.8 TO 1.1 μm).

FIGURE 4.8 Landsat-1 MSS images of the Los Angeles region acquired October 21, 1972. Images are 185 km wide. Images were digitally processed by Jet Propulsion Laboratory. Courtesy Jet Propulsion Laboratory, California Institute of Technology.

an IR color image. Bands 4, 5, and 7 of Figure 4.8 were combined in this fashion to produce the color image in Plate 4. Spectral characteristics and color signatures of Landsat MSS color images are comparable to those of IR color aerial photographs. Typical signatures are as follows:

Healthy vegetation	Red
Clear water	Dark blue to black
Silty water	Light blue
Red beds	Yellow
Bare soil, fallow fields	Blue
Windblown sand	White to yellow
Cities	Blue
Clouds and snow	White
Shadows	Black

For many of the MSS scenes, EDC has prepared color composite images that one may purchase at scales and prices described later in the chapter. If a color composite does not exist for a particular scene, it may be possible to have one generated for an additional fee.

Land-Use and Land-Cover Interpretation

The location map (Figure 4.9) shows the major physiographic and geographic features of the Los Angeles MSS scene. Urban areas are concentrated in the Los Angeles Basin, which has a grid pattern of major traffic arteries. Central commercial areas, such as the cities of Los Angeles and Long Beach, have blue signatures caused by pavement, roofs, and an absence of vegetation. The suburbs are pink to red, depending on density and condition of lawns, trees, and other landscape vegetation. Small, bright red areas are parks, golf courses, cemeteries, and other concentrations of vegetation.

Agriculture is concentrated in the Central Valley, the Ventura Basin, the eastern part of the Los Angeles Basin, and scattered patches in the Antelope Valley. Vegetation has a rectangular bright red (growing crops) and blue-gray (fallow fields) pattern. In the extreme northeast portion of the image are red circles formed by alfalfa fields irrigated by centerpoint irrigation sprinklers.

Rangeland occurs in the rolling hills of the eastern part of the Los Angeles Basin and has a red-brown signature in this fall season image. Forest and brush cover mountainous terrain of the Sierra Nevada and the Transverse Ranges: lower elevations are covered by chaparral and higher elevations by pine trees. A large dark patch

in the Santa Susana Mountains is the scar of an old brush fire.

Water is represented by the Pacific Ocean and scattered reservoirs. The dark blue color is typical of the Pacific much of the year, but during the winter rainy season, muddy water from various rivers forms light-colored plumes that are carried southward by the California current. A few scattered clouds occur along the coast and along the southeast flank of the Transverse Ranges. Some patches of early snow cover a few peaks.

The Mojave Desert and its western extension, the Antelope Valley have a light yellow signature that is typical of arid land. Along the southern margin of the Antelope Valley are several light gray to very dark gray triangles, which are alluvial fans of gravel eroded from the bedrock of the Transverse Ranges. The most striking is the Sheep Canyon fan (at the eastern margin of the image), which has a conspicuous dark gray color caused by dark minerals eroded from schist bedrock. Edwards and Rosamond dry lakes have white signatures caused by silt and clay deposits. Incidentally, Edwards dry lake is a landing site for the Space Shuttle.

Major geologic features are also recognizable in the Landsat image. The San Andreas fault, which separates the Antelope Valley from the Transverse Ranges, is expressed as linear scarps and canyons. This active right-

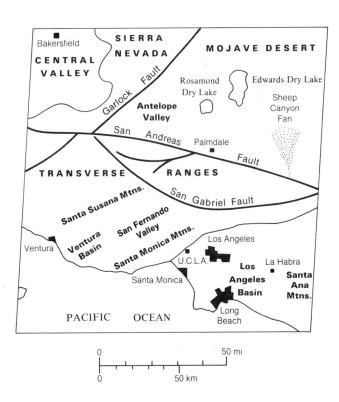

FIGURE 4.9 Location map for Landsat MSS images of the Los Angeles region.

lateral, strike-slip fault is the boundary between the Pacific plate on the southwest and the North America plate on the northeast. The left-lateral Garlock fault, which separates the Antelope Valley and the Sierra Nevada, has an expression similar to that of the San Andreas fault.

RETURN-BEAM VIDICON SYSTEM

Return-beam vidicons (RBV) are framing systems that are essentially television cameras. Landsats 1 and 2 carried three RBVs that recorded green, red, and photographic IR images of the same area on the ground. These images can be projected in blue, green, and red to produce infrared color images comparable to MSS images. There were problems with the color RBV system, and the images were inferior to MSS images; for these reasons, only a few color RBV images were acquired. Landsat 3 deployed an extensively modified version of RBV.

Landsat 3 RBV System

In addition to an MSS system, Landsat 3 carried an RBV system that acquired a single black-and-white image in the wavelength region from 0.50 to 0.75 μm. The main reason for deploying this new RBV was to acquire images with a 40-m ground resolution cell, which is superior to the 79-m cell of MSS.

RBV (Figure 4.10) employs a lens and shutter similar to those of cameras to form an image at the focal plane of the system. Instead of film, however, the image forms on an electrically charged photosensitive tube. The photons of light reflected from the scene and focused on the tube cause changes in the electrical charge of the tube. A beam of electrons from an electron gun scans the tube in a *raster pattern* of parallel lines similar to that of a television screen. The electron beam, which returns from the tube to a detector, changes in intensity proportional to the electrical charge of the image on the tube. The return beam, therefore, carries a signal that can be digitized and transmitted to ground receiving stations, where it is recorded on magnetic tape. After the image has been scanned and transmitted, the tube is electronically erased and is ready to receive the next image. The recorded images are played back onto film to produce black-and-white images.

Landsat 3 employed two RBV cameras, each of which covered adjacent ground areas that are 99 by 99 km in size, with 15 km of sidelap (Figure 4.11). Successive pairs of RBV images have 17 km of forward overlap. Two pairs of RBV images cover an area 181 by 183 km, which is equivalent to the area of the corresponding MSS image.

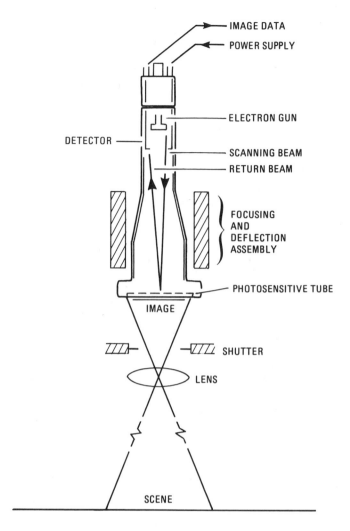

FIGURE 4.10 Return-beam vidicon system.

Typical RBV Images

Figure 4.12 shows typical RBV images of the Los Angeles region. The array of small crosses, called *reseau marks,* are used for geometric control. The 1:1,000,000 scale is the same as that of the MSS image of the region (Plate 4) to which these RBV frames may be compared. This comparison illustrates the advantages of the higher spatial resolution of RBV. For example, in the urban area of the San Fernando Valley, just north of the Santa Monica Mountains (Figure 4.9), the grid of secondary streets is recognizable on the RBV image (Figure 4.12D), but not on the MSS image (Plate 4).

Landsat 3 collected RBV images of many areas around the world. Where RBV and MSS images are available, it is useful to obtain both data sets in order to have the advantages of higher spatial resolution (from RBV) plus IR color spectral information (from MSS).

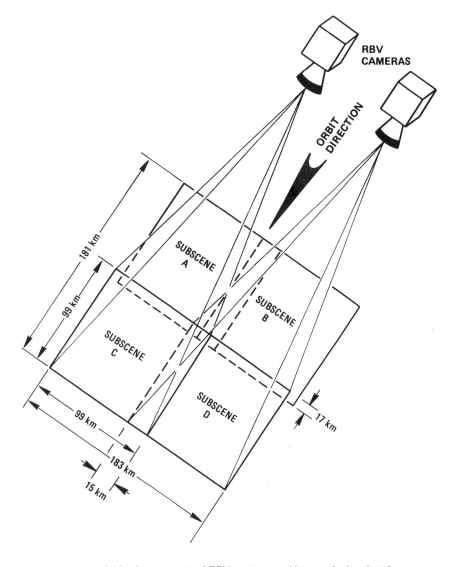

FIGURE 4.11 Arrangement of RBV systems and images for Landsat 3.

THEMATIC MAPPER SYSTEM

The thematic mapper (TM), deployed on Landsats 4 and 5, is a major improvement over MSS. However, because of problems with Landsat 4 and with the TDRS system, a relatively limited number of images has been acquired since TM was deployed in 1982. In time, the amount of coverage may be comparable to that from MSS.

System Characteristics

Figure 4.13 illustrates the TM system, and Table 4.2 lists its characteristics. TM is a cross-track scanner similar to MSS with an oscillating scan mirror and arrays of detectors, but TM is an improvement in the following respects:

1. TM has a spatial resolution of 30 m, compared with 79 m for MSS.

2. TM has seven spectral bands, whereas MSS has only four.

3. TM has an extended spectral range in the visible and reflected IR regions.

4. The addition of blue spectral data (TM band 1) enables production of normal color composite images.

5. TM has a thermal IR band.

6. TM has improved detector sensitivity and radiometric resolution.

TM records on both east-bound and west-bound sweeps in order to reduce the scan rate and provide a longer

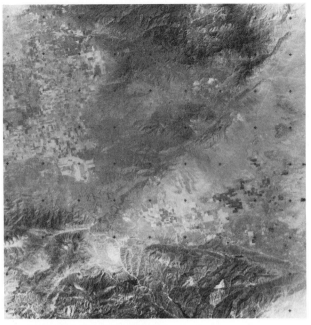

A. NORTHWEST QUADRANT.

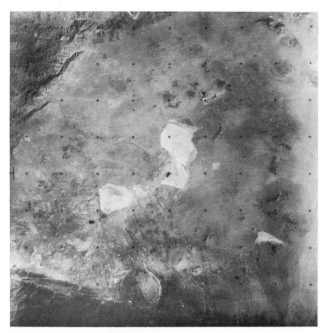

B. NORTHEAST QUADRANT

C. SOUTHWEST QUADRANT.

D. SOUTHEAST QUADRANT.

FIGURE **4.12** Landsat-3 RBV images of the Los Angeles region acquired February 18, 1981. Images are 99 km wide.

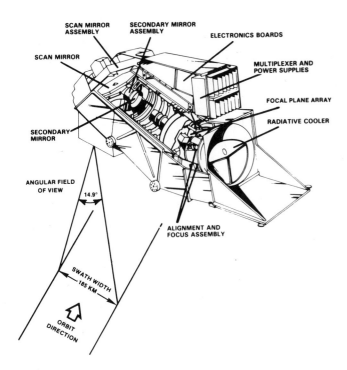

FIGURE 4.13 Thematic-mapper system.

dwell time, which improves radiometric accuracy. The TM mirror sweeps east and west seven times per second. Each TM band uses an array of 16 detectors (with the exception of band 6, which uses only 4).

Figure 4.14 shows the spectral ranges of the six visible and reflected IR TM bands together with spectral reflectance curves of vegetation, hydrothermally altered rock, and unaltered rock. Spectral ranges of the four MSS bands are also shown. Band 6 of TM (10.4 to 12.5 μm) lies in the thermal IR region, beyond the range shown in Figure 4.14. As originally configured, TM included bands 1 through 5 in the visible and reflected IR regions and band 6 in the thermal IR region. When geological users pointed out the value of information from the 2.1-to-2.4-μm region, band 7 was added to acquire these data. The original numbering system remained the same, however, which is why band 7 is out of sequence on a spectral basis.

Typical Thematic Mapper Images

Table 4.4 summarizes the characteristics of TM spectral bands. Two scenes were selected to illustrate the characteristics of TM images in different types of ter-

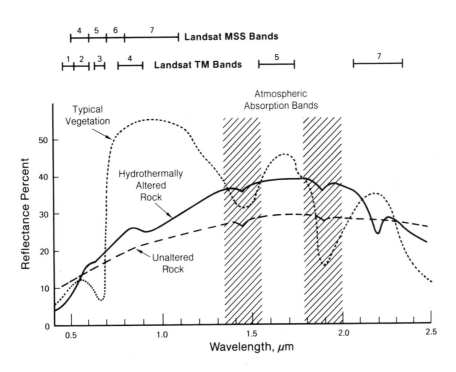

FIGURE 4.14 Spectral bands for TM and MSS systems. Reflectance curves for vegetation, unaltered rocks, and hydrothermally altered rocks. From Sabins (1983, Figure C-5).

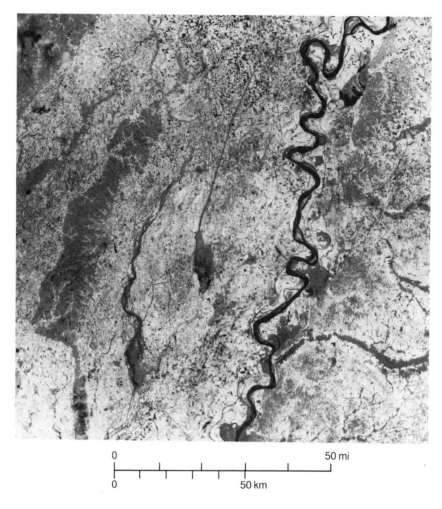

0   50 mi

0   50 km

FIGURE 4.15   Landsat TM band-4 image of northeast Arkansas acquired August 22, 1982.

TABLE 4.4   Thematic-mapper spectral bands

Band	Wavelength, μm	Characteristics
1	0.45 to 0.52	Blue-green—no MSS equivalent. Maximum penetration of water, which is useful for bathymetric mapping in shallow water. Useful for distinguishing soil from vegetation and deciduous from coniferous plants.
2	0.52 to 0.60	Green—coincident with MSS band 4. Matches green reflectance peak of vegetation, which is useful for assessing plant vigor.
3	0.63 to 0.69	Red—coincident with MSS band 5. Matches a chlorophyll absorption band that is important for discriminating vegetation types.
4	0.76 to 0.90	Reflected IR—coincident with portions of MSS bands 6 and 7. Useful for determining biomass content and for mapping shorelines.
5	1.55 to 1.75	Reflected IR. Indicates moisture content of soil and vegetation. Penetrates thin clouds. Good contrast between vegetation types.
6	10.40 to 12.50	Thermal IR. Nighttime images are useful for thermal mapping and for estimating soil moisture.
7	2.08 to 2.35	Reflected IR. Coincides with an absorption band caused by hydroxyl ions in minerals. Ratios of bands 5 and 7 are potentially useful for mapping hydrothermally altered rocks associated with mineral deposits.

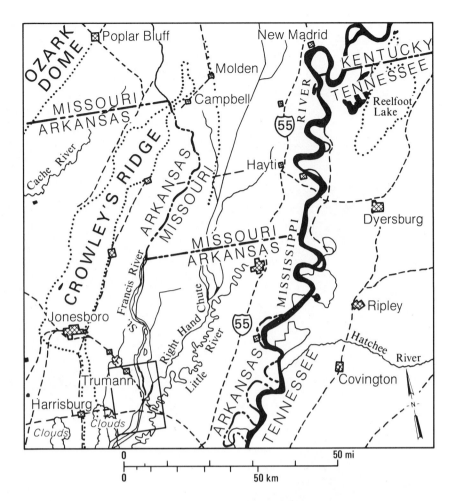

FIGURE 4.16 Location map for the TM image of northeast Arkansas. Rectangle shows location of enlarged subscene of the St. Francis River area.

rain. A scene of northeast Arkansas and vicinity covers agricultural terrain in the humid Mississippi River Valley. A scene in central Wyoming covers semiarid rangeland and mountains.

Northeast Arkansas The band-4 TM image in Figure 4.15 covers a stretch of the Mississippi River and portions of Arkansas, Missouri, and Tennessee. Figure 4.16 indicates major features; the rectangle in the southwest corner shows the location of the St. Francis River subscene enlarged and illustrated in Figure 4.17. The subscene shows features typical of this region: agriculture, native vegetation, open water and marsh, urban areas, transportation, and drainage networks. Plate 5A,B illustrates a normal color image and an IR color image of the subscene. The normal color image is a composite of bands 1, 2, and 3 in blue, green, and red. The IR color image is a composite of bands 2, 3, and 4 in blue, green, and red. Cotton and soybeans are major crops in this region. An extensive marsh along the St. Francis River is covered with dense native vegetation. On either side of the marsh is a network of flood-control canals. Urban areas are represented by the towns of Trumann and Marked Tree (Figure 4.17H). A few thin clouds are present.

While evaluating the individual TM images in Figure 4.17, it is useful to refer to Figure 4.14 and compare each spectral band with the reflectance curve for vegetation. In the visible region represented by TM bands 1, 2, and 3, vegetation has a low reflectance as shown by the dark signature of the densely vegetated marsh. Band 4 coincides with the major reflectance peak for vegetation and the marsh and vegetated fields have very bright tones (Figure 4.17D). Vegetation reflectance be-

A. BAND 1 (0.45 TO 0.52 μm).

B. BAND 2 (0.52 TO 0.60 μm).

C. BAND 3 (0.63 TO 0.69 μm).

D. BAND 4 (0.76 TO 0.90 μm).

E. BAND 5 (1.55 TO 1.75 μm).

F. BAND 6 (10.40 TO 12.50 μm).

G. BAND 7 (2.08 TO 2.35 μm).

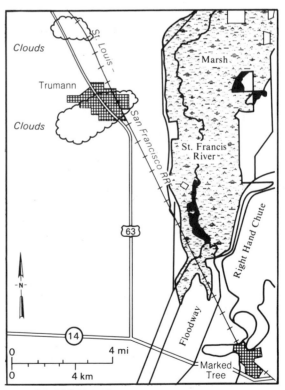

H. LOCATION MAP.

FIGURE 4.17 St. Francis River subscene of enlarged Landsat-4 TM images in northeast Arkansas.

FIGURE **4.18** Landsat TM band-4 image of central Wyoming acquired November 21, 1982. Image digitally processed by Chevron Oil Field Research Company.

comes progressively lower for bands 5 and 7, and the marsh and cultivated fields are increasingly darker in these images. The silty water of the St. Francis River reflects visible wavelengths and has a bright signature in these bands. The river and other water is dark in the reflected IR images because water absorbs these wavelengths. Thin clouds conceal the town of Trumann in the visible images due to scattering, but the longer-wavelength energy of band 4 penetrates the clouds, producing a clear image of the town. Such penetration occurs only for very thin clouds, however, and denser clouds effectively scatter reflected IR wavelengths.

Band 6 (Figure 4.17F) records heat radiated from the surface. Warm areas are bright, and cool areas are dark.

In this daytime image, water, vegetation, and moist soil are relatively cool, while bare, dry soil and urban areas are relatively warm. Chapter 5 explains the reasons for these signatures. Band-6 detectors record a large ground resolution cell of 120 by 120 m, which explains the lower spatial resolution of this image. The low contrast ratio is attributed to the humid atmosphere, which absorbs and reradiates thermal IR energy.

Central Wyoming The band-4 TM image (Figure 4.18) and map (Figure 4.19) of central Wyoming include several broad topographic basins (Wind River, Bighorn, and Powder River basins) bordered by mountain ranges (Wind River, Owl Creek, and Bighorn mountains). In this late-

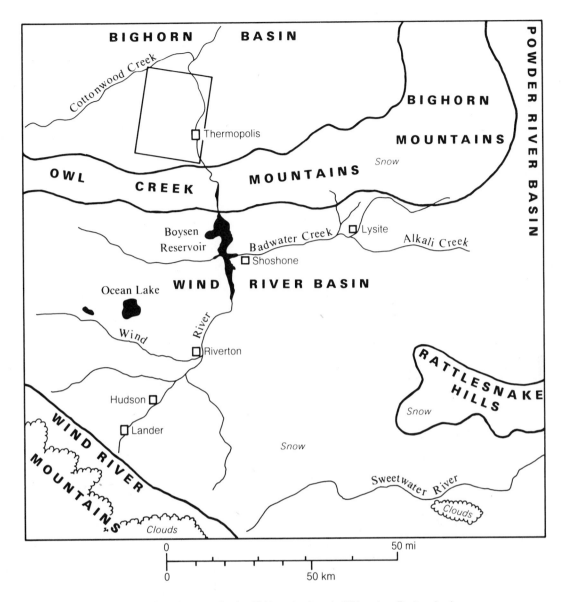

FIGURE 4.19 Location map for the TM image of central Wyoming. Rectangle shows location of enlarged subscene of the Thermopolis subarea.

fall image (November 21, 1982), snow covers the crests of the ranges. The Wind River Mountains are largely concealed by clouds. Because of the high elevation, northern latitude, and limited rainfall, agriculture is restricted to irrigated fields along the major streams. The basins and foothills are primarily used for cattle ranching. Granitic and metamorphic rocks occur in the crests of the mountains; these are flanked by folded and faulted sedimentary rocks, which form traps for a number of oil fields. Figure 4.19 shows the location of the Thermopolis subscene on the northern flank of the Owl Creek Mountains, which was selected to illustrate enlarged images of the seven TM bands. Figure 4.20 shows the bands and a map of the subscene, which cover a stretch

of the Wind River and the town of Thermopolis. Sedimentary rocks are well exposed and are folded into four anticlines, two of which form oil fields (Figure 4.20H). Plate 5C,D are normal color and IR color composite images of the subscene.

Clear water of the Wind River is dark in all TM bands. Cultivated vegetation is confined to the valleys of the Wind River and Owl Creek, where it is dark in the visible bands (Figure 4.20A,B,C). The season, environment, and types of vegetation in Wyoming are completely different from those in the Arkansas subscene (Figure 4.17). Despite these differences, the relative reflectance of vegetation in the TM bands is similar in both areas. As in the St. Francis River subscene, signatures of vegetation

A. BAND 1 (0.45 TO 0.52 μm).

B. BAND 2 (0.52 TO 0.60 μm).

C. BAND 3 (0.63 TO 0.69 μm).

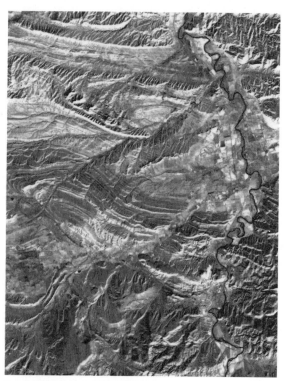

D. BAND 4 (0.76 TO 0.90 μm).

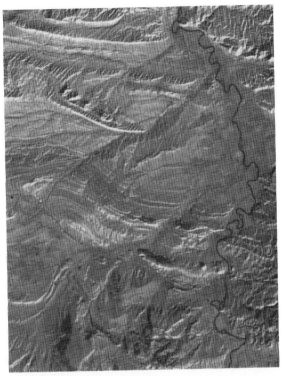

E. BAND 5 (1.55 TO 1.75 μm).

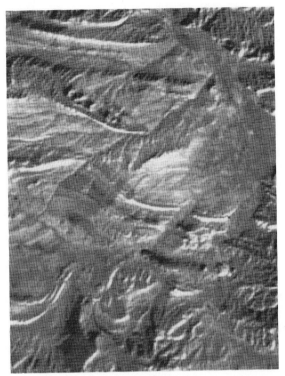

F. BAND 6 (10.40 TO 12.50 μm).

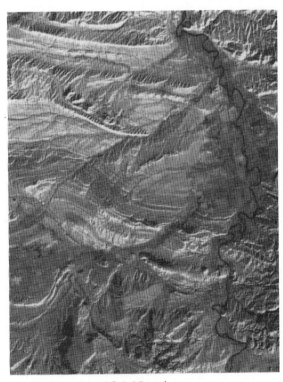

G. BAND 7 (2.08 TO 2.35 μm).

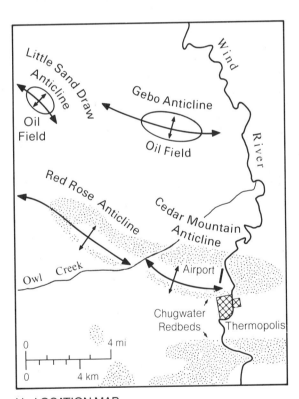

H. LOCATION MAP.

FIGURE **4.20** Thermopolis subscene of enlarged Landsat-4 TM images in central Wyoming. Images digitally processed by Chevron Oil Field Research Company.

are brightest in reflected IR band 4, and progressively decrease in bands 5 and 7. In the normal color image of the Thermopolis area (Plate 5C), vegetation has a very dark green signature, and in the IR color image (Plate 5D), the typical red signature. A striking color shift between the normal color and IR color images is shown by the Chugwater Formation, which consists of reddish orange siltstone. The Chugwater Formation crops out around the flanks of the Red Rose and Cedar Mountain anticlines and in the south part of the subscene (Figure 4.20H). In the normal color image, the Chugwater Formation has a reddish orange signature, which is typical of the outcrop color. In the IR color image, however, the Chugwater rocks have a bright yellow signature, which is characteristic for red materials in these images.

The thermal IR image (Figure 4.20F) is marred by stripes caused by the scanning system. Despite this and the low spatial resolution (120 m), topographic details are recognizable. The dark signatures record relatively cool temperatures and are associated with shadowed, north-facing slopes. Bright signatures are relatively warm features and are associated with sunlit, south-facing slopes. Contrast ratio and spatial resolution in the thermal IR image of Thermopolis are superior to those in the Arkansas image because the dry atmosphere of Wyoming transmits thermal IR wavelengths more readily than does the humid atmosphere of Arkansas.

THERMOPOLIS AREA, WYOMING

The comparison of individual TM bands, normal color, and IR color images is a useful introduction to a detailed interpretation of an enlarged IR color TM image of the Thermopolis area (Plate 6). Land use, geomorphology, rock types, and geologic structure are recognizable.

Land Use and Geomorphology

As Figure 4.19 shows, the Thermopolis area is located in the south flank of the Bighorn Basin directly north of the Owl Creek Mountains. The area is drained by the north-flowing Wind River (which farther north becomes the Bighorn River) and by Owl Creek (Figure 4.20H). The floodplains of the river and creek are used for pasture land and irrigated agriculture, as shown by the rectangular red and brown pattern of crops and fallow fields respectively. The town of Thermopolis, on the west bank of the river, has the blue signature characteristic of urban areas. The thin, dark strip north of town is the asphalt runway of an airport that is bordered on the west by the bright red fairways of a golf course. Except for the town, the area is sparsely populated and used primarily for ranching.

Topography of the northern two-thirds of the area is dominated by alternating ridges and broad valleys that trend northwest. In the fall season, when the image in Plate 6 was acquired, the cover of annual grass is dormant, giving the area the light gray signature of the bedrock. Some north-facing slopes support evergreen pinyon trees, which have red signatures.

South of Owl Creek, the terrain is dominated by the broad slopes of the northern flank of the Owl Creek Mountains. Elevations are higher, which results in greater precipitation and denser vegetation. These dark-red and red-brown vegetation signatures contrast with the white and light gray signatures of the lower, barren terrain to the north.

Rock Types

As the geologic map (Figure 4.21) shows, sedimentary rocks, ranging from Permian to Tertiary in age, underlie the Thermopolis area. In the image (Plate 6) the stratified nature of the terrain clearly expresses the sedimentary bedrock. Parallel ridges are formed by rocks that are resistant to erosion, such as sandstone, conglomerate, and carbonate rocks. The ridges are separated by valleys formed on less resistant shale and siltstone. The legend of Figure 4.21 lists the sequence of formations by geologic age; their image signatures are summarized below.

Alluvial deposits These form floodplains of the Wind River and Owl Creek and show no topographic relief. Red and brown signatures are caused by crops and fallow fields.

Fort Union Formation Resistant sandstone with some shale that forms dissected ridges with white and light gray signatures.

Lance and Meeteetse formations Predominantly nonresistant siltstone and soft sandstone that form a broad valley with a light-gray to medium-gray signature.

Mesaverde Formation Resistant sandstone with interbeds of shale and coal that forms dissected ridges with a light gray signature. The Mesaverde and Fort Union formations have similar expressions in the image because of their similar lithology.

Cody Shale Nonresistant shale that forms a broad valley with light gray to medium gray signature. The Cody Shale signature is very similar to that of the Lance and Meeteetse formations.

Frontier Sandstone Alternating beds of resistant sandstone and nonresistant shale that form relatively narrow ridges and valleys. Dip slopes of the sandstone support native vegetation that causes red signatures.

Cloverly, Mowry, and Thermopolis formations Shale beds with some sandstone that form a valley with a few ridges. These formations are treated as a single mapping unit.

Undifferentiated Early Jurassic formations Sandstone and shale.

Chugwater Formation Reddish orange siltstone with a conspicuous yellow signature on the IR color image.

Phosphoria Formation Carbonate unit that crops out only in the crest of the Red Rose and Cedar Mountain anticlines. The lack of apparent stratification distinguishes this massive unit from the other formations.

Now that the formations have been described and their contacts have been mapped in Figure 4.21, the next step is to map the structure of the area.

Structural Geology

Geologic structure is dominated by the regional dip northward from the Owl Creek Mountains into the Bighorn Basin. The recognition of strike and dip attitudes is based on geomorphology, drainage patterns, and the orientation of highlights and shadows. In this mid-morning November image, the sun is shining from the southeast at an elevation of about 25° above the horizon. For north-dipping beds, the dip slopes are shadowed and the narrow antidip scarps produce linear highlights. The pattern of highlights and shadows reverses for south-dipping beds. The regional dip is interrupted by several folds. As shown in the map (Figure 4.21) and cross section (Figure 4.22), the Red Rose and Cedar Mountain anticlines are bounded on the south by reverse faults and are deeply eroded to expose the Phosphoria Formation. In the northern part of the area, the Gebo and Little Sand Draw anticlines are topographic depressions eroded into nonresistant Cody Shale. The overlying Mesaverde Sandstone was stripped from the crest of these structures and forms a resistant rimrock. Both the Gebo and Little Sand Draw anticlines are oil fields that produce from the Phosphoria Formation (Figure 4.22). In the images the folds are recognizable from the arcuate outcrop patterns and by the attitudes of the beds, as described in the Appendix.

COMPARISON OF LANDSAT IMAGE SYSTEMS

Images from the three Landsat systems (MSS, RBV, and TM) have been described and illustrated. With this background, the characteristics of the images (Table 4.2) can be compared. Images of the Palmdale area in the Antelope Valley of southern California were photographically enlarged to a uniform scale of approximately 1:80,000 (Figure 4.23). For the location of the town of Palmdale, see Figure 4.9. Before discussing the different spatial resolutions, some other differences among the images must be mentioned.

The images of the Palmdale area were acquired on three different dates over a span of 8 years (1976 to 1984), during which time urban development occurred in this desert area. For example, on the 1984 TM image, there is an artificial lake directly south of the Palmdale airport (Figure 4.23D) that is not present on the 1981 RBV image. The images were acquired at slightly different spectral bands—MSS band 5 at 0.60 to 0.70 μm, RBV at 0.50 to 0.75 μm, and TM band 4 at 0.76 to 0.90 μm—which accounts for minor tonal differences among the images. For example, the concrete surfaces of the Antelope Valley Freeway and the Palmdale airport have bright signatures in the visible bands of MSS and RBV, but the albedo is much lower in the reflected IR band of the TM image, where these features are less apparent.

The images differ primarily in their spatial resolution, which is determined by the ground resolution cell: 79 by 79 m for MSS, 40 by 40 m for RBV, and 30 by 30 m for TM. At the enlargement factor used for Figure 4.23, the individual picture elements, or ground resolution cells, of the MSS image are visible and cause a blocky appearance. The RBV image has a smoother appearance, and additional details are visible. The TM image has the maximum spatial detail as the pattern of streets and individual buildings in the city of Lancaster reveals. Some of this improved resolution in TM may be due to improved recording techniques that have produced a better contrast ratio.

AVAILABILITY OF IMAGES

Beginning in late 1985, the EOSAT Company assumed responsibility for distributing all the Landsat data that are archived at EDC. Inquiries and orders should be sent to

EOSAT Company
EROS Data Center
Sioux Falls, SD 57198

EOSAT is not responsible for non-Landsat data at EDC. Images for individual continents and countries are also available from the facilities listed in Table 4.5. For many areas, these local receiving stations may have recorded images that are not available from EDC. At EDC, all images are catalogued in a computer system, using their identification number, as described in the earlier subsection "Image Annotation." In late 1985 EDC had the following inventory of images:

MSS	500,000
RBV	154,000
TM	10,000
Total	664,000

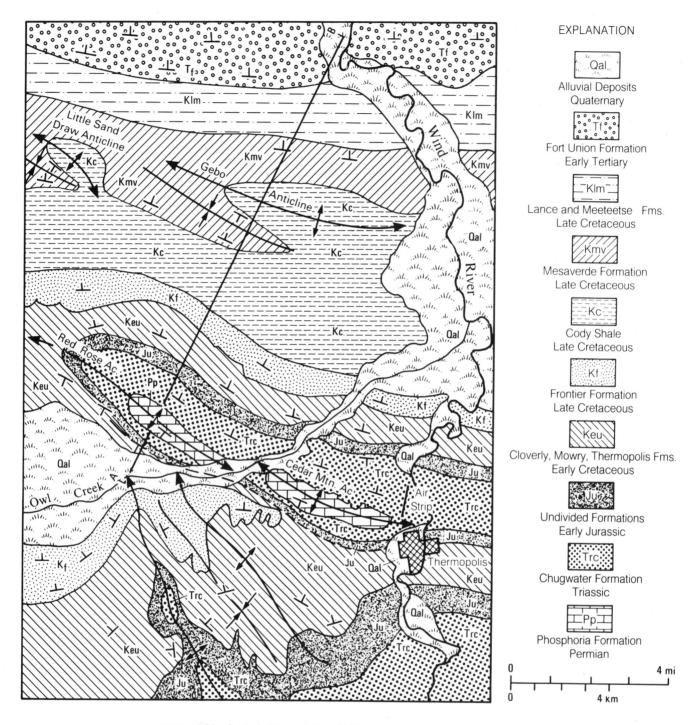

EXPLANATION

Qal
Alluvial Deposits
Quaternary

Tf
Fort Union Formation
Early Tertiary

Klm
Lance and Meeteetse Fms.
Late Cretaceous

Kmv
Mesaverde Formation
Late Cretaceous

Kc
Cody Shale
Late Cretaceous

Kf
Frontier Formation
Late Cretaceous

Keu
Cloverly, Mowry, Thermopolis Fms.
Early Cretaceous

Ju
Undivided Formations
Early Jurassic

Trc
Chugwater Formation
Triassic

Pp
Phosphoria Formation
Permian

0   4 mi
0   4 km

FIGURE 4.21 Geologic interpretation of TM image of the Thermopolis subscene.

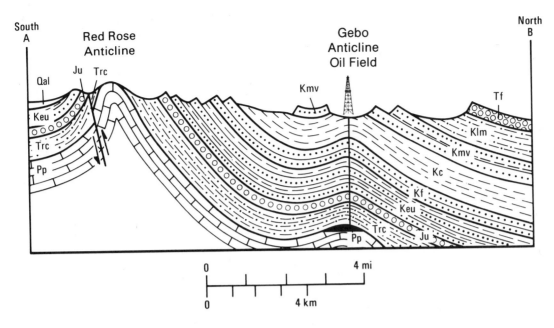

FIGURE 4.22 Geologic cross section of the Thermopolis subscene. Formation symbols are explained in Figure 4.21.

The images are referenced to a series of *path-and-row index maps*.

Annotated collections of Landsat images have been published by Short and others (1976), Williams and Carter (1976), and Slaney (1981). EDC periodically publishes ''Landsat Data Users Notes,'' which provides additional useful information.

The USSR has operated a series of earth-orbiting satellites (Meteor, Salut, and Soyuz) that gather remote sensing images. Unlike Landsat images, however, these USSR data are not available and only rare samples have been published. The few samples illustrated by Trifonov (1984) appear comparable in spatial resolution to Landsat MSS images.

Worldwide Path-and-Row Index Maps

Path-and-row index maps of all the continents are available from EDC and are valuable for locating images. The index map for the western United States (Figure 4.24) shows the centerpoints at which MSS images (and RBV images for Landsat 3) were acquired on the repeated 18-day cycles of Landsats 1, 2, and 3. *Center-points* are defined by the intersection of the south-bound orbit path and east-west rows that are spaced approximately 80 km apart. For example, all images of the Los Angeles area, including the example in Figure 4.8, are referenced to path 45, row 36. The centers of individual images may deviate as much as a few tens of kilometers

from the nominal center, but this is rarely a problem in determining coverage. A different set of index maps are used for MSS and TM images acquired by Landsats 4 and 5 because their orbit paths are more widely spaced than the orbit paths of the first generation of Landsats.

Selection of Images

To identify specific images, a user gives EDC the latitude and longitude boundaries or path and row coordinates of an area of interest. The user may also specify type of image, image quality, and percentage of cloud cover. EDC will then provide a computer printout listing the identification numbers of all images that cover the area of interest, their locations and dates of acquisition, image quality, and percentage of cloud cover. From this information, the user orders the optimum images for his application. The regional centers (Table 4.5) provide similar services.

Formats and Prices of Images

Table 4.6 lists the standard formats and prices of images at EDC. For each centerpoint in the United States, EDC has selected and computer-enhanced a high-quality MSS image. These enhanced images are available at standard prices and are recommended unless the user requires a specific image.

A. MSS BAND 5, MARCH 7, 1976.

B. RBV IMAGE, FEBRUARY 18, 1981.

C. TM BAND 4, OCTOBER 4, 1984.

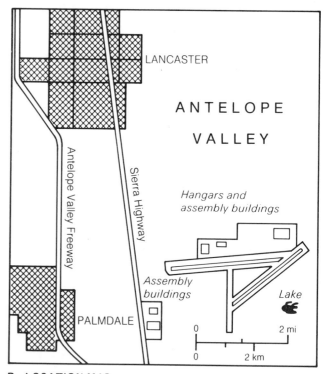

D. LOCATION MAP.

FIGURE 4.23 Comparison of enlarged Landsat MSS, RBV, and TM images of the Palmdale area, California.

TABLE 4.5   International Landsat data distribution centers

1. EOSAT Company
    EROS Data Center
    Sioux Falls, SD 57198
    USA
    Telephone: 605-594-6511
    TWX: 910-668-0310

2. Instituto de Pesquisas Espaciais (NPE)
    Departamento de Producao de Imagens
    ATUS–Banco de Imagens terrestres
    Rodovia Presidente Dutra, Km 210
    Cachoeira Paulista-CEP 12.630
    Sao Paulo
    Brazil
    Telephone: (0125)611507

3. Canadian Centre for Remote Sensing
    User Assistance and Marketing Unit
    717 Belfast Road
    Ottawa, Ontario K1A 0Y7
    Canada
    Telephone: (613)994-1210
    Telex: 053-3777

4. ESA-ESRIN
    Earthnet User Services
    Via Galileo Galilei
    000 44 Frascati
    Italy
    Telephone: 39-6-9401360 or 39-6-9401216
    Telex: 611295 or 610637

5. Remote Sensing Tech. Center of Japan
    Uni-Roppongi Bldg.
    7-15-17 Roppongi Minato-ku
    Tokyo 106
    Japan
    Telephone: Tokyo 3-403-1761
    Telex: 0242780 RESTECJ

6. National Remote Sensing Agency
    Balanagar, Hyderbad 500037
    Andhra Pradesh
    India
    Telephone: 262572 Ext. 67
    Telex: 0155-522

7. Australia Landsat Station
    14-16 Oatley Court
    P.O. Box 28
    Belconen, A.C.T. 2616
    Australia
    Telephone: 062-515411
    Telex: 61510

8. Comision Nacional de Investigaciones Espaciales
    Dorrego 4010
    (1425) Buenos Aires
    Argentina
    Telephone: 772-5108
    Telex: 17511 LANBA AR

9. National Institute for Telecommunications Research
    ATTN: Satellite Remote Sensing Ctr.
    P.O. Box 3718
    Johannesburg 2000
    South Africa
    Telephone: (012)26-5271
    Telex: 3-21005 SOUTH AFRICA

10. Remote Sensing Division
    National Research Council
    196 Phahonyothin Road
    Bangkok 10900
    Thailand
    Telephone: 579-0017
    Telex: 82213 NRCTRSD
    Cable: NRC Bangkok

11. Academia Sinica
    Landsat Ground Station
    Beijing
    People's Republic of China
    Telephone: 285093 (Beijing, China)
    Telex: 22474 ASCHI CN

12. Chairman
    Indonesian National Institute of Aeronautics and Space
    Jln. Pemuda Persil No. 1
    P.O. Box 3048
    Jakarta
    Indonesia
    Telex: 49175

LANDSAT MOSAICS

The broad regional coverage of an individual Landsat image (34,000 km^2) can be extended by combining adjacent images into a mosaic (Figure 4.25). The sidelap of adjacent orbit swaths and the 10 percent forward overlap of consecutive images greatly facilitate mosaic compilation. The uniform scale and minimal distortion of Landsat images also make mosaic compilation easier; anyone who has ever made mosaics of aerial photographs and attempted to match radially distorted prints will appreciate these two features. The 18-day repetition cycles have enabled Landsats 1, 2, and 3 to acquire essentially cloud-free MSS coverage of most of the world.

Aside from locating the needed images, the major problem is matching the photographic density and contrast of the images to produce a uniform mosaic. This is particularly difficult when mosaics are compiled from images taken at different seasons of the year.

Mosaic Compilation

The following suggestions are useful in compiling mosaics:

1. Select images for the same day along an orbit path, if available.

TABLE 4.6 Formats and prices of Landsat images available from EOSAT

Image size, cm	Scale	Format	Price,* $ MSS and RBV	TM
Black-and-white				
18.5	1:1,000,000	Film positive	80	100
18.5	1:1,000,000	Film negative	90	160
18.5	1:1,000,000	Paper	50	150
37.1	1:500,000	Paper	100	170
74.2	1:250,000	Paper	150	250
Color composite†				
18.5	1:1,000,000	Film positive	150	360
18.5	1:1,000,000	Paper	100	300
37.1	1:500,000	Paper	200	400
74.2	1:250,000	Paper	350	500

*These prices were current in 1986 but are subject to change. Current prices may be obtained from EOSAT.
†If a master color composite does not exist, the fee to generate a master is $200 for MSS and $305 for TM.

2. Select images from the same season to minimize differences caused by snow cover, shadows, and vegetation cycles.

3. For black-and-white mosaics, bands 5 or 7 are generally superior, for reasons given later.

4. Obtain negatives, rather than prints, from EDC. One can control density and contrast in the darkroom to provide uniform prints for the mosaic.

5. First match the images along each orbit path. Then fit the adjacent orbit paths together and distribute any mismatch along the entire length.

6. In making a large mosaic, begin with the central orbit path and work outward rather than starting at one edge.

Available Mosaics

Figure 4.25 is a greatly reduced copy of the mosaic of the United States assembled by the U.S. Soil Conservation Service Photographic Laboratory from approximately 600 band-5 Landsat MSS images. Photographic reproductions of this mosaic and mosaics of Alaska are available at scales from 1:500,000 to 1:10,000,000. For information write

Agricultural Stabilization and Conservation Service
APFO
P.O. Box 300010
Salt Lake City, UT 84130

Ryder (1981) published a convenient version of the U.S. mosaic. The 1:1,000,000 scale images are bound in a folio and annotated with political boundaries, cities, highways, and geographic information.

A number of color and black-and-white mosaics of foreign areas and individual states in the United States are available from EDC and from the Agricultural Stabilization and Conservation Service (ASCS). Both organizations will provide lists of available mosaics.

INTERPRETATION METHODS

The objectives of the interpreter determine the interpretation methods for Landsat images and other forms of remote sensing imagery. Geographers, agronomists, and geologists would produce different interpretation maps of the Los Angeles image (Plate 4). The terminology employed by most disciplines is adequate for describing features on Landsat images, and there is little need for additional terms. The Appendix summarizes basic geologic terms; however, lineaments are a common feature on Landsat and other images that require additional explanation.

Lineaments and Related Features

In the early 1900s the American geologist William H. Hobbs (1904, 1912) recognized the existence and significance of linear geomorphic features that are the sur-

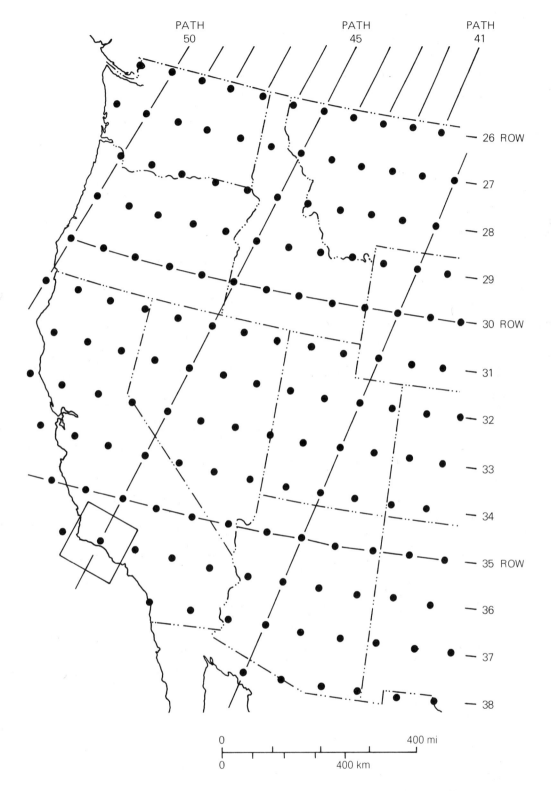

PATH 50   PATH 45   PATH 41

— 26 ROW
— 27
— 28
— 29
— 30 ROW
— 31
— 32
— 33
— 34
— 35 ROW
— 36
— 37
— 38

0   400 mi

0   400 km

FIGURE 4.24 Portion of the path-and-row index map for Landsats 1, 2, and 3. The MSS image of the Los Angeles area is indicated by the square at path 45, row 36. A different index map is used for Landsats 4 and 5.

FIGURE 4.25 Landsat mosaic of the conterminous United States compiled from Landsat MSS band-5 images acquired in the summer season. Mosaic compiled by the U.S. Department of Agriculture Soil Conservation Service.

face expression of zones of weakness or structural displacement in the crust of the earth. Hobbs defined lineaments as "the significant lines of landscape which reveal the hidden architecture of the rock basement. . . . They are character lines of the earth's physiognomy" (1912, p. 227). Over the years, additional terms have been misused as synonyms for *lineament,* and the resulting confusion has tended to obscure the geologic significance of lineaments. O'Leary, Friedman, and Pohn (1976) reviewed the origin and usage of the terms *linear, lineation,* and *lineament.* Their work and definitions are the basis for the following discussion.

Linear is an adjective that describes the linelike character of an object or an array of objects. Some geologists misuse the term as a noun substitute for *lineament,* which is grammatically incorrect. If we accept *linear* and *linears* as nouns, we would likewise have to accept planars, circulars, and so on as real objects. *Linear* is an indispensable descriptive word (linear valleys, linear escarp-

ment, and so forth) and should be used in this sense. *Linear feature* is a good, informal term to describe objects in terms of their geometry without causal or structural implications.

Lineation is the one-dimensional fabric of the internal components of a rock, such as the parallel orientation of elongate crystals in a metamorphic rock. Some geologists have used this petrographic term as a synonym for *lineament,* but this is incorrect and unacceptable.

A *lineament* is a mappable simple or composite linear feature of a surface whose parts align in a straight or slightly curving relationship and that differs distinctly from the patterns of adjacent features. Presumably a lineament expresses a subsurface phenomenon. The surface features making up a lineament may be *geomorphic* (caused by relief) or *tonal* (caused by contrast differences). The surface features may be landforms, the linear boundaries between different types of terrain, or breaks within a uniform terrain. Straight stream valleys and aligned segments of valleys are typical geomorphic expressions of lineaments. A tonal feature may occur as a straight boundary between areas of contrasting tone or as a stripe against a background of contrasting tone. Differences in vegetation, moisture content, and soil or rock composition account for most tonal contrasts. Lineaments may be continuous or discontinuous. An uninterrupted linear scarp is an example of a *continuous lineament*. In *discontinuous* lineaments, the separate features align in a consistent direction and are relatively closely spaced. Lineaments may be simple or composite. *Simple* lineaments consist of a single type of feature, such as a linear stream valley or a series of aligned topographic escarpments. *Composite* lineaments consist of more than one type of feature, such as a combination of aligned tonal features, stream segments, and ridges.

Although many lineaments are controlled at least in part by faults, *structural displacement* (faulting) is not a requirement in the definition of a lineament. On Landsat images of Precambrian shield areas, for example, long lineaments are clearly defined by aligned straight segments of valleys and by linear tonal contrasts. On 1:250,000-scale images, no offset of the lithologic contacts transected by the lineaments is detectable. These and similar lineaments throughout the world are thought to represent zones of weakness in the crust. Although displacement has apparently not occurred, the rocks may be more highly fractured and susceptible to stream erosion. Lineaments that are continuous lines or zones of structural offset are called *faults* or *fault zones*. On the Landsat image (Plate 4) and map (Figure 4.9) of the Los Angeles region, the San Andreas, San Gabriel, and Garlock faults qualify as lineaments but are designated as faults because they are known to be zones of structural displacement. Features newly discovered on images are initially called lineaments; if field-checking establishes the presence of structural offset, they can then be designated as faults. This procedure is illustrated in the example of the Peninsular Ranges, California, given later in this chapter.

There is no minimum length for lineaments, but significant crustal features are typically measured in tens or hundreds of kilometers. Lineaments may also be recognized on gravity, magnetic, and seismic contour maps by such features as aligned highs and lows, steep contour gradients, and linear offsets of trends. Lineaments are well expressed on Landsat images because of the low sun angle, the suppression of distracting spatial details, and the regional coverage. Note, however, that lineaments are but one of many types of geologic features recognized on Landsat and other remote sensing images. Faults, folds, rock types, and other features are readily identifiable as shown by the examples from Thermopolis and elsewhere.

Analysis of Landsat Images

Landsat images are interpreted in much the same manner as small-scale aerial photographs or images and photographs acquired from manned satellites. However, there are some differences and potential advantages of Landsat images. Linear features caused by topography may be enhanced or suppressed on Landsat images depending on orientation of the features relative to sun azimuth. Linear features trending normal, or at a high angle, to the sun azimuth are enhanced by shadows and highlights. Those trending parallel with the azimuth are suppressed and difficult to recognize, as are linear features parallel with the MSS scan lines.

Scratches and other film defects may be mistaken for natural features, but these defects are identified by determining whether the questionable features appear on more than a single band of imagery. Shadows of aircraft contrails may be mistaken for tonal linear features but are recognized by checking for the parallel white image of the contrail. Many questionable features are explained by examining several images acquired at different dates. With experience, an interpreter learns to recognize linear features of cultural origin, such as roads and field boundaries.

The recommended interpretation procedure is to plot lineaments as dotted lines on the interpretation map. Field checking and reference to existing maps will identify some lineaments as faults; for these the dots are connected by solid lines on the interpretation map. The remaining dotted lines may represent (1) previously unrecognized faults, (2) zones of fracturing with no displacement, or (3) lineaments unrelated to geologic structure.

The repeated coverage of Landsat enables interpreters to select images from the optimum season for their purpose. Several examples of seasonal effects on images occur later in the chapter. Winter images provide minimum sun elevations and maximum enhancement of suitably oriented topographic linear features. Larger structural and topographic features are commonly enhanced on images of snow-covered terrain because the snow eliminates or suppresses tonal differences and minor terrain features, such as small lakes. Areas with wet and dry seasonal climates should be interpreted from images acquired at the different seasons. In southern California and South Africa, cloud-free rainy-season images are best for most applications, but this selection may not apply everywhere.

Significance of colors on Landsat IR color images was described earlier in the section on MSS images. For special interpretation objectives, black-and-white images of individual bands are useful. Table 4.4 gives some specific applications of TM bands.

Mosaics are valuable for interpreting regional structural trends that extend across more than a single image. It is also helpful to project a Landsat image on a screen and study it from a distance. Almost invariably some previously unrecognized features become apparent.

PENINSULAR RANGES, CALIFORNIA

The structure of California south of the Transverse Ranges is dominated by the southwest-trending San Andreas, San Jacinto, and Elsinore strike-slip faults. On Landsat images, however, Sabins (1973) noted prominent lineaments trending north and northeast that did not correspond to mapped faults. Lamar and Merifield (1975) interpreted the Landsat image of Figure 4.26 to produce a lineament map (Figure 4.27) that was checked and evaluated in the field, which is a difficult task in the rugged Peninsular Ranges with their limited access and dense cover of brush. Bedrock consists of Late Mesozoic plutonic rocks and roof pendants of Paleozoic and Mesozoic metamorphic rocks. Breccia and gouge zones

were the main criteria for recognizing faults in the field. Displaced or terminated lithologic contacts were also used to recognize faults, but these features are scarce in this region.

Based on fieldwork, the lineaments were assigned to three categories:

1. lineaments that correlate with previously mapped faults (two lineaments)

2. lineaments that correlate with faults that previously were unmapped or were of controversial origin (seven)

3. lineaments that do not correlate with faults or have an unknown origin (seven)

Figure 4.27 plots the lineament categories with different symbols. Two lineaments that correlate with previously mapped faults (category 1 above) require no additional explanation. Table 4.7 explains representative examples of the other two categories. The seven lineaments in category 2 are now classified as faults; characteristics of four of these are summarized in the table. The Witch

TABLE 4.7 Field evaluation of lineaments on a Landsat MSS image of the Peninsular Ranges, California

Lineament	Geologic setting	Description	Remarks
Lineaments that correlate with previously unmapped or controversial faults			
San Ysidro Creek fault	Fault occurs along contact between granitic intrusive and schist.	Fault is a zone of crushed and sheared gouge up to 7 m wide. Striations indicate predominantly horizontal movement.	Alignment with San Diego River fault to the southwest is probably fortuitous.
San Diego River fault zone	Fault transects terrain of schist and plutonic rocks.	Fault is a zone of gouge, sheared, and brecciated rock, with up to 630 m of right-lateral separation of lithologic contacts.	Fault zone controls straight course of the San Diego River.
Chariot Canyon fault	Fault follows contact between schist on the west and granitic rocks on the east.	Fault is a broad shear zone with slickensides that separates granite and schist.	Fault is discontinuous and requires additional field mapping.
Lineaments that do not correlate with faults or have unknown origins			
Thing Valley fault	Bedrock is predominantly granite with inclusions of schist.	Fault is over 20 km long with a gouge and breccia zone up to 35 m wide.	Possible right-lateral separation ranges from 100 m to 1 km.
Witch Creek lineament	Bedrock is diorite and schist. Jointing and foliation are not aligned with the lineament.	Aligned, remarkably straight canyons parallel the San Diego River fault zone.	Lineament may correlate with fault, but field evidence is lacking.
Pamo Valley lineament	Granitic rocks	Lineament consists of discontinuous straight stream segments and is second only to the San Diego River lineament in prominence on images.	No fault control. Alignment of straight segments of diverse origin is possibly fortuitous.

Source: From Lamar and Merifield (1975).

Creek and Pamo Valley lineaments (Table 4.7) are examples of category 3, lineaments for which no evidence of faulting was found in the field. They are controlled by jointing and foliation directions in the bedrock or by aligned stream segments of diverse origin. It is significant that 9 of the 16 Landsat lineaments correlate with faults that were known prior to Landsat or with faults recognized on the basis of Landsat interpretation. The northwest-trending Elsinore fault (Figure 4.27) is an active fault that has moved during the past 10,000 years and caused earthquakes. The fresh appearance of the fault is evidence for the recent movement. The north- and northeast-trending lineaments and faults, however, are deeply eroded and lack recent offsets, indicating that they are not active faults. Additional criteria for recognizing active faults are given in Chapter 11.

ADIRONDACK MOUNTAINS, NEW YORK

An instructive contrast to the arid, high-relief terrain of California is provided by the Adirondack Mountains of northeast New York, which were studied by Isachsen, Fakundiny, and Forster (1973, 1974).

Geologic Setting

This humid area of subdued relief is shown on the Landsat mosaic (Figure 4.28) where the Adirondack Mountains are the dark, highly fractured oval area occupying most of the scene. The Adirondacks are a glaciated dome of Precambrian gneiss, quartzite, and marble with a core of metamorphosed anorthosite. This dome is surrounded by gently dipping sandstones and carbonate rocks of Lower Paleozoic age that are lighter in tone and much less fractured than the crystalline rocks. Physiographically the Adirondack Mountains are bordered on the north and east by the St. Lawrence–Champlain lowlands, on the south by the Mohawk lowlands, and on the west by the Tug Hill Plateau.

Landsat Investigation

Isachsen, Fakundiny, and Forster (1974) plotted all linear features visible on Landsat images at the 1:1,000,000 scale. These lineaments were screened to eliminate

1. artificial features such as highways, railroads, power lines, and canals

2. linear features coincident with foliation trends and rock contacts not caused by faulting

This screening eliminated approximately 20 percent of the original features; Figure 4.29 shows the results of comparing the remaining lineaments with the major faults and topographic lineaments on the geologic map of New York state (1:250,000 scale). Many of the lineaments correlate with previously mapped faults and are indicated by a solid line through the dotted pattern. Most of the previously mapped faults that are not visible on the mosaic are short and may prove to be visible on larger-scale images. The lineaments that do not correspond to previously mapped faults are shown with lines of unconnected dots.

Analysis of Lineaments

The major problem encountered in field checking was the difficulty of locating the Landsat lineaments on the ground. Low-level aircraft flights were the most effective and economical method of location, after which key localities were investigated on the ground. The lineaments were then classified as shown in Table 4.8. Straight stream valleys are the predominant category.

The radial diagram of Figure 4.30A plots the lengths of all the faults and linear topographic features shown on the geologic map of New York for the Adirondacks region. Figure 4.30B is a plot of all the lineaments on the Landsat interpretation map of Figure 4.29 (connected and unconnected dotted lines). Aside from the greater number of Landsat lineaments, the two radial diagrams are similar; the major difference is that the pronounced maximum of N15°E for the mapped faults is less prominent in the Landsat diagram.

TABLE 4.8 Lineaments on Landsat mosaic of the Adirondack Mountains, New York

Lineaments	Number	Percent
Straight stream valleys	130	41
Elongate lakes or straight shorelines of lakes	4	1
Topographic escarpments	6	2
Dark vegetation strips	8	3
Natural vegetation borders	6	2
Combinations of linear features	129	40
Unexplained	36	11
Total	319	100

Source: Summarized from Isachsen, Fakundiny, and Forster (1974).

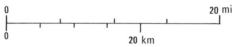

FIGURE 4.26 Landsat MSS band-5 image of the Peninsular Ranges, southern California. See Figure 4.27 for identification of major faults and lineaments. From Lamar and Merifield (1975, Figure 4). Courtesy P. M. Merifield, California Earth Science Company.

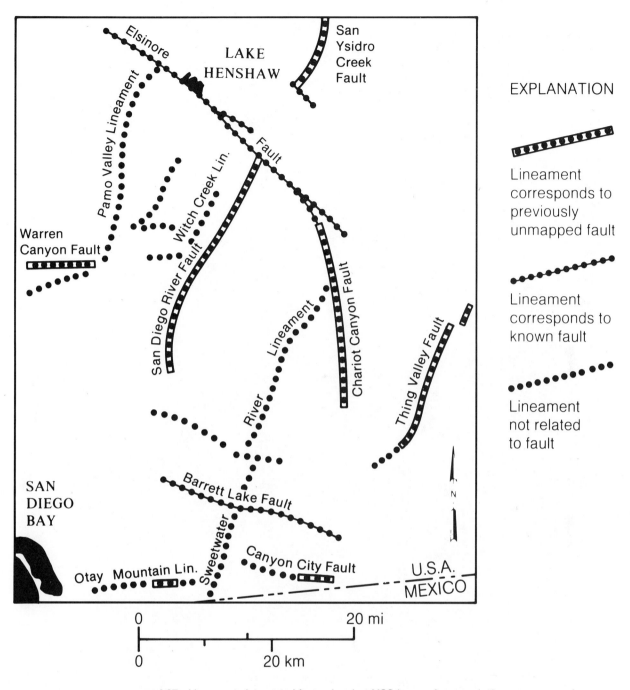

FIGURE 4.27 Lineaments interpreted from a Landsat MSS image of an area in the Peninsular Ranges, California. From Lamar and Merifield (1975, Figure 2).

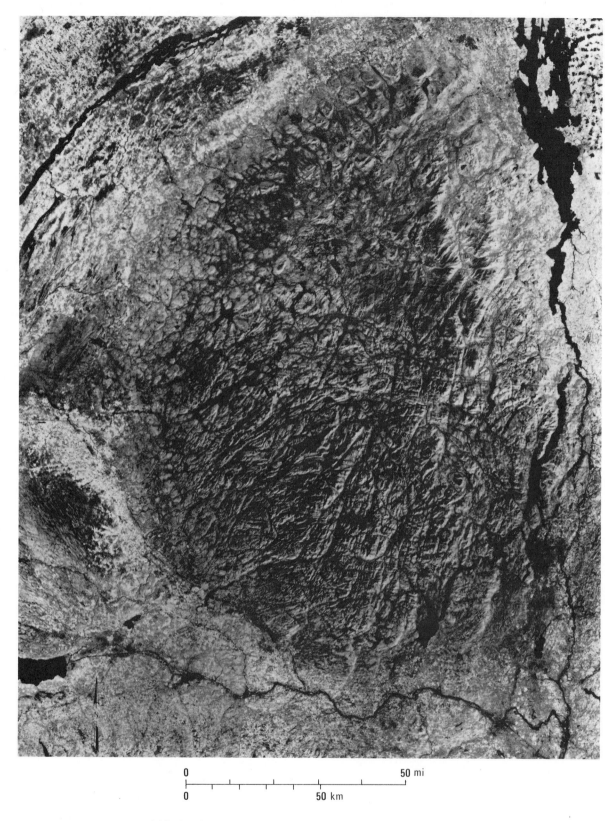

0   50 mi

0   50 km

FIGURE 4.28 Landsat mosaic of the Adirondack Mountains, New York. Compiled from MSS band-7 images acquired October 10 and 11, 1972. Courtesy Y. W. Isachsen, New York State Geological Survey.

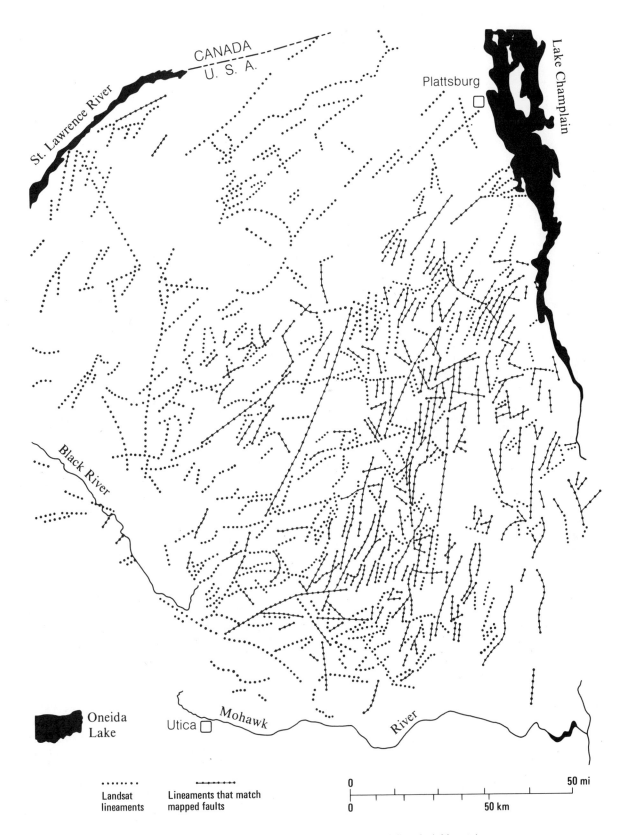

CANADA
U. S. A.

Plattsburg

Lake Champlain

St. Lawrence River

Black River

Oneida Lake

Utica

Mohawk River

· · · · · · · ·
Landsat
lineaments

·—·—·—·
Lineaments that match
mapped faults

0   50 mi
0   50 km

FIGURE 4.29 Interpretation of Landsat mosaic of the Adirondack Mountains,
New York. From Isachsen, Fakundiny, and Forster (1973, Figure 2). Courtesy
Y. W. Isachsen, New York State Geological Survey.

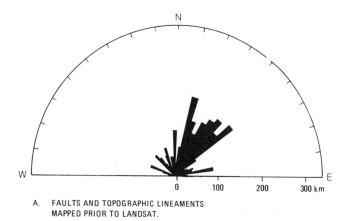

A. FAULTS AND TOPOGRAPHIC LINEAMENTS
MAPPED PRIOR TO LANDSAT.

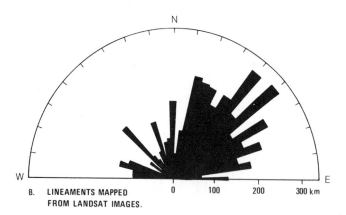

B. LINEAMENTS MAPPED
FROM LANDSAT IMAGES.

FIGURE 4.30 Radial diagrams for faults and lineaments of the Adirondack Mountains. Summed lengths are shown at 5° intervals. From Isachsen, Fakundiny, and Forster (1974, Figures 13B and 15).

In summary, the close coincidence between the Landsat lineaments and the pattern of previously mapped faults is very strong evidence that the Landsat lineaments represent faults and fractures. Because bedrock is so poorly exposed in the Adirondacks, Isachsen, Fakundiny, and Forster (1973) suggested that Landsat images may be the best data available for the regional mapping of fracture zones.

STEREO VIEWING OF LANDSAT IMAGES OF THE PETEN BASIN, MEXICO AND GUATEMALA

Stereoscopic viewing of overlapping aerial photographs is possible because of the displacement of targets (Chapter 2). The geometry of the Landsat multispectral scan-

ner and of other cross-track scanner systems causes displacement in the cross-track scan direction but not in the flight-line direction. The satellite or aircraft is directly over the center of each scan line, which means that the stereo base (see Figure 3.12) is zero in the flight direction and stereo viewing is not possible in this direction. There is sidelap, however, between images acquired on adjacent Landsat orbit paths (Figure 4.3), which provides some capability for stereo viewing.

The altitude of Landsats 1, 2, and 3 is 920 km. The sidelap of adjacent orbit paths increases from 14 percent at the equator to 85 percent at the 80° latitude. Base–height ratios are calculated from this information and range from 0.031 at the 80° latitude to 0.174 at the equator. These low values result in vertical exaggeration factors that are small, even for images at the equator. MSS and TM images acquired by Landsats 4 and 5 have less sidelap than the first generation of Landsats (Table 4.1), which reduces the area available for stereo viewing. Despite the limited coverage and minimal vertical exaggeration, stereo viewing can be useful for interpreting Landsat images. The French SPOT satellite, described later in this chapter, is designed to provide worldwide coverage of stereo images.

Figure 4.31 is a Landsat stereo pair taken from the sidelapping portions of adjacent MSS swaths. The area is the Peten Basin of Mexico and Guatemala, where several large oil fields have been discovered. The 10-month time lapse in acquiring the adjacent images was caused by cloud cover and priorities for using the Landsat tape recorders. Note that the image on the right was acquired by Landsat 1 and the image on the left by Landsat 2. To derive maximum benefit from this discussion, view the figure with a pocket-lens stereoscope. At this 17°N latitude, the base–height ratio is only 0.17, resulting in minimal vertical exaggeration. Nevertheless, the stereo model is valuable, which you can verify by interpreting the image first without and then with stereo. The procedure for interpreting a Landsat stereo model is the same as that described earlier for stereo aerial photographs. The first step is to map drainage patterns to provide a geographic reference. The next step is to define major rock units and map their boundaries. Then one records structural features, such as fold axes, faults, lineaments, and strike and dip to produce a geologic map (Figure 4.32). The final step is to produce a geologic cross section.

Despite the vegetation cover of this tropical area, structural features show up with remarkable clarity on the stereo model. Structure of the area is dominated by east-west trending symmetrical anticlines and synclines. The following features of the folds can be interpreted on the stereo model:

1. Anticlines may be distinguished from synclines by their topographic expression and by the attitude of the dipping strata.

2. Symmetry and relative amplitude of the folds.

3. Plunge of the folds is mapped by tracing strata that curve around the plunging noses of the structures.

Some major faults trend parallel with the fold axes, and others cut diagonally across them. On a single image, many of the faults can be interpreted only as possible lineaments. On the stereo model, however, one can estimate the topographic expression and stratigraphic offset for many of the faults. A major west-trending fault at the south part of the image is particularly well expressed. On a single image, this fault appears as a bright lineament against a darker background; in the stereo model, however, the bright lineament is a steep south-facing scarp illuminated by the sun. The scarp is almost surely a steeply dipping fault with the south side downthrown relative to the north side. Approximately 15 km to the north is a second west-trending lineament. The eastern portion is expressed as a dark linear feature that is interpreted as a shadow on a north-facing scarp. In the northern part of the area, several northwest-trending faults cause lateral offset of stratigraphic units and may have a component of strike-slip displacement.

The sparse stratigraphic information available for this region indicates that the outcrops are marine strata of Cretaceous and Tertiary age. On the Landsat images, three major rock units can be distinguished. The oldest unit, which crops out in the crests of the major anticlines, consists of poorly stratified rocks with a distinctive, fine-textured, mottled pattern of irregular dark spots against a bright background. The unit is interpreted as massive carbonate rocks with well-developed karst topography. *Karst topography* forms by solution and collapse of carbonate rocks in tropical climates and is characterized by steep-sided, irregular pinnacles and depressions. Shadows cast by the pinnacles and depressions cause the dark mottled pattern against the bright background of high vegetation reflectance on the Landsat MSS band-7 images of Figure 4.31. Overlying the carbonate unit is a sequence of well-stratified clastic rocks, probably sandstone and shale, exposed on the flanks of anticlines and in the troughs of synclines. Stratification is expressed by variations in tone and resistance to weathering of the beds. Most of the strike and dip information on Figure 4.32 is interpreted from outcrops of this clastic unit. The third rock unit consists of alluvial deposits in the stream valleys of the northern part of the area. According to W. V. Trollinger (personal communication), geologic interpretation of Landsat stereo models was a major factor in the successful oil exploration campaign in the Peten Basin.

SEASONAL INFLUENCE ON IMAGES

The repetitive coverage by Landsat has demonstrated that images acquired at different seasons may enhance geologic features. In regions with seasonal rainfall patterns, images acquired in the wet season are markedly different from those in the dry season. At high latitudes there are major differences between summer and winter images.

Wet-Season and Dry-Season Images in South Africa

Plate 7 shows two Landsat MSS images of the south flank of the Transvaal Basin; one image was acquired during the winter dry season and the other in the summer rainfall season. The area is a grass-covered plateau with conspicuous ridges of resistant rock units.

Geologic Setting The geologic map of Figure 4.33 was interpreted from the rainy-season image. Rock types distinguished on the image were identified by referring to the maps of Grootenboer (1973). The northern two-thirds of the image covers the southwest flank of the Transvaal Basin where the strata dip gently north. A northeast-trending anticline separates the Transvaal Basin from the more complex Potchefstroom Basin in the southeast corner of the image. The oldest rocks in the area are granites of Archean age exposed along the axis of the anticline. The granites are overlain by the Witwatersrand quartzite and shale and the Ventersdorp andesites and sediments. The overlying Transvaal System consists of a thin basal clastic unit (Black Reef Series) overlain by massive dolomitic limestone (Dolomite Series), which is overlain by alternating quartzites and shales with some volcanic layers (Pretoria Series). Intrusive mafic sills are abundant near the top of the Pretoria Series.

The Bushveld igneous complex that occupies the center of the Transvaal Basin occurs in the northern part of the area. The Layered Sequence at the base consists of mafic igneous rocks and is overlain by the Bushveld Granite. At the north edge of the image, the Bushveld Complex is intruded by the Pilansberg Complex, a circular structure of silicic intrusive rocks. Most of the area is covered by residual soil that conceals much of the bedrock. The only significant outcrops are the Pilansberg Complex, quartzites of the Witwatersrand and

A. LANDSAT 2317–15522,
    DECEMBER 5, 1975.

B. LANDSAT 1572–15554,
    FEBRUARY 15, 1974.

FIGURE 4.31 Stereo pair of Landsat MSS band-7 images, Peten Basin, Mexico and Guatemala.

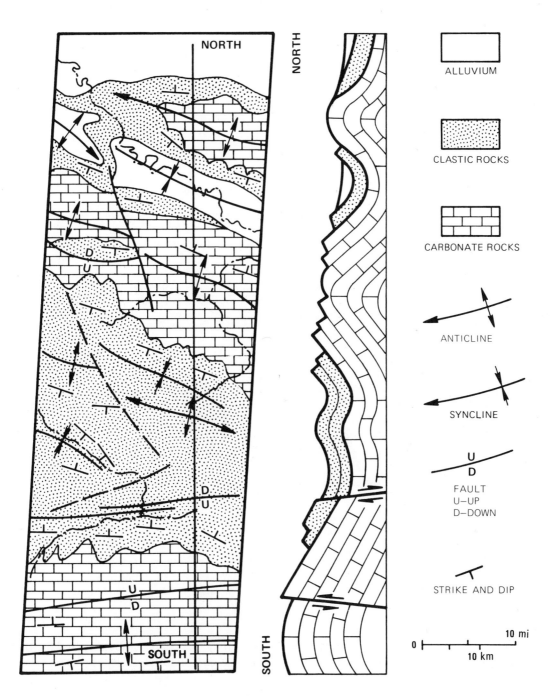

FIGURE 4.32 Geologic interpretation map and cross section of the Landsat stereo pair of an area in the Peten Basin. Modified from an unpublished interpretation by W. V. Trollinger, Denver, Colorado.

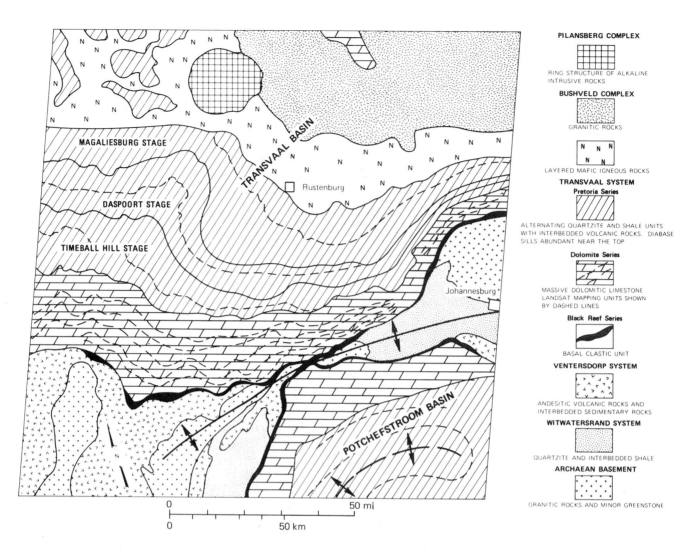

PILANSBERG COMPLEX

RING STRUCTURE OF ALKALINE
INTRUSIVE ROCKS

BUSHVELD COMPLEX

GRANITIC ROCKS

LAYERED MAFIC IGNEOUS ROCKS

TRANSVAAL SYSTEM

Pretoria Series

ALTERNATING QUARTZITE AND SHALE UNITS
WITH INTERBEDDED VOLCANIC ROCKS. DIABASE
SILLS ABUNDANT NEAR THE TOP

Dolomite Series

MASSIVE DOLOMITIC LIMESTONE
LANDSAT MAPPING UNITS SHOWN
BY DASHED LINES

Black Reef Series

BASAL CLASTIC UNIT

VENTERSDORP SYSTEM

ANDESITIC VOLCANIC ROCKS AND
INTERBEDDED SEDIMENTARY ROCKS

WITWATERSRAND SYSTEM

QUARTZITE AND INTERBEDDED SHALE

ARCHAEAN BASEMENT

GRANITIC ROCKS AND MINOR GREENSTONE

FIGURE 4.33 Geologic map of the south flank of the Transvaal Basin, South Africa, interpreted from a Landsat MSS image acquired in the rainy season. From Grootenboer (1973, Figure 1).

Transvaal systems, and scattered exposures of the Bushveld Complex.

Comparison of Images When the dry-season image (Plate 7A) was acquired, the area was covered with dry, brown grass, the indigenous vegetation was leafless, and the corn fields were fallow. Black patches mark areas of recent burning. Slight tonal variations in the image enable recognition of the major stratigraphic units to a degree comparable to that on 1:1,000,000 geologic maps published prior to Landsat (Grootenboer, 1973). The wet-season image (Plate 7B) was acquired at the height of the rainy summer season, when the area was covered by green annual grasses and the perennial vegetation was in full leaf. The strong tonal variations are directly related to bedrock lithology, particularly in the area un-

derlain by the Transvaal System and the Bushveld Complex (Figure 4.33). Locally the lithologic detail on the image is comparable to that on 1:250,000-scale published geologic maps of the area.

Of particular interest are the seven zones of tonal variations within the outcrop of the Dolomite Series. Field checks by Grootenboer, Eriksson, and Truswell (1973) established that the four darker zones correspond to dark, chert-free dolomite and the three lighter zones to light-toned dolomite with abundant chert. During the previous 90 years of geologic investigation in the area, no such stratigraphic subdivisions had been recognized in the Dolomite Series.

Analysis of Seasonal Differences Several factors contribute to the superiority of the wet-season image:

1. Windblown dust causes atmospheric haze, which severely scatters light in the dry season. During the wet season, rainfall removes dust from the air, producing a clearer atmosphere and good image contrast. It also washes away the surface dust layer.

2. Greater soil moisture enhances tonal and color differences between rock types.

3. Vegetation grows preferentially on belts of soil with higher moisture. The red stripes of vegetation in the wet-season image help delineate geologic trends.

These advantages of wet-season images have also been observed in other areas.

Winter and Summer Images in Arctic Canada

At high and moderate latitudes, winter and summer images differ in sun elevation and in snow cover. The advantages of lower sun elevation for geologic mapping were demonstrated in Chapter 2 and also apply to Landsat images. The two Landsat MSS band-7 images (Figure 4.34) near Bathurst Inlet in Arctic Canada were acquired in the summer and late winter. On the summer image (Figure 4.34A), there is no snow; vegetation growth is vigorous, as shown by the bright signatures; and most of the small lakes have thawed, as revealed by their dark signatures. Few geologic features are recognizable. Most of the lakes are only a few hundred meters in size and tend to obscure the geologic features that are thousands of meters in size. The snow cover of the winter image (Figure 4.34B) conceals the frozen lakes, thereby enhancing the appearance of geologic structures.

The relatively high sun angle (45°) of the summer image causes only minimal highlights and shadows. The low sun angle (27°) of the winter image, however, causes highlights and shadows that emphasize subtle topographic features expressing strike ridges of sedimentary rocks, folds, lineaments, faults, and igneous dikes. Figure 4.35 shows these features, which were interpreted from the winter image.

A major fold, outlined by strike ridges of sedimentary rocks, is surrounded by highly fractured, unstratified crystalline basement rocks. The fold is a syncline with younger strata preserved in the center. The strata are argillites, sandstones, and quartzites of Proterozoic age that overlie older crystalline rocks of Archean age. The lineament trending northwest across the northern part of the image is a major fault. North of the fault is a small anticline of Proterozoic strata. Also north of the fault at the east margin of the map and images is a prominent ridge that corresponds to an igneous dike. At the western boundary of the map, a distinct circular drainage

anomaly 8 km in diameter marks a ring dike or an igneous plug. These geologic features are mappable only from the winter image.

This Arctic example demonstrates the advantages of snow cover and low sun angles for geologic mapping. These conditions also occur in winter images of areas at intermediate latitudes.

PLATE-TECTONIC ANALYSIS: AFAR REGION, ETHIOPIA

The geologic discipline of *plate tectonics* deals with the dynamics of the lithologic plates that move over the earth. Of particular interest are the boundaries where plates diverge (*spreading centers*), collide (*subduction zones*), or shift laterally (*transform faults*). Many of these boundaries occur in the oceans and cannot be imaged by Landsat or other satellites. Where the boundary features are exposed in the continents, they are ideal subjects for Landsat analysis because of their great extent and lack of adequate maps in some areas. The MSS image of the Los Angeles region (Plate 4) includes a portion of the San Andreas fault, which is the transform boundary between the Pacific and North American plates.

Plate-tectonic features are also exposed in the Afar region, or Danakil depression, of northern Ethiopia. This area covers the triple junction at the intersection of the spreading axes of the Red Sea, Gulf of Aden, and Ethiopian rifts (Figure 4.36, index map). The Arabia, Nubia, and Somalia plates are separating along these spreading axes. Much of the Afar region is below sea level, but it is separated from the Red Sea and Gulf of Aden by uplifted fault blocks. The region is inhospitable and poorly accessible, but excellent Landsat images are available.

On the MSS image (Figure 4.37), Lake Abbe marks the present position of the triple junction, which has migrated through time. The Gulf of Tadjura is the western end of the Gulf of Aden. Light-toned areas are recent deposits of clay, silt, and evaporite in the depressions. Basalt (darker tone) and andesite (lighter tone) of Tertiary age form the outcrops. Mount Damahali is a recent volcano with associated lava flows. Geologic structure is dominated by normal faults that form the boundaries of numerous horsts and grabens (Appendix). The faults are accentuated by shadows on the image that indicate the topographic relief and the sense of throw. In the southwest part of the image, the north-trending faults belong to the Ethiopian rift system. The northwest-trending faults are parallel with the Red Sea spreading axis. Prominent west-trending faults in the southeast are parallel with the Gulf of Aden spreading axis. Except for an apparent strike-slip fault northwest of Lake Asal,

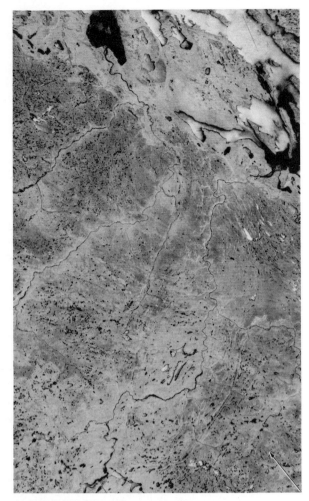 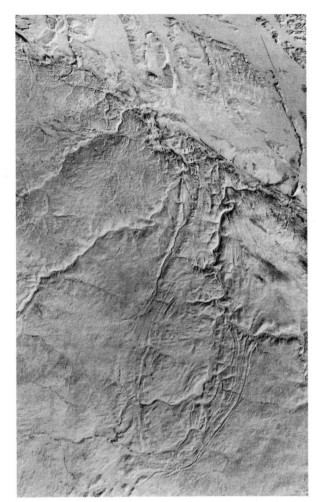

A. SUN ELEVATION 45°, JUNE 18, 1973.   B. SUN ELEVATION 27°, APRIL 2, 1974.

FIGURE 4.34 Seasonal effect of snow cover and different sun elevations. Landsat MSS band-7 images of Bathurst Inlet, Arctic Canada.

there is no suggestion of transform faulting. The sinuous Gawa graben in the north part of the image is a departure from the regional pattern.

Landsat images have been used effectively to understand patterns of plate tectonics in many other regions. A good example is the collision zone resulting from the northward drift of the Indian subcontinent toward Eurasia. Following the initial collision of the two crustal plates 40 to 60 million years ago, India has moved northward another 2000 km toward Eurasia. A major problem in tectonic theory is accounting for the vast land area that was displaced by the convergence of the plates. The problem is complicated by two factors: (1) the geology of Eurasia is complex and appears to present a chaotic jumble of landforms, and (2) the geology is poorly known. When Molnar and Tapponnier (1977) viewed Eurasia as a whole on Landsat MSS images, they recognized a simple, coherent geologic pattern attributed to the collision. They interpreted structural information that was previously unavailable or unknown. Major strike-slip faults were recognized, and for some the sense of displacement was evident. Their Landsat interpretation was combined with earthquake data to determine the sense of motion associated with the principal faults. This analysis indicated that west-trending faults with left-lateral displacement are predominant. Thus China has been displaced laterally eastward out of the way of India, which is reasonable because material displaced westward would encounter the resistance of the Eurasian land mass. Eastward motion, however, is easily accommodated by thrusting of China over the oceanic plates along the margins of the Pacific.

A. NORMAL COLOR PHOTOGRAPH.

B. IR COLOR PHOTOGRAPH.

PLATE 1 Normal color and infrared color aerial photographs of the UCLA and Westwood area, Los Angeles, California. Each photograph covers a width of 5 km.

A. PORTION OF NHAP IR COLOR PHOTOGRAPH OF COEUR D'ALENE, IDAHO. PHOTOGRAPH COVERS A WIDTH OF 10 km.

B. NORMAL COLOR COMPOSITE SCANNER IMAGE, SAN PABLO BAY, CALIFORNIA. IMAGE COVERS A WIDTH OF 13 km.

C. IR COLOR COMPOSITE SCANNER IMAGE, SAN PABLO BAY, CALIFORNIA.

PLATE 2   NHAP photograph and aircraft multispectral scanner images.

A. GREAT BARRIER REEF AND COAST OF QUEENSLAND, AUSTRALIA.

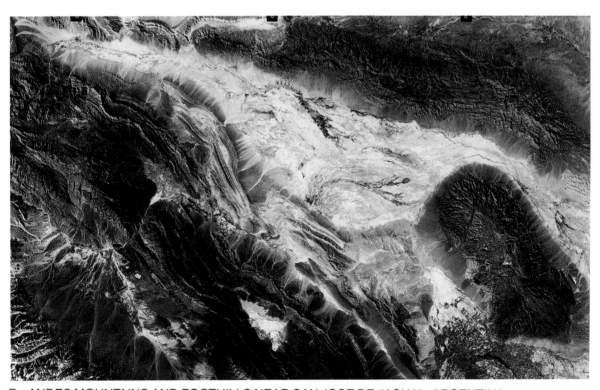

B. ANDES MOUNTAINS AND FOOTHILLS NEAR SAN JOSE DE JACHAL, ARGENTINA.

PLATE 3 IR color photographs acquired by the large-format camera from the Space Shuttle, October, 1984. Each photograph covers a width of 240 km.

PLATE 4 IR color Landsat MSS image of the Los Angeles region. Image covers a width of 185 km.

A. NORMAL COLOR IMAGE, ST. FRANCIS
RIVER, ARKANSAS, SUBSCENE.

B. IR COLOR IMAGE, ST. FRANCIS RIVER,
ARKANSAS, SUBSCENE.

C. NORMAL COLOR IMAGE, THERMOPOLIS,
WYOMING, SUBSCENE.

D. IR COLOR IMAGE, THERMOPOLIS,
WYOMING, SUBSCENE.

PLATE 5 Landsat TM images. Each image covers a width of 20 km.

PLATE 6 IR color Landsat TM image of the Thermopolis, Wyoming, subscene. Image covers a width of 20 km.

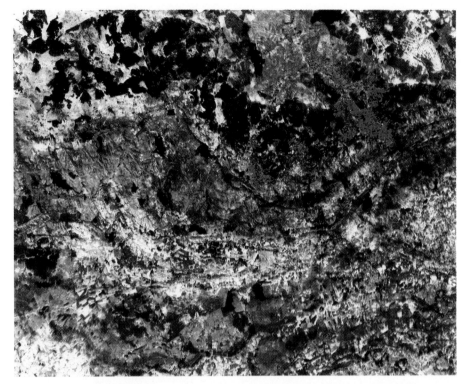

A. DRY-SEASON IMAGE.

B. WET-SEASON IMAGE.

PLATE 7 Landsat MSS images of western Transvaal Province, South Africa. Images processed by Environmental Research Institute of Michigan. Each image covers a width of 100 km.

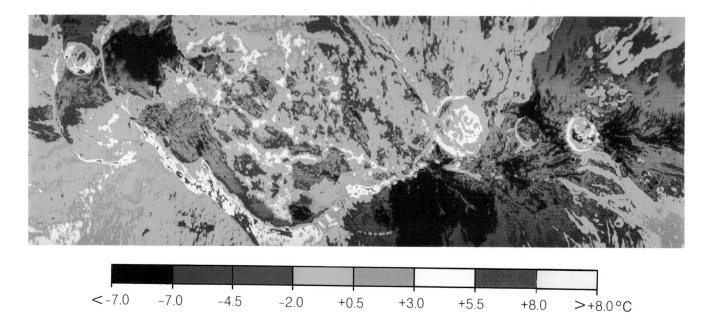

| < -7.0 | -7.0 | -4.5 | -2.0 | +0.5 | +3.0 | +5.5 | +8.0 | > +8.0 °C |

A. SUMMIT OF MAUNA LOA, HAWAII. IMAGE COVERS A WIDTH OF 10 km.

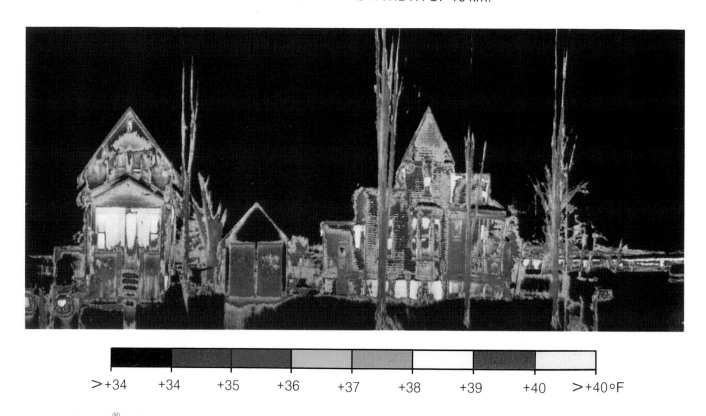

| > +34 | +34 | +35 | +36 | +37 | +38 | +39 | +40 | > +40 °F |

B. VANSCAN® WINTER IMAGE IN PLYMOUTH, MICHIGAN.

PLATE 8 Color density-sliced nighttime thermal IR images (8 to 14 μm). Courtesy Daedalus Enterprises, Incorporated.

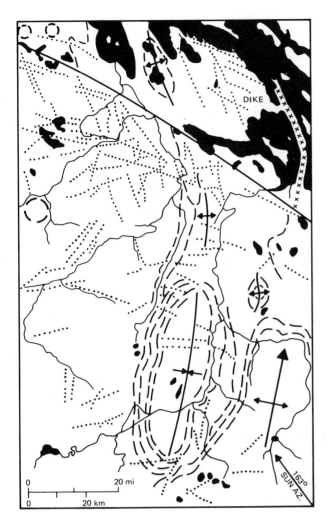

FIGURE 4.35 Interpretation map of Bathurst Inlet. Dashed lines are resistant strata. Other symbols are conventional. Reproduced with permission from A. F. Gregory and H. D. Moore, *Recent advances in geologic applications of remote sensing from space*, copyright 1976, Figure 5, Pergamon Press, Ltd.

FUTURE LANDSAT MISSIONS

The U.S. government operated and funded the Landsat program from its inception. Income from the sale of image data was only a small fraction of the cost of the program. In 1985 the U.S. government transferred ownership and operation of the Landsat program to the private sector for the following reasons: (1) to reduce government expenditures, (2) to transfer control of operating Landsat from taxpayers to users of the data, and (3) to remove the government from a commercial activity. The EOSAT Company (a joint venture of Hughes Aircraft Corp. and RCA) now operates the Landsat program and will launch replacement satellites.

SPOT PROGRAM

In 1986, the French space agency plans to launch SPOT (Système Probatoire d'Observation de la Terre) satellite on an Arianne rocket from their facility in French Guiana. The orbital altitude of SPOT is 832 km, and the sun-synchronous orbit is similar to that of Landsat. The south-bound orbits cross latitude 40°N at 10:00 a.m. local sun time, and ground tracks are repeated at 26-day intervals.

Imaging Systems

SPOT employs *high-resolution visible* (HRV) imaging systems, which are illustrated in Figure 4.38. HRV is an along-track scanner of the type described in Chapter 1. The tiltable mirror does not move during the scanning process but reflects energy into the optical system, which focuses the image onto the linear arrays of *charge-coupled detectors* (CCD). The HRV system operates in either a multispectral mode or a high-resolution panchromatic mode (Table 4.9). The 6000 CCDs of the high-resolution mode have a 10-by-10-m ground resolution cell and record a single panchromatic image in the 0.51-to-0.73-μm spectral band. The multispectral mode of HRV employs three arrays of 3000 CCDs with a ground resolution cell of 20 by 20 m. The multispectral mode records green, red, and reflected IR images, from which IR color images may be prepared. In either mode, the HRV ground swath is 60 km wide. SPOT employs two HRV systems that may be operated in the nadir position to acquire parallel strips of imagery with a total width of 117 km and 3 km of sidelap (Figure 4.39).

An innovation in SPOT is the off-nadir viewing capability (Figure 4.40). The mirror is tiltable up to 27° on either side of vertical to obtain image strips at distances up to 475 km away from the nadir. Because of the greater viewing distance at the extreme angles, the ground resolution cells are larger and the image strips are up to 80 km wide. Off-nadir viewing provides considerable flexibility in scheduling times of image acquisition. Without this capability, an orbit path would be revisited at 26-day intervals, which is not adequate for monitoring dynamic events such as floods, volcanic eruptions, and growth cycles of crops. Figure 4.41 illustrates SPOT's capability to image the same swath from several orbits during a 26-day cycle. Localities at the equator can be imaged 7 times during a cycle; those at latitude 45° can be imaged 11 times.

Repeated off-nadir viewing of a ground swath produces overlapping image pairs that are suitable for stereo viewing. The base–height ratio is determined by the distance between the two satellite positions (Figure 4.41) divided by the altitude of 832 km. Base–height ratios range from

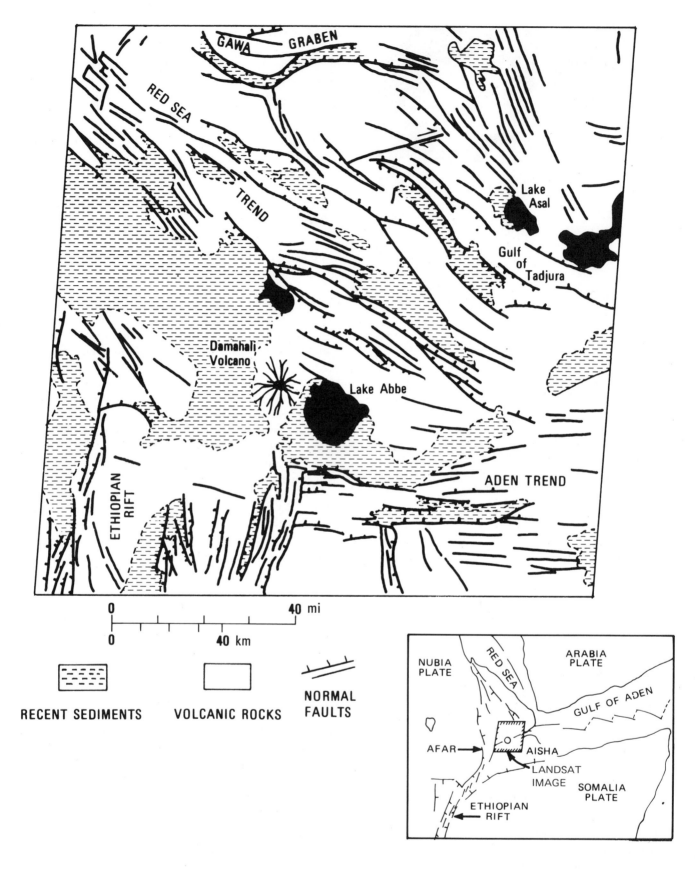

FIGURE 4.36 Geologic interpretation of the Landsat image of an area in the Afar region, Ethiopia.

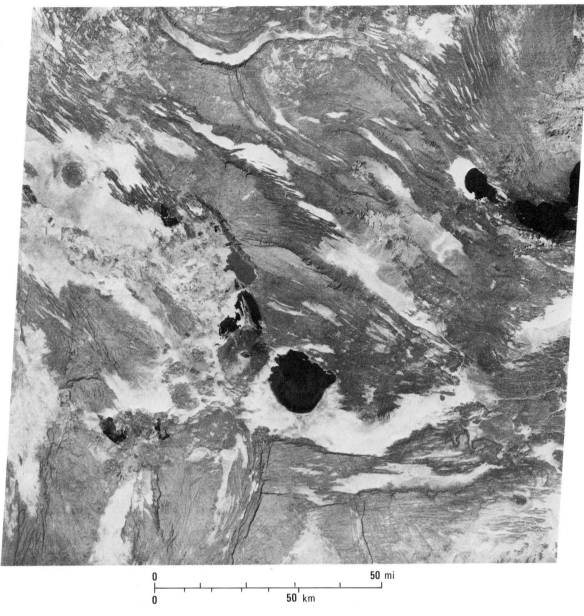

FIGURE 4.37 Landsat MSS band-6 image of an area in the Afar region, Ethiopia. Image acquired November 18, 1972.

0.75 at the equator to 0.50 at latitude 45°, which provide stereo models with strong vertical exaggeration.

The primary SPOT receiving station is located at Toulouse, France, and additional stations are planned. SPOT lacks the Landsat network of ground receiving stations and data-relay satellites; therefore, two on-board tape recorders store image data for later transmission to the receiving station.

Availability of Images

Information on the availability of images may be obtained from

SPOT Image
18 Avenue Edouard-Belin
F 31055 Toulouse Cedex, France

SPOT Image Corporation
Suite 307
1150 17th Street NW
Washington, DC 20036
USA

SPOT images have better spatial resolution than Landsat TM images and may be acquired as stereo pairs. SPOT lacks the spectral range of TM, however. The

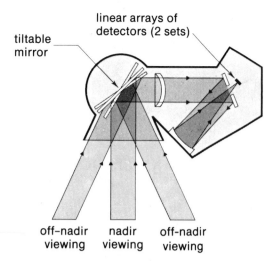

FIGURE 4.38 SPOT imaging system. Courtesy G. Weill, SPOT Image Corporation.

TABLE 4.9 Characteristics of SPOT high-resolution visible system

	Multispectral mode	Panchromatic mode
Spectral bands:		
green	0.50 to 0.59 μm	0.51 to 0.73 μm
red	0.61 to 0.68 μm	—
reflected IR	0.79 to 0.89 μm	—
Angular field of view	4.13°	4.13°
Ground resolution cell (nadir viewing)	20 by 20 m	10 by 10 m
Detectors per spectral band	3000	6000
Ground-swath width (nadir viewing)	60 km	60 km

availability and cost of Landsat data are more favorable for the user than are SPOT data.

COMMENTS

The Landsat program is a major advance in remote sensing. The advantages of Landsat images are as follows:

1. Cloud-free images are available for most of the world with no political or security restrictions.

2. The low to intermediate sun angle enhances many subtle geologic features.

3. Long-term repetitive coverage provide images at different seasons and illumination conditions.

4. The images are low in cost.

5. IR color composites are available for many of the scenes. With suitable equipment, color composites may be made for any image.

6. Synoptic coverage of each scene under uniform illumination aids recognition of major features. Mosaics extend this coverage.

7. There is negligible image distortion.

8. Images are available in a digital format suitable for computer processing.

9. Limited stereo coverage is available.

10. TM provides images with improved spatial resolution, extended spectral range, and additional spectral bands.

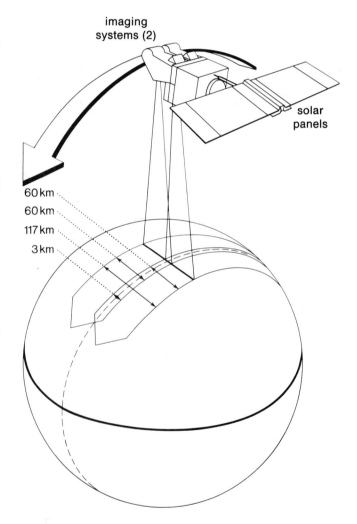

FIGURE 4.39 SPOT satellite platform and image swaths acquired in the nadir-viewing mode. Courtesy G. Weill, SPOT Image Corporation.

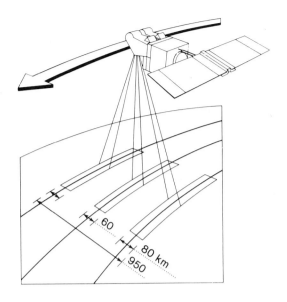

FIGURE 4.40 Off-nadir viewing capability of SPOT using tiltable mirror. Courtesy G. Weill, SPOT Image Corporation.

In addition to the applications shown in this chapter, Landsat images are valuable for resource exploration, environmental monitoring, land-use analysis, and evaluating natural hazards (as illustrated in Chapters 8, 9, 10, and 11).

Another major contribution of Landsat is the impetus it has given to digital image processing (described in Chapter 7). The availability of low-cost multispectral image data in digital form has encouraged the application and development of computer methods for image processing, which are increasing the usefulness of the data for interpreters in many disciplines.

Since the first launch in 1972, Landsat has evolved from an experiment into an operational system. There has been a steady improvement in the quality and utility of the image data. Many users throughout the world now rely on Landsat images as routinely as they do on weather and communication satellites. It is essential that the Landsat program continue to provide images.

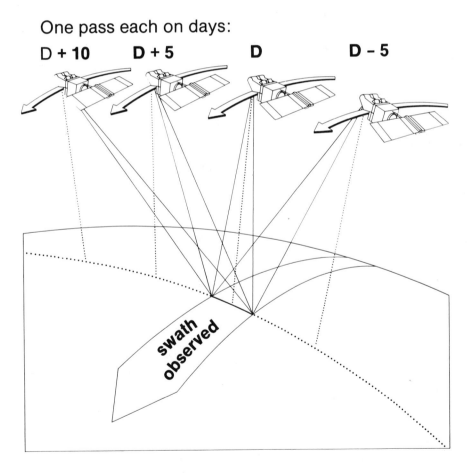

FIGURE 4.41 Capability of SPOT to revisit localities during a 26-day cycle. Localities at this equator can be observed seven times; localities at latitude 45° can be observed 11 times. Courtesy G. Weill, SPOT Image Corporation.

QUESTIONS

1. Why are path-and-row index maps for Landsats 1, 2, and 3 different from index maps for Landsats 4 and 5?

2. Generate the identification number for a Landsat-1 MSS image acquired of your area at 9:45 a.m. standard time one year after launch.

3. Assume that your eyes have normal spatial resolution (Chapter 1) and that you are an astronaut traveling on the Landsat-1 platform. What would be the dimension of the ground resolution cell you observe? How does the resolving power of your eyes compare with that of MSS?

4. Calculate the ground resolution cell of your eyes as if you were an observer on the Landsat-5 platform. How does this resolving power compare with that of TM?

5. Compare the spectral range of the eye with the spectral ranges of MSS and TM.

6. Based on the analyses for questions 4 and 5, discuss the relative merits of earth observations made by an astronaut versus those made by Landsat TM.

7. The oscillating mirror of TM completes 14 scans each second. Calculate the dwell time for each ground resolution cell.

8. For obtaining images of areas outside the United States, what is the major advantage of checking with regional facilities (Table 4.5) on image availability?

9. Critique and correct this statement made by an image interpreter: "In the Landsat image, straight and aligned stream segments form lineations (less than 10 km long) and linears (more than 10 km long) that are undoubtedly faults, although no geologic maps are available and I have not field-checked the area."

10. Prepare a table comparing significant characteristics of SPOT HRV with the three Landsat imaging systems. Analyze the advantages and disadvantages of each system for your applications.

REFERENCES

Freden, S. C., and F. Gordon, 1983, Landsat satellites *in* Colwell, R. N., ed., Manual of remote sensing, second edition: ch. 12, p. 517–570, American Society for Photogrammetry and Remote Sensing, Falls Church, Va.

Gregory, A. F., and H. D. Moore, 1976, Recent advances in geologic applications of remote sensing from space: Proceedings of 24th International Astronautical Congress, Pergamon Press, p. 153–170.

Grootenboer, J., 1973, The influence of seasonal factors on the recognition of surface lithologies from ERTS imagery of the western Transvaal: Third ERTS Symposium, NASA SP-351, v. 1, p. 643–655.

Grootenboer, J., K. Eriksson, and J. Truswell, 1973, Stratigraphic subdivision of the Transvaal Dolomite from ERTS imagery: Third ERTS Symposium, NASA SP-351, v. 1, p. 657–664.

Hobbs, W. H., 1904, Lineaments of the Atlantic border region: Geological Society of America Bulletin, v. 15, p. 483–506.

Hobbs, W. H., 1912, Earth features and their meaning—an introduction to geology for the student and general reader: Macmillan Publishing Co., N.Y.

Isachsen, Y. W., R. H. Fakundiny, and S. W. Forster, 1973, Evaluation of ERTS-1 imagery for geological sensing over the diverse geological terranes of New York State: Symposium on Significant Results Obtained from ERTS-1, NASA SP-327, v. 1, p. 223–230.

Isachen, Y. W., R. H. Fakundiny, and S. W. Forster, 1974, Assessment of ERTS-1 imagery as a tool for regional geological analysis in New York State: NASA contract NAS 5-21764, U.S. Technical Information Service Document E74-10363, Springfield, Va.

Lamar, D. L., and P. M. Merifield, 1975, Application of Skylab and ERTS imagery to fault tectonics and earthquake hazards of Peninsular Ranges, southwestern California: California Earth Science Corporation, Technical Report 75-2, Santa Monica, Calif.

Molnar, P., and P. Tapponnier, 1977, The collision between India and Eurasia: Scientific American, v. 236, p. 30–41.

NASA, 1976, Landsat data users handbook: Goddard Space Flight Center Document No. 76 SDS 4258, Greenbelt, Md.

O'Leary, D. W., J. D. Friedman, and H. A. Pohn, 1976, Lineaments, linear, lineation—some proposed new standards for old terms: Geological Society of America Bulletin, v. 87, p. 1463–1469.

Ryder, N. G., 1981, Ryder's standard geographic reference—the United States of America: Ryder Geosystems, Denver, Co.

Sabins, F. F., 1973, Geologic interpretation of radar and space imagery of California (abstract): American Association Petroleum Geologists Bulletin, v. 57, p. 802.

Sabins, F. F., 1983, Remote sensing laboratory manual, second edition: Remote Sensing Enterprises, La Habra, Calif.

Short, N. M., P. D. Lowman, S. C. Freden, and W. A. Finch, 1976, Mission to Earth, Landsat views the world: NASA Publication SP-360, U.S. Government Printing Office, stock no. 033-000-00659-4, Washington, D.C.

Slaney, V. R., 1981, Landsat images of Canada—a geological appraisal: Geological Survey of Canada paper 80-15, Ottawa, Ont.

Trifonov, V. G., 1984, Applications of space images for neotectonic studies *in* Teleki, P., and C. Weber, eds. Remote sensing for geological mapping: p. 41–56, Bureau de Recherchés Géologiques et Minières no. 82, Orleans, France.

Williams, R. S., and W. D. Carter, eds., 1976, ERTS-1, a new window on our planet: U.S. Geological Survey Professional Paper 929.

Bodechtel, J., and H. G. Gierloff-Emden, 1974, Weltrambildung die dritte Entdeckung der Erde: Paul List Verlag KG, Munich.

McCracken, K. G., and C. E. Astley-Boden, eds., 1982, Satellite images of Australia: Harcourt, Brace, and Jovanovich Group, Sydney, Australia.

Sheffield, C., 1981, Earth watch: Macmillan Publishing Co., N.Y.

Short, N. M., 1982, The Landsat tutorial workbook: NASA Reference Publication 1078, Washington, D.C.

Southworth, C. S., 1985, Characteristics and availability of data from earth-imaging satellites: U.S. Geological Survey Bulletin 1631.

SPOT Image Corporation, 1984, SPOT simulation applications handbook: American Society for Photogrammetry and Remote Sensing, Falls Church, Va.

Thermal Infrared Images

All matter radiates energy at thermal IR wavelengths (3 to 15 μm) both day and night. The ability to detect and record this thermal radiation in image form at night takes away the cover of darkness and has obvious reconnaissance applications. Accordingly, the early development of thermal IR imaging technology, beginning in the 1950s, was funded by government agencies and classified for security purposes. Military interpreters recognized that geologic and terrain features greatly influenced the background against which strategic targets were displayed. Word of these potential nonmilitary applications created interest in the civilian geologic community, and in the mid-1960s some manufacturers received approval to acquire images for civilian clients using the classified systems. In 1968 the government declassified systems that did not exceed certain standards for spatial resolution and temperature sensitivity. Today excellent scanner systems are available for unrestricted use. The term *thermography* has been suggested as a replacement for *thermal IR imagery* but is not adopted in this text. Thermography has long been associated with medical applications of thermal IR imagery; any change in the accepted use of this term would be confusing.

Satellite acquisition of thermal IR images began in 1960 with the U.S. Meteorologic Television IR Operational Satellites (TIROS) and has continued with subsequent programs. The coarse spatial resolution of these images is optimum for monitoring cloud patterns and ocean temperatures; large terrain features are also recognizable. In 1978 the NASA Heat Capacity Mapping Mission (HCMM) obtained daytime and nighttime thermal IR images with a 600-m spatial resolution for geologic applications. The thematic mapper of Landsats 4 and 5 (launched in 1982 and 1984) records a thermal IR image (band 6) with a 120-m resolution that has potential applications for terrain studies.

Until recently, thermal IR images were acquired at broad spectral bands, typically 8 to 14 μm for aircraft and 10.5 to 12.5 μm for satellites. In 1980 the thermal infrared multispectral scanner (TIMS) was developed to acquire six bands of imagery at wavelength intervals of 1 μm or less. These aircraft images are proving useful for discriminating rock types on the basis of variations in silica content.

THERMAL PROCESSES AND PROPERTIES

In order to interpret thermal IR images, one must understand the basic physical processes that control the interactions between thermal energy and matter, as well as the thermal properties of matter that determine the rate and intensity of the interactions. The following sections discuss these properties.

Heat, Temperature, and Radiant Flux

Kinetic heat is the kinetic energy of particles of matter in random motion. The random motion causes particles to collide, resulting in changes of energy state and the emission of electromagnetic radiation from the surface of materials. The internal, or kinetic, heat energy of matter is thus converted into *radiant energy*. The amount of heat is measured in calories. A *calorie* is the amount of heat required to raise the temperature of 1 g of water 1°C. *Temperature* is a measure of the concentration of heat. On the Celsius scale, 0°C and 100°C are the temperatures of melting ice and boiling water. On the Kelvin, or absolute, temperature scale, 0°K is *absolute zero*, the point at which all molecular motion ceases. The Kelvin and Celsius scales correlate as follows: 0°C = 273°K, and 100°C = 373°K. The electromagnetic energy radiated from a source is called *radiant flux* (F) and is measured in watts per square centimeter ($W \cdot cm^{-2}$).

The concentration of kinetic heat of a material is called the *kinetic temperature* (T_{kin}) and is measured with a thermometer placed in direct contact with the material. The concentration of the radiant flux of a body is the *radiant temperature* (T_{rad}). Radiant temperature may be measured remotely by devices that detect electromagnetic radiation in the thermal IR wavelength region. The radiant temperature of materials is always less than the kinetic temperature because of a thermal property called emissivity, which will be defined later.

Heat Transfer

Heat energy is transferred from one place to another by three mechanisms:

1. *Conduction* transfers heat through a material by molecular contact. The transfer of heat through a frying pan to cook food is one example.

2. *Convection* transfers heat through the physical movement of heated matter. The circulation of heated water and air are examples of convection.

3. *Radiation* transfers heat in the form of electromagnetic waves. Heat from the sun reaches the earth by radiation. In contrast to conduction and convection, which can only transfer heat through matter, radiation can transfer heat through a vacuum.

Materials at the surface of the earth receive thermal energy primarily by radiation from the sun and to a much lesser extent by conduction of heat from the interior of the earth. There are daily and annual cyclic variations in the duration and intensity of solar energy. Energy from the interior of the earth is transferred to the surface primarily by conduction and is relatively constant at any locality, although there are regional variations in this heat flow. Locally the energy is transferred by convection at hot springs and volcanoes. The regional heat flow patterns may be altered by local geologic features such as salt domes and faults.

IR Region of the Electromagnetic Spectrum

The IR region is that portion of the electromagnetic spectrum ranging in wavelength from 0.7 to 300 μm. The terms *short, middle,* and *long* have been used to subdivide the IR region but are not used here because of confusion about the boundaries. The *reflected IR region* ranges from wavelengths of 0.7 to 3 μm and includes the *photographic IR band* (0.7 to 0.9 μm) that may be detected directly by IR-sensitive film. On both IR color photographs and Landsat color composite images, the red signature records IR energy that is strongly reflected by vegetation and is not related to thermal radiation. IR radiation at wavelengths from 3 to 14 μm is called the *thermal IR region* (Figure 5.1). Thermal IR radiation is absorbed by the glass lenses of conventional cameras and cannot be detected by photographic films. Special detectors and optical-mechanical scanners are used to detect and record images in the thermal IR spectral region. Unfortunately the term *IR energy* connotes heat to many people; therefore it is important to recognize the difference between reflected IR energy and thermal IR energy.

Atmospheric Transmission Not all wavelengths of thermal IR radiation are transmitted uniformly through the atmosphere. Carbon dioxide, ozone, and water vapor absorb energy at certain wavelengths. As shown in Figure 5.1, IR radiation at wavelengths from 3 to 5 μm and from 8 to 14 μm is readily transmitted through the atmosphere; these regions are referred to as *atmospheric windows*. A number of detection devices have been designed that are sensitive to radiation of these wavelengths. The narrow absorption band from 9 to 10 μm (shown as a dashed curve in Figure 5.1) is caused by the ozone layer at the top of the earth's atmosphere. To avoid the effects of this absorption band, satellite thermal IR systems typically operate in the 10.5-to-12.5-μm band. Systems on aircraft, which fly beneath the ozone layer, are not affected and may record the full 8-to-14-μm band.

Radiant Energy Peaks and Wien's Displacement Law For an object at a constant kinetic temperature, the radiant energy, or flux, varies as a function of wavelength. The *radiant energy peak* (λ_{max}) is the wavelength

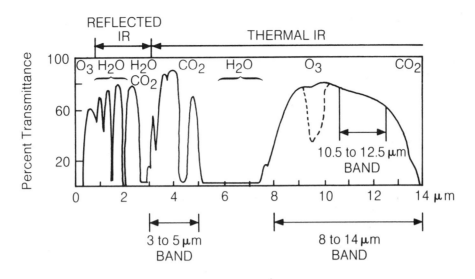

REFLECTED
IR

THERMAL IR

O_3 H_2O H_2O CO_2 H_2O O_3 CO_2
CO_2

10.5 to 12.5 μm
BAND

3 to 5 μm
BAND

8 to 14 μm
BAND

FIGURE 5.1 Electromagnetic spectrum showing spectral bands used in the thermal IR region. Gases responsible for atmospheric absorption are indicated. From Santa Barbara Research Center.

at which the maximum amount of energy is radiated. Figure 5.2 shows radiant energy curves for objects ranging in temperature from 300° to 700°K. With increasing temperature the total amount of radiant energy (the area under each curve) increases and the radiant energy peak shifts to shorter wavelengths (as indicated in Figure 5.2 by the dotted line, which passes through the radiant energy peaks of each curve). This shift to shorter wavelengths with increasing temperature is described by *Wien's displacement law*, which states that

$$\lambda_{\max} = \frac{2897 \ \mu m \cdot °K}{T_{\mathrm{rad}}} \qquad (5.1)$$

where T_{rad} is radiant temperature in degrees Kelvin and 2897 μm · °K is a physical constant. The wavelength of the radiant energy peak of an object may be determined by substituting the value of T_{rad} into Equation 5.1. For example, the average radiant temperature of the earth is approximately 300°K (27°C or 80°F). Substituting this temperature into Equation 5.1 results in

$$\lambda_{\max} = \frac{2897 \ \mu m \cdot °K}{T_{\mathrm{rad}}}$$

$$= \frac{2897 \ \mu m \cdot °K}{300°K}$$

$$= 9.7 \ \mu m$$

which is the value shown in Figure 5.2 and is the radiant power peak of the earth. Figure 5.2 will be used later

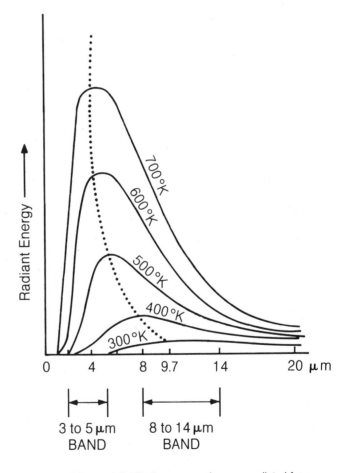

FIGURE 5.2 Spectral distribution curves of energy radiated from objects at different temperatures. Note position of 3-to-5-μm and 8-to-14-μm bands. From Colwell and others (1963, Figure 2).

to evaluate optimum wavelength bands for detecting targets at various temperatures.

Wien's displacement law also applies to hot objects that glow at visible wavelengths, such as an iron poker in a fire. As the poker heats up, the color progresses from dark red through bright red, orange, and yellow at successively shorter wavelengths. The colors represent λ_{max} of Equation 5.1.

Thermal Properties of Materials

Radiant energy striking the surface of a material is partly reflected, partly absorbed, and partly transmitted through the material. Therefore,

$$\text{Reflectivity + absorptivity + transmissivity} = 1 \tag{5.2}$$

Reflectivity, absorptivity, and transmissivity are determined by properties of matter and also vary with the wavelength of the incident radiant energy and with the temperature of the surface. As discussed in Chapter 2, reflectivity may be expressed as albedo (A), which is the ratio of reflected energy to incident energy. For materials in which transmissivity is negligible, Equation 5.2 reduces to

$$\text{Reflectivity + absorptivity} = 1 \tag{5.3}$$

The absorbed energy causes an increase in the kinetic temperature of the material.

Blackbody Concept, Emissivity, and Radiant Temperature

The concept of a blackbody is fundamental to understanding heat radiation. A *blackbody* is a material that absorbs all the radiant energy that strikes it, which means

$$\text{Absorptivity} = 1 \tag{5.4}$$

A blackbody also radiates all of its energy in a wavelength distribution pattern that is dependent only on the kinetic temperature. According to the *Stefan-Boltzmann law*, the radiant flux of a blackbody (F_b) at a kinetic temperature of T_{kin} is

$$F_b = \sigma T_{kin}^4 \tag{5.5}$$

where σ is the *Stefan-Boltzmann constant*,

$$5.67 \times 10^{-12} \text{ W} \cdot \text{cm}^{-2} \cdot {}^\circ\text{K}^{-4}$$

For a blackbody with a T_{kin} of 10°C (283°K), the radiant flux may be calculated from Equation 5.5 as

$$
\begin{aligned}
F_b &= \sigma T_{kin}^4 \\
&= (5.67 \times 10^{-12} \text{ W} \cdot \text{cm}^{-2} \cdot {}^\circ\text{K}^{-4})(283{}^\circ\text{K})^4 \\
&= (5.67 \times 10^{-12} \text{ W} \cdot \text{cm}^{-2} \cdot {}^\circ\text{K}^{-4}) \\
&\quad (6.41 \times 10^9 \cdot {}^\circ\text{K}^4) \\
&= 3.6 \times 10^{-2} \text{ W} \cdot \text{cm}^{-2}
\end{aligned}
$$

A blackbody is a physical abstraction, for no material has an absorptivity of 1 and no material radiates the full amount of energy given in Equation 5.5. For real materials a property called *emissivity* (ϵ) has been defined as

$$\epsilon = \frac{F_r}{F_b} \tag{5.6}$$

where F_r is radiant flux from a real material. The emissivity for a blackbody is 1, but for all real materials it is less than 1. Emissivity is wavelength-dependent, which means that the emissivity of a real material will be different when measured at different wavelengths of radiant energy.

Table 5.1 lists the emissivities of various materials in the 8-to-12 μm wavelength region, which is widely used in remote sensing. Most of the emissivities of materials in Table 5.1 fall within the relatively narrow range from 0.81 to 0.96. Note that water has a high emissivity and that a thin film of petroleum lowers the emissivity, which is a significant relationship for remote sensing of oil slicks.

TABLE 5.1 Emissivity of materials measured in the 8-to-12-μm wavelength region

Material	Emissivity (ϵ)
Granite, typical	0.815
Dunite	0.856
Obsidian	0.862
Feldspar	0.870
Granite, rough	0.898
Silicon sandstone, polished	0.909
Sand, quartz, large-grain	0.914
Dolomite, polished	0.929
Basalt, rough	0.934
Dolomite, rough	0.958
Asphalt paving	0.959
Concrete walkway	0.966
Water, with a thin film of petroleum	0.972
Water, pure	0.993

Source: From Buettner, K. J. K., and C. D. Kern, Journal of Geophysical Research, v. 70, p. 1333, 1965, copyrighted by American Geophysical Union.

Combining Equations 5.5 and 5.6 produces the following equation for radiant flux of a real material:

$$F_r = \epsilon \sigma T_{kin}{}^4 \tag{5.7}$$

where ϵ is the emissivity for that material. Emissivity is a measure of the ability of a material both to radiate and to absorb energy. Materials with a high emissivity absorb large amounts of incident energy and radiate large quantities of kinetic energy. Materials with low emissivities absorb and radiate lower amounts of energy. Figure 5.3 illustrates the effect of different emissivities on radiant flux from an aluminum block with a uniform kinetic temperature of 10°C (283°K). The portion of the block that is painted dull black has an emissivity of 0.97. The radiant flux for this material may be calculated from Equation 5.7 as

$$F_r = \epsilon \sigma T_{kin}{}^4$$
$$= 0.97 \, (5.67 \times 10^{-12} \, \text{W} \cdot \text{cm}^{-2} \cdot {}^\circ\text{K}^{-4})(283°\text{K})^4$$
$$= 3.5 \times 10^{-2} \, \text{W} \cdot \text{cm}^{-2}$$

For the shiny portion of the aluminum block with an emissivity of 0.06, the radiant flux may be calculated from Equation 5.7 as

$$F_r = \epsilon \sigma T_{kin}{}^4$$
$$= 0.06 \, (5.67 \times 10^{-2} \, \text{W} \cdot \text{cm}^{-2} \cdot {}^\circ\text{K}^{-4})(283°\text{K})^4$$
$$= 2.2 \times 10^{-3} \, \text{W} \cdot \text{cm}^{-2}$$

Although the aluminum block has a uniform kinetic temperature of 283°K, the radiant flux from the surface with high emissivity is more than 10 times greater than from the surface with low emissivity.

Most thermal IR remote sensing systems record the radiant temperature (T_{rad}) of terrain rather than radiant flux. In order to determine T_{rad}, consider a blackbody and a real material that have different kinetic temperatures but the same radiant flux, so that $F_b = F_r$. For a blackbody, $T_{rad} = T_{kin}$; therefore Equation 5.5 may be written as

$$F_b = \sigma T_{rad}{}^4$$

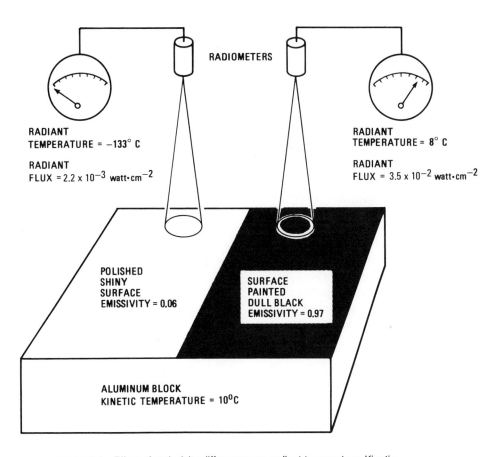

RADIOMETERS

RADIANT
TEMPERATURE = −133° C

RADIANT
FLUX = 2.2 x 10^{-3} watt·cm^{-2}

RADIANT
TEMPERATURE = 8° C

RADIANT
FLUX = 3.5 x 10^{-2} watt·cm^{-2}

POLISHED
SHINY
SURFACE
EMISSIVITY = 0.06

SURFACE
PAINTED
DULL BLACK
EMISSIVITY = 0.97

ALUMINUM BLOCK
KINETIC TEMPERATURE = 10°C

FIGURE 5.3 Effect of emissivity differences on radiant temperature. Kinetic (internal) temperature of aluminum block is uniformly 10°C. Different emissivities cause different radiant temperatures, which are measured by radiometers.

This equation and Equation 5.7 may then be combined as follows:

$$F_r = \epsilon\sigma T_{kin}^4$$

$$F_b = \sigma T_{rad}^4$$

$$F_b = F_r$$

$$\sigma T_{rad}^4 = \epsilon\sigma T_{kin}^4$$

$$T_{rad} = \epsilon^{1/4} T_{kin} \qquad (5.8)$$

For a real material of known emissivity and kinetic temperature, Equation 5.8 may be used to calculate the radiant temperature. Radiant temperatures may be measured with nonimaging remote sensing devices called *radiometers*. For the portion of the aluminum block in Figure 5.3 with an emissivity of 0.97, radiant temperature is calculated from Equation 5.8 as

$$T_{rad} = 0.97^{1/4} \times 283°K$$

$$= 281°K$$

For the portion of the block with an emissivity of 0.06,

$$T_{rad} = 0.06^{1/4} \times 283°K$$

$$= 140°K$$

which is 141°K lower than the radiant temperature for the portion of the aluminum block with high emissivity. An alternate way to understand the low radiant temperature of the shiny surface involves Equation 5.3. Because the emissivity is low, the absorptivity is also low and the reflectivity is therefore high. In the out-of-doors with no clouds, the very low temperature of outer space is reflected by the shiny aluminum surface. For this reason, metallic objects such as airplanes and metal-roofed buildings have cold radiant temperatures.

Thermal Conductivity *Thermal conductivity* (*K*) is the rate at which heat will pass through a material and is expressed as calories per centimeter per second per degree centigrade. *K* is the number of calories that will pass through a 1-cm cube of the material in 1 sec when two opposite faces are maintained at a 1°C difference in temperature. Table 5.2 gives thermal conductivities for geologic materials. For any rock type the thermal conductivity may vary up to 20 percent from the value given. Thermal conductivities of porous materials may vary up to 200 percent depending on the nature of the substance that fills the pores. Rocks and soils are relatively poor conductors of heat. The average thermal conductivity of the materials in Table 5.2 is 0.006 cal $\cdot$ cm^{-1} $\cdot$ sec^{-1}

$\cdot$ °C^{-1}, which is two orders of magnitude lower than the thermal conductivity of such metals as aluminum, copper, and silver.

Thermal Capacity *Thermal capacity* (*c*) is the ability of a material to store heat. Thermal capacity is the number of calories required to raise the temperature of 1 g of a material by 1°C and is expressed in calories per gram per degree centigrade. In Table 5.2 note that water has the highest thermal capacity of any substance. Figure 5.4 shows the difference between thermal capacity and kinetic temperature. Spheres of the same volume made from rhyolite, limestone, and sandstone are placed in boiling water to reach a uniform temperature of 100°C. In this experiment the rocks have zero porosity and no water is absorbed. The thermal capacity and density, from Table 5.2, are multiplied to determine the number of calories per cubic centimeter per degree centigrade that each rock stores. The rocks are assumed to have a uniform density of 2.5 g $\cdot$ cm^{-3}; therefore the different values are determined solely by differences in thermal capacity. As Figure 5.4A shows, sandstone stores the greatest amount of heat and rhyolite stores the least. The heated spheres are simultaneously placed on a sheet of paraffin. Melting ceases when the spheres and paraffin have reached a uniform temperature. As Figure 5.4B shows, the amount of melting is related to the thermal capacity of the rocks and not to their temperature.

Thermal Inertia *Thermal inertia* (*P*) is a measure of the thermal response of a material to temperature changes and is given in calories per square centimeter per second square root per degree centigrade (cal $\cdot$ cm^{-2} $\cdot$ sec$^{-1/2}$ $\cdot$ °C) Thermal inertia is expressed as

$$P = (K\rho c)^{1/2} \qquad (5.9)$$

where *K* is thermal conductivity, ρ is density, and *c* is thermal capacity. Of the three properties that determine thermal inertia, density is the most important. For the most part, thermal inertia increases linearly with increasing density. This linear trend is shown in Figure 5.5, which plots the values for the materials listed in Table 5.2.

Figure 5.6 illustrates the effect of differences in thermal inertia on surface temperatures. The difference between maximum and minimum temperature occurring during a diurnal solar cycle is called ΔT. Materials with low thermal inertia, such as shale and volcanic cinders, have low resistance to temperature change and have a relatively high ΔT. They reach a high maximum surface temperature in the daytime and a low minimum temperature at night. Materials with high thermal inertia, such as sandstone and basalt, strongly resist tempera-

TABLE 5.2 Thermal properties of geologic materials and water at 20°C

	Thermal conductivity (K), cal · cm^{-1} sec^{-1} · °C^{-1}	Density (ρ), g · cm^{-3}	Thermal capacity (c), cal · g^{-1} · °C^{-1}	Thermal diffusivity (k), cm^2 · sec^{-1}	Thermal inertia (P), cal · cm^{-2} · sec$^{-1/2}$ · °C^{-1}
1. Basalt	0.0050	2.8	0.20	0.009	0.053
2. Clay soil, moist	0.0030	1.7	0.35	0.005	0.042
3. Dolomite	0.012	2.6	0.18	0.026	0.075
4. Gabbro	0.0060	3.0	0.17	0.012	0.055
5. Granite	0.0075	2.6	0.16	0.016	0.052
6. Gravel	0.0030	2.0	0.18	0.008	0.033
7. Limestone	0.0048	2.5	0.17	0.011	0.045
8. Marble	0.0055	2.7	0.21	0.010	0.056
9. Obsidian	0.0030	2.4	0.17	0.007	0.035
10. Peridotite	0.011	3.2	0.20	0.017	0.084
11. Pumice, loose, dry	0.0006	1.0	0.16	0.004	0.009
12. Quartzite	0.012	2.7	0.17	0.026	0.074
13. Rhyolite	0.0055	2.5	0.16	0.014	0.047
14. Sandy gravel	0.0060	2.1	0.20	0.014	0.050
15. Sandy soil	0.0014	1.8	0.24	0.003	0.024
16. Sandstone, quartz	0.0120	2.5	0.19	0.013	0.054
17. Serpentine	0.0063	2.4	0.23	0.013	0.063
18. Shale	0.0042	2.3	0.17	0.008	0.034
19. Slate	0.0050	2.8	0.17	0.011	0.049
20. Syenite	0.0077	2.2	0.23	0.009	0.047
21. Tuff, welded	0.0028	1.8	0.20	0.008	0.032
22. Water	0.0013	1.0	1.01	0.001	0.037

Source: From Janza and others (1975, Table 4.1).

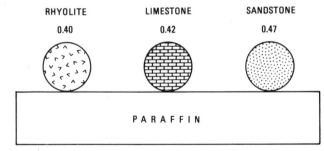

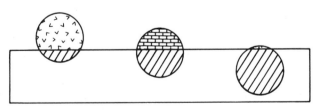

A. SPHERES OF ROCK HEATED TO 100°C AND PLACED ON A SHEET OF PARAFFIN. THE VALUE FOR EACH ROCK IS THE PRODUCT OF ITS THERMAL CAPACITY (c) AND DENSITY ρ IN cal · cm^{-3} · °C^{-1}.

B. AFTER THE ROCKS AND PARAFFIN HAVE REACHED THE SAME TEMPERATURE.

FIGURE 5.4 The effect of differences in thermal capacity of various rock types.

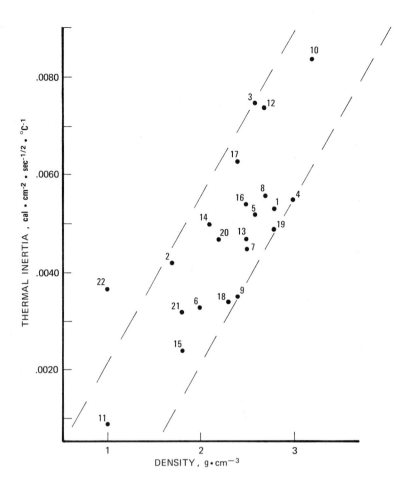

FIGURE 5.5 Relationship of thermal inertia to density for rocks and water. Numbers refer to the materials listed in Table 5.2.

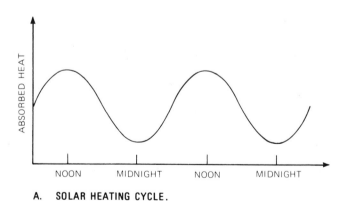

A. SOLAR HEATING CYCLE.

FIGURE 5.6 Effect of differences in thermal inertia on surface temperature during diurnal solar cycles. Note differences in ΔT for materials with high and low thermal inertia. From K. Watson (unpublished), U.S. Geological Survey.

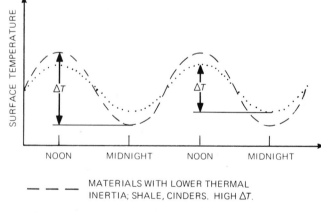

⎯ ⎯ MATERIALS WITH LOWER THERMAL INERTIA; SHALE, CINDERS. HIGH ΔT.

· · · · · · · MATERIALS WITH HIGHER THERMAL INERTIA; SANDSTONE, BASALT. LOW ΔT.

B. VARIATIONS IN SURFACE TEMPERATURE.

ture changes and have a relatively low ΔT. These materials are relatively cool in the daytime and warm at night.

Apparent Thermal Inertia

Apparent Thermal Inertia Thermal inertia cannot be determined by remote sensing methods because conductivity, density, and thermal capacity must be measured by contact methods. Maximum and minimum radiant temperature, however, may be measured from digitally recorded daytime and nighttime images. For corresponding ground resolution cells ΔT is determined by subtracting the nighttime temperature from the daytime temperature. The fact that ΔT is low for materials with high thermal inertia and high for those with low thermal inertia (Figure 5.6) may be used to determine a property called *apparent thermal inertia* (*ATI*) by the relationship

$$ATI = \frac{1 - A}{\Delta T} \qquad (5.10)$$

where A is the albedo in the visible band. Albedo is employed to compensate for the effects that differences in absorptivity have on radiant temperature. During the day, dark materials (low albedo) absorb more sunlight than light materials. The absorbed solar energy increases kinetic temperature, which increases the radiant thermal energy. Therefore, a dark material typically has a higher ΔT than an otherwise identical material that has a light color. The term $1 - A$ corrects for some of these effects. A typical *ATI* image produced from a visible (albedo) image and daytime and nighttime thermal images is illustrated and interpreted in the section later in the chapter describing the Heat Capacity Mapping Mission.

ATI images must be interpreted with caution because ΔT may be influenced by factors other than thermal inertia. Consider an area of uniform material, such as granite, that has high topographic relief. In a daytime IR image the shadowed areas have a lower radiant temperature than sunlit areas. In a nighttime image the sunlit and shadowed areas have similar temperatures. As a result the shadowed granite has a lower ΔT and higher *ATI* than the same granite that is sunlit. Some *ATI* computer programs compensate for shadows by using topographic data together with information on solar elevation and azimuth. Water and vegetation typically have *ATI* values that are determined by factors other than their thermal inertia, as discussed in the section "Influence of Water and Vegetation."

ATI may be measured with a thermal-inertia meter, which employs a radiometer and standard rock samples of known emissivity and thermal inertia (Kahle and others, 1981). Electric lamps heat the target material and the standards to a uniform level, the radiometer measures radiant temperatures, and the meter then calculates the *ATI* for the target material. Table 5.3 lists *ATI* values determined in the field using this meter. These relative values are useful for discriminating among different materials and are not intended as absolute measures of thermal inertia.

Thermal Diffusivity

Thermal Diffusivity The same values used to determine thermal inertia may be used to determine *thermal diffusivity* (k), which is given as

$$k = \frac{K}{c\rho} \qquad (5.11)$$

Thermal diffusivity (given in centimeters squared per second) governs the rate at which temperature changes within a substance. More specifically, it states the ability of a substance during a period of solar heating to transfer heat from the surface to the interior and at night during a period of cooling to transfer stored heat to the surface.

Diurnal Temperature Variations

Typical diurnal variations in radiant temperature are shown diagrammatically in Figure 5.7. Note that the most rapid temperature changes, shown by steep curves, occur near dawn and sunset. At the points where two curves intersect (*thermal crossover*), there is no radiant temperature difference between the materials. Quantitative data on diurnal changes in radiant temperature

TABLE 5.3 Apparent thermal inertia values measured in the field with a thermal inertia meter

Material	ATI*
1. Sandy alluvium	0.014
2. Sand, windblown	0.015
3. Rhyolite tuff	0.022
4. Clay-silt playa	0.024
5. Basalt, pahoehoe	0.039
6. Basalt, olivine	0.042
7. Basalt, aa	0.042
8. Barite	0.043
9. Andesite	0.044
10. Rhyodacite, silicified	0.048
11. Chert	0.053

*ATI values are relative to a dolomite standard with thermal inertia of 0.984 $cal \cdot cm^{-2} \cdot sec^{-1/2} \cdot {}^{\circ}C^{-1}$. Error limits are approximately ± 10 percent.
Source: From Kahle and others (1981, Table 4).

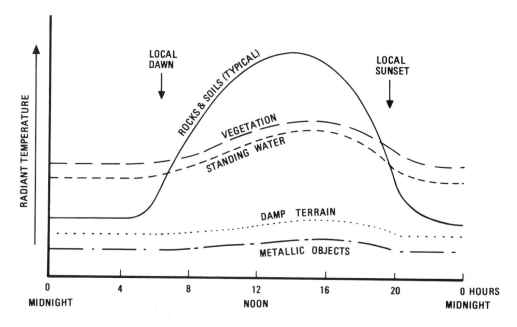

FIGURE 5.7 Diurnal radiant temperature curves (diagrammatic) for typical materials.

for various materials are given later in the section describing the Indio Hills.

Thermal Models

There are two basic approaches to understanding the significance of temperature signatures on thermal IR images. The first approach is to make an empirical correlation between the image signatures and the corresponding ground features. Warm and cool areas on the images can be matched with localities on the ground to determine the material responsible for the signature. This empirical method is rapid and direct but does not consider the underlying physical causes for the thermal expression of different materials.

The second approach uses a mathematical model to relate surface temperatures to the physical properties of materials. The advantage of the model approach is that it enables the investigator to understand the interactions between materials and thermal processes. Watson (1971) calculated a mathematical model to predict ground temperature based on physical properties of materials and the diurnal solar heating cycle. It is assumed that each material is a semi-infinite solid with homogeneous thermal properties. The two equations of the model are

$$F = [I_0 (1 - A) \cos Z] - [\sigma T_{\text{kin}}^4] \text{ (day)} \quad (5.12)$$

$$= -\sigma T_{\text{kin}}^4 \quad \text{(night)} \quad (5.13)$$

where

F = radiant flux of the sun

I_0 = the solar constant, a measure of radiation from the sun

A = albedo of the surface

Z = zenith angle of the sun

σ = the Stefan-Boltzmann constant, 5.67×10^{-12} $\text{W} \cdot \text{cm}^{-2} \cdot {}^\circ\text{K}^{-4}$

These equations are programmed into a computer to calculate curves showing diurnal variations of surface temperature. Figure 5.8A shows diurnal curves for materials with different thermal inertia values. Curves for materials with lower thermal inertia have greater extremes (higher ΔT); materials with higher thermal inertia have lower extremes (lower ΔT). The curves cross over (intersect) at sunrise and sunset. These calculated model curves show the same basic relationships as the empirical curves in Figure 5.6B. Figure 5.8B shows curves for materials of different albedo but uniform thermal inertia. Darker materials (low albedo) have higher day and night temperatures and higher ΔT values than do lighter colored materials (high albedo). The albedo curves are essentially parallel and do not cross over at sunrise or sunset.

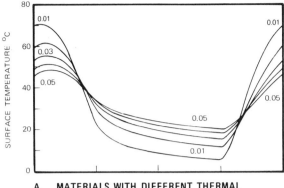

A. MATERIALS WITH DIFFERENT THERMAL
INERTIAS.

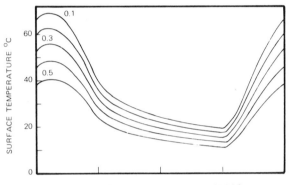

B. MATERIALS WITH DIFFERENT ALBEDOS.

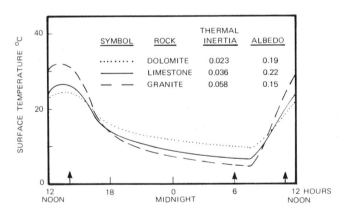

SYMBOL	ROCK	THERMAL INERTIA	ALBEDO
....	DOLOMITE	0.023	0.19
————	LIMESTONE	0.036	0.22
– – – –	GRANITE	0.058	0.15

C. LIMESTONE, DOLOMITE, AND GRANITE OUTCROPS
FROM MILL CREEK, OKLAHOMA.

FIGURE 5.8 Mathematical models showing surface temperature calculated for materials with different thermal inertia and albedo values and for different rock types. From Watson (1971).

Equations 5.12 and 5.13 were used to calculate diurnal curves for dolomite, limestone, and granite with known thermal inertias and albedos (Figure 5.8C). The solar components of the equation were based on a December date, a site latitude of 34.5°N, and a solar declination of −23.3°. From these curves one can predict the temperature signature of these materials throughout the diurnal cycle.

IR DETECTION AND IMAGING TECHNOLOGY

The pattern of radiant temperature variations of the terrain may be recorded as an image by *airborne infrared scanners.*

Airborne IR Scanners

Thermal IR images are produced by airborne scanner systems that consist of three basic components: (1) optical-mechanical scanning system, (2) thermal IR detector, and (3) image recording system. As shown in Figure 5.9, the cross-track scanning system consists of an electric motor mounted in the aircraft with the rotating shaft oriented parallel with the aircraft fuselage and hence the flight direction. A front-surface mirror with a 45° facet mounted on the end of the shaft sweeps the terrain at a right angle to the flight path. The angular field of view is typically 90°. IR energy radiated from the terrain is focused from the scan mirror onto the detector, which converts the radiant energy into an electrical signal that varies in proportion to the intensity of the IR radiation. Detectors are cooled to 73°K to reduce the electronic noise caused by the molecular vibrations of the detector crystal. In scanners such as the one in Figure 5.9 the detector is enclosed by a vacuum bottle, or *dewar,* filled with liquid nitrogen. Other detectors employ a closed-cycle mechanical cooler to maintain a low detector temperature. In the simplest scanners of the mid-1960s the amplified signal modulates the intensity of a small light source. A second mirror rotating synchronously with the scanning mirror sweeps the image of this modulated light source across a strip of recording film. The film advances at a rate proportional to the aircraft ground speed divided by the altitude, so that each scan line on the ground is represented by a scan line on the film.

Since the late 1960s, magnetic tape recording has largely replaced direct film recording. The tapes are played back onto film in the laboratory to produce images. Magnetic data are also suitable for digital image processing. The controlled radiant temperature reference sources shown in Figure 5.9 are incorporated into the scanner housing so that the scanner mirror views one at the beginning

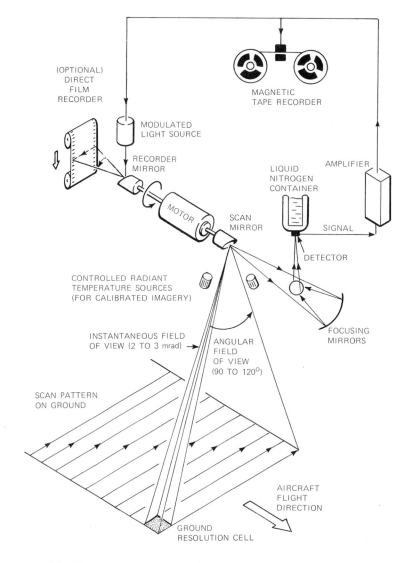

FIGURE 5.9 Diagram of a thermal IR scanner system. After Sabins (1969, Figure 2).

and end of each scan. The known temperatures from the reference sources are used to produce calibrated images as discussed later. Figure 5.10 illustrates the components of a typical scanner system. The gyroscopically controlled stabilization device provides data for correcting for the effect of aircraft roll. The control unit enables the operator to monitor and modify the operation of the scanner.

In some scanners the faceted mirror rotates about a vertical axis to generate a conical sweep. These forward-looking scanners record a segment of the circular ground path in advance of the aircraft. Additional information on thermal IR detectors and systems is given in Norwood and Lansing (1983). IR scanners in satellites employ the same principles as those in aircraft.

Stationary IR Scanners

Airborne scanners are designed for use in aircraft or spacecraft where the forward movement of the vehicle provides coverage along the flight path and the rotating scan mirror provides coverage at right angles to that direction. *Stationary scanners* are designed to acquire images from a fixed position. In these scanners the faceted scan mirror rotates about a vertical axis to provide coverage in the horizontal direction. Coverage in the vertical direction is provided by a plane mirror that tilts about a horizontal axis. Another system employs a television-type camera that electronically scans the scene. On both types of stationary scanners the radiation transmitted through an IR filter is focused onto a detector

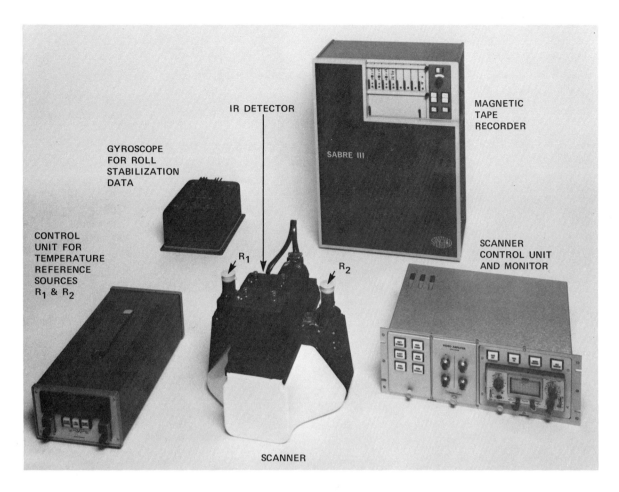

GYROSCOPE
FOR ROLL
STABILIZATION
DATA

IR DETECTOR

MAGNETIC
TAPE
RECORDER

SABRE III

CONTROL
UNIT FOR
TEMPERATURE
REFERENCE
SOURCES
R₁ & R₂

R₁

R₂

SCANNER
CONTROL UNIT
AND MONITOR

SCANNER

FIGURE 5.10 Airborne thermal IR scanner system. Courtesy Daedalus Enterprises, Incorporated.

that converts the radiation into an electrical signal. The resulting image is displayed in real time on a small television-type screen that may be photographed to produce a permanent record. Stationary scanners are mounted on a tripod stand for most applications, although some systems are sufficiently light and compact to be held in the operator's hand.

One use of stationary scanners is to record the pattern of heat radiating from the human body. As mentioned at the beginning of the chapter, the term *thermography* is generally associated with this medical application of thermal IR images, which are called thermograms. Tumors and impaired blood circulation are physiological disorders that have been detected on thermograms. Stationary scanners are also used to monitor industrial facilities for hot spots that may indicate potential problems. Anomalous hot spots on the exterior of industrial furnaces may be areas where the fire brick lining has eroded and failure is imminent. In electrical transmission facilities, faulty transformers and insulators have been detected by their high radiant temperatures. Rail-

roads use IR scanners to detect overheated wheel bearings in moving trains.

CHARACTERISTICS OF IR IMAGES

On most thermal IR images the brightest tones represent the warmest radiant temperatures and the darkest tones represent the coolest temperatures (Figure 5.11). The apparent similarity of IR images to black-and-white aerial photographs results from the fact that both are recorded as gray-scale variations on film. In photography, film acts as the medium for detecting, recording, and displaying reflected energy in the 0.4-to-0.9-μm wavelength region. For thermal IR images, however, a semiconductor device detects the energy, and film serves only as a medium to display radiant temperatures.

Geometric Distortion and Correction

Cross-track scanner systems produce a characteristic geometric distortion on the images. The distance from

A. IMAGE WITH NORMAL
   SCANNER DISTORTION.

B. RECTILINEAR IMAGE.

FIGURE 5.11 Distorted and rectilinear versions of a thermal IR scanner image. Flight direction is from bottom to top of each image. From Sabins (1973, page 839).

the scanner to the ground is greater at either end of the scan line than at the nadir (Figure 5.9); therefore the ground resolution cells of the detector are larger at either end of the scan line than in the center (Figure 5.12A). The scanner mirror rotates at a constant angular rate, but the image is recorded at a constant linear rate. This difference between angular and linear rates causes compression toward the edges of the image. This compression results in the geometric distortion shown diagrammatically in Figure 5.12B and with a real image in Figure 5.11A. The S-shaped curvature of straight roads trending diagonally across the flight path is a typical expression of all cross-track scanner distortion. Images recorded on magnetic tape may be played back onto film with an electronic correction to produce *rectilinear images* free from distortion, as shown in Figure 5.11B. There are several advantages to rectilinear images, particularly for compiling mosaics. The full width of a rectilinear image strip may be used for interpretation; there-

fore fewer flight lines with wider spacing can cover a survey area.

Effects of Weather on Images

Clouds typically have the patchy warm and cool pattern illustrated in Figure 5.13A, where the dark signatures are relatively cool and the bright signatures are relatively warm. Scattered rain showers produce a pattern of streaks parallel with the scan lines on the image. A heavy overcast layer reduces thermal contrasts between terrain objects because of reradiation of energy between the terrain and cloud layer. Images may be acquired by flying below the cloud layer, but the resulting thermal contrast is relatively low.

Surface winds produce characteristic patterns of smears and streaks on images. *Wind smears* (Figure 5.13B) are parallel curved lines of alternating lighter and darker signatures that may extend over wide expanses of the

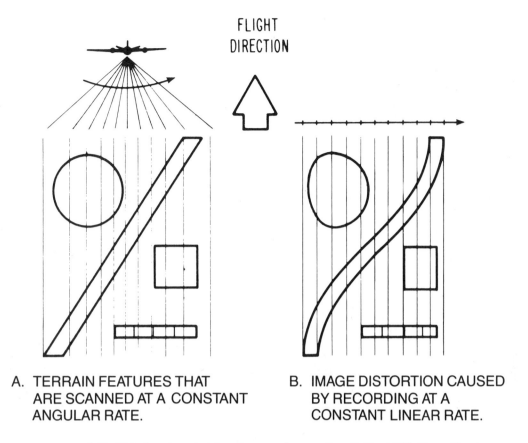

FLIGHT
DIRECTION

A. TERRAIN FEATURES THAT
ARE SCANNED AT A CONSTANT
ANGULAR RATE.

B. IMAGE DISTORTION CAUSED
BY RECORDING AT A
CONSTANT LINEAR RATE.

FIGURE 5.12   Distortion characteristics of scanner images. From Sabins (1969, Figure 3).

image. *Wind streaks* occur downwind from obstructions on flat terrain and typically appear as the warm (bright) patterns shown in Figure 5.13C. On this example the obstructions are clumps of trees with warm signatures. Wind velocity is lower downwind from obstructions, which reduces the cooling effect; thus terrain in the sheltered areas is warmer than terrain exposed to the wind. Wind smears and streaks may be avoided by acquiring images only on calm nights, but in many regions, surface winds persist for much of the year and their effects must be endured. Interpreters must be alert to avoid confusing wind-caused signatures with terrain features.

Penetration of Smoke Plumes

Clouds consist of finely divided particles of ice or water that have the same temperature as the surrounding air. As shown in Figure 5.13A, images acquired from aircraft or satellites above cloud banks record the radiant temperature of the clouds. Energy from the earth's surface does not penetrate the clouds, but is absorbed and reradiated. Smoke plumes, however, consist of ash particles and other combustion products so fine that they are readily penetrated by the relatively long wavelengths of thermal IR radiation.

Figure 5.14 shows a visible image and a thermal IR image that were acquired simultaneously during a daytime flight over a forest fire. The smoke plume completely conceals the ground in the visible image, but terrain features are clearly visible in the IR image where the burning front of the fire has a bright signature. The U.S. Forest Service uses aircraft equipped with IR scanners that produce image copies in flight, which are dropped to fire fighters on the ground. These images provide information about the fire location that cannot be obtained by visual observation through the smoke plumes. IR images are also acquired after fires are extinguished in order to detect hot spots that could reignite.

Influence of Water and Vegetation

The thermal inertia of water is similar to that of soils and rocks (Table 5.2), but in the daytime, water bodies have a cooler surface temperature than soils and rocks. At night the relative surface temperatures are reversed,

A. CLOUDS.

B. SURFACE WIND SMEAR.

C. SURFACE WIND STREAKS.

FIGURE 5.13 Effects of weather on thermal IR images. From Sabins (1973, Figure 2).

so that water is warmer than soils and rocks (Figure 5.7). This reversal in relative temperatures is apparent when day and night images are compared, as with HCMM images, which are described later in the chapter. Convection currents maintain a relatively uniform temperature at the surface of a water body. Convection does not operate to transfer heat in soils and rocks; therefore heat from solar flux is concentrated near the surface of these solids in daytime, causing a higher surface temperature. At night this heat radiates into the atmosphere and is not replenished by convection currents in these solid materials, causing surface temperatures to be lower than in adjacent water bodies (K. Watson, personal communication). Some images may not be annotated for the time of day at which they were acquired. The thermal signatures of water bodies are a reliable index to the time of image acquisition. If water bodies have warm signatures relative to the adjacent terrain, the image was acquired at night. Relatively cool water bodies indicate daytime imagery.

As shown in the diurnal temperature curves (Figure 5.7), damp soil is cooler than dry soil, both day and night. As absorbed water evaporates, it cools the soil. Figure 5.15 clearly shows this *evaporative cooling* effect. The aerial photograph and IR image were simultaneously acquired over an orchard of immature trees where adjacent rows were irrigated on successive days. In the photograph only the wettest, most recently irrigated soil has a discernibly darker tone than the surroundings. In the IR image, however, the moisture pattern is clearly visible as variations in radiant temperature. The gray scale of the image is calibrated to show radiant temperature, which in turn correlates with moisture content. Some researchers have noted that adding water to dry soil increases the thermal inertia of the soil to values comparable to those of rocks. The effect of evaporative cooling, however, dominates the radiant temperature signature of damp ground. Many geologic faults and fractures are recognizable in IR images because of evaporative cooling. Examples from California and South Africa are described later in the chapter.

Green deciduous vegetation has a cool signature on daytime images and a warm signature on nighttime images. During the day, transpiration of water vapor lowers leaf temperature, causing vegetation to have a cool signature relative to the surrounding soil. At night the insulating effect of leafy foliage and the high water content retain heat, which results in warm nighttime temperatures. The relatively high nighttime and low daytime radiant temperature of conifers, however, does not appear to be related to their water content. The composite emissivity of the needle clusters making up a whole tree approaches that of a blackbody. Dry vegetation, such as crop stubble in agricultural areas, appears warm on

A. VISIBLE IMAGE.

B. THERMAL IR IMAGE.

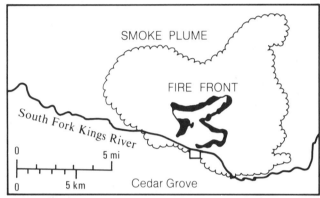

C. LOCATION MAP.

FIGURE 5.14 Thermal IR image and visible image of a forest fire in King's Canyon, Sequoia National Forest, California. Courtesy NASA Ames Research Center.

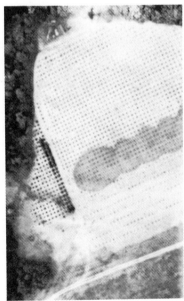

A. AERIAL PHOTOGRAPH.

B. DAYTIME THERMAL IR IMAGE (8 to 14 μm).

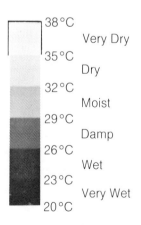

C. TEMPERATURE SCALE.

FIGURE 5.15 Relationship between soil moisture and radiant temperature in an irrigated orchard. Courtesy Daedalus Enterprises, Incorporated.

nighttime imagery in contrast to bare soil, which is cool. The dry vegetation insulates the ground to retain heat and cause the warm nighttime signature.

For reasons discussed above, water, moist soil, and vegetation have relatively low values of ΔT when their signatures are compared on daytime and nighttime IR images. When these ΔT values are used in Equation 5.10, the resulting high *ATI* values differ significantly from the actual thermal inertias for these materials, which have low to intermediate values. Therefore one must be cautious when interpreting areas of water, moist soil, or vegetation in *ATI* images.

Temperature Calibration of Images

Most of the older IR scanners lacked temperature calibration and produced images in which the gray tones recorded relative rather than absolute radiant temperatures. These qualitative images were satisfactory for many purposes. In the early 1970s manufacturers began to equip scanners with internal temperature calibration sources that are now standard on most systems. The scanner in Figure 5.9 has temperature calibration sources mounted on either side of the angular field of view. The scanner records the radiant temperature of the first calibration source, then sweeps the terrain, and finally records the temperature of the second source. The resulting signal is recorded on magnetic tape and has the appearance shown in the upper part of Figure 5.16A. Calibration source BB1 was set at a temperature of 84°F and source BB2 was set at 102°F. These reference temperatures provide a scale for determining the temperature at any point along the magnetic tape record of the terrain temperature. During playback of the magnetic tape, the temperature range may be divided into intervals and displayed quantitatively. One playback system employs six colors, plus black and white, to display the temperature values. In Figure 5.16B each color has been assigned a temperature interval of 3°F. The boundary between any two color areas is a temperature contour,

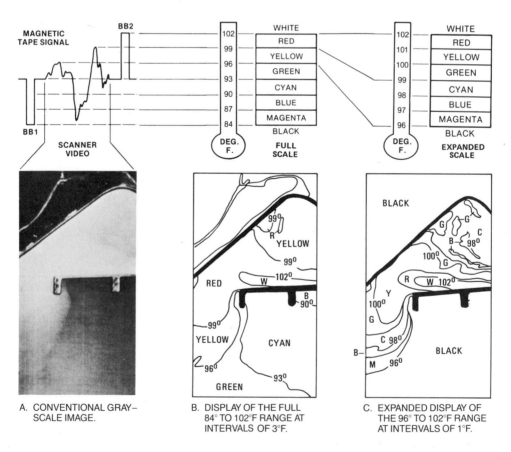

A. CONVENTIONAL GRAY-SCALE IMAGE.

B. DISPLAY OF THE FULL 84° TO 102°F RANGE AT INTERVALS OF 3°F.

C. EXPANDED DISPLAY OF THE 96° TO 102°F RANGE AT INTERVALS OF 1°F.

FIGURE 5.16 Temperature-calibrated thermal IR image of power plant discharge. Courtesy Daedalus Enterprises, Incorporated.

or *isotherm*. Finer temperature detail may be displayed by assigning each color to a 1°F interval (Figure 5.16C). Comparison of the calibrated displays with the conventional gray-scale image (Figure 5.16A) illustrates the advantage of quantitative images. Plate 8A shows color displays of calibrated images for Hawaiian volcanoes. Chapter 9 shows an isotherm map.

CONDUCTING AIRBORNE IR SURVEYS

Aircraft thermal IR surveys have not been made for most areas; therefore airborne surveys must be conducted over an area of interest. The acquisition cost of IR images is approximately three times that of aerial photographs, although the cost per square kilometer decreases as the area of the survey increases. IR surveys are of two general types: single flight lines and mosaic coverage. A single flight line over a small test site is simple to plan and conduct. The flight line may be repeated at different times of day to evaluate diurnal thermal variations. Figure 5.17 illustrates repeated images of the Caliente and Temblor ranges, California. Mosaic coverage with parallel sidelapping image strips provides regional coverage and is employed for reconnaissance surveys. Figure 5.18 shows a typical mosaic of thermal IR image strips. For any IR survey the following factors must be considered: time of day, wavelength band, and orientation and altitude of flight lines.

Time of Day

Nighttime images are essential for most applications because the thermal contrasts due to solar heating and shadowing are greatly reduced. On daytime images, topography is typically the dominant expression because of these differential solar effects. The predawn and postsunrise images of the Caliente and Temblor ranges, California, illustrate these differences (Figure 5.17). On the postsunrise image, the ridges and canyons in the mountain ranges are clearly defined by differential solar heating and shadowing, as is the anticlinal hill in the Carrizo Plain. On the predawn image (Figure 5.17B), topographic features are largely eliminated and geologic features are emphasized, as shown by comparison with the map (Figure 5.17C). The narrow warm signatures in the Caliente Range are basalt outcrops. In the Temblor Range the bands with cool signatures are outcrops of shale and siltstone. The broad belts of warm signature are sandstone and conglomerate outcrops. These geologic features are obscure or invisible in the daytime image. Another advantage of nighttime images is that radiant temperatures of a particular target are more constant

than in the daytime, as shown in the diurnal temperature curves of Figure 5.7.

Wavelength Bands

Thermal IR images may be acquired at wavelength bands of 3 to 5 μm and 8 to 14 μm, which are atmospheric windows (Figure 5.1). Wien's displacement law enables us to determine the temperatures at which the maximum energy will radiate for each of these bands. Figure 5.2 shows that the 3-to-5-μm band corresponds to the radiant energy peak for temperatures of 600°K and greater, which are associated with fires, lava flows, and other hot features. The 8-to-14-μm band spans the radiant energy peak for a temperature of 300°K; this is the ambient temperature of the earth, which has a radiant energy peak at 9.7 μm. Images in the 3-to-5-μm band are commonly acquired with an indium-antimonide detector; in the 8-to-14-μm band, mercury-cadmium-telluride detectors are employed. Norwood and Lansing (1983) discuss the characteristics of these and other detectors.

Figure 5.19 shows nighttime images in central Michigan acquired by a multispectral scanner equipped with both an indium-antimonide detector and a mercury-cadmium-telluride detector. The area on the images consists of pastures, fields, and woodlands cut by a network of roads and a few streams. In the 8-to-14 μm image (Figure 5.19B), terrain features are well expressed and have the following signatures: trees and freeways are relatively warm, fields have intermediate temperatures, and marshy areas along the left margin of the image are cool. The overall radiant temperature level of the 3-to-5-μm image (Figure 5.19A) is lower and the thermal contrasts among terrain features are much lower than in the 8-to-14-μm image. Of special interest are localities A through D in the location map, which represent the following features:

A. Three small fires of glowing charcoal briquets are located within a grove of trees. On the 8-to-14-μm image the warm signature of the trees effectively masks the fires, but on the 3-to-5-μm image the fires are clearly visible and the signature of the trees is subdued.

B. A large campfire in an open field is visible on both images.

C. In an open field a pit containing a small charcoal fire is concealed beneath a pile of brush. On the 8-to-14-μm image this target could be mistaken for vegetation, but on the 3-to-5-μm target it is clearly recognizable as a hot target.

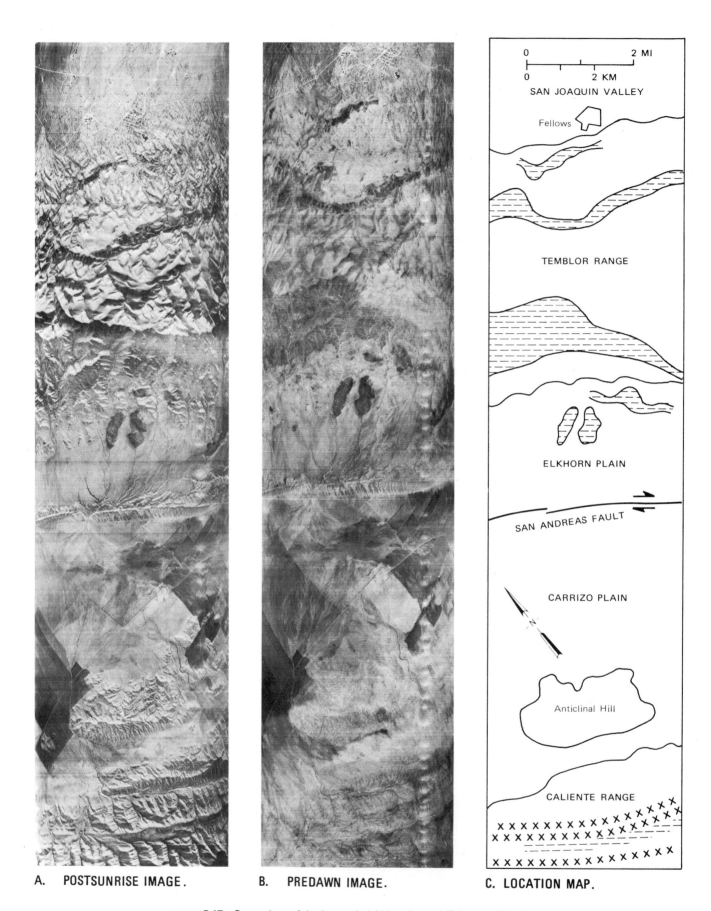

A. POSTSUNRISE IMAGE.   B. PREDAWN IMAGE.   C. LOCATION MAP.

FIGURE 5.17 Comparison of daytime and nighttime thermal IR images (8 to 14 μm) of the Caliente and Temblor ranges, California. The crosses on the map are basalt outcrops, and the dashes are shale outcrops. From Wolfe (1971, Figures 3 and 4). Courtesy E. W. Wolfe, U.S. Geological Survey.

A. MOSAIC OF IMAGE STRIPS (8 TO 14 μm).

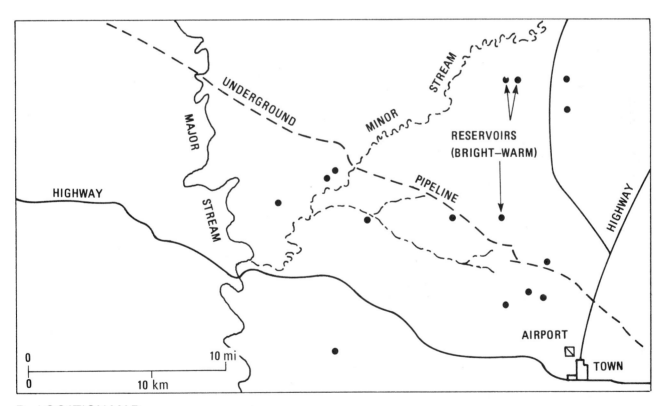

B. LOCATION MAP.

FIGURE 5.18 Mosaic constructed of nighttime thermal IR images (8 to 14 μm), South Dakota.

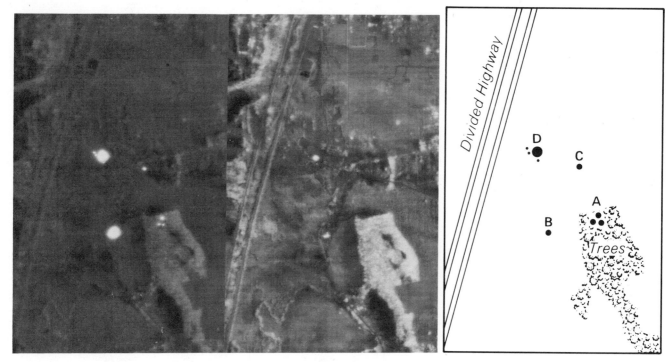

A. 3 TO 5 μm BAND.   B. 8 TO 14 μm BAND.   C. LOCATION MAP.

FIGURE 5.19 Thermal IR images from different spectral bands. These images in central Michigan were acquired October 30, 1974, at 8:30 p.m. Central standard time. Letters on the map are explained in the text. Courtesy Daedalus Enterprises, Incorporated.

D. A large campfire and three vehicles with warm engines are located in an open field. The four targets are resolvable on the 3-to-5-μm image, but the campfire signature conceals the other targets on the 8-to-14-μm image.

This example illustrates that 8-to-14-μm images are optimum for terrain mapping, whereas 3-to-5-μm images are optimum for mapping hot targets, such as fires.

Orientation and Altitude of Flight Lines

For geologic projects, it is useful to know the regional structural strike or tectonic "grain" of the area in advance of an IR survey. This information helps determine optimum orientation of flight lines. If flight lines are oriented normal to the regional strike, the scan-line pattern will be parallel with the strike and may mask linear geologic features. To avoid this, it is preferable to orient flight lines parallel with, or at an acute angle to, the regional strike. On very high quality images acquired with modern scanner systems, however, the scan lines are not a severe problem.

Flight altitude influences image scale, lateral ground coverage, and spatial resolution. Flight altitude is mean-

ingful only when expressed as height above average terrain, which is called *flight elevation*. For a typical scanner system with 90° angular field of view, the *lateral ground coverage,* which is normal to the flight line, is related to flight elevation as shown on the chart in Figure 5.20. At a typical elevation of 2 km, for example, the lateral coverage of the image strip is 4 km. The resulting image is typically displayed on film at a width of 50 mm, or a scale of 1:80,000.

Maximum length of flight lines is determined by the capacity of the magnetic recording tape. The tape recorder shown in Figure 5.10 can record continuously for 1 hour.

Spatial Resolution and Information Content of Images

Spatial resolution of IR images is determined by the flight elevation and by the instantaneous field of view (IFOV) of the detector, which typically ranges from 2 to 3 mrad. At a distance of 1000 m, 1 mrad subtends an arc of 1 m. At an elevation of 2 km, a 3-mrad detector records a ground resolution cell that is 6 by 6 m in size (Figure 5.20). On the resulting images, one should be able to resolve targets separated by more than 6 m,

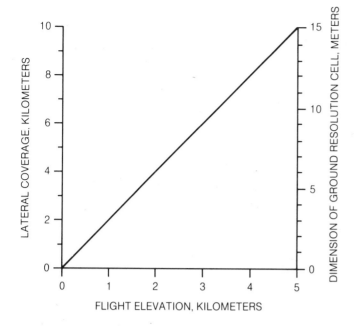

FIGURE 5.20 Lateral coverage and spatial resolution of scanner images as a function of elevation above terrain. For this scanner, the angular field of view is 90° and the instantaneous field of view is 3.0 mrad.

provided there is adequate thermal contrast between targets and background. Resolution becomes lower (poorer) toward the margins of the image because the greater viewing distance increases the size of the ground resolution cell. Spatial resolution, however, is not the sole measure of image quality. Detection of thermal patterns is determined primarily by differences in radiant temperature, which are expressed as tonal contrast on the image. Aircraft images acquired at an elevation of approximately 2 km provide suitable coverage and adequate spatial resolution for most applications. For regional investigations, however, images acquired from satellite altitudes are useful, as demonstrated later in this chapter.

Ground Measurements

As described earlier, weather and surface conditions play a large role in determining terrain expression on IR images. It may be useful to collect ground information on weather conditions, soil moisture, and vegetation at the time of the IR survey. In the early days of remote sensing, the term *ground truth* was coined for these measurements, but most investigators have abandoned the term. There is no reason that measurements made on the ground are more truthful than those made remotely. Ground measurements are most practical and useful for surveys of relatively small areas that can be

covered with a single flight line. If repeated flights are made, ground data on changing weather conditions and solar flux may be valuable in comparing and interpreting the images.

For larger areas, such as the one shown in Figure 5.18, ground measurements can be made at only a limited number of localities during the 3 to 4 hours required to acquire the images. Ground measurements are most valuable if they are made at localities that have anomalous image signatures. The measurements may help explain the anomalies or may eliminate possible causes. In practice, however, it is virtually impossible to anticipate where anomalies will occur; therefore most regional surveys omit contemporaneous ground measurements. The availability of calibrated thermal IR images also reduces the need for ground measurements.

Ground measurements to help explain image signatures may be made some time after the image flight, as in the Indio Hills example shown later in the chapter. Ground information is useful for understanding other forms of remote sensing imagery in addition to thermal IR images. The type of ground measurements will depend on the wavelength region of the airborne sensor. Dozier and Strahler (1983) describe typical ground measurements recorded in support of remote sensing surveys.

LAND USE AND LAND COVER— ANN ARBOR, MICHIGAN

The discussion of image interpretation begins with a scene familiar to all readers: a city and its surroundings (Figure 5.21). The city of Ann Arbor is located west of Detroit and includes the University of Michigan. As shown in the IR image and aerial photograph, the business district and university are surrounded by residential suburbs, which in turn are surrounded by open land. The IR image was acquired shortly after sunrise when radiant temperatures were rising rapidly, as seen in the diurnal curves of Figure 5.7. The road and highway network has a conspicuous warm signature because of the high thermal inertia of the relatively dense concrete and asphalt. Building roofs are cool because of low emissivity of these materials. The inner city is noticeably warmer than the surrounding area because of the concentration of heat-generating activities. Water in the ponds and in the Huron River is warmer than the surrounding soil, which has not yet absorbed sufficient solar energy to reach the typical high radiant temperatures of midday. Meadows and fields are relatively cool, but wooded areas are warm. The stadium in the left central part of the scene has a warm grandstand and cool playing field, which is covered with artificial turf. IR images acquired at lower elevations with higher spatial resolution are

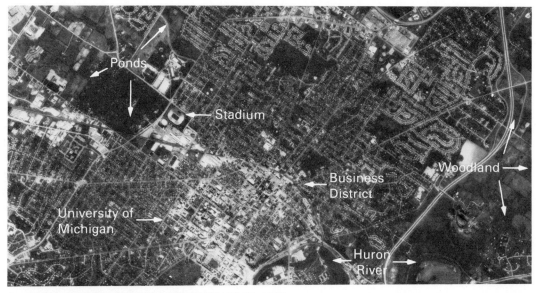

A. AERIAL PHOTOGRAPH ACQUIRED MAY 5, 1983.

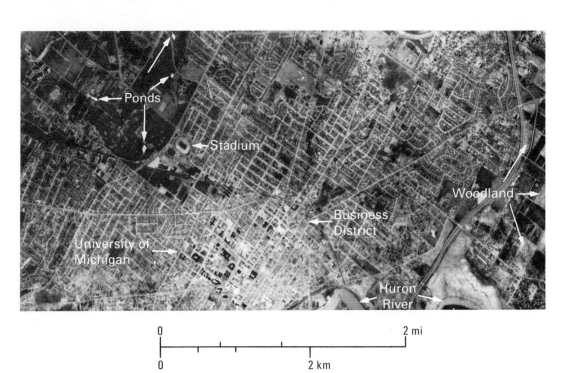

B. THERMAL IR IMAGE (8 TO 14 μm) ACQUIRED MAY 9,
1979, AT 6:30 A.M. COURTESY DAEDALUS ENTERPRISES, INC.

FIGURE 5.21 Ann Arbor, Michigan.

used for heat-loss surveys of urban and industrial areas, as described in the following section.

HEAT-LOSS SURVEYS

An obvious application of thermal IR technology is to survey heated buildings, factories, and buried steam lines for anomalous hot spots that may indicate heat leakage and poorly insulated roofs. Airborne IR surveys and image interpretations are relatively inexpensive. By locating and correcting heat losses, the fuel saved in a few months can repay the cost of a survey. In the northern and central United States, many building complexes, such as university campuses and industrial complexes, are heated by steam distributed through buried pipelines from a central generating station. As the steam heats the buildings, it condenses into hot water, which returns via condensate lines to the steam plant, where it is used to generate more steam. Many of the pipeline systems are several decades old and have developed leaks that are difficult to detect because the lines are buried and many stretches are covered by sidewalks, streets, and parking lots.

Brookhaven National Laboratory in Long Island, New York, fits the above description, so it contracted for a heat-loss survey. Aerial photographs and nighttime thermal IR images were acquired in November 1976. The aerial photograph (Figure 5.22A) includes an overlay of buried steam and condensate lines and manholes that was provided by the Brookhaven maintenance staff. The nighttime IR image (Figure 5.22B) has characteristic signatures. Trees, standing water, and pavement are relatively warm. Roofs of well-insulated buildings are cool with warm spots formed by exhaust ventilators. Sides of buildings are relatively warm because of heat radiated from windows. The buried heating lines and manholes form bright lines and spots. In some surveys the IR images revealed locations of lines for which the engineering records had been lost. Localities A, B, and C in the image and photograph are anomalous hot spots that proved to be major leaks or areas where pipe insulation had deteriorated. Localities D and E are building roofs that are significantly warmer than other roofs in the laboratory complex. Significant savings in energy costs could probably be made by improving the ceiling insulation in these buildings. A number of other facilities have been surveyed with results similar to those in Brookhaven.

Aerial surveys detect heat losses from pipelines and roofs, but much heat is lost from inadequately insulated walls of buildings. The *Vanscan® technique*, devised by Daedalus Enterprises, monitors these heat losses. An IR scanner is mounted on its side in a van with the axis of rotation parallel with the direction of travel. As the van moves forward, the angular field of view of the scanner is directed horizontally through an opening in the side of the vehicle to record images of building walls along the route of travel.

Plate 8B is a typical nighttime Vanscan® image of a residential area acquired in the winter. The radiant temperatures have been digitally subdivided into eight intervals with black representing less than 34°F and white representing more than 40°F. The range from 34° to 40°F is displayed in six colors at intervals of 1°F. The garage between the two residences is unheated and shows essentially no energy loss. Both houses have heat loss from the windows, as shown by the white tones. The ground floor of the house on the left also has high heat loss from the walls which have white signatures. Walls of the house on the right are well insulated and have lower heat loss, shown by the green and blue signatures. The house on the left also radiates much heat from the walls of the attic, shown by the red and yellow tones beneath the roof. The attic of the house on the right has little heat loss, as shown by the blue tones in the image. Utility companies have commissioned surveys of this type to educate their customers on the advantages of installing adequate insulation for their houses.

Vanscan® surveys have also been made of oil refineries and other industrial facilities to record temperature patterns of the walls of catalytic cracking towers, heat exchangers, and other vessels. Anomalous hot spots in the images can indicate potential problems, such as internal corrosion, fractures, and damaged insulation. Preventative maintenance can then be done before failure occurs.

INDIO HILLS, CALIFORNIA

The Indio Hills, in the eastern part of the Coachella Valley, are a low range of deformed clastic sedimentary rocks of Late Tertiary age trending southeastward parallel with the San Andreas fault zone. A geologic map (Figure 5.23) and an aerial photograph and a nighttime IR image (Figure 5.24) cover the south end of the hills. This arid terrain, with little vegetation and well-exposed bedrock, is ideal for acquiring and analyzing IR images.

Thermal IR Signatures of Rock Types

The alluvium surrounding the hills has a relatively cool and featureless signature on the nighttime image, which is consistent with the low thermal inertia for this low-density material. Two types of bedrock are readily

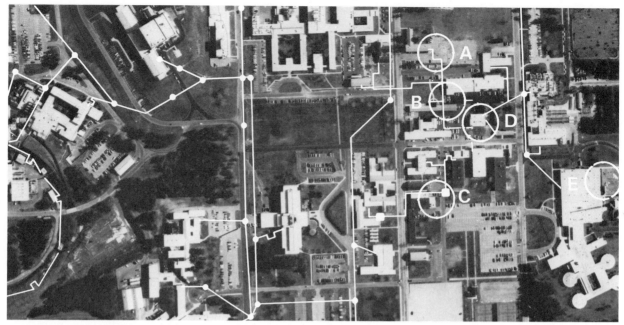

A. AERIAL PHOTOGRAPH WITH OVERLAY OF HEATING LINES.

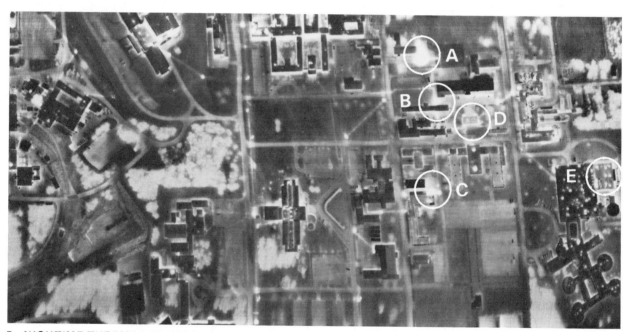

B. NIGHTIME THERMAL IR IMAGE (8 to 14 μm).

FIGURE 5.22 Heat loss survey of Brookhaven National Laboratories, Long Island, New York. Localities are explained in the text. Courtesy Daedalus Enterprises, Incorporated.

distinguished on the image. One type has a relatively warm and uniform signature and consists of poorly stratified, moderately to poorly consolidated conglomerate of the Ocotillo, Canebrake, and Mecca formations (Figure 5.23). The outlying hill in the lower center of the image (location 1.3, A.8 to D.5 of Figure 5.24B) is a good example.

The other bedrock type is the Palm Spring Formation, consisting of well-stratified alternating beds of resistant conglomeratic sandstone and nonresistant siltstone up to 12 m thick that erode to form the ridges and slopes illustrated in Figure 5.25. The Palm Spring Formation has a distinctive pattern of alternating warm and cool bands on the nighttime IR image. Careful correlation of the image with outcrops in the field established that the warm signatures are sandstones and the cool signatures are siltstones. These signatures were later verified by radiometer studies.

Radiometer Investigations

A portable radiometer operating at 8 to 14 μm was used to measure daytime and nighttime radiant temperatures of 8 sandstone and 10 siltstone outcrops at locality 3.5, B.7 of Figure 5.24B. In Figure 5.26 the daytime temperatures are plotted in the upper diagram and nighttime temperatures in the lower diagram. The average temperature values are summarized below.

	Average Temperature, °C		
Rock type	Day	Night	ΔT
Sandstone	21	10	11
Siltstone	25	8	17

These data show that relative to siltstones, the sandstones are cooler in daytime, warmer at night, and have a lower ΔT, which indicates that the sandstone has a higher thermal inertia than the siltstone. These field results may be compared with the thermal properties given in Table 5.2. Siltstone is not listed in Table 5.2, but shale (no. 18) is a similar rock type. Thermal inertia is lower for shale (0.034) than for sandstone (0.054), which is consistent with field measurements at the Indio Hills.

In the nighttime IR image of the Indio Hills (Figure 5.24B), the sparse vegetation has a distinct warm signature and soil and alluvium are cool. In daytime IR images of similar areas (not illustrated), the signatures are reversed. In order to evaluate these relationships

quantitatively, radiometer measurements were made at locality 3.1, F.3 in Figure 5.24B, where three salt cedar trees have distinctly warmer signatures than the surrounding bare soil. Radiant temperatures were measured during a diurnal cycle for the salt cedars plus three smaller creosote bushes, and the values were averaged. For each observation period, radiant temperature measurements of six soil exposures were also averaged. Figure 5.27 plots the results, together with air temperature readings. This diagram shows that vegetation at night is consistently warmer than the soil, with a maximum temperature difference of 4°C. The temperature relationships are reversed during the day, when the soil is much warmer than the vegetation. Note that the thermal crossovers of the various curves occur within less than 1 hour both in the evening and morning. These diurnal temperature relationships of soil and vegetation have since been confirmed on daytime and nighttime IR images at many localities.

Folds and Faults

Geologic structures shown in the map (Figure 5.23) are well expressed on the IR image. The plunging anticlines and synclines in the Palm Spring Formation are shown by the pattern of the warm and cool signatures. The San Andreas fault borders the west side of the Indio Hills; to the south it passes along the east side of the outlier of Ocotillo Conglomerate; farther south the fault trace is concealed by alluvium and has no topographic expression. On the aerial photograph, the concealed trace of the fault is marked on the northeast side by a vegetation anomaly, which is a concentration of vegetation that abruptly terminates at the fault trace (1.6, D.2 to 1.3, G.3 in Figure 5.24A). On the IR image the fault trace is expressed by an alignment of very cool (dark) anomalies along the northeast side (1.6, E.1 to 1.6, G.3 in Figure 5.24B). The cool anomalies are not related to the vegetation distribution because vegetation is warm on the nighttime image. The cool anomalies are probably related to the barrier effect of the San Andreas fault on groundwater movement. In the spring of 1961, which was a few months before the IR survey, Cummings (1964, p. 34) observed that the water table on the east side of the fault was 15 m shallower than on the west side. The shallower water table causes a high near-surface moisture content. This near-surface moisture results in evaporative cooling, which explains the cool anomaly on the east side of the fault. It also allows for the greater density of vegetation on the east side. On IR images of the northern Indio Hills (not illustrated) the Mission Creek fault is indicated by cool anomalies where the fault is a barrier to groundwater movement. Similar thermal

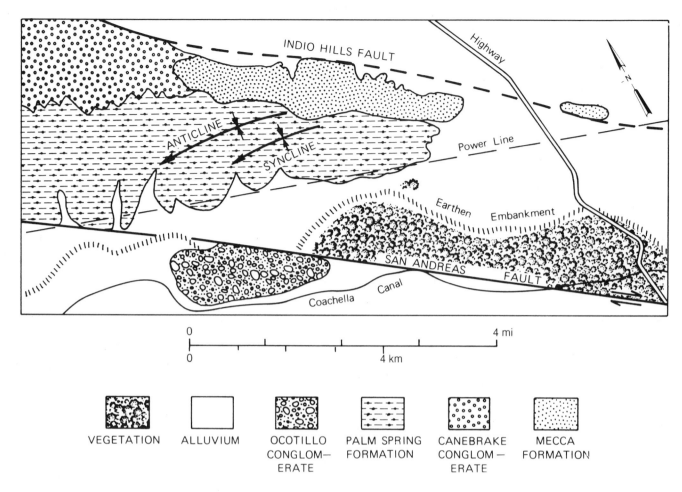

VEGETATION ALLUVIUM OCOTILLO CONGLOM— ERATE PALM SPRING FORMATION CANEBRAKE CONGLOM— ERATE MECCA FORMATION

FIGURE 5.23 Geologic map of the southern part of the Indio Hills, Riverside County, California.

anomalies appear on nighttime IR images of the San Andreas fault in the Carrizo Plain 320 km to the northwest.

Some anomalies on thermal IR images are difficult to explain. An example is the trough of a syncline at locality 3.3, A.0 to 3.7, A.9 in Figure 5.24B, which is marked by a distinct tonal anomaly that is cool on the south side and warm on the north. The straight interface coincides with the axis of the syncline. Careful field checks show that there is no faulting or unusual fracturing, nor is there any apparent stratigraphic or topographic cause of this anomaly. One possible explanation is that the synclinal structure influences moisture content in the rocks, but this is difficult to verify.

IMLER ROAD AREA, CALIFORNIA

The Imler Road area (Figure 5.28 and 5.29) on the west margin of the Imperial Valley has some similarities to the Indio Hills. Both areas are in the southern California desert and have bedrock of deformed siltstone and sandstone of Late Tertiary age. There are major differences, however. The Indio Hills have rugged topography with well-exposed bedrock. The Imler Road area is a featureless plain where bedrock is partially covered by gravel and windblown sand which largely conceal geologic structures. The featureless nature of the area is shown in the aerial photograph (Figure 5.28B). The nighttime thermal IR image (Figure 5.28A), however, displays a wealth of geologic and terrain information that is explained in the following sections.

Terrain Expression

The cultivated fields in the south part of the area are irrigated by the Fillaree Canal, which has the typical warm nighttime signature of water. The very warm field north of the canal (1.8, C.1 in Figure 5.28A) was prob-

A. AERIAL PHOTOGRAPH ACQUIRED MAY 5, 1953.

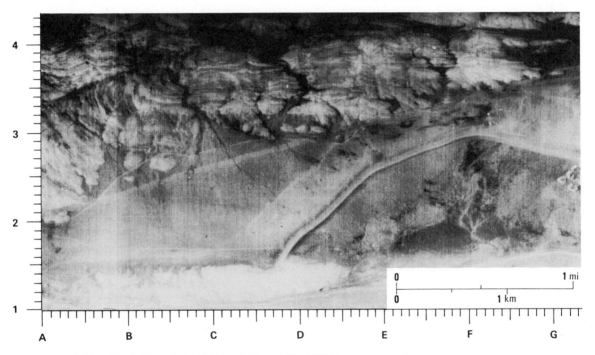

B. NIGHTTIME THERMAL IR IMAGE (8 TO 14 μm) ACQUIRED
OCTOBER 1963.

FIGURE 5.24 Nighttime thermal IR image and aerial photograph of southern part of
the Indio Hills, California. From Sabins (1967, Figures 3 and 4).

FIGURE 5.25 Typical ledge-forming sandstones and slope-forming siltstones of the Palm Spring Formation at locality 3.4, B.7 of Figure 5.24B. Radiant temperatures of these outcrops are shown in Figure 5.26.

ably flooded with standing water to leach salt from the soil at the time the image was acquired. The very cool fields were probably damp from recent irrigation, resulting in evaporative cooling.

Most of the area is flat desert terrain with sparse clumps of vegetation. The map (Figure 5.29) shows a number of sand dunes that are stabilized by mesquite trees, which appear warm on the image and dark on the aerial photograph. The very warm, Y-shaped feature at locality 3.2, B.1 in Figure 5.28A is a thick accumulation of windblown sand lodged against an earthen embankment. Imler Road in the northern part of the area is actually straight, but appears curved because of distortion caused by the IR scanner. The road was surfaced with hard-packed sand when the image was acquired but today is paved with asphalt.

Bedrock in the area is the Borrego Formation (Pleistocene age), which consists of brownish gray, lacustrine siltstone with thin interbeds of well-cemented, brown sandstone. Slabs and concretions of sandstone litter the surface where it crops out. Light-colored, nodular, thin layers in the siltstone help define bedding trends within this monotonous sequence. On the nighttime image the siltstone is relatively cool and the sandstone is warm,

corresponding to thermal signatures of similar rock types at the Indio Hills.

Anticline

An east-plunging anticline forms a conspicuous arcuate feature in the center of the IR image (4.1, C.0 in Figure 5.28A). Had this structure not been observed first on the image, one could walk across the anticline in the field without recognizing it, for there are no conspicuous lithologic or topographic patterns. After the anticline was located in the field by referring to the image, the limbs and plunge were defined by walking along the outcrops of individual beds. Structural attitudes are obscure in the siltstone, but dips up to 45° were measured in isolated outcrops of the sandstones and the dip-reversal across the fold axis was located.

In Figure 5.29 the anticline is mapped as solid bedrock, but there are numerous thin patches of windblown sand that cause the local gray tones on the image. The core of the anticline consists of contorted siltstone with a very cool signature. The pattern of alternating warm and cool bands outlining the anticline correlates with outcrops of sandstone and siltstone respectively. The

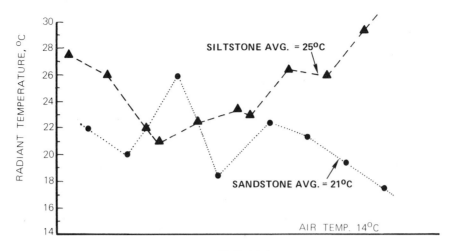

A. DAYTIME TEMPERATURES (8:45 A.M.). THE SILTSTONES ARE WARMER THAN THE SANDSTONES, WITH ONE EXCEPTION.

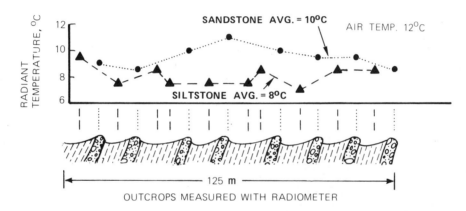

B. MIDNIGHT TEMPERATURES. THE SANDSTONES ARE WARMER THAN THE SILTSTONES, WITH TWO EXCEPTIONS.

FIGURE 5.26 Daytime and nighttime radiant temperatures of sandstones and siltstones of the Palm Spring Formation at locality 3.4, B.7 of Figure 5.24B.

west end of the anticline is truncated by the southeastward projection of the Superstition Hills fault. The inferred trace of the fault is obscured by windblown sand, but siltstone outcrops in the immediate vicinity of the fault are strongly deformed, probably as a result of fault movement.

Superstition Hills Fault

This right-lateral, strike-slip fault was named for exposures in the Superstition Hills, 14 km to the northwest

and projected into this area on the El Centro sheet of the California Geologic Atlas. The fault alignment shown in Figure 5.29 differs from that on the El Centro sheet. In addition to truncating the anticline, the fault is marked in the southeast part of the image by a southeast-trending linear feature that is cooler on the east and warmer on the west (2.2, C.5 to 1.2, D.0 in Figure 5.28A). The trend of the linear tonal anomaly is parallel with, and about 0.2 km to the east of, the row of prominent sand dunes. On April 9, 1968, the Borrego Mountain earthquake caused surface breaks along the trace of the Su-

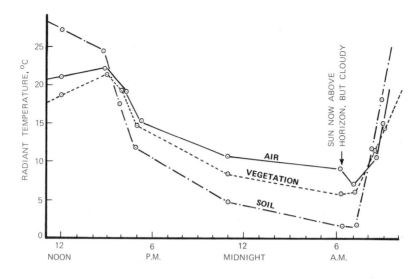

FIGURE 5.27 Diurnal radiant temperatures of vegetation and soil at the Indio Hills. Locality 3.1, F.3 of Figure 5.24B.

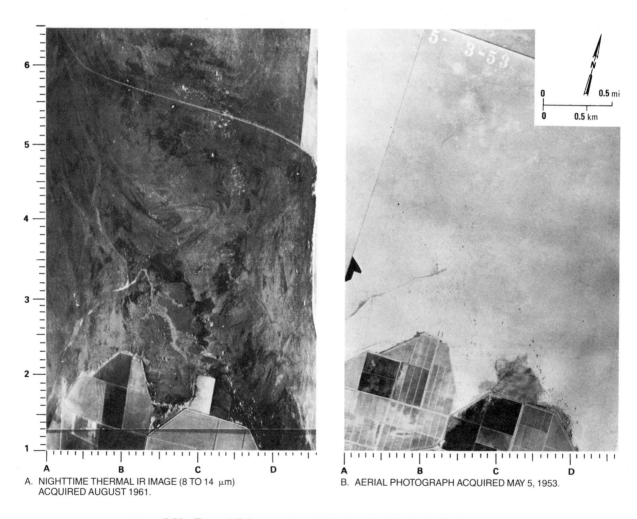

A. NIGHTTIME THERMAL IR IMAGE (8 TO 14 μm) ACQUIRED AUGUST 1961.

B. AERIAL PHOTOGRAPH ACQUIRED MAY 5, 1953.

FIGURE 5.28 Thermal IR image and aerial photograph of the Imler Road area, Imperial County, California. From Sabins (1969, Plate 1).

perstition Hills fault that were mapped by A. A. Grantz and M. Wyss (Allen and others, 1972, Plate 2). In the area of Figure 5.28A, their map shows a series of breaks with less than 2.5 cm of right-lateral displacement. The trend of the breaks closely coincides with the linear anomaly on the thermal IR image (Figure 5.28A). The image, which was acquired 7 years prior to the earthquake, located an important structural feature that is obscure both on aerial photographs and in the field.

Comparison of IR Image and Aerial Photograph

The striking difference in tonal contrast and geologic information content between the IR image and the aerial photograph (Figure 5.28) is not caused by the 8-year difference in the dates when the images were acquired. This desert area is a relatively stable environment in which natural changes occur very slowly. During annual field trips over a period of 10 years, no significant changes were noted in the area. The contrast and resolution of

the aerial photograph are good and accurately record the low contrast of this area in the visible spectral region. Color and IR color aerial photographs of the anticline are not significantly better than the black-and-white aerial photographs.

In the visible band there is little reflectance difference between the various rocks and surface materials. In the IR band, however, there are marked differences in the thermal inertia of these materials, which explains the higher information content of the IR image.

WESTERN TRANSVAAL, SOUTH AFRICA

South Africa is well suited for thermal IR surveys because of the dry climate and sparse vegetation, as shown by an example from the Stilfontein area in western Transvaal. The aerial photograph (Figure 5.30B) shows a featureless surface of low relief with a thin soil cover of 0.5 m or less and scattered trees with dark signatures.

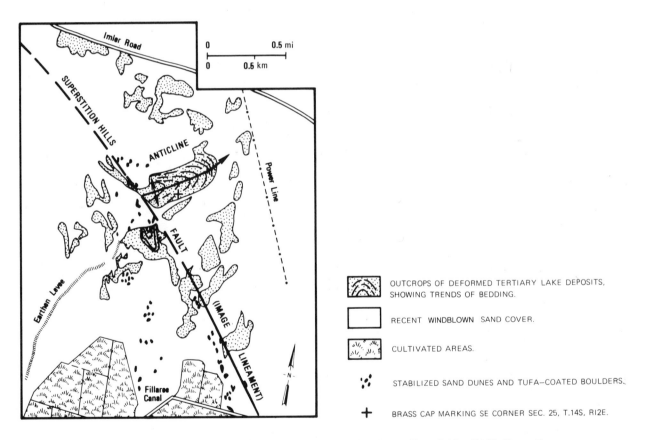

FIGURE 5.29 Interpretation map of IR image of the Imler Road area. From Sabins (1969, Figure 5).

A. NIGHTTIME THERMAL IR IMAGE (8 TO 14 μm).

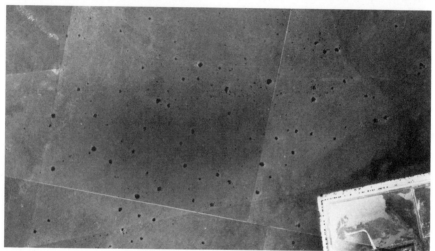

B. AERIAL PHOTOGRAPH.

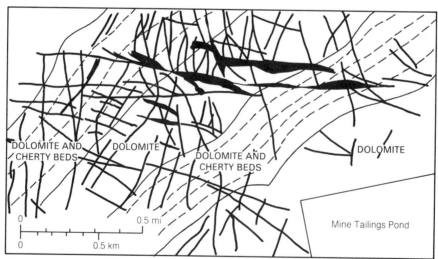

C. INTERPRETATION MAP OF THERMAL IR IMAGE.

FIGURE 5.30 Stilfontein area, western Transvaal, South Africa. From Warwick, Hartopp, and Viljoen (1979, Figures 8 and 9).

A tailings pond for a gold mine occurs in the southeast part of the image. No significant geologic information can be extracted from the photograph. The nighttime thermal IR image (Figure 5.30A), however, contains a wealth of information on geologic structure and lithology that are shown in the interpretation map (Figure 5.30C). The area is underlain by dolomite that includes a number of beds rich in chert, which is a siliceous sedimentary rock. The beds strike northeast and dip 10° to 15° to the southeast. In the IR image the dolomite has a bright (warm) signature, which is attributed to its relatively high density and high thermal inertia. The chert-rich beds have distinctly darker (cooler) signatures caused by the lower density and lower thermal inertia of these rocks. A belt of alternating dolomite beds and chert-rich beds trends northeastward across the image and is bounded on the northwest and southeast by broad areas of dolomite with uniform warm radiant temperature.

The numerous linear features with very dark signatures are the expression of faults and joints that have been enlarged by erosion and filled with moist soil. Evaporative cooling causes the dark signatures. The two major sets of fracture trend approximately north to south and east to west. In much of the area, the bedrock is cut by closely spaced joints that produce a fine network of cool lines in the image. South African geologists interpret IR images to locate areas of thicker soil cover that may be excavated to build earth-fill dams.

The location and geology of the Stilfontein area differ from those of the Indio Hills and Imler Road areas, but in all these examples the IR image is superior to the aerial photograph for geologic interpretation. In these areas the different rock units have little contrast in the visible region, but there are easily detectable differences in their thermal properties. Fractures and faults in the Stilfontein area have cool signatures due to evaporative cooling of moist soil concentrated along the breaks. In the Indio Hills, evaporative cooling also marks the trace of the San Andreas fault.

MAUNA LOA, HAWAII

Mauna Loa is a classic shield volcano with a broad, convex-upward profile and a large depression, Moku-aweoweo caldera, at its summit. The calibrated night-time IR image shown in color in Plate 8A and the aerial photomosaic (Figure 5.31A) cover the caldera and part of the Southwest Rift Zone including the pit craters, South Pit, Lua Hohonu, and Lua Hou. These features are identified on the geologic map of Figure 5.31B, which also shows the dates of the historic lava flows. Pit craters form as the surface subsides and are not primarily vents

for lava. Lua Hohonu is the youngest pit crater, having formed after 1841 and partially filled with lava in 1940. The north end of the summit caldera coalesces with another pit crater, North Pit. East of the junction between these is Lua Poholo, a small pit crater that probably formed between 1874 and 1885. The cliffs bounding the summit caldera are slightly eroded fault scarps. The main faults along which the floors of the caldera and the pits have subsided are covered by young lava flows. Dashed lines in Figure 5.31B show the inferred position of these faults. Smaller faults and fractures are common. Historic lava flows shown on the geologic map originated from fissures on the floor of the summit caldera and in the Southwest Rift Zone. Lava from different flows has spilled into the pit craters. Near the fissures the lava flows are the smooth, ropy pahoehoe type, changing character downslope to the rough, fragmented aa type.

The IR image was acquired in February 1973, some 23 years after the latest eruption in June 1950. Activity resumed with the eruption of July 1975. On the thermal IR image (Plate 8A) the temperature range from $-7°$ to $+8°$C has been digitally subdivided into six intervals. Each color represents a 2.5°C interval. Radiant temperatures cooler than $-7°$C are shown in black; those warmer than 8°C are in white. The range of $-7°$ to $+8°$C was selected for color display because preliminary study of the image data indicated that most features of geologic interest had radiant temperatures within this range. The low overall temperature level was caused by the 4200-m elevation of the summit of Mauna Loa. Individual lava flows cannot be distinguished on the basis of radiant temperature; the flows of the 1940s within the summit caldera have the same temperature ranges as the pre-historic flows on the flanks of the summit. The flows are more readily mapped from tonal differences on the aerial photograph (Figure 5.31A). After the flows have cooled, their radiant temperatures are determined by albedo and thermal properties of the rocks. Geologic descriptions indicate that all the flows have similar rock composition, which suggests that the thermal properties are also similar. This similarity explains the lack of unique thermal signatures in Plate 8A for the different flows.

Some of the warmest radiant temperatures occur at the walls of the scarps bounding the summit caldera, North Pit, South Pit, and Lua Hou. The dense interior portions of lava flows that are exposed in the scarps have higher thermal inertias than the porous vesicular surfaces of the flows exposed on the flanks of the volcano and in the floors of the caldera and pit craters. The reticulate pattern of very warm signatures on the floor of the summit crater appears to be concentrated within the area of the 1940 flow (Figure 5.31B). Fumes of steam and sulfur dioxide issue from cracks in the surface of

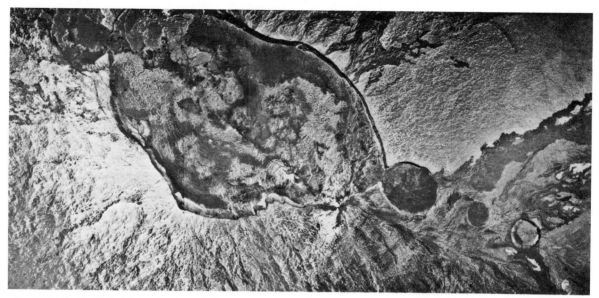

A. AERIAL PHOTOMOSAIC ACQUIRED JANUARY 30, 1965,
   BY THE U.S. DEPARTMENT OF AGRICULTURE.

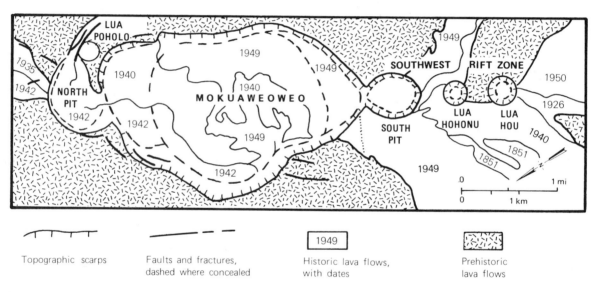

Topographic scarps

Faults and fractures,
dashed where concealed

| 1949 |

Historic lava flows,
with dates

Prehistoric
lava flows

B. GEOLOGIC MAP.

FIGURE 5.31   Summit of Mauna Loa and upper part of Southwest Rift Zone, Hawaii.

the 1940 flow (Macdonald, 1971) and probably are responsible, at least in part, for the warm pattern.

The floors of the three pit craters along the Southwest Rift Zone have marked differences in radiant temperature. South Pit is the warmest (greater than 3°C), Lua Hohonu the coolest (less than −2°C), and Lua Hou has intermediate temperatures (−2° to +0.5°C). The warmer radiant temperatures are probably caused by higher rates

of convective heat transfer at South Pit and Lua Hou. The relative temperatures of the craters were confirmed in early 1975 when the summit of Mauna Loa was blanketed by 2 m of snow. Aerial photographs acquired in April 1975 (Lockwood and others, 1976, p. 12) showed that the snow had melted on the floors of South Pit and Lua Hou while the cooler floor of Lua Hohonu was completely snow covered.

HEAT CAPACITY MAPPING MISSION

Thermal IR imaging systems have been orbited in several unmanned satellite programs, including the Heat Capacity Mapping Mission (HCMM) and band 6 of Landsat's thematic mapper. Both systems are described in the following sections and summarized in Table 5.4. Thermal IR systems are also used in environmental and meteorologic satellites, which are described in Chapter 9.

NASA launched HCMM in 1978, and it functioned into 1980. South-bound segments of orbits occurred at night and north-bound segments during the day. Localities along latitude 40°N were covered at 1:30 p.m. (local sun time) and 13 hours later at 2:30 a.m., a time interval that provided maximum ΔT. In the northern and southern hemispheres, areas between latitudes 22° and 33° had a 36-hour interval between corresponding day and night coverage. HCMM did not have an onboard tape recorder, so image acquisition was limited by the availability of ground receiving stations. The following areas were covered: coterminous United States, Alaska, western Canada, western Europe, northern Africa, and eastern Australia. An HCMM user's guide (Price, 1980) describes the system, orbit paths, and image format, and Short and Stuart (1983) provides numerous examples and descriptions of HCMM images.

Images, digital tapes, and user's guides are available from

National Space Science Data Center
Code 601
NASA Goddard Space Flight Center
Greenbelt, MD 20771

The National Space Science Data Center (NSSDC) does not provide computer searches for specific HCMM images, but it will provide catalogs from which the user may select images. The standard HCMM film format is a 241-by-241-mm image (scale of 1:31 million) that is available as prints and or as positive or negative transparencies.

California and Vicinity

Figure 5.32 is an HCMM nighttime thermal IR image reproduced at approximately the original scale. Fifteen Landsat MSS images would be needed to cover this region, which extends over most of California, portions of Nevada, and the Pacific Ocean. Figure 5.33 is a location map of the region. As shown by the gray scale, dark tones represent relatively cool radiant temperatures and bright tones represent warm temperatures. The belt of cooler water (darker tone) along the Pacific

TABLE 5.4 Characteristics of images from HCMM and Landsat TM satellites

	HCMM	TM
Operational period	1978 to 1980	1982 to present
Orbital altitude	620 km	705 km
Image coverage	700 by 700 km	185 by 170 km
Acquisition time, day	1:30 p.m.	10:30 a.m.
Acquisition time, night	2:30 a.m.	9:30 p.m.
Visible and reflected IR detectors		
Number of bands	1	6
Spectral range	0.5 to 1.1 μm	0.4 to 2.35 μm
Ground resolution cell	500 by 500 m	30 by 30 m
Thermal IR detector		
Spectral range	10.5 to 12.5 μm	10.5 to 12.5 μm
Ground resolution cell	600 by 600 m	120 by 120 m

coast results from offshore winds shifting the warmer surface water seaward, which allows upwelling of deeper, cooler water near the shore. Warmer water occurs farther offshore, and numerous circulation patterns are visible. The shallow water in San Francisco Bay and Monterey Bay is relatively warm. The various lakes have conspicuous warm signatures, which are characteristic of standing water in nighttime IR images.

The Coast Ranges geologic province consists of northwest-trending linear ridges and valleys. The Salinas Valley extends southeast from Monterey Bay and has a marked cool (dark) signature, possibly due to evaporative cooling of damp soil in the extensive irrigated fields. The Central Valley extends from the southeast to northwest corners of the image. The relatively cool signature of the valley floor may be due to irrigated fields and to low thermal inertia of the soils. Foothills and slopes are underlain by bedrock with relatively high thermal inertia and warm nighttime temperatures. The Sierra Nevada is the prominent cool belt along the east flank of the Central Valley. The Sierra Nevada, which has summit elevations of several thousand meters, was free of snow when this image was acquired. The cool signature is related to *adiabatic cooling*, the phenomenon that causes air temperature to decrease by 6.5°C per 1000 m of altitude gained. The northeast portion of the image covers part of the Basin and Range province of eastern California and adjacent Nevada, which consists of high mountain ranges surrounded by broad, semiarid valleys. In this and other HCMM nighttime IR images of the Basin and Range province, the mountain crests are very cool, the slopes are warm, and the valleys

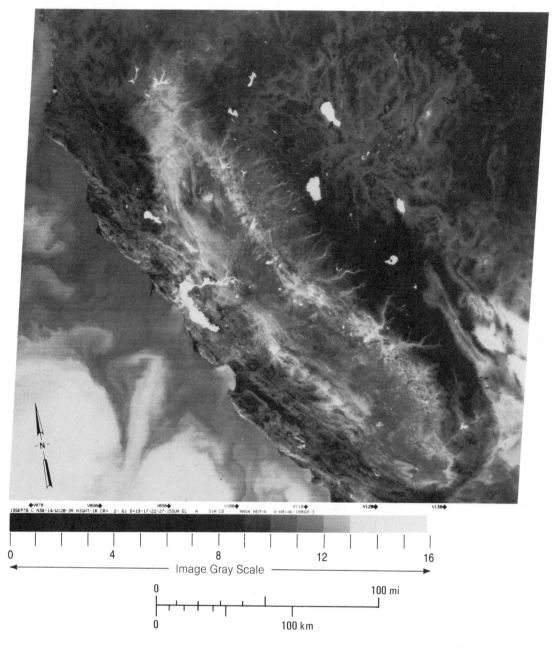

FIGURE 5.32 HCMM nighttime thermal IR image (10.5 to 12.5 μm) of California and vicinity acquired September 19, 1978, at 2:06 a.m. Pacific Standard Time. From Sabins (1983, Figure 5.11).

have intermediate to cool temperatures. Adiabatic cooling may account for the cool crests and warm slopes but cannot explain the cooler temperature of the valley floors, which are at low elevations relative to the mountains. Possible explanations include (1) meteorologic inversion, in which a layer of warmer air confines underlying cooler air at lower elevations; (2) drainage of cooler, denser air into topographic depressions; (3) lower thermal inertia of unconsolidated soil in the valley floors; and (4) evaporative cooling of moist soil.

Daytime and Nighttime Images Compared

Figure 5.34 compares the three types of images acquired by HCMM: daytime visible, daytime thermal IR, and nighttime thermal IR images. As shown in the lo-

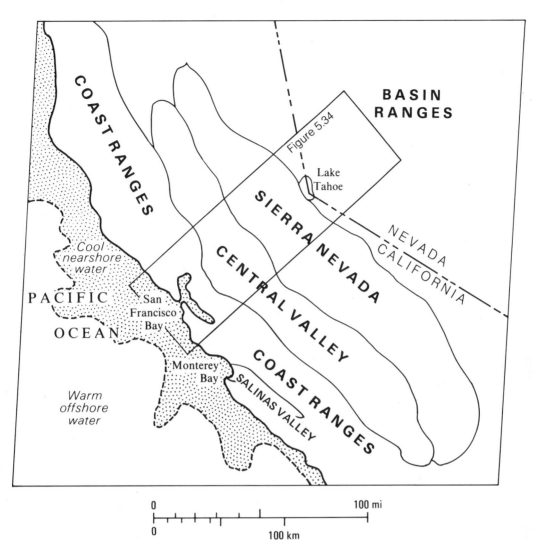

FIGURE 5.33 Location map for HCMM image of California and vicinity.

cation map (Figure 5.35), these enlarged images extend from San Francisco Bay northeast to Pyramid Lake, Nevada. The daytime images were acquired 13 hours earlier than the nighttime image. Gray tones of the daytime IR image represent a different range of temperatures than the tones of the nighttime image. Figure 5.36 compares the temperature calibration of gray scales for a typical pair of day and night IR images acquired by HCMM. In this diagram the gray scale with 16 tones from dark to light refers to the gray scale that accompanies each HCMM image, such as Figure 5.32. The gray scale of the daytime IR image in Figure 5.36 has a temperature range of 280° to 340°K, whereas the nighttime range is from 260° to 300°K. An image tone at the center of the gray scale (value 8) represents a nighttime temperature of 280°K and a daytime temperature of 310°K.

If one assumes that a lake, such as Lake Tahoe, has approximately the same day and night temperature of 20°C, Figure 5.36 shows the water to have a dark tone of 3 in the daytime image (Figure 5.34B) and a bright tone of 13 in the nighttime image (Figure 5.34C).

Apparent Thermal Inertia Image of San Rafael Swell, Utah

Kahle and others (1981) used daytime and nighttime HCMM images (Figure 5.37) of the San Rafael swell in central Utah to prepare an image showing apparent thermal inertia (*ATI*). Digital data for the daytime and nighttime images were geometrically registered. For each ground resolution cell the nighttime radiant temperature was subtracted from the daytime temperature to deter-

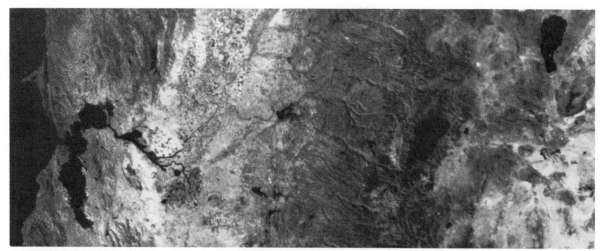

A. DAYTIME VISIBLE IMAGE ACQUIRED SEPTEMBER 20,
   1978, AT 1:17 P.M. PST.

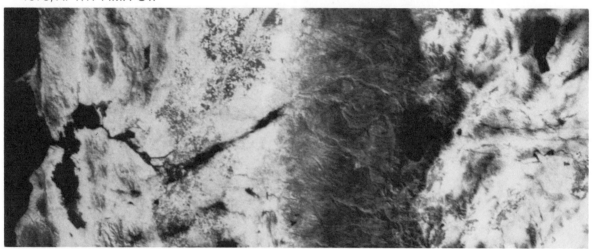

B. DAYTIME THERMAL IR IMAGE ACQUIRED SEPTEMBER 20,
   1978, AT 1:17 P.M. PST.

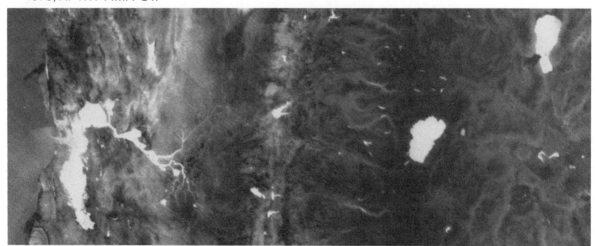

C. NIGHTTIME THERMAL IR IMAGE ACQUIRED SEPTEMBER 19,
   1978, AT 2:06 A.M. PST.

FIGURE 5.34   Enlarged HCMM images of central California and Nevada. From Sabins (1983, Figure 5.14).

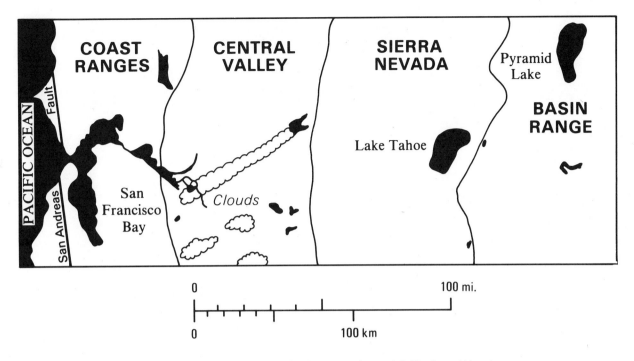

FIGURE 5.35 Location map for HCMM images of central California and Nevada.

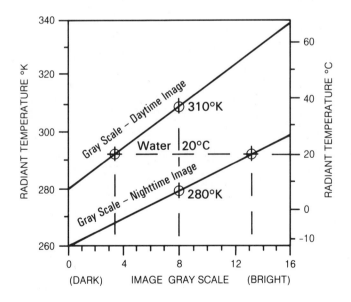

FIGURE 5.36 Relationship of image gray scale to radiant temperature for typical HCMM daytime and nighttime thermal IR images. From Short and Stuart (1983, Figure 3.8).

mine ΔT. Albedo values from the daytime visible image were used to calculate the term $1 - A$ for each cell. Equation 5.10 was then used to determine ATI. The ATI values are converted into an image (Figure 5.37D) in which dark tones display low values of ATI and bright tones display high values.

Figure 5.38 is a Landsat MSS band-5 (visible red) image with a 79-m spatial resolution, which may be compared with the HCMM visible image (500-m resolution) and thermal IR images (600-m resolution). The Landsat image was a valuable aid in preparing the interpretation map (Figure 5.39) for the ATI image. The San Rafael swell is a low domal feature bounded by the Wasatch Plateau on the northwest, the Buckhorn Plateau on northeast, and the San Rafael Desert of windblown sand on the southeast. The San Rafael swell is semiarid with little vegetation, and bedrock is well-exposed except where it is covered by sand and gravel. Areas of vegetation occur along the east slope of the Wasatch Plateau and are seen in the Landsat image as dark patches surrounded by soil with a bright signature. Additional vegetation occurs at the higher elevations of the plateaus. Standing water occurs in three small reservoirs in the northwest part of the area.

The interpretation map (Figure 5.39) was prepared by delineating areas of high, intermediate, and low values of ATI from Figure 5.37D. The Landsat image and pub-

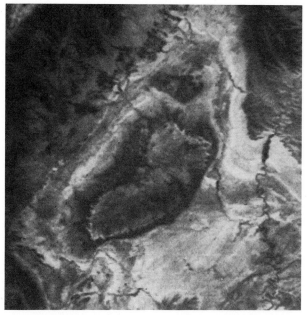

A. DAYTIME THERMAL IR IMAGE ACQUIRED AUGUST 28, 1978.

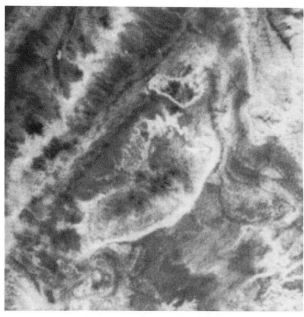

B. NIGHTTIME THERMAL IR IMAGE ACQUIRED AUGUST 27, 1978.

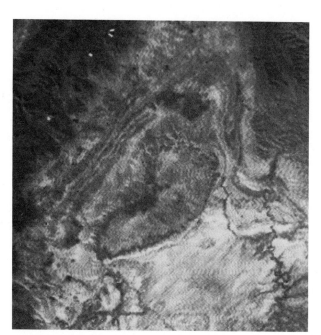

C. VISIBLE (ALBEDO) IMAGE ACQUIRED AUGUST 28, 1978.

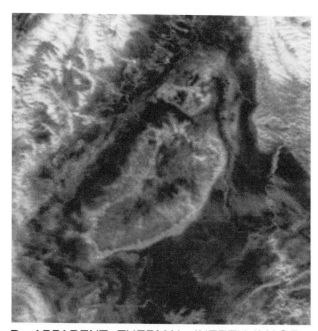

D. APPARENT–THERMAL–INERTIA IMAGE.

FIGURE 5.37 Enlarged HCMM images of the San Rafael swell, Utah. From Kahle and others (1981). Courtesy A. B. Kahle, Jet Propulsion Laboratory.

FIGURE 5.38 Landsat MSS band-5 image of the San Rafael swell, Utah, acquired September 29, 1972.

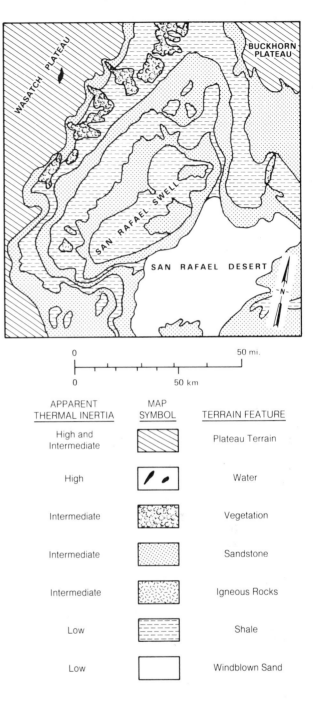

FIGURE 5.39 Interpretation map of the apparent-thermal-inertia image (Figure 5.37D) of the San Rafael swell, Utah.

lished images were used to determine the features that cause the *ATI* signatures. Water, vegetation, and plateau terrain have the following signatures and explanations.

Water (high *ATI*)   Has a nearly constant day and night temperature, which for reasons described earlier, results in a very low ΔT and a corresponding high *ATI*.

Vegetation (high to intermediate *ATI*)   Has a low ΔT, relative to soil, as shown in the earlier example from the Indio Hills. Therefore vegetation should have a high *ATI*, but most of the vegetation along the eastern slope of the Wasatch Plateau has only a moderate *ATI*. The Landsat image shows that much of this vegetation occurs along drainage channels that are separated by areas of bare soil. Each 600-by-600-m ground resolution cell of HCMM includes areas of vegetation and soil, which results in intermediate values of *ATI*.

Plateau terrain (high and intermediate *ATI*)   Because of their high elevation, the Wasatch and Buckhorn plateaus are relatively cool both day and night. These areas also support a moderate vegetation cover. Both elevation and vegetation contribute to a relatively low ΔT and high *ATI*. In the Wasatch Plateau, the ridges have high *ATI* values and the valleys have intermediate values.

The San Rafael swell anticline has eroded to expose sedimentary rocks ranging in age from Late Cretaceous to Late Paleozoic. The rocks are sandstone, shale or siltstone, and minor limestone. Two bodies of mafic ig-

neous rock occur at the southwest plunge of the arch. The Landsat image (Figure 5.38) shows the concentric belts of strata exposed around the swell, where sandstones form ridges and shales form valleys. The large ground resolution cell of the HCMM images does not resolve these details, and only some broad zones of contrasting signature are visible. In the *ATI* image these signatures are light gray (intermediate *ATI*) and very dark gray (low *ATI*). The following correlation of *ATI* signatures with rock units was made by comparing the image with a geologic map of the region.

Sandstone (intermediate *ATI*) The zones of intermediate *ATI* correlate with sandstones, which have relatively high thermal inertias.

Shale and siltstone (low *ATI*) The zones of low *ATI* correlate with shale and siltstone, which have low thermal inertia values.

Mafic igneous rocks (intermediate *ATI*) The two exposures of igneous rock have intermediate signatures that correspond to published values of thermal inertia for these rocks.

Windblown sand (low *ATI*) The windblown sand of the San Rafael Desert has a dark signature in the *ATI* image. The very low density of this unconsolidated sand explains the low thermal inertia.

This example from Utah demonstrates that, despite the low spatial resolution, HCMM image data may be processed to express thermal characteristics of terrain and geologic features.

THERMAL IR BAND OF LANDSAT THEMATIC MAPPER

Landsat TM records thermal IR images (10.5 to 12.5 μm) on band 6, with a 120-by-120-m ground resolution cell. At mid-latitudes, daytime images are acquired on southbound orbit segments at 10:30 a.m. and nighttime images are acquired at 9:30 p.m. on northbound segments. Path-and-row locations for the nighttime images are completely different from locations for daytime images. Index maps for the nighttime images are available from EDC.

Lake Ontario and Lake Erie

The TM band-6 nighttime IR image shown in Figure 5.40 covers portions of Lake Ontario and Lake Erie. A few scattered clouds cause the cool (dark) spots along the south shore of Lake Ontario. The cities of Toronto, Hamilton, and Buffalo are warm relative to the surrounding woodlands and agricultural areas. The warmer water of Lake Erie flows northward through the Niagara River and over Niagara Falls into Lake Ontario, where the thermal patterns are clearly visible. The very warm water concentrated at the western extremity of Lake Ontario is Hamilton Harbor, which is confined behind a dike. A small plume of warm water flows eastward through a channel in the dike into the lake.

Appalachian Mountains and Cumberland Plateau

Figure 5.41 illustrates TM daytime IR and visible images acquired simultaneously and a nighttime IR image acquired 8 days earlier. These images in West Virginia cover portions of the Appalachian Mountains in the east and the Cumberland Plateau in the west. In the two daytime images, note the difference in spatial resolution between the thermal IR image (Figure 5.41A) with 120-m resolution and the visible image (Figure 5.41B) with 30-m resolution. Thin clouds in the northeast and snow cover in the northwest portions of these images produce dark (cool) signatures in the IR image and bright signatures in the visible image. In both images topography is accentuated by solar illumination from the southeast. In the visible image (Figure 5.41B) the southeast-facing slopes are highlighted and northwest-facing slopes are shadowed. In the daytime IR image, southeast-facing slopes are warm and those facing northwest are cool. In the nighttime IR image (Figure 5.41C) the effects of differential heating and shadowing have dissipated. In the Appalachians the ridges are warm and the adjacent valleys are cool. Similar patterns occur in other HCMM nighttime IR images of the Appalachians acquired in the fall season. Ridges are formed on resistant rocks, such as quartzite and conglomerate, which have high thermal inertia values. Valleys are eroded in less-resistant limestone and shale and are covered with soil, which has a low thermal inertia. These differences in materials and thermal inertia explain the nighttime temperature patterns. The elevation difference between ridges and valleys is only a few hundred meters; therefore adiabatic cooling is not a significant factor. Water in the rivers and lakes has the typical warm signature on the daytime IR image and cool signature on the nighttime image.

TM band 6 acquires thermal IR images with the highest spatial resolution of any satellite system. As shown in the West Virginia example and in images from aircraft, thermal IR images must be acquired at night to be useful for most interpretations. However, most TM images are acquired in the daytime, because six of the seven TM bands only function at this time. It is recommended that additional IR images be acquired during the nighttime orbits.

THERMAL IR MULTISPECTRAL SCANNER

The preceding images in this chapter have been acquired by scanners that detect a single, broad band of thermal IR energy, which is typically 8 to 14 μm for aircraft systems and 10.5 to 12.5 μm for satellites. In the early 1980s, Jet Propulsion Laboratory developed a new air-

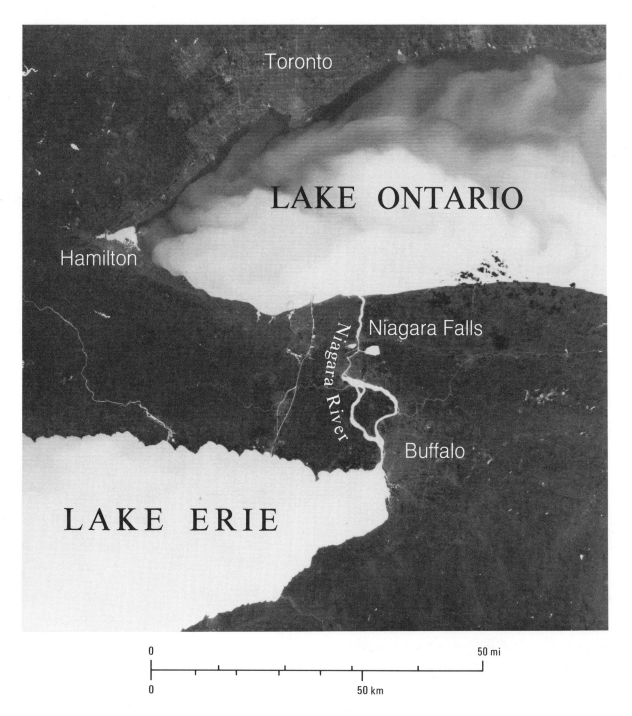

Toronto

LAKE ONTARIO

Hamilton

Niagara Falls

Niagara River

Buffalo

LAKE ERIE

0   50 mi

0   50 km

FIGURE 5.40 Landsat TM band-6 thermal IR image of Lake Erie and Lake Ontario acquired August 22, 1982, at 9:24 p.m. Eastern Standard Time.

A. DAYTIME THERMAL IR IMAGE (BAND 6)
   ACQUIRED NOVEMBER 16, 1982.

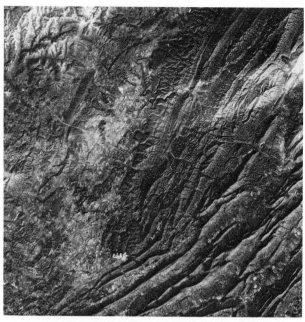

B. DAYTIME VISIBLE IMAGE (BAND 3) ACQUIRED
   NOVEMBER 16, 1982.

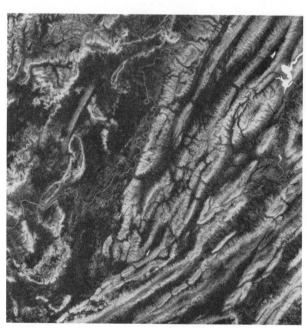

C. NIGHTTIME THERMAL IR IMAGE (BAND 6)
   ACQUIRED NOVEMBER 8, 1982.

D. LOCATION MAP.

FIGURE 5.41   Thermal IR images and visible image acquired by Landsat 4 TM of
Greenbriar River, West Virginia.

borne instrument, the Thermal Infrared Multispectral Scanner (TIMS) (Kahle and Goetz, 1983). TIMS records six bands of image data in the 8-to-12-μm region. The IFOV is 2.5 mrad, and the angular field of view is 80°. The nominal spectral bands recorded are 8.2 to 8.6 μm, 8.6 to 9.0 μm, 9.0 to 9.4 μm, 9.4 to 10.2 μm, 10.2 to 11.2 μm, and 11.2 to 12.2 μm.

Thermal IR Spectra

Figure 5.42 shows the positions of the TIMS bands together with transmission spectral curves for typical igneous rocks. Emission spectra would be more applicable to remote sensing, but emission spectra are difficult to measure and interpret. The transmission curves adequately describe spectral characteristics in the thermal IR region. In the 8-to-14-μm region, the curves in Figure 5.42 each show a broad absorption feature, which is known as a *Reststrahlen band* in reflectance spectra of polished samples. In silicate rocks this absorption effect is due to stretching vibrations between silicon and oxygen atoms bonded in the silicate crystal lattice. The position and depth of the absorption feature are related to the crystal structure of the constituent minerals and are especially sensitive to the quartz content of the rocks. For the rocks in Figure 5.42 the silica content decreases

from granite through peridotite and the absorption band shifts toward longer wavelengths. An arrow marks the center of each absorption feature, which is not necessarily the position of maximum absorption. The center of the absorption band occurs at the wavelength of TIMS band 4 for granite and shifts to band 6 for peridotite.

Figure 5.43 shows IR spectra of additional silicate rocks, along with clay minerals and carbonate rocks. The silicate rocks show the typical absorption feature. From the top of the diagram, their spectra are arranged in order of decreasing silica content from quartzite through basalt and show the shift of the absorption feature toward longer wavelengths. In the clay minerals (kaolinite and montmorillonite), spectral features in the 8-to-14-μm region are attributed to various Si-O-Si and Si-O stretching vibrations and to an Al-O-H bending mode. The spectral features in montmorillonite are less distinct than in kaolinite because the numerous exchangeable cations and water molecules in the montmorillonite structure allow many different vibrations.

Spectra of the carbonate rocks—limestone and dolomite—in Figure 5.43 have a major absorption feature in the 6-to-8-μm region due to internal vibrations in the carbonate anion. This wavelength region, however, coincides with an atmospheric absorption band and is not usable in remote sensing of earth. The spectra of pure carbonate rocks are featureless in the 8-to-12-μm region recorded by TIMS, but the addition of clay and quartz to limestone causes an absorption feature near 9 μm.

Panamint Mountains, California

TIMS images with an 18-by-18-m ground resolution cell were acquired of the east flank of the Panamint Mountains, in Death Valley, California, near noon on August 17, 1982. The digital image data were processed at JPL using a technique that suppresses temperature information while exaggerating emissivity features (Kahle, Madura, and Soha, 1980). The resulting images for TIMS bands 1, 3, and 5 were composited in blue, green, and red, respectively, to produce the color image in Plate 9A. An IR color image prepared from Landsat TM data is shown for comparison in Plate 9B. The interpretation map for this area (Figure 5.44) is a generalized version of the detailed map by Gillespie, Kahle, and Palluconi (1984). Vegetation is very sparse in this arid region. The image includes three terrain categories; bedrock, alluvial fans, and valley deposits.

Bedrock Bedrock of the Panamint Mountains is exposed in the western portion of the image and consists of volcanic rocks (of Miocene age) and slightly metamorphosed quartzite, carbonate rocks, and shale (of Precambrian and Paleozoic age). In the interpretation

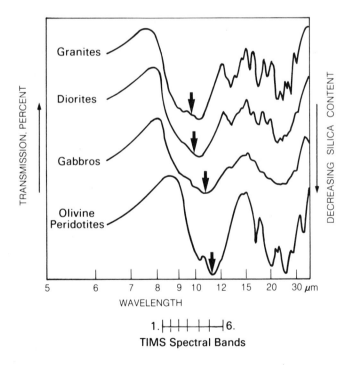

FIGURE 5.42 Thermal IR spectra of igneous rocks. Arrow indicates the center of each absorption band, which does not necessarily coincide with absorption minimum. The spectra are offset vertically to avoid overlap. Spectral bands of TIMS are shown.

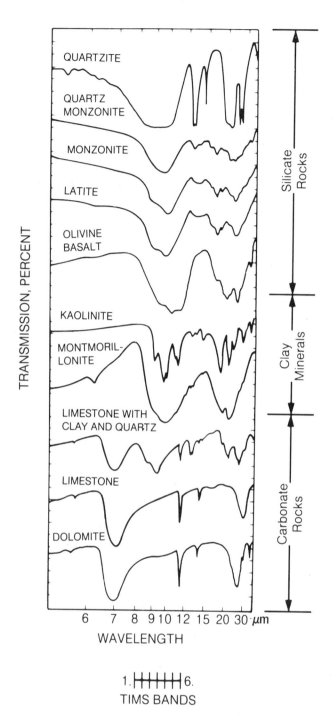

FIGURE 5.43 Thermal IR spectra of rocks and minerals. The spectra are offset vertically to avoid overlap. From Kahle (1984, Figure 4).

readily distinguished in the TIMS image than in the Landsat image.

Alluvial Fans The broad alluvial fans along the east flank of the Panamint Mountains consist of gravel eroded from bedrock and transported eastward by the washes (ephemeral streams). Lithology of the gravel in each fan is determined by the type of bedrock exposed along the washes. In the Landsat image the fans are seen to consist of two types of gravel: older deposits and younger deposits. Gravels of the older deposits are generally coated with desert varnish, which is a thin layer of clay, iron oxide, and manganese oxide that forms in arid climates and imparts a dark tone to the surface. The younger gravels, which occur in active stream channels, lack this desert varnish and have bright tones. In the TIMS image, some differences in development of desert varnish are discernible, but the color signatures are predominantly related to the lithology of the gravel. The Trail Canyon and Blackwater Wash fans have both red and magenta tones; the red tones are associated with quartzite gravel, and the magenta tones represent mixtures of quartzite and shale. The Tucki Wash fan is a mixture of dolomite and shale fragments. Between the Trail Canyon and Blackwater Wash fans are two small fans that consist of carbonate gravel and have the same green tone as the carbonate bedrock. These green fans are separated by a patch of volcanic gravel with a lilac signature that was derived from adjacent volcanic bedrock with a purple signature.

Valley Deposits Saline deposits on the floor of Death Valley occur along the east margin of the fans. A distinctive yellow belt in the TIMS image represents deposits of sand cemented with sulfate and carbonate salts. The green signature represents halite (rock salt) deposits. The very dark signature represents floodplain deposits of clay and silt that are saturated with brine and have a thin salt crust. Evaporative cooling of these damp floodplain deposits causes a low radiant temperature in all TIMS bands, which results in the dark signature.

COMMENTS

Thermal IR images record the radiant temperature of materials. Radiant temperature is determined by a material's kinetic temperature and by its emissivity, which is a measure of its ability to radiate and to absorb thermal energy. The diurnal temperature range (ΔT) is a function of thermal inertia, which is directly related to density of materials. Thermal IR images are useful for many applications, including

map (Figure 5.44), three types of bedrock are shown together with the closely associated detritus. Quartzite outcrops and the adjacent detritus have a conspicuous red color. Carbonate rocks are blue-green. Shale and volcanic rocks are purple. These bedrock units are more

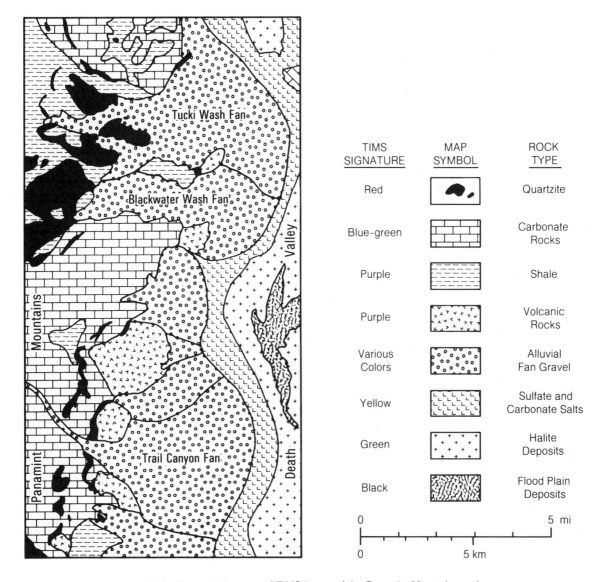

TIMS SIGNATURE	MAP SYMBOL	ROCK TYPE
Red		Quartzite
Blue-green		Carbonate Rocks
Purple		Shale
Purple		Volcanic Rocks
Various Colors		Alluvial Fan Gravel
Yellow		Sulfate and Carbonate Salts
Green		Halite Deposits
Black		Flood Plain Deposits

0   5 mi

0   5 km

FIGURE 5.44 Interpretation map of TIMS image of the Panamint Mountains and Death Valley, California. After Gillespie, Kahle, and Palluconi (1984, Figure 3).

1. Differentiation of rock types. Denser rocks, such as basalt and sandstone, have higher thermal inertias than less dense rocks, such as cinders and siltstone. On nighttime thermal IR images, the rocks with higher thermal inertia values have warmer signatures.

2. Mapping surface moisture. Damp ground has a cool signature on thermal IR images that is caused by evaporative cooling.

3. Mapping geologic structure. Faults may be marked by cool linear anomalies caused by evaporative cooling of moisture concentrated along the fault zone. Folds may be indicated by the thermal pattern caused by the outcrops of different rock types.

4. Measuring surface temperatures of active volcanoes.

5. Aiding in monitoring forest fires. Thermal IR wavelengths penetrate smoke, but not clouds, and the images are valuable.

6. Mapping geothermal areas, sea ice, subsurface coal fires, and for detecting oil films, thermal plumes, and current patterns in water bodies. Later chapters describe these uses of thermal IR images.

Thermal IR images are affected by environmental factors that the interpreter must consider. These include

1. Clouds and surface winds that produce confusing thermal patterns.

2. Time of day. Daytime images record the differential solar heating and shadowing of topographic features. Geologic and other interpretations require nighttime images.

3. Surface moisture and dense vegetation. These effectively mask other features on thermal IR images; therefore thermal IR images are most useful in arid and semiarid terrain.

The HCMM experiment demonstrated the value of low-resolution IR images acquired at night from satellites. Band 6 of TM has demonstrated the value of images with medium spatial resolution, but additional nighttime data should be acquired. The TIMS aircraft system records images at narrow spectral bands in the 8-to-14-μm atmospheric window. When they are suitably processed, these images discriminate different rock types, primarily on the basis of variations in silica content.

QUESTIONS

1. An iron poker in a fire glows dull red, which corresponds to a visible wavelength of 0.65 μm and represents λ_{max}. Use Wein's displacement law (Equation 5.1) to calculate the radiant temperature of the iron.

2. The filament in a light bulb is heated to a radiant temperature of 7000°K. What is the wavelength at which the maximum energy radiates from the filament?

3. Calculate the radiant flux (F_b) for a blackbody with a kinetic temperature of 21°C.

4. Calculate radiant flux (F_r) from a block of rough granite (Table 5.1) with a kinetic temperature (T_{kin}) of 15°C.

5. Calculate radiant temperature (T_{rad}) of a block of rough dolomite (Table 5.1) with a kinetic temperature (T_{kin}) of 13°C.

6. Calculate the thermal inertia of a rock with the following properties: a thermal conductivity (K) of 0.005 cal · cm^{-1} · sec^{-1} · °C^{-1}, a density (ρ) of 2.7 g · cm^{-3}; and a thermal capacity (c) of 0.19 cal · g^{-1} · °C^{-1}.

7. You are interpreting daytime and nighttime HCMM thermal IR images and a visible image of an area. A rock outcrop has a daytime radiant temperature of 20°C and a nighttime temperature of 10°C. Albedo from the visible image is 0.50. Calculate the apparent thermal inertia (ATI) for this rock.

8. Describe the appearance of clouds and of smoke plumes on thermal IR images. Explain the difference in appearance.

9. List the signature (warm or cool) of the following targets on daytime and nighttime IR images: dry soil, damp soil, and standing water. Explain these signatures.

10. In arid terrain such as southern California and South Africa, thermal IR images commonly portray more geologic information than do aerial photographs. Explain this advantage of IR images.

REFERENCES

Allen, C. R., M. Wyss, J. N. Brune, A. Grantz, and R. E. Wallace, 1972, Displacements on the Imperial, Superstition Hills and San Andreas faults triggered by the Borrego Mountain earthquake *in* The Borrego Mountain earthquake of April 9, 1968: U.S. Geological Survey Professional Paper 787, p. 87–104.

Buettner, K. J. K., and C. D. Kern, 1965, Determination of infrared emissivities of terrestrial surfaces: Journal of Geophysical Research, v. 70, p. 1329–1337.

Colwell, R. N., W. Brewer, G. Landis, P. Langley, J. Morgan, J. Rinker, J. M. Robinson, and A. L. Sorem, 1963, Basic matter and energy relationships involved in remote reconnaissance: Photogrammetric Engineering, v. 29, p. 761–799.

Dozier, J., and A. H. Strahler, 1983, Ground investigations in support of remote sensing *in* Colwell, R. N., ed., Manual of remote sensing, second edition: ch. 23, p. 969–989, American Society of Photogrammetry, Falls Church, Va.

Gillespie, A. R., A. B. Kahle, and F. D. Palluconi, 1984, Mapping alluvial fans in Death Valley, California, using multichannel thermal infrared images: Geophysical Research Letters, v. 11, p. 1153–1156.

Janza, F. J., and others, 1975, Interaction mechanisms *in* Reeves, R. G., ed., Manual of remote sensing, first edition: ch. 4, p. 75–179, American Society of Photogrammetry, Falls Church, Va.

Kahle, A. B., 1984, Measuring spectra of arid lands *in* El-Baz, F., ed., Deserts and arid lands: ch. 11, p. 195–217, Martinus Nijhoff Publishers, The Hague, Netherlands.

Kahle, A. B., and A. F. H. Goetz, 1983, Mineralogic information from a new airborne thermal infrared multispectral scanner: Science, v. 222, p. 24–27.

Kahle, A. B., D. P. Madura, and J. M. Soha, 1980, Middle infrared multispectral aircraft scanner data—analysis for geologic applications: Applied Optics, v. 19, p. 2279–2290.

Kahle, A. B., J. P. Schieldge, M. J. Abrams, R. E. Alley, and C. J. LeVine, 1981, Geologic applications of thermal inertia imaging using HCMM data: Jet Propulsion Laboratory Publication 81–55, Pasadena, Calif.

Lockwood, J. P., R. Y. Koyanagi, R. I. Tilling, R. T. Holcomb, and D. W. Peterson, 1976, Mauna Loa threatening: Geotimes, v. 21, p. 12–15.

Macdonald, G. A., 1971, Geologic map of the Mauna Loa Quadrangle, Hawaii: U.S. Geological Survey Geologic Quadrangle Map GQ-897.

Norwood, V. T., and J. C. Lansing, 1983, Electro-optical imaging systems *in* Colwell, R. N., ed., Manual of remote sensing, second edition: ch. 8, p. 335–367, American Society of Photogrammetry, Falls Church, Va.

Price, J. C., 1980, Heat Capacity Mapping Mission (HCMM) data users handbook for Applications Explorer Mission (AEM): NASA Goddard Space Flight Center, Greenbelt, Md.

Sabins, F. F., 1967, Infrared imagery and geologic aspects: Photogrammetric Engineering, v. 29, p. 83–87.

Sabins, F. F., 1969, Thermal infrared imagery and its application to structural mapping in Southern California: Geological Society of America Bulletin, v. 80, p. 397–404.

Sabins, F. F., 1973, Recording and processing thermal IR imagery: Photogrammetric Engineering, v. 39, p. 839–844.

Sabins, F. F., 1983, Remote sensing laboratory manual, second edition: Remote Sensing Enterprises, La Habra, Calif.

Short, N. M., and L. M. Stuart, 1983, The Heat Capacity Mapping Mission (HCMM) anthology: NASA SP 465, U.S. Government Printing Office, Washington, D.C.

Warwick, D., P. G. Hartopp, and R. P. Viljoen, 1979, Application of the thermal infrared linescanning technique to engineering geological mapping in South Africa: Quarterly Journal of Engineering Geology, v. 12, p. 159–179.

Watson, K., 1971, Geophysical aspects of remote sensing: Proceedings of the International Workshop on Earth Resources Survey Systems, NASA SP 283, v. 2, p. 409–428.

Wolfe, E. W., 1971, Thermal IR for geology: Photogrammetric Engineering, v. 37, p. 43–52.

ADDITIONAL READING

Estes, J. E., E. J. Hajic, and L. R. Tinney, 1983, Fundamentals of image analysis—analysis of visible and thermal infrared data *in* Colwell, R. N., ed., Manual of remote sensing, second edition: ch. 24, p. 987–1124, American Society for Photogrammetry and Remote Sensing, Falls Church, Va.

Kahle, A. B., 1980, Surface thermal properties *in* Siegal, B. S., and A. R. Gillespie, eds., Remote sensing in geology: ch. 8, p. 257–273, John Wiley & Sons, N.Y.

Sabins, F. F., 1980, Interpretation of thermal infrared images *in* Siegal, B. S., and A. R. Gillespie, eds., Remote sensing in geology: ch. 9, p. 275–295, John Wiley & Sons, N.Y.

Radar Images

Radar is an *active* remote sensing system because it provides its own source of energy. The system "illuminates" the terrain with electromagnetic energy, detects the energy returning from the terrain (called *radar return*), and then records it as an image. *Passive* remote sensing systems, such as photography and thermal IR, detect the available energy reflected or radiated from the terrain. Radar systems operate independently of lighting conditions and largely independently of weather. In addition, the terrain can be "illuminated" in the optimum direction to enhance features of interest.

Radar is the acronym for radio detection and ranging; it operates in the radio and microwave bands of the electromagnetic spectrum ranging from a meter to a few millimeters in wavelength. The reflection of radio waves from objects was noted in the late 1800s and early 1900s. Definitive investigations of radar began in the 1920s in the United States and Great Britain for the detection of ships and aircraft. Radar was developed during World War II for navigation and target location and used the familiar rotating antenna and circular *cathode-ray-tube* (CRT) display. The continuous-strip mapping capability of *side-looking airborne radar* (SLAR) was developed in the 1950s to acquire reconnaissance images without the necessity of flying over politically unfriendly regions. Fischer (1975, p. 41–43) has prepared a comprehensive history of radar development.

AIRCRAFT RADAR SYSTEMS

To aid in understanding the image-forming process, radar systems aboard aircraft are described here, but the principles of aircraft systems are also applicable to satellite systems. Moore (1983) and Moore and others (1983) give details of radar theory and practice.

Radar Components

Figure 6.1 is a diagram of the components of a typical SLAR system. The timing pulses of electromagnetic energy from the pulse-generating device serve two purposes: (1) they control the bursts of energy from the transmitter, and (2) they trigger the sweep of the CRT film-recording device. Film and CRTs are being replaced by magnetic tape recorders, but the principles are the same. The bursts of electromagnetic energy from the transmitter are of specific wavelength and duration, or pulse length. (Pulse length is further defined later in the chapter.)

The same antenna transmits the radar pulse and receives the return from the terrain. An electronic switch, or *duplexer,* prevents interference between transmitted and received pulses by blocking the receiver circuit during transmission and blocking the transmitter circuit during reception. The *antenna* is a reflector that focuses

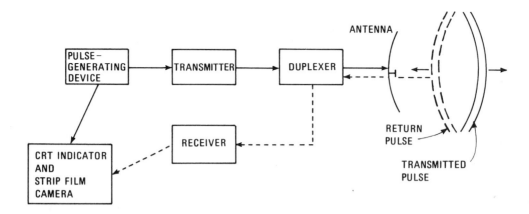

FIGURE 6.1 Block diagram of a radar imaging system.

the pulse of energy into the desired form for transmission and also collects the energy returning from the terrain. A *receiver,* similar to a home radio, amplifies the weak energy waves collected by the antenna. At the same time it preserves the variations in intensity of the returning pulse. The receiver also records the timing of the return pulse, which determines the position of terrain features on the image. The return pulse may be displayed as a line sweep on a CRT and simultaneously recorded on film; it may be recorded as nonimage data on *signal film* for later optical processing into images; or it may be recorded on magnetic tape for later computer processing into images.

Airborne Imaging System

The SLAR antenna illustrated in Figure 6.2 is housed in a cylindrical pod mounted with its long axis parallel with the aircraft fuselage. Newer, solid-state antennas are flat plates. Pulses of energy transmitted from the antenna illuminate strips of terrain in the *look direction* (also called the *range direction*). The look direction is oriented normal to the *azimuth direction* (aircraft flight direction). Figure 6.3 illustrates such a strip of terrain and the shape and timing of the energy pulse that it returns to the antenna. The return pulse is displayed as a function of two-way travel time on the horizontal axis with the shortest times at the right, or *near range,* closest to the aircraft flight path. The longest travel times are at the *far range.* Travel time is converted to distance by multiplying the time by c, the speed of electromagnetic radiation (3×10^8 m · sec^{-1}). The amplitude of the return pulse is a complex function of the interaction between the terrain and the transmitted pulse.

The return pulse shown in Figure 6.3 is converted to a scan line on a CRT by assigning the darkest tones of

the CRT gray scale to returns of the lowest intensity and the brightest tones to those of the highest intensity. The resulting scan line is recorded on a film strip in which the long dimension is parallel with azimuth direction. The CRT trace, which represents look direction, is recorded across the film width, normal to the look direction. The film strip moves past the CRT synchronously with the aircraft ground speed. As the aircraft moves forward, successive scan lines are generated to form images such as those in Figure 6.4.

In addition to being an active system, radar differs from other remote sensing systems such as cameras and optical-mechanical scanners because it records data on the basis of time rather than angular distance. Time can be much more precisely measured and recorded than angular distance can; hence radar images can be acquired at longer ranges with higher resolution. Also, atmospheric absorption and scattering are minimal except at the shortest microwave wavelengths.

Table 6.1 lists the terrain features illustrated in Figure 6.3 together with their signatures and tones in a radar image. The table also summarizes the causes of the signatures and tones. The following section illustrates and describes the appearance of these terrain features on a typical radar image.

Typical Image

Figure 6.4 is an aircraft radar image and map of Weiss Lake and vicinity in northeast Alabama that includes examples of the terrain features diagrammed in Figure 6.3. The image was acquired with the look direction toward the west, which is toward the left margin of the image; thus the look direction has the same orientation for Figures 6.3 and 6.4A. The east-facing slopes of the mountains and ridges in Figure 6.4A face the radar look

FIGURE 6.2 Radar survey aircraft. Antenna for radar imaging system is housed in the pod beneath the fuselage. Courtesy Aero Service Division, Western Geophysical Corporation of America.

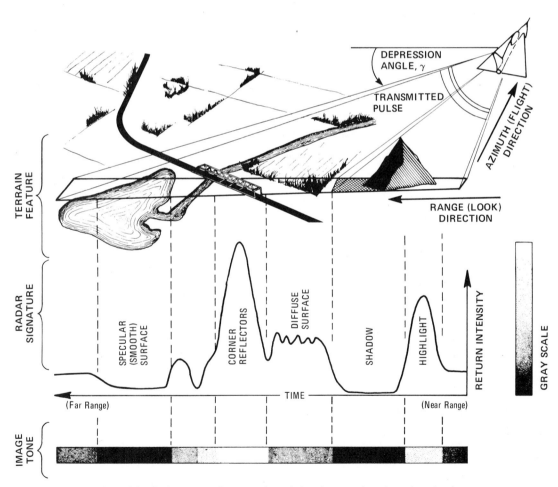

FIGURE 6.3 Radar returns from terrain and signal processing of a pulse of radar energy. From Sabins (1973, Figure 7). Depression angle is explained later in text.

A. RADAR IMAGE.

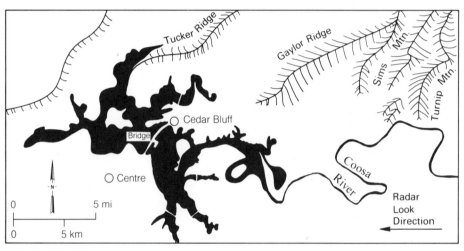

B. LOCATION MAP.

FIGURE 6.4 Aircraft radar image of Weiss Lake and vicinity, northeast Alabama.

direction and form bright signatures, or *highlights,* caused by the strong radar returns. The west-facing slopes are oriented away from the radar look direction and form dark *shadows* because no energy from the transmitted pulses reached these areas.

The flat terrain adjacent to Weiss Lake and the Coosa River is covered with various types of crops and native vegetation. These irregular surfaces cause *diffuse* signatures with intermediate gray tones. The bridge southwest of Cedar Bluff (Figure 6.4B) has a bright signature caused by *corner reflectors,* which are formed by intersecting planar structures. Energy encountering corner reflectors is strongly reflected toward the antenna. The towns of Cedar Bluff and Centre likewise have very bright signatures because of abundant corner reflectors. The calm water of Weiss Lake and the Coosa River is a *specular,* or smooth, mirrorlike surface. Specular surfaces reflect all incident energy such that the angle of reflection is equal to but opposite the angle of incidence. Hence no transmitted energy returns to the antenna and a very dark signature forms.

From the preceding descriptions, it is obvious that tone alone (bright, intermediate, or dark) is insufficient to identify terrain features on radar images. Topographic scarps facing away from the antenna and specular surfaces both have dark tones but are completely different features. Radar signatures, however, are determined not only by tone but also by size, shape, texture, and associations of the image feature. The size and shape of specular features are different from radar shadows. Radar shadows are generally associated with highlights, which are absent from specular features. By utilizing all

TABLE 6.1 Typical features and signatures on radar images

Image signature	Image tone	Terrain feature	Cause of signature
Highlights	Bright	Steep slopes and scarps facing *toward* antenna	Much energy is reflected back to antenna.
Shadows	Very dark	Steep slopes and scarps facing *away* from antenna	No energy reaches terrain; hence there is no return to antenna.
Diffuse surfaces	Intermediate	Vegetation	Vegetation scatters energy in many directions, including returns to antenna.
Corner reflectors	Very bright	Bridges and cities	Intersecting planar surfaces strongly reflect energy toward antenna.
Specular surfaces	Very dark	Calm water, pavement, dry lake beds	Smooth, horizontal surfaces totally reflect energy, with angle of reflectance opposite to angle of incidence.

information available in an image, remarkably accurate interpretations are possible.

The orientation of the radar look direction relative to the interpreter's perspective is important because it determines the orientation of the resulting highlights and shadows. For most interpreters, the optimum look direction is toward the lower margin of an image. In images where this orientation is reversed (look direction toward the upper margin), many interpreters find that topography seems to be reversed, with ridges appearing as valleys and valleys appearing as ridges. This phenomenon, called *topographic inversion,* can be corrected by rotating the image 180° so the look direction is toward the lower margin of the image.

Radar Wavelengths

Table 6.2 lists the various radar wavelengths and corresponding frequencies. Frequency is a more fundamental property of electromagnetic radiation than is wavelength because, as radiation passes through media of different densities, frequency remains constant whereas velocity and wavelength change. Most interpreters, however, comprehend wavelengths more readily than frequencies; also, wavelengths are used to describe the visible and infrared spectral regions. Therefore wavelengths are used here to designate various radar systems. Equation 1.1, (Chapter 1) enables any frequency (ν) to be converted into wavelength (λ) in the following manner:

$$c = \lambda \nu \qquad (1.1)$$

$$\lambda = \frac{3 \times 10^8 \, \text{m} \cdot \text{sec}^{-1}}{\nu}$$

where c is the speed of electromagnetic radiation. A convenient version of Equation 1.1 for converting radar frequencies into wavelength equivalents is

$$\lambda \text{ (in cm)} = \frac{30}{\nu \text{ (in GHz)}}$$

TABLE 6.2 Radar wavelengths and frequencies used in remote sensing

Band designation*	Wavelength (λ), cm	Frequency (ν), GHz (10^9 cycles · sec^{-1})
Ka (0.86 cm)	0.8 to 1.1	40.0 to 26.5
K	1.1 to 1.7	26.5 to 18.0
Ku	1.7 to 2.4	18.0 to 12.5
X (3.0 cm, 3.2 cm)	2.4 to 3.8	12.5 to 8.0
C	3.8 to 7.5	8.0 to 4.0
S	7.5 to 15.0	4.0 to 2.0
L (23.5 cm, 25.0 cm)	15.0 to 30.0	2.0 to 1.0
P	30.0 to 100.0	1.0 to 0.3

*Wavelengths commonly used in imaging radars are shown in parentheses.

A random-letter code (Table 6.2) was assigned to different frequencies during the early classified stages of radar development to avoid mention of the wavelength regions under investigation. In the 1960s and early 1970s, the first unclassified airborne radar was a Ka-band system that produced numerous images, a few of which are illustrated in this chapter. This system was discontinued by the mid-1970s, and today commercial aircraft radar systems operate at X-band wavelengths. Aircraft radar experiments are also conducted at various other wavelengths, and images from satellites are currently acquired at L-band wavelengths. Future systems will employ additional wavelengths, as described toward the end of this chapter.

Depression Angle and Spatial Resolution

Resolution in the range (look) direction and azimuth (flight) direction is determined by the engineering characteristics of the radar system. An important characteristic is the *depression angle* (γ), defined as the angle between a horizontal plane and a beam from the antenna to a target on the ground (Figure 6.5A). The depression angle is steeper at the near-range side of an image strip and shallower at the far-range side. Average depression angle of an image is measured for a beam to the midline of an image strip. An alternate geometric term is *incidence angle* (θ), defined as the angle between a radar beam and a line perpendicular to the surface. For a horizontal surface, θ is the complement of γ, but for an inclined surface there is no correlation between the two

angles (Figure 6.5B). The incidence angle more correctly describes the relationship between a radar beam and a surface than does depression angle; however, in actual practice the surface is usually assumed to be horizontal and the incidence angle is taken as the angle between the radar beam and a vertical plane passing through the antenna. Several other terms have been applied to this angle, including *look angle* and *off-nadir angle;* using the unambiguous term *depression angle* avoids this confusion. Recall, however, that depression angle is the complement of incidence angle.

The combination of range resolution and azimuth resolution determines the dimensions of the ground resolution cell, which in turn determines the spatial resolution of a radar image.

Range Resolution *Range resolution* (R_r), or resolution in the range direction, is determined by the depression angle and by the pulse length. Range resolution is theoretically equal to one-half the pulse length (Figure 6.6). *Pulse length* (τ) is the duration of the transmitted pulse and is measured in microseconds (μsec, or 10^{-6} sec). It is converted from time into distance by multiplying by the speed of electromagnetic radiation ($c = 3 \times 10^8$ m · sec^{-1}). The resulting distance is measured in the *slant range,* or direction in which the energy propagates from the antenna to the target. Range resolution, however, is expressed in *ground range,* which is the distance measured on the terrain (Figure 6.6). Dividing the slant-range distance by the cosine of the depression (γ) angle converts slant-range distance into ground-range distance.

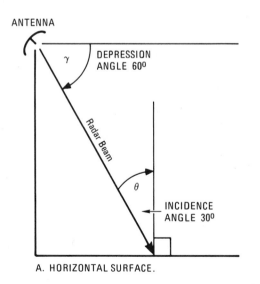

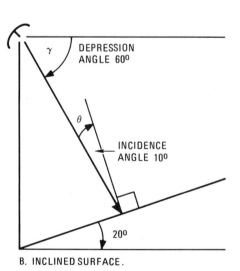

FIGURE 6.5 Depression angle and incidence angle.

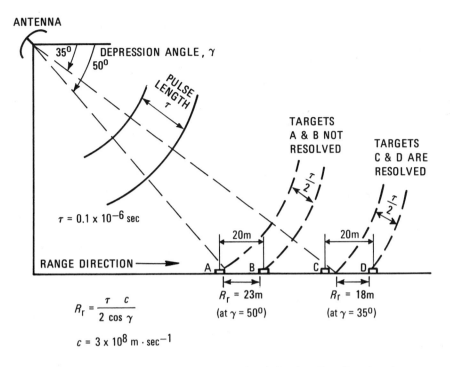

$$R_r = \frac{\tau \ c}{2 \cos \gamma}$$

$$c = 3 \times 10^8 \ m \cdot sec^{-1}$$

FIGURE 6.6 Radar resolution in the range (look) direction. From Barr (1969).

The equation for range resolution is

$$R_r = \frac{\tau c}{2 \cos \gamma} \qquad (6.1)$$

For a depression angle of 50° and a pulse length of 0.1 μsec, range resolution is calculated as

$$R_r = \frac{(0.1 \times 10^{-6} \ sec)(3 \times 10^8 \ m \cdot sec^{-1})}{2 \cos 50°}$$

$$= \frac{30 \ m}{2 \times 0.64}$$

$$= 23.4 \ m$$

Therefore, at a depression angle of 50° and pulse length of 0.1 μsec, targets must be separated by more than 23 m in the range direction to be resolved. At a depression angle of 35°, however, range resolution improves to 18.3 m. In Figure 6.6, target pairs A–B and C–D are both separated by 20 m. Targets A and B are located in the near-range position where $\gamma = 50°$ and $R_r = 23$ m; therefore A and B are not resolved as separate targets in the image. Targets C and D are located in the far-range position where $\gamma = 35°$ and $R_r = 18$ m; therefore

C and D are resolved as two separate features in the image.

One method of improving range resolution is to shorten the pulse length, but this reduces the total amount of energy in each transmitted pulse. The energy and pulse length cannot be reduced below the level required to produce a sufficiently strong return from the terrain. Electronic techniques have been developed for shortening the apparent pulse length while providing adequate signal strength.

Azimuth Resolution *Azimuth resolution* (R_a), or resolution in the azimuth direction, is determined by the width of the terrain strip illuminated by the radar beam. To be resolved, targets must be separated in the azimuth direction by a distance greater than the beam width as measured on the ground. As shown in Figure 6.7, the fan-shaped beam is narrower in the near range than in the far range, causing azimuth resolution to be smaller in the near-range portion of the image. *Angular beam width* is directly proportional to wavelength of the transmitted energy; therefore azimuth resolution is higher for shorter wavelengths, but the short wavelengths lack the desirable weather penetration capability. Angular beam width is inversely proportional to *antenna length;* there-

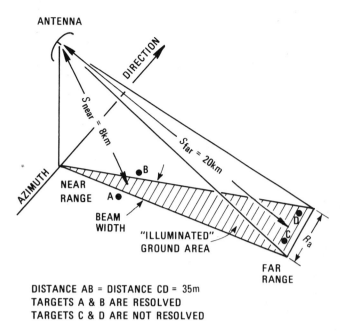

DISTANCE AB = DISTANCE CD = 35m
TARGETS A & B ARE RESOLVED
TARGETS C & D ARE NOT RESOLVED

FIGURE 6.7 Radar beam width and resolution in the azimuth direction. From Barr (1969).

Real-Aperture and Synthetic-Aperture Systems

The two basic systems are real-aperture radar and synthetic-aperture radar, which differ primarily in the method each uses to achieve resolution in the azimuth direction. The *real-aperture,* or "brute force," system uses an antenna of the maximum practical length to produce a narrow angular beam width in the azimuth direction, as illustrated in Figure 6.7.

The *synthetic-aperture radar* (SAR) employs a small antenna that transmits a relatively broad beam (Figure 6.8A). The Doppler principle and special data-processing techniques are employed to synthesize the azimuth resolution of a very narrow beam. Using the familiar

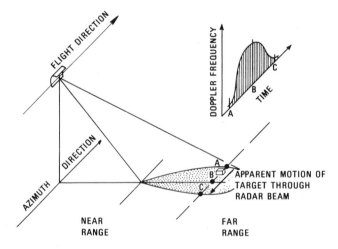

A. DOPPLER FREQUENCY SHIFT DUE TO RELATIVE MOTION OF TARGET THROUGH RADAR BEAM.

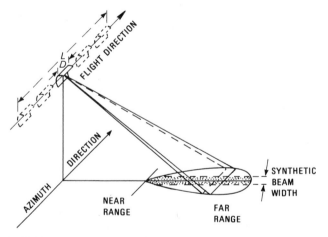

B. RESOLUTION OF SYNTHETIC–APERTURE RADAR IN THE AZIMUTH DIRECTION. NOTE THAT THE PHYSICAL ANTENNA LENGTH D IS SYNTHETICALLY LENGTHENED TO L.

FIGURE 6.8 Synthetic-aperture radar system. From Craib (1972, Figures 3 and 5).

fore resolution improves with longer antennas, but there are practical limitations to the maximum antenna length.

The equation for azimuth resolution (R_a) is

$$R_a = \frac{0.7\,S\,\lambda}{D} \qquad (6.2)$$

where S is slant-range distance, and D is the antenna length. For a typical X-band system, $\lambda = 3.0$ cm and $D = 500$ cm. At the near-range position, the slant-range distance (S_{near} in Figure 6.7) is 8 km and R_a is calculated from Equation 6.2 as

$$R_a = \frac{0.7(8 \text{ km} \times 3.0 \text{ cm})}{500 \text{ cm}}$$

$$= 33.6 \text{ m}$$

Therefore targets in the near range, such as A and B in Figure 6.7, must be separated by approximately 34 m to be resolved. At the far-range position, the slant-range distance (S_{far} in Figure 6.7) is 20 km and R_a is calculated as 84 m; therefore targets C and D, also separated by 35 m, are not resolved. Values for range resolution in modern radar systems far greater than those in Equation 6.2 are obtained using synthetic-aperture methods, as described in the following section.

example of sound, the Doppler principle states that the frequency (pitch) of the sound heard differs from the frequency of the vibrating source whenever the listener and the source are in motion relative to one another. The rise and drop in pitch of the siren as an ambulance approaches and recedes is a familiar example. This principle is applicable to all harmonic wave motion, including the microwaves employed in radar systems.

Figure 6.8A illustrates the apparent motion of a target through the successive radar beams from points A to C as a consequence of the forward motion of the aircraft. As shown on the Doppler frequency diagram, the frequency of the energy pulse returning from the target increases from a minimum at point A to a maximum at point B normal to the aircraft. As the target recedes from B to C, the frequency decreases.

Synthetic-aperture data are recorded by a laser system that produces a holographic film record of the amplitude and phase history of the returns from each target as the repeated radar beams pass across the target from A to C. The holographic film is later played back to produce an image film, as described by Jensen and others (1977). The record of Doppler frequency changes enables the target to be resolved on the image film as though it had been observed with an antenna of length

L, as shown in Figure 6.8B. This synthetically lengthened antenna produces the effect of a very narrow beam with constant width in the azimuth direction, shown by the shaded area in Figure 6.8B. This pattern can be compared with the fan-shaped beam of a real-aperture SLAR system shown in Figure 6.7. For both real-aperture and synthetic-aperture systems, resolution in the range direction is determined by pulse length and depression angle (Figure 6.6).

Figure 6.9 compares real-aperture and synthetic-aperture images of a volcanic area in the Alaska Peninsula. The depression angle is shallower for the real-aperture image, which results in longer shadows; otherwise the images were acquired under similar circumstances. The higher spatial resolution of the SAR image is apparent, particularly when examining the hummocky lava flows in the lower portion of the images. In the SAR image, spatial resolution is constant from the near range (top margin of image) to far range (bottom of image), but in the real-aperture image, resolution becomes poorer in the far-range direction.

Real-aperture systems are simple in design and do not require sophisticated data recording and processing. However, coverage in the range direction is relatively limited and only shorter wavelengths can be employed

A. REAL-APERTURE IMAGE.

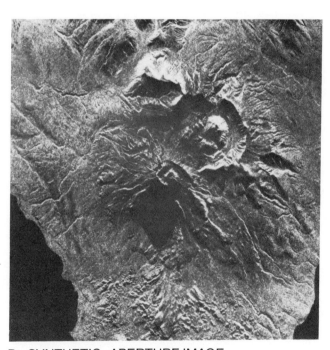

B. SYNTHETIC-APERTURE IMAGE.

FIGURE 6.9 Real-aperture and synthetic-aperture X-band radar images of Mount Peulik, Alaska Peninsula. Look direction is from upper to lower margin of image. Images cover an area of 18 by 18 km.

if high resolution is required. SAR systems maintain high azimuth resolution at long distances in the range direction at both long and short wavelengths. The added complexity and cost of SAR are the tradeoffs for these advantages. The development of SAR systems has made it possible to acquire radar images from orbital platforms such as Seasat and the Space Shuttle.

Stereo Images

Adjacent radar flight lines may be spaced to obtain overlapping images such as the example in Figure 6.10 from Venezuela. These images should be viewed with a pocket stereoscope to obtain the three-dimensional effect. Stereo viewing is particularly useful for areas of low relief. Stereo radar coverage may also be obtained by imaging the same area from two different flight altitudes or from two opposing antenna look directions.

Mosaics

Image strips acquired from aircraft may be hundreds of kilometers in length and may vary from 10 to 30 km in width depending on the range of depression angles and the flight altitudes. Parallel overlapping strips may be composited into a mosaic to provide more extensive coverage. Figure 6.11 is a portion of a mosaic of the Appalachian Mountains in Pennsylvania. This image is reproduced at the original 1:250,000 scale of the mosaic. Look direction is toward the northwest, as shown by highlights and shadows formed by the ridges. The mosaic was so skillfully constructed that boundaries be-

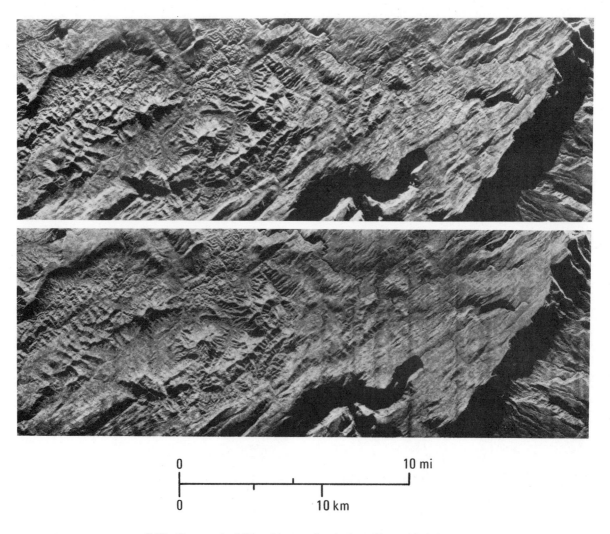

FIGURE 6.10 Stereo pair of X-band images for viewing with a pocket stereoscope. Look direction is from upper to lower margin of images. Cerro Duida area, south-central Venezuela. Courtesy Aero Service Division, Western Geophysical Corporation of America.

FIGURE 6.11 Portion of a radar mosaic of Harrisburgh, Pennsylvania, acquired by an X-band, synthetic-aperture, aircraft system. Look direction is toward the northwest.

tween the northeast-trending image strips are not discernible.

Acquisition of images and construction of a mosaic are exacting tasks. Aircraft navigation must be precise to provide proper sidelap of image strips. Altitude and speed must be controlled to provide uniform scale on the images. The radar look direction must be uniformly oriented in all image strips; otherwise shadows and highlights will be reversed in different parts of the mosaic. Mosaics have been made for portions of Alaska and conterminous United States, as well as most of Brazil and Venezuela.

Available Aircraft Systems and Images

Table 6.3 lists the currently operational nonmilitary, aircraft radar systems. The JPL system was used for research; the other systems are available for commercial surveys. Radar mosaics of portions of the United States (scale 1:250,000) have been acquired by the U.S. Geological Survey and are available from EDC, which can provide index maps and price lists.

GEOMETRY OF RADAR IMAGES

The nature of radar illumination causes specific geometric characteristics in the images that include shadows and highlights, slant-range distortion, and image displacement.

Shadows and Highlights

As illustrated in Figures 6.3 and 6.4, the oblique illumination of radar produces strong returns from the sides of buildings, ridges, and peaks facing the antenna. These topographic obstructions also create shadows in the look direction, as shown on the image of the Santa Rosa Mountains (Figure 6.12A). The outline of the radar shadow matches the topographic profile of the ridge crest, as shown by comparison with the contour map (Figure 6.12B). The importance of radar highlights is illustrated by the southeast-trending linear feature marked by a very bright highlight at F on the image. This feature is a topographic scarp along the active Clark fault that is part of the San Jacinto fault zone. Southeastward from this scarp the Clark fault can be traced into the badlands topography at the south end of the Santa Rosa Mountains.

In aerial photographs the elevation angle of solar illumination is constant throughout the scene. On radar images the depression angle becomes smaller in the far-range direction and shadows are proportionately longer (Figure 6.13) in the far-range portion of the image. For terrain with low relief, it is desirable to acquire radar images with a small depression angle in order to produce maximum highlights and shadows. In terrain with high relief, an intermediate depression angle is desirable, because the extensive shadows caused by a small angle obscure much of the image.

Slant-Range and Ground-Range Images

Depending on design of the recording system, images are presented either as slant-range or as ground-range displays. On *slant-range displays* the scale in the near-range portion of the image is compressed relative to the far-range portion. The upper part of Figure 6.14 is a vertical section showing two targets on the ground that are of equal size, A in the near range and B in the far range. In the plane of the slant-range display, however, the image A' is greatly compressed relative to image B'. This geometric compression in the near-range portion of an image is called *slant-range distortion* and is shown in the image of Figure 6.14A, which covers an agricultural area with rectangular fields and orthogonal roads. In this slant-range display the fields are distorted into

TABLE 6.3 Operational aircraft radar

Operator	Equipment	Remarks
Aero Service Division, Western Geophysical Corporation of America Houston, Texas	X-band, synthetic-aperture system.	Radar system built by Goodyear Aerospace Corporation.
Intera Technologies, Ltd. Calgary, Alberta	X-band, synthetic-aperture system.	Images digitally recorded.
Jet Propulsion Laboratory Pasadena, California	Simultaneous C-band and L-band, synthetic-aperture system. Aircraft and radar destroyed in July 1985 but scheduled for replacement.	Parallel- and cross-polarized images acquired at either wavelength.

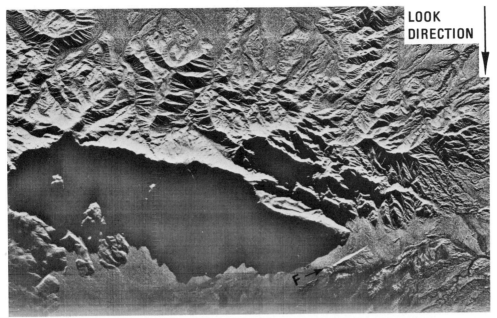

LOOK DIRECTION

A. Ka–BAND, REAL–APERTURE IMAGE WITH LOOK DIRECTION TO THE SOUTHWEST. SCARP OF ACTIVE CLARK FAULT IS INDICATED AT F.

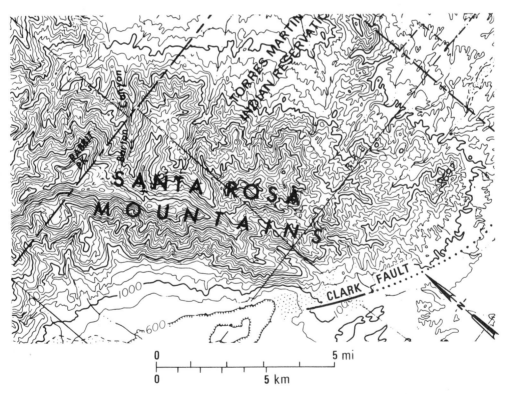

B. TOPOGRAPHIC MAP WITH CONTOUR INTERVAL OF 60 m (200 ft). COMPARE PROFILE OF THE MOUNTAIN CREST WITH THE RADAR SHADOW.

FIGURE 6.12 Radar shadows and highlights in the Santa Rosa Mountains, Riverside County, California.

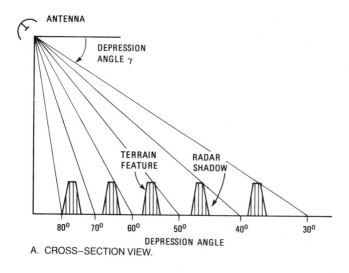

ANTENNA

DEPRESSION
ANGLE γ

TERRAIN
FEATURE

RADAR
SHADOW

80° 70° 60° 50° 40° 30°
DEPRESSION ANGLE

A. CROSS–SECTION VIEW.

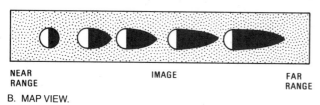

NEAR
RANGE

IMAGE

FAR
RANGE

B. MAP VIEW.

FIGURE 6.13 Radar illumination and shadows at different depression angles.

rhomboid outlines and the roads are warped in the near-range portion of the image.

Modern radar systems include a circuit that corrects slant-range distortion and produces ground-display images. Figure 6.14B is the ground-range display version of the distorted image. The fields are rectangular and the roads are orthogonal.

Image Layover

The curvature of a transmitted radar pulse causes the top of a tall vertical target to reflect energy in advance of its base, which results in displacement of the top toward the near range on the image. This distortion is called *image layover*. On the diagram of Figure 6.15A the curved wavefront encounters the mountain crest at A in advance of the base B. The image of the crest is offset toward the near range relative to the base. The ridges of the Grapevine Mountains are relatively symmetrical, but on the radar image (Figure 6.15A) they appear *foreshortened*, "leaning" in the near-range direction because of the layover effect. Comparison with the undistorted Skylab photograph (Figure 6.15B) illustrates the extent of displacement on the radar image. The JPL aircraft radar system provides a topographic

profile along the aircraft flight path, which is shown on the left side of Figure 6.15A.

The amount of layover is influenced by the following factors:

1. *Height of targets.* Taller targets are displaced more than shorter targets.

2. *Radar depression angle.* Images acquired with steep (large) depression angles have more displacement than those acquired with shallow (small) depression angles.

3. *Location of targets.* For targets of the same height, those located in the near-range are displaced more on the image than are those in the far range because depression angle is steeper in the near range.

Because of the complex three-dimensional factors that cause layover, it is not practical to correct for this distortion. Layover can be minimized by acquiring images at relatively shallow depression angles, but radar shadows may be excessive for terrain with high topographic relief.

SATELLITE RADAR SYSTEMS

Since 1978, NASA has launched three earth-orbiting radar satellite systems, one Seasat and two Shuttle imaging radars (SIR–A and SIR–B), whose characteristics are summarized in Table 6.4. The preceding description of aircraft systems and images applies equally to satellite systems. Seasat and SIR are both synthetic-aperture systems operating at an L-band wavelength of 23.5 cm. Figure 6.16 shows images of the Santa Barbara coast, California, acquired by the two systems. Seasat and SIR–A images are recorded in a slant-range format, but variations in geometric distortion across each image are minimal because there is only a 6° difference in depression angle from near range to far range for both systems. The major difference is the steeper overall depression angle of Seasat and its correspondingly greater image layover. Figure 6.17 is an index map of images acquired by SIR and Seasat.

Seasat

As the name implies, Seasat was designed primarily to investigate oceanic phenomena (such as roughness, current patterns, and sea ice conditions), but the images have also proven valuable for terrain observations. The unmanned satellite (Figure 6.18) was launched in June 1978 and prematurely ceased operation in October 1978 because of a major electrical failure. As shown in Figure 6.19, the depression angle ranges from 67° to 73°; at the

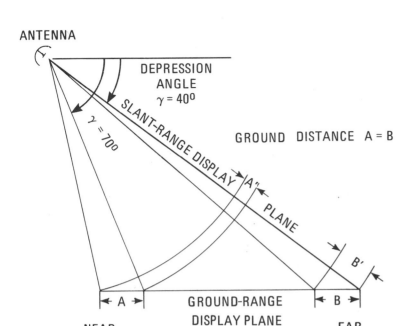

ANTENNA

DEPRESSION
ANGLE
$\gamma = 40^0$

$\gamma = 70^0$

SLANT-RANGE DISPLAY

GROUND DISTANCE A = B

A''

PLANE

B'

A

GROUND-RANGE
DISPLAY PLANE

B

NEAR
RANGE

FAR
RANGE

← LOOK DIRECTION →

A. SLANT–RANGE DISPLAY.

B. GROUND–RANGE DISPLAY.

FIGURE 6.14 Slant-range display corrected into ground-range display. Aircraft
L-band image in central Illinois. Each image covers a ground area of 10 by 10 km.
Images courtesy J. P. Ford, Jet Propulsion Laboratory.

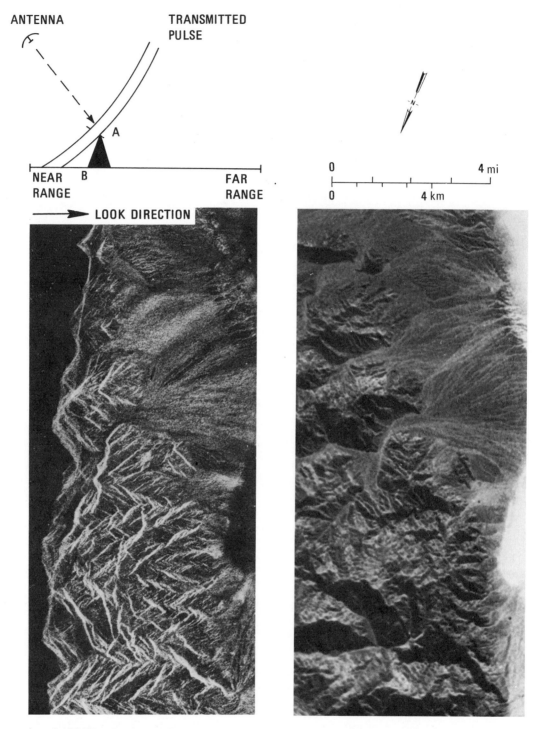

ANTENNA   TRANSMITTED
            PULSE

A

B

NEAR   FAR
RANGE   RANGE

LOOK DIRECTION

0   4 mi

0   4 km

A. L-BAND RADAR IMAGE ACQUIRED
    BY NASA AND JET PROPULSION
    LABORATORY.

B. ENLARGED SKYLAB-4 PHOTOGRAPH.

FIGURE 6.15 Layover of topographic features on a radar image of the Grapevine Mountains on the east side of Death Valley, California. Images courtesy G. C. Schaber, U.S. Geological Survey.

A. SEASAT IMAGE ACQUIRED 1978.

B. SIR–A IMAGE ACQUIRED NOVEMBER 1981.

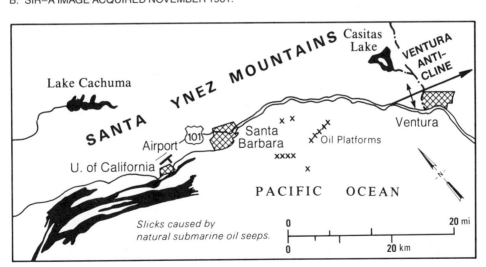

C. LOCATION MAP.

FIGURE 6.16 Seasat and SIR–A images of the Santa Barbara coast, California. From Sabins (1983b, Figure D-3).

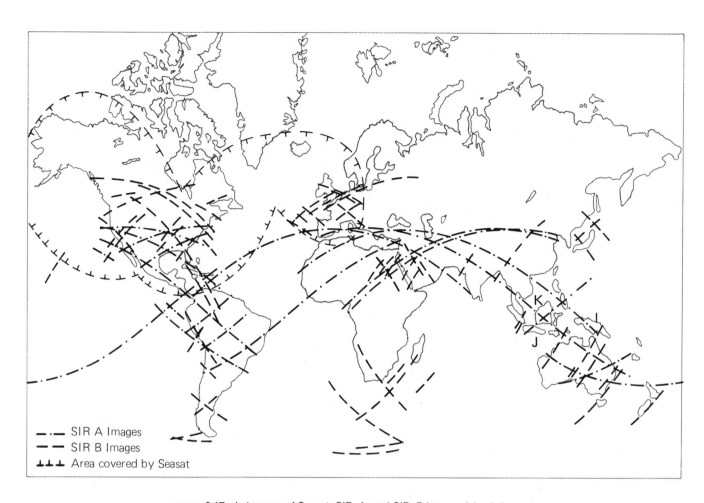

— · — SIR A Images
— — SIR B Images
⊥⊥⊥ Area covered by Seasat

FIGURE 6.17 Index map of Seasat, SIR–A, and SIR–B images. Islands in Indonesia are indicated as follows: I = Irian Jaya; J = Java, K = Kalimantan. From Ford, Cimino, and Elachi (1983, Figure 1). Courtesy J. P. Ford, Jet Propulsion Laboratory.

TABLE 6.4 Characteristics of Seasat, SIR–A, and SIR–B systems

Characteristics	Seasat (1978)	SIR–A (1981)	SIR–B (1984)
Orbit inclination	108°	38°	57°
Wavelength	L-band (23.5 cm)	L-band (23.5 cm)	L-band (23.5 cm)
Spatial resolution	25 m	38 m	25 m
Latitude coverage	72°N to 72°S	50°N to 35°S	58°N to 58°S
Altitude	790 km	250 km	225 km
Image-swath width	100 km	50 km	40 km
Depression angle, γ	67° to 73°	37° to 43°	30° to 75° (variable)
At average γ	70°	40°	52°
Smooth criterion	$h < 1.0$ cm	$h < 1.5$ cm	$h < 1.2$ cm
Rough criterion	$h > 5.7$ cm	$h > 8.3$ cm	$h > 6.8$ cm
Polarization	HH	HH	HH

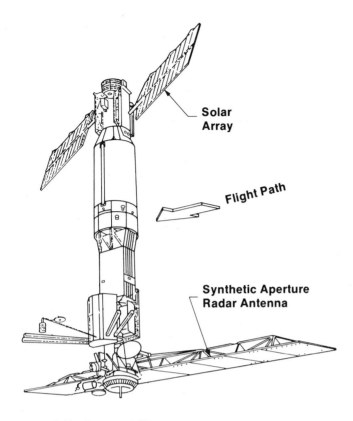

FIGURE 6.18 Seasat satellite.

pretation map (Figure 6.16C), the linear dark patterns in the Pacific Ocean in the western part of the image are streaks of calm water caused by slicks from submarine oil seeps. Metal structures of oil production platforms cause small, bright signatures that are detectable where the adjacent water is calm and has a dark signature. Ships have similar bright signatures. In areas of rough water, however, platforms and ships are virtually undetectable. The cities of Ventura and Santa Barbara along the coast have bright signatures. The calm surfaces of Lake Cachuma and Casitas Lake have distinctive dark signatures.

SIR–A System

The first Shuttle imaging radar experiment (SIR–A) was launched in November 1981. Figure 6.20 shows the configuration of the SIR–A system, and Table 6.4 summarizes its characteristics. The synthetic-aperture antenna is stowed inside the Shuttle cargo bay and operates when the Shuttle is in an inverted attitude (Figure 6.21). The index map (Figure 6.17) shows location of the 50-km-wide strips of SIR–A imagery, which cover a surface area of about 10 million km². The image data were recorded as holographic film on board the Shuttle.

790-km altitude this geometry produces an image swath that is 100 km wide. Image data were telemetered to earth receiving stations in Alaska, California, Florida, Newfoundland, and Great Britain, which recorded the coverage shown in Figure 6.17. Data for each 100-km-wide image swath were optically processed to produce four film strips, each of which covers a width of 25 km and a length of several thousand kilometers. Selected scenes were digitally processed to produce images that cover 100 by 100 km. Both optically processed and digitally processed images may be purchased from

Environmental Data and Information Service
NOAA
Room 606, World Weather Building
Washington, DC 20233

Catalogs and price lists of Seasat images are available from this facility. Ford and others (1980) have prepared a useful atlas of Seasat images.

The image of the Santa Barbara coast (Figure 6.16A) was acquired with a look direction toward the northeast. The steep depression angle (70° average) of Seasat causes the pronounced layover of mountain ridges toward the west (left margin of image). The extensive bright patches in the Pacific Ocean are rough water that contrasts with the dark signature of calm water. As shown in the inter-

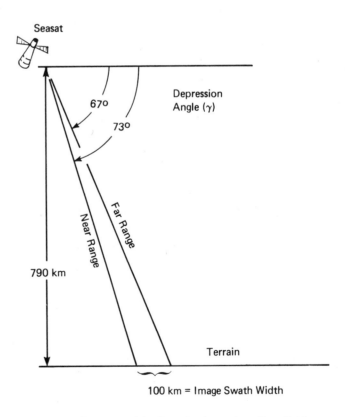

FIGURE 6.19 Geometry of the Seasat radar system. From Sabins (1983a, Figure 6.19).

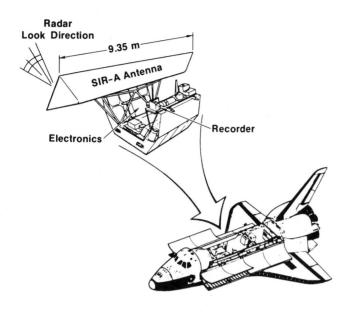

FIGURE 6.20 Shuttle imaging radar (SIR–A) system. Essentially the same system was employed for SIR–B in 1984 with an additional device to change depression angle of the antenna.

The data film was developed and then processed at Jet Propulsion Laboratory to produce the original image film at a scale of 1:500,000. The films are archived at

National Space Science Data Center
Code 601.4
NASA Goddard Space Flight Center
Greenbelt, MD 20771

A "SIR–A Information Packet," including index maps and ordering information, is available from NSSDC in their document number 81-111A-01A. Extensive collections of SIR–A images were published by Cimino and Elachi (1982) and by Ford, Cimino, and Elachi (1983).

In comparing the images of the Santa Barbara coast (Figure 6.16), note that the look direction for the Seasat image was toward the northeast and for the SIR–A image toward the north, which accounts for the different orientation of shadows and highlights. The moderate depression angle (40°) of SIR–A causes less topographic displacement than in Seasat; therefore terrain features are more readily interpreted in Shuttle images. The SIR–A image of the Pacific Ocean lacks the expression of currents and *sea state* (variations in roughness) recorded in the Seasat image. Some of this difference may be due to differences in sea state between the summer of 1978 (when the Seasat image was acquired) and November 1981 (when the SIR–A image was acquired); however, most of the difference is attributed to the steeper depression angle of Seasat, which is optimal for viewing oceanic conditions. Careful photographic processing and reproduction of the SIR–A film of the Santa Barbara coast

reveals the dark oil slicks, but these are less obvious than in Seasat images.

SIR–B System

A second Shuttle radar experiment (SIR–B) was conducted in October 1984 to acquire images of the areas shown in the index map (Figure 6.17). Wavelength, spatial resolution, and polarization were the same as in the earlier mission. For SIR–B the antenna was modified, however, to permit the depression angle to be changed during the mission within a range of 30° to 75°. Multiple images of certain areas were acquired at different depression angles for two purposes: (1) to evaluate changes in radar return as a function of depression angle for different terrain types, and (2) to produce stereo images based on the parallax provided by difference in viewing geometry. Most of the SIR–B images were recorded digitally. Some of the data were transmitted to ground receiving stations via the TDRS communication satellite described in Chapter 4. Other SIR–B images were recorded by on-board tape recorders and returned to earth with the Shuttle vehicle.

On three orbits over northern Florida, SIR–B acquired images at depression angles of 62°, 45°, and 32°. A subscene of Fisher Lake and vicinity was enlarged and analyzed by Hoffer, Mueller, and Lozano-Garcia (1985), who also collected information on land-cover types. Figure 6.22 shows images at the three depression

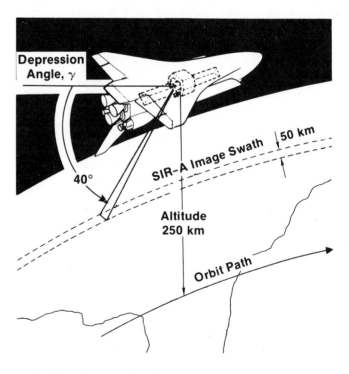

FIGURE 6.21 Geometry of the SIR–A system.

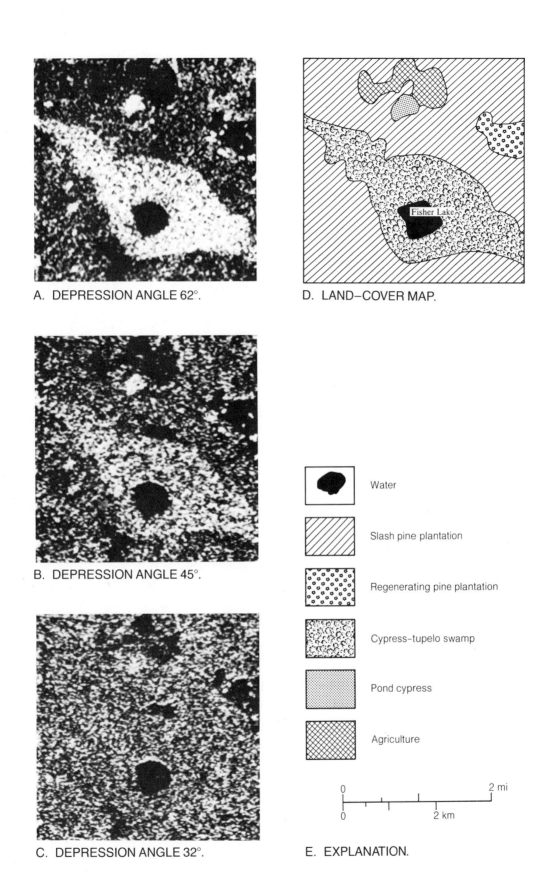

A. DEPRESSION ANGLE 62°.

D. LAND–COVER MAP.

B. DEPRESSION ANGLE 45°.

C. DEPRESSION ANGLE 32°.

Fisher Lake

Water

Slash pine plantation

Regenerating pine plantation

Cypress–tupelo swamp

Pond cypress

Agriculture

0   2 mi

0   2 km

E. EXPLANATION.

FIGURE 6.22 Enlarged SIR–B images acquired at three depression angles. Fisher Lake and vicinity, northern Florida. From Hoffer, Mueller, and Lozano-Garcia (1985, Figure 4). Images courtesy R. M. Hoffer, Purdue University.

angles, together with a map and legend that show land cover. The area is flat and largely covered with different types of forest. The ability to distinguish different types of land cover improves with increasing (steeper) depression angles. In Figure 6.22A, acquired at the maximum 62° depression angle, the cypress-tupelo swamp surrounding Fisher Lake has a very bright signature that contrasts with the dark gray signature of adjacent plantations of slash pine. At the 32° depression angle (Figure 6.22C), however, the swamp and plantation have identical signatures. The smooth surface of Fisher Lake has a dark signature at all depression angles. Surfaces of intermediate roughness (agriculture and regenerating pine plantations) have relatively brighter signatures at shallower depression angles. A color image (not shown) was prepared by projecting each of the three black-and-white images in red, green, or blue. The resulting color image provided good discrimination of land-cover types (Hoffer, Mueller, and Lozano-Garcia, 1985).

RADAR RETURN AND IMAGE SIGNATURES

Stronger radar returns produce brighter signatures on an image than do weaker returns, as shown diagrammatically in Figure 6.3 and in the image of Figure 6.4. Intensity of the radar return, for both aircraft and satellite systems, is determined by the following properties:

1. Radar system properties
   Wavelength
   Depression angle
   Polarization

2. Terrain properties
   Dielectric properties (including water content)
   Surface roughness
   Feature orientation

The following sections discuss these properties in more detail. The effects of depression angle and wavelength of the radar system are described in the section on surface roughness of the terrain; polarization is discussed in the section "Image Characteristics"; and feature orientation is described in the section "Look Direction and Terrain Features."

Dielectric Properties and Water Content

One electrical property of matter that influences its interaction with electromagnetic energy is called the *dielectric constant*. At radar wavelengths the dielectric constant of dry rocks and soils ranges from about 3 to 8, while water has a value of 80. Therefore as the mois-

ture content of a material increases, the dielectric constant can reach values of 20 or more, as shown for sand and clay in Figure 6.23.

Backscatter coefficient is a quantitative measure of the intensity of energy returned to the antenna. Figure 6.24 plots the relationship between backscatter coefficient and moisture content for fields of corn, bare soil, and milo, a grain resembling millet. As moisture increases, the backscatter coefficient also increases, which in turn indicates an increasing brightness in image tone.

These experimental relationships between moisture content and image tone are confirmed by a Seasat image in Iowa (Figure 6.25) where a rainstorm moved northeastward across an agricultural region 12 hours before the image was acquired. The map of rainfall measurements shows that the bright signature of the western portion of the image correlates with areas that received

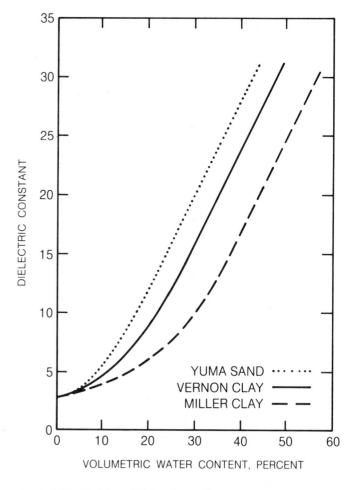

FIGURE 6.23 Variation of dielectric constant (at 27-cm wavelength) as a function of moisture content in sand and clay. From Wang and Schmugge (1980, Figure 3).

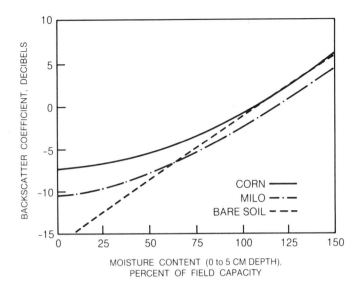

FIGURE 6.24 Variation of backscatter (at 6.7-cm wavelength) as a function of moisture content. Milo is a grain resembling millet. From Ulaby, Aslam, and Dobson (1982, Figures 4 and 7).

A. SEASAT IMAGE.

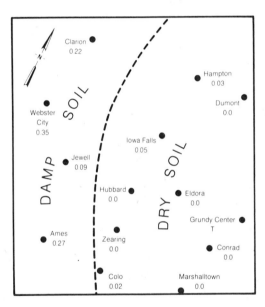

B. MAP OF RAIN–GAUGE DATA.

FIGURE 6.25 Seasat image and map of rain-gauge data for Ames, Iowa, and vicinity acquired August 16, 1978, 12 hours after a rainfall. From Ulaby, Brisco, and Dobson (1983, Figure 2). Image courtesy J. P. Ford, Jet Propulsion Laboratory.

up to 0.35 in. (0.89 cm) of rain, whereas the dark eastern area was dry. The bright streaks in the otherwise dark eastern portion of the image are attributed to small rain storms that moved northeastward in advance of the main weather front (Ford and others, 1980, p. 118). Similar signatures of damp and dry agricultural areas occur in Seasat images acquired at other dates and localities (Ulaby, Brisco, and Dobson, 1983).

At a test area in Oklahoma, soil moisture was measured at a number of sites at the same time an image was acquired by Seasat (Blanchard and others, 1981). For each site the Seasat scattering coefficient was plotted as a function of soil moisture. The resulting graph (Figure 6.26) shows that radar backscatter, or image brightness, increases linearly with increasing soil moisture. The test sites plotted in Figure 6.26 included bare soil, alfalfa, and milo. Backscatter measurements were also made for cut and standing cornfields, which showed no correlation with soil moisture. Apparently both the cut and standing corn effectively masked the underlying soil from any interaction with the Seasat radar beam. Increasing amounts of soil moisture reduces the penetration of radar energy beneath the surface.

These image signatures and experimental data suggest that radar images may be used to estimate soil moisture, which would be valuable information for hydrology and agronomy. Radar backscatter, however, is also strongly influenced by other characteristics of the scene, such as surface roughness and type of vegetation, both of which are discussed later in this chapter. A number of re-

searchers are working at sorting out the effects of these various characteristics of a scene.

The preceding comments refer to absorbed water. Standing water (fresh or salty) is highly reflective of radar energy and has dark or bright signatures depending on whether the surface is calm or agitated.

Surface Roughness

Surface roughness is the terrain property that most strongly influences the strength of radar returns. Surface roughness is distinct from *topographic relief,* which is measured in meters and hundreds of meters. Topographic relief features include hills, mountains, valleys, and canyons that are expressed on the imagery by highlights and shadows. *Surface roughness* is measured in centimeters and is determined by textural features comparable in size to the radar wavelength, such as leaves and twigs of vegetation and sand, gravel, and cobble particles.

The average surface roughness within a ground resolution cell determines the intensity of the return for that cell. Ground resolution cells are 10 by 10 m for typical airborne systems and 25 by 25 m for Seasat. Surface roughness is a composite of the vertical and horizontal dimensions and spacing of the small-scale features, together with the geometry of the individual features (leaves, twigs, sand, and gravel particles). Because of the complex geometry of most natural surfaces, it is difficult to characterize them mathematically, particularly for the large area of a resolution cell. For most surfaces the *vertical relief,* or average height of surface irregularities is an adequate approximation of surface relief.

Surfaces may be grouped into the following three roughness categories:

1. A *smooth surface* reflects all the incident radar energy with the angle of reflection equal and opposite to the angle of incidence (Snell's law).

2. A *rough surface* diffusely scatters the incident energy at all angles. The rays of scattered energy may be thought of as enclosed within a hemisphere, the center of which is located at the point where the incident wave encounters the surface.

3. A *surface of intermediate roughness* reflects a portion of the incident energy and diffusely scatters a portion.

Roughness of a surface return is determined by the relationship of surface relief, at the scale of centimeters, to radar wavelength and to the depression angle of the antenna.

Roughness Criteria The *Rayleigh criterion* considers a surface to be smooth if

$$h < \frac{\lambda}{8 \sin \gamma} \qquad (6.3)$$

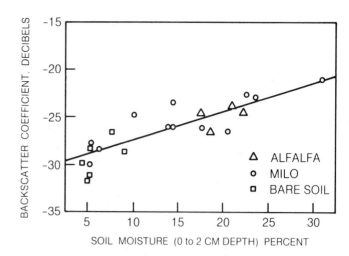

FIGURE 6.26 Variation of Seasat backscatter coefficient as a function of soil moisture for individual fields in a Guymon, Oklahoma, test area. From Blanchard and others (1981).

where h is the vertical relief, λ is the radar wavelength, and γ is the depression angle. Both h and λ are given in the same units, usually centimeters. For Seasat with a wavelength of 23.5 cm and a depression angle of 70°, the surface relief below which the surface will appear smooth is determined by substituting into Equation 6.3

$$h < \frac{23.5 \text{ cm}}{8 \sin 70°}$$

$$< \frac{23.5 \text{ cm}}{8 \times 0.94}$$

$$< 3.1 \text{ cm}$$

Therefore a vertical relief of 3.1 cm is the theoretical boundary between smooth and rough surfaces for the given wavelength and depression angle. The Rayleigh criterion does not consider the important category of surface relief that is intermediate between definitely smooth and definitely rough surfaces.

Peake and Oliver (1971) modified the Rayleigh criterion by defining upper and lower values of h for surfaces of intermediate roughness. By their "smooth" criterion, a surface is smooth if

$$h < \frac{\lambda}{25 \sin \gamma} \qquad (6.4)$$

Substituting Seasat's λ of 23.5 cm and γ of 70° results in

$$h < \frac{23.5 \text{ cm}}{25 \sin 70°}$$

$$< \frac{23.5 \text{ cm}}{25 \times 0.94}$$

$$< 1.0 \text{ cm}$$

Therefore a vertical relief of 1.0 cm is the boundary between smooth surfaces and surfaces of intermediate roughness for the given wavelength and depression angle.

Peake and Oliver (1971) also derived a "rough" criterion that considers a surface to be rough if

$$h > \frac{\lambda}{4.4 \sin \gamma} \qquad (6.5)$$

Substituting for Seasat results in

$$h > \frac{23.5 \text{ cm}}{4.4 \sin 70°}$$

$$> \frac{23.5 \text{ cm}}{4.4 \times 0.94}$$

$$> 5.7 \text{ cm}$$

Therefore a vertical relief of 5.7 cm is the boundary between intermediate surfaces and rough surfaces for the given radar wavelength and depression angle. Note that the value determined earlier from the Rayleigh criterion ($h < 3.1$ cm) is intermediate between those derived for the smooth criterion and the rough criterion.

Figure 6.27 illustrates the interaction between a transmitted pulse of Seasat energy and surfaces with smooth, intermediate, and rough relief. The smooth surface (Figure 6.27A) reflects all the energy and no energy is returned to the antenna. The surface with intermediate roughness reflects part of the energy and scatters the remainder (Figure 6.27B). The waves that are backscattered will produce a return of intermediate brightness on the image. The rough surface (Figure 6.27C) diffusely scatters all the energy, causing a relatively strong backscattered component that produces a bright signature on the image.

Equations 6.4 and 6.5 were used to calculate the limiting values of vertical relief (h) that define smooth, intermediate, and rough surfaces for four radar systems listed in Table 6.5. This information is shown in a different fashion in Table 6.6, where the *roughness response* for different values of h is given for different radar wavelengths. For example, a surface with vertical relief of 0.5 cm, typical of medium-to-coarse sand, ap-

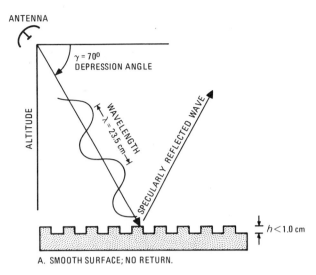

A. SMOOTH SURFACE; NO RETURN.

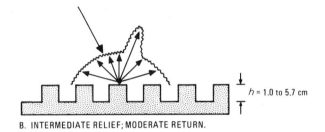

B. INTERMEDIATE RELIEF; MODERATE RETURN.

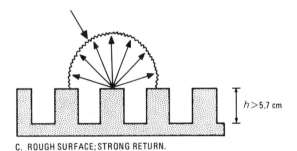

C. ROUGH SURFACE; STRONG RETURN.

FIGURE 6.27 Models of surface roughness criteria and return intensity for Seasat images (23.5-cm wavelength).

pears rough on Ka-band images, intermediate on X-band, and smooth on L-band images. The image signature of this surface will be bright, medium gray, and dark, at these respective wavelengths. Table 6.6 also illustrates the advantage of acquiring radar images at more than one wavelength for terrain analysis. By comparing the image signatures at two or more wavelengths, one can estimate the surface relief more accurately than by looking at the image signature of a single wavelength. Images acquired at different wavelengths are compared later in this section.

TABLE 6.5 Surface roughness categories for typical radar systems

Roughness category	Aircraft Ka-band, cm ($\lambda = 0.86$ cm, $\gamma = 40°$)	Aircraft X-band, cm ($\lambda = 3$ cm, $\gamma = 40°$)	SIR–A L-band, cm ($\lambda = 23.5$ cm, $\gamma = 40°$)	Seasat L-band, cm ($\lambda = 23.5$ cm, $\gamma = 70°$)
Smooth	$h < 0.05$	$h < 0.19$	$h < 1.46$	$h < 1.00$
Intermediate	$h = 0.05$ to 0.30	$h = 0.19$ to 1.06	$h = 1.46$ to 8.35	$h = 1.00$ to 5.68
Rough	$h > 0.30$	$h > 1.06$	$h > 8.35$	$h > 5.68$

TABLE 6.6 Response of different values of vertical relief (h) at different radar wavelengths. Radar depression angle (γ) is 40°

h, cm	Aircraft Ka-band ($\lambda = 0.86$ cm, $\gamma = 40°$)	Aircraft X-band ($\lambda = 3$ cm, $\gamma = 40°$)	SIR–A L-band ($\lambda = 23.5$ cm, $\gamma = 40°$)	Seasat L-band ($\lambda = 23.5$ cm, $\gamma = 70°$)
0.05	Smooth	Smooth	Smooth	Smooth
0.10	Intermediate	Smooth	Smooth	Smooth
0.5	Rough	Intermediate	Smooth	Smooth
1.5	Rough	Rough	Intermediate	Intermediate
10	Rough	Rough	Rough	Rough

Depression Angle and Surface Roughness As discussed earlier, the depression angle (γ) affects the smooth and rough criteria. Figure 6.28A shows that at low to intermediate depression angles the specular reflection from a smooth surface returns little or no energy to the antenna. At very high depression angles (80° to 90°), however, the specularly reflected wave may be received by the antenna and produce a strong return. A rough surface produces diffuse scattering of relatively uniform strong intensity for a wide range of depression angles (Figure 6.28B), which results in strong returns at all depression angles. Figure 6.28C compares the relative return intensity for smooth and rough surfaces at different depression angles. The relatively uniform return from the rough surface decreases somewhat at low depression angles because of the greater two-way travel distance. Smooth surfaces produce strong returns at depression angles near vertical, but little or no return at lower angles. Images acquired at different depression angles by SIR–B were described earlier.

Radar Signatures and Surface Roughness at Cottonball Basin

The relationship between radar backscatter and surface roughness was initially based on theoretical anal-

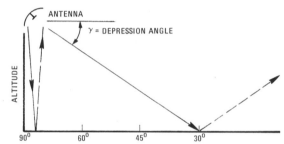

A. SMOOTH SURFACE WITH SPECULAR REFLECTION.

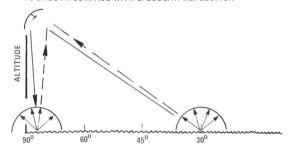

B. ROUGH SURFACE WITH DIFFUSE SCATTERING.

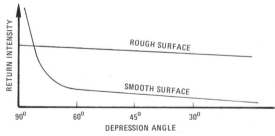

C. RETURN INTENSITY AS A FUNCTION OF DEPRESSION ANGLE.

FIGURE 6.28 Radar return from smooth and rough surfaces as a function of depression angle.

yses supplemented by laboratory studies of artificial surfaces. Subsequently, however, field studies in Death Valley by the U.S. Geological Survey and by Jet Propulsion Laboratory established the relationship between radar roughness criteria and natural surfaces with different roughness. Key papers are by Schaber, Berlin, and Brown (1976) and Schaber, Berlin, and Pitrone (1975). Cottonball Basin, at the north end of Death Valley, is one of a number of closed evaporation basins in the Basin and Range physiographic province. The floor of the basin is a flat saltpan with a wide range of surface roughness. The gravel surfaces of the alluvial fans bordering the basin provide additional degrees and types of surface roughness. The diversity of surface materials and scarcity of vegetation make this an excellent site for correlating radar signatures with materials of different surface roughness.

Figure 6.29 shows a Seasat image and a Landsat TM image of Cottonball Basin, and Figure 6.30 provides a geologic map of the area. Table 6.7 correlates the Seasat image signatures with roughness categories and with values of vertical relief calculated earlier from the smooth and rough criteria. The steps in this analysis are as follows:

1. Correlate the image signature with actual surface materials.

2. Determine vertical relief for these materials in the field.

3. Compare the measured and calculated values.

The materials listed in Table 6.7 and mapped in Figure 6.30 are illustrated in Figure 6.31 and are described below.

Coarse Gravel (h = 12.0 cm) Cobbles and boulders eroded from the mountains surrounding Death Valley form alluvial fans on the east and west margins of Cottonball Basin. Tucki Wash fan in the southwest part of the map (Figure 6.30) is a good example. The gravel is deposited by intermittent streams and slope wash during the infrequent rainstorms in the mountains. The gravel consists of a wide range of rock types (Figure 6.31A) that vary according to the lithology of the bedrock. In the Seasat image (Figure 6.29A) the gravel has a bright signature that contrasts with the dark signatures of desert pavement and sand that also occur in the fan.

Rough Halite (h = 29.0 cm) This unit is the roughest of all materials in the Cottonball Basin. At the end of the glacial period a salt lake that filled Death Valley evaporated, depositing a layer of halite (rock salt). The salt crystals have dissolved and recrystallized, causing stresses that break the salt into jumbled slabs a meter or more in diameter. The slabs are covered with sharp

pinnacles formed by solution during the infrequent rains (Figure 6.31B). The areas of rough halite that are slightly elevated above the floor of the basin are not subject to periodic flooding. This material has the brightest signature in the Seasat image (Figure 6.29A). In the Landsat image (Figure 6.29B) the signature of rough halite ranges from bright to dark, depending on the amount of silt contained in the salt.

Intermediate Halite (h = 6.0 cm) Halite deposits in the lower elevations of the basin are periodically flooded, a process that reduces surface relief to a hummocky surface of intermediate roughness (Figure 6.31C). This surface has an intermediate gray signature in the Seasat image and a similar gray signature in the Landsat image.

Carbonate and Sulfate Deposits (h = 6.0 cm) Carbonate and sulfate deposits form a belt around the margin of the basin, where they are mixed with sand (Figure 6.31D). Periodic wetting and drying produces a puffy surface with intermediate relief that produces a gray signature in the Seasat image. In the TM image the carbonate and sulfate deposits have a light gray tone.

Desert Pavement (h = 1.0 cm) Desert pavement forms on alluvial fans where older gravel deposits have been subjected to prolonged weathering, which disintegrates the surface layer of cobbles and boulders into slabs and chips. These fragments form a smooth surface that resembles a tile mosaic (Figure 6.31E). Areas of desert pavement are common at Tucki Wash fan and other fans on the west side of Death Valley. Much of the desert pavement is coated with desert varnish, which causes a dark signature in the TM image that contrasts with the brighter signature of the younger deposits of coarse gravel. In the Seasat image the smooth desert pavement at Tucki Wash Fan has a distinctive dark signature.

Sand and Fine Gravel (h = 1.0 cm) A narrow belt of sand and fine gravel (not illustrated in Figure 6.31) occurs at the margin of Tucki Wash Fan and has a narrow dark signature in the Seasat image. The signature is consistent with the relatively fine grain size and low relief of this material. The belt of sand and fine gravel has a distinctive gray signature in the TM image.

Floodplain Deposits (h = 0.2 cm) The ephemeral streams in Cottonball Basin have formed floodplains of silt that are saturated with brine and are extensively coated with thin salt crusts (Figure 6.31F). These floodplains are the smoothest surfaces in the basin and have distinctive dark signatures in the Seasat image. In the TM image the tone ranges from bright to medium gray

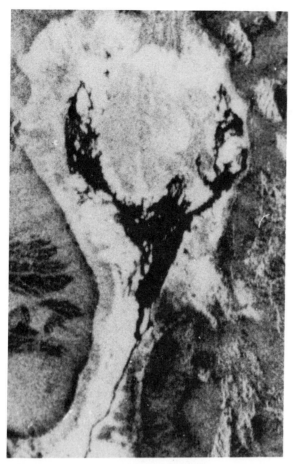 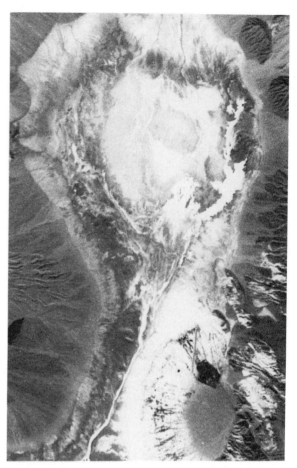

A. SEASAT IMAGE WITH LOOK DIRECTION TOWARD THE EAST.

B. LANDSAT TM BAND–3 (RED) IMAGE.

FIGURE 6.29 Seasat and Landsat images of Cottonball Basin, Death Valley, California. From Sabins (1984, Figure 5).

depending on the thickness of the white salt crust and the degree of water saturation.

Bedrock Bedrock, consisting of deeply eroded sedimentary rocks, is exposed in the eastern portion of the Seasat image. Relief is measured in tens of meters, and radar signatures are determined by the relationship between topography and the eastern look direction of the radar. West-facing slopes are bright and east-facing slopes are shadowed and dark.

Conclusions The Cottonball Basin example demonstrates that calculated roughness criteria correlate with surface relief of materials in the field and with signatures on a radar image. This close correlation between radar signature and surface relief points out another important fact—namely that radar signatures alone cannot be used to identify the composition of materials. For example,

coarse gravel and rough halite have completely different compositions, but both have similar radar signatures. Floodplain deposits, sand, desert pavement, and standing water are very different materials, but all have dark signatures because of their smooth surfaces. In the image these materials can be distinguished because of their patterns but not on the basis of their radar signature.

The correlations between roughness and radar signature demonstrated at Cottonball Basin have been confirmed on Seasat images covering all of Death Valley and other regions.

X-Band and L-Band Images Compared

The Copper Canyon alluvial fan on the east side of Death Valley is an instructive site for comparing the signatures of images acquired at different wavelengths. The materials at Copper Canyon fan (Figure 6.32C) are

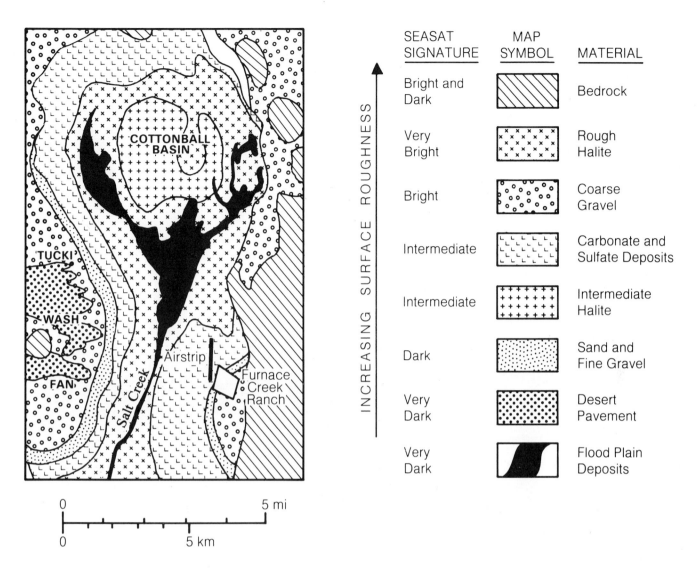

FIGURE 6.30 Map showing distribution of surface materials at Cottonball Basin. Based on a geologic map by Hunt and Mabey (1966, Plate 1).

TABLE 6.7 Characteristics of a Seasat image of Cottonball Basin, Death Valley, California

Image signature	Roughness category	Calculated vertical relief, cm	Materials
Dark	Smooth	$h < 1.0$	Desert pavement, sand and fine gravel, flood-plain deposits
Intermediate	Intermediate	$h = 5.7$ to 1.0	Intermediate halite, carbonate and sulfate deposits
Bright	Rough	$h > 5.7$	Coarse gravel, rough halite

A. COARSE GRAVEL, $h = 12.0$ cm.

B. ROUGH HALITE, $h = 29.0$ cm.

C. INTERMEDIATE HALITE, $h = 6.0$ cm.

D. CARBONATE AND SULFATE DEPOSITS,
$h = 6.0$ cm.

E. DESERT PAVEMENT, $h = 1.0$ cm.

F. FLOODPLAIN DEPOSITS, $h = 0.2$ cm.

FIGURE 6.31 Oblique ground photographs of surface materials at Cottonball Basin, with values for typical vertical relief. Areas are approximately 50 cm wide.

similar to those at Cottonball Basin. The images in Figure 6.32 were acquired by an aircraft X-band system (3-cm wavelength) and the Seasat L-band system (23.5-cm wavelength). The comparison is subjective because the two images were acquired and processed separately and were not calibrated to a known standard.

For the aircraft image the look direction is toward the west, which results in pronounced shadows from the high ridge of bedrock east of the fan. Details of this radar mission are not available; spatial resolution of the image is estimated at 5 m, and the depression angle is estimated at 25°. The smooth criterion is approximately 0.3 cm and the rough criterion 1.6 cm. For the Seasat image the look direction is toward the east, which results in very bright signatures for the west-facing bedrock scarp and strong layover toward the west. Spatial resolution is 25 m and depression angle is 70°. The smooth criterion is 1.0 cm, and the rough criterion is 5.7 cm.

The aircraft image has much more spatial detail (5-m resolution) than does the Seasat image (25-m resolution), but distribution of materials is similar in both images as seen by comparison with the map (Figure 6.32C); the map explanation (Figure 6.32D) lists the signatures in both images. Rough halite exceeds the rough criterion of both images and has bright signatures. Floodplain deposits have less relief than either of the smooth criteria and are dark in both images. Coarse gravel of Copper Canyon fan is finer grained than that at the Tucki Wash Fan; its relief of 5 to 6 cm is rough for the aircraft image (bright signature) but intermediate for the Seasat image (gray signature). The belt of sand and fine gravel near the margin of the fan has a vertical relief of 1 cm, which is intermediate for the aircraft image (gray) and smooth for Seasat (dark). A belt of carbonate and sulfate deposits, with a vertical relief of 6 cm, occurs between the sand and the floodplain deposits. The carbonate and sulfate deposits are bright on the aircraft image and intermediate on the Seasat image. This example demonstrates the relationship of radar wavelength to roughness of materials and to their resulting signatures on images. Images acquired at different wavelengths provide better definitions of surface relief. Consider the gravel at Copper Canyon Fan, which has an indicated relief greater than 1.6 cm in the aircraft (X-band) image and 1 to 5.7 cm in the Seasat (L-band) image. The combination of images provides an estimated roughness range of 1.6 to 5.7 cm for this gravel.

Durmid Hills, California

The lessons learned at Death Valley about the relationship between surface roughness and radar signatures were valuable for interpreting a Seasat image of the Durmid Hills (Figure 6.33A). The Durmid Hills are a broad, gentle arch of arid terrain along the eastern shore of the Salton Sea, which is a large salty lake in southern California. The grainy appearance of the Seasat image is due to its enlargement from the small-scale original image. An enlarged Skylab photograph (Figure 6.33B) is included for comparison, together with a Seasat interpretation map (Figure 6.33C). Choppy water of the Salton Sea has a bright signature in the Seasat image and a dark signature in the infrared photograph because infrared energy at these short wavelengths is absorbed by water. Inspection of the Seasat image reveals a lineament extending the length of the hills. The lineament is formed by the linear contact between the bright tones on the southwest and the dark tones on the northeast and is an example of a tonal lineament. According to Table 6.5 the dark tones on the Seasat image indicate a smooth surface with relief less than 1.0 cm, and the bright tones indicate a rough surface with relief greater than 5.7 cm.

The next step is to visit the Durmid Hills in order to (1) compare the actual terrain roughness with predicted values of vertical relief and (2) determine the significance of the Seasat lineament. Figure 6.34A illustrates terrain with bright Seasat tones that is underlain by the Borrego Formation (Pleistocene), consisting of poorly consolidated siltstone with beds and concretions of hard, resistant sandstone. The infrequent rains erode the soft siltstone to form a surface littered with gravel and boulders with a relief of 10 to 20 cm. Figure 6.34B illustrates terrain with dark Seasat tones that is covered with sand and silt deposited in Lake Cahuila, which is now extinct. Relief on this fine-grained material is less than the 1.0-cm limit for smooth surfaces. The linear contact between the Cahuila deposits and the Borrego outcrops corresponds to the tonal lineament of the Seasat image and to the trace of the San Andreas fault. The fault is more clearly shown in the Seasat image than in the Skylab photograph or in Landsat images (not illustrated). The Durmid Hills example illustrates the value of radar signatures for geologic mapping based on variations in surface roughness.

IMAGE CHARACTERISTICS

In addition to the properties described above, characteristics of radar images are influenced by polarization, look direction, and image irregularities. These factors are discussed below.

Polarization

The electrical field vector of the transmitted energy pulse may be polarized (or vibrating) in either the vertical or horizontal plane. On striking the terrain, most of the energy returning to the antenna usually has the

A. X–BAND AIRCRAFT IMAGE.

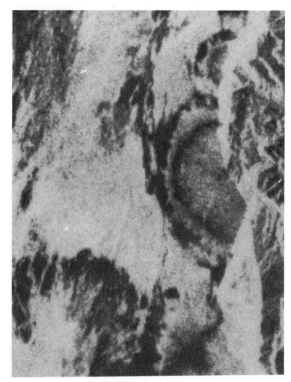

B. L–BAND SEASAT IMAGE.

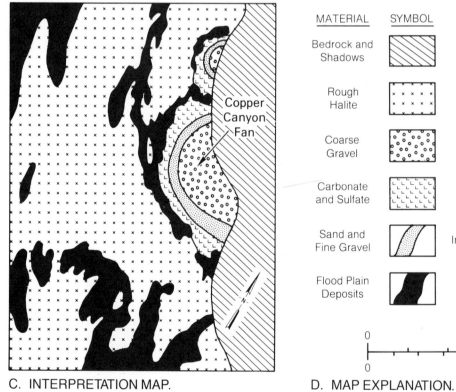

C. INTERPRETATION MAP.

Copper Canyon Fan

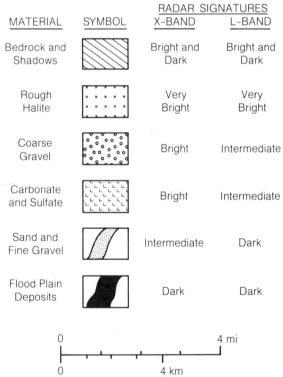

D. MAP EXPLANATION.

| MATERIAL | SYMBOL | RADAR SIGNATURES | |
		X-BAND	L-BAND
Bedrock and Shadows		Bright and Dark	Bright and Dark
Rough Halite		Very Bright	Very Bright
Coarse Gravel		Bright	Intermediate
Carbonate and Sulfate		Bright	Intermediate
Sand and Fine Gravel		Intermediate	Dark
Flood Plain Deposits		Dark	Dark

0   4 mi

0   4 km

FIGURE 6.32   X-band and L-band images of the Copper Canyon alluvial fan and vicinity, Death Valley. From Sabins (1984, Figures 9 and 10).

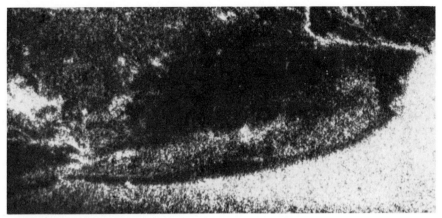

A. SEASAT RADAR IMAGE.

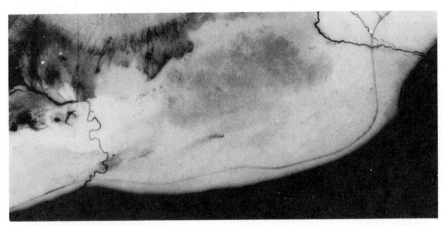

B. SKYLAB IR PHOTOGRAPH.

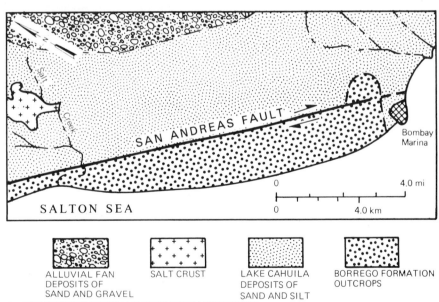

C. INTERPRETATION MAP OF THE SEASAT IMAGE.

FIGURE 6.33 Seasat and Skylab images of the Durmid Hills, southern California. From Sabins, Blom, and Elachi (1980, Figures 5 and 9).

A. OUTCROPS OF THE BORREGO FORMATION ON THE SOUTHWEST SIDE OF THE SAN ANDREAS FAULT. THIS ROUGH SURFACE HAS A BRIGHT SIGNATURE ON THE SEASAT IMAGE.

B. SILT AND SAND DEPOSITS OF LAKE CAHUILA ON THE NORTHEAST SIDE OF THE SAN ANDREAS FAULT. THIS SMOOTH SURFACE HAS A DARK SIGNATURE ON THE SEASAT IMAGE.

FIGURE 6.34 Terrain features associated with the tonal lineament on a Seasat image of the Durmid Hills. From Sabins, Blom, and Elachi (1980, Figures 6 and 7).

same polarization as the transmitted pulse. This energy is recorded as *parallel-polarized* (or like-polarized) imagery and is designated HH (*horizontal transmit, horizontal return*) or VV (*vertical transmit, vertical return*). A portion of the returning energy has been *depolarized* by the terrain surface and vibrates in various directions. The mechanisms responsible for depolarization are not definitely known, but the most widely accepted theory

attributes it to multiple reflections at the surface. This theory is supported by the fact that depolarization effects are much stronger from vegetation than from bare ground. Leaves, twigs, and branches are believed to cause the multiple reflections responsible for depolarization. Some experimental radar systems have a second antenna element that receives the depolarized energy vibrating at right angles to the plane of the transmitted pulse. The resulting imagery is termed *cross-polarized* and may be either HV (*horizontal transmit, vertical return*) or VH (*vertical transmit, horizontal return*). Most mapping radar systems operate in the HH mode because this mode produces the strongest return signals.

Jet Propulsion Laboratory operated an airborne L-band (25-cm wavelength), synthetic-aperture system that simultaneously acquired HH, HV, VV, and VH images on a digital recording system. The aircraft and radar were destroyed in an accident in July 1985. Figure 6.35, of Furnace Creek Ranch and vicinity in Death Valley, is a typical set of these images. Digital processing of the cross-polarized images increased their brightness to levels comparable to those of the parallel-polarized images (Figure 6.35). Figure 6.36 is an aerial photograph and geologic map of the Furnace Creek Ranch area.

Despite the differences in polarization, all four radar images are remarkably similar. On all the images, smooth surfaces (airport runway, floodplain, and the sand and fine gravel at the margins of the alluvial fan) have dark signatures. Rough surfaces (rough halite) have bright signatures. Surfaces with intermediate relief (carbonate and sulfate deposits) have intermediate signatures. Bedrock has similar highlights and shadows on all four images. For these materials the backscattering processes are apparently similar in both the parallel- and cross-polarized modes.

The major polarization differences in the radar signatures are caused by vegetation, which consists of small mesquite trees and shrubs of creosote bush. Distribution of the sparse plant community is shown in the aerial photograph and map (Figure 6.36A,B). Native vegetation is concentrated in the dry washes that radiate across the carbonate and sulfate deposits south and west of Furnace Creek Ranch (Figure 6.36B). In the photograph (Figure 6.36A) native vegetation forms dark stripes that contrast with the bright tone of the carbonate and sulfate deposits. Furnace Creek Ranch is enclosed by a windbreak of tamarisk trees and inside includes a date-palm grove, a golf course, and scattered tamarisk trees.

In the parallel-polarized images (Figure 6.35A,C), vegetation is indistinguishable from the background in the VV mode and is only faintly brighter than the background in the HH mode. In both cross-polarized images (HV and VH), however, the cultivated vegetation at Furnace Creek Ranch and the native vegetation are distinctly brighter than the background. The twigs and branches cause multiple reflections of incident radar energy (sometimes referred to as *volume scattering*), which depolarizes a significant proportion of the energy. For a horizontally transmitted pulse of energy, the major return from the scene is HH-polarized, but returns from vegetation will also have a significant HV-polarized signal. Unvegetated terrain, however, does not cause multiple reflections and does not depolarize the incident energy; thus the HV return from bare surfaces is weak relative to the HV return from vegetation. The same relationships apply to the VV and VH images (Figure 6.35C,D) and account for the bright signature of vegetation in the VH image. This correlation between vegetation and relatively bright signatures on cross-polarized images has been observed in numerous images. Donovan, Evans, and Held (1985) published multiple-polarized images of several test sites.

Look Direction and Terrain Features

Many natural and cultural features of the terrain have a strong *preferred orientation,* which is commonly expressed as parallel linear features, such as fractures or roads. The geometric relationship between the preferred orientation and the look direction may influence the radar signature.

Geologic Features, Venezuela Lineaments, faults, and outcrops of layered rocks are linear features that may be enhanced or subdued in radar images depending on their orientation relative to the look direction. Features trending normal or at an acute angle to the look direction are enhanced by highlights and shadows. Features trending parallel with the look direction produce no highlights or shadows and are suppressed in an image. These relationships are illustrated by an area in Venezuela (Figure 6.37) where aircraft X-band images (3-cm wavelength) were acquired with look directions toward the south and west. As shown in the geologic map (Figure 6.37C), the area includes a high mesa capped by relatively horizontal beds of resistant quartzite. The mesa is surrounded by lowlands of folded metamorphosed sedimentary rock that clearly express the bedding trends.

In the large fold in the eastern part of the area, one limb trends northward and the second limb trends northwest. The north-trending limb is enhanced by the west look direction (Figure 6.37B). The northwest-striking limb trends at an acute angle to either look direction and is equally enhanced in both images. In the western part of the area, a major lineament (probably a fault) strikes south through the metamorphic terrain and cuts the west portion of the mesa where it forms a linear valley. The lineament is strongly enhanced by the west look direction and subdued by the south look direction (Figure 6.37A). Other lineaments in the mesa that trend a few degrees east or west of north are enhanced and subdued in the same fashion on the two images. In the southeast part of the area, a number of parallel lineaments trend slightly north of east in the metamorphic terrain; these are clearly visible with the south look direction but are not recognizable with the west look direction. In Figure 6.37B, the west-trending, parallel light and dark bands are minor defects in the image. If this area were to be imaged with only a single look direction, an orientation toward the southwest or northeast would enhance most of the linear features.

Cultural Features, New Orleans, Louisiana In urban areas a strong preferred orientation is formed by street patterns and by buildings with their walls parallel and normal to the streets. Radar energy with a look direction that is normal or parallel to a street direction will thus be oriented perpendicular to many walls, resulting in a strong return and a bright signature. A look direction that is oblique to a street direction will be oriented obliquely to walls, resulting in a weak return and an intermediate signature.

The New Orleans, Louisiana, region is well suited to evaluate the relationship between the radar look direction and the orientation of cultural features. Figure 6.38 illustrates Seasat images acquired with look directions toward the northwest and northeast, together with an aerial photograph. The map of Figure 6.39 shows the major categories of land cover and land use. Calm water in the lakes and in the Mississippi River is dark in all three images; bright spots on the river are ships and barges. Forested areas are bright in both radar images because the trees have no preferred orientation and produce strong volume scattering. Much of the area is covered with marsh vegetation, which has a medium gray signature on both Seasat images.

As shown in the aerial photograph (Figure 6.38C), there is a wide diversity of street patterns and orientations in the area. The more recently developed areas around the shore of Lake Ponchartrain and the towns of Arabi and Chalmette (Figure 6.39) have rectangular street grids. The older part of New Orleans, which is located within a bend of the Mississippi River, has a radial street pattern.

In most of the urban areas the orientation of streets and building walls is oblique to both the northwest and northeast Seasat look directions, which results in dark to intermediate signatures in both images (Figure 6.38A,B). In the map (Figure 6.39) these areas are shown with a blank pattern ("No Preferred Orientation" category). The map also shows the orientation of the two

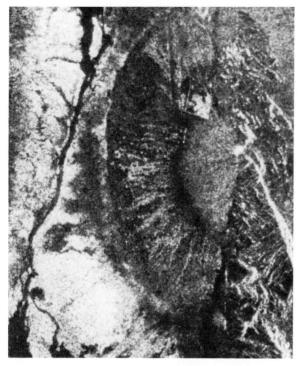

A. HH (PARALLEL–POLARIZED) IMAGE.

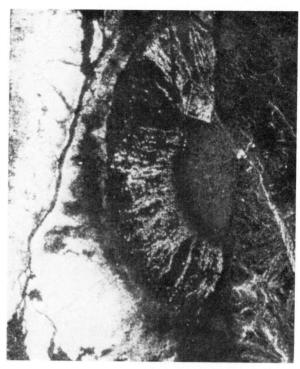

B. HV (CROSS–POLARIZED) IMAGE.

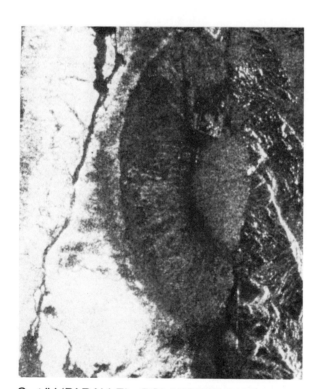

C. VV (PARALLEL–POLARIZED) IMAGE.

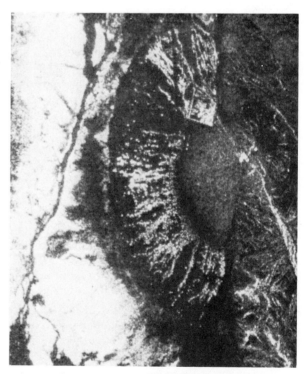

D. VH (CROSS–POLARIZED) IMAGE.

FIGURE 6.35 Parallel-polarized and cross-polarized, L-band (25-cm wavelength), aircraft images of Furnace Creek Ranch, Death Valley, California. Courtesy Jet Propulsion Laboratory.

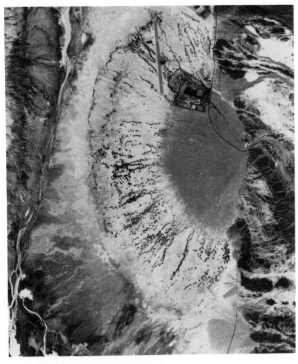

A. AERIAL PHOTOGRAPH.

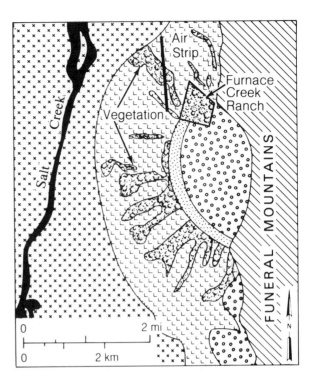

B. GEOLOGIC MAP. SYMBOLS ARE
EXPLAINED IN FIGURE 6.30.

FIGURE 6.36   Aerial photograph and map of Furnace Creek Ranch and vicinity.

Seasat look directions, which are not orthogonal to each other. Of special interest are urban areas that are bright on one image, but dark or intermediate on the other. On the map these areas are labeled NW or NE to indicate the look direction of the image in which the signature is bright. The radial street pattern in the bend of the river produces these different returns, which are known as *cardinal point effects*. In Figure 6.38A (northwest look direction) there are two wedges with bright signatures in this radial street pattern. The bright wedge on the western side is caused by the radial streets oriented normal to the northwest look direction. Parallel streets oriented normal to the northwest look direction cause the bright wedge in the eastern part. In Figure 6.38B (northeast look direction) there is a single bright wedge from the radial streets in the central part of the bend. The street directions are used for convenient reference; actually the buildings are responsible for the bright signatures in both images. In the Arabi-Chalmette area, streets are oriented parallel and perpendicular to the look direction, and this produces the bright signature in Figure 6.38A and the dark signature in Figure 6.38B. Similar explanations apply to the other NW and NE

areas indicated on the map (Figure 6.39). This relationship between orientation of cultural features and radar look direction has also been observed in the Los Angeles area (Bryan, 1979) and elsewhere.

Significance of the Look Direction   The X-band (aircraft) images of Venezuela and the L-band (Seasat) images of New Orleans demonstrate that, irrespective of radar wavelength, there is an important relationship between radar look direction, image signatures, and orientation of linear features in the terrain. Linear geologic features such as faults, outcrops, and lineaments that are oriented at a normal or oblique angle to the radar look direction are enhanced by highlights and shadows in the image. Features oriented parallel with look direction are suppressed and are difficult to recognize in the image. The importance of radar look direction is comparable to the importance of sun azimuth in aerial photographs and in Landsat images acquired at low sun angles. When planning an aircraft radar survey, one should orient the look direction at a normal or acute angle relative to the known geologic trends of the survey area.

LOOK
DIRECTION

A. LOOK DIRECTION TOWARD SOUTH.

LOOK
DIRECTION

B. LOOK DIRECTION TOWARD WEST.

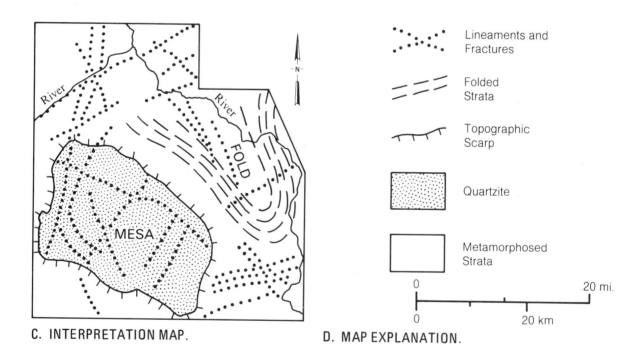

C. INTERPRETATION MAP.

D. MAP EXPLANATION.

- ⋯ Lineaments and Fractures
- – – – Folded Strata
- Topographic Scarp
- Quartzite
- Metamorphosed Strata

0   20 mi.

0   20 km

FIGURE 6.37 Expression of linear geologic features on images with different look directions. Synthetic-aperture, X-band, aircraft images in southeast Venezuela. Courtesy Aero Service Division, Western Geophysical Corporation of America.

In urban areas the orientation of streets and buildings relative to the look direction also influences the signatures. In addition, linear metal features, such as fences and power lines, have bright signatures where they are oriented normal to look directions but are not recognizable where they are oriented parallel with the look direction. Interpreters of radar images of geologic or urban scenes must be aware of these geometric relationships.

SIR–A IMAGES OF INDONESIA

The SIR–A mission acquired images of portions of Indonesia, including parts of Irian Jaya (New Guinea), Kalimantan (Borneo), Java, and some of the smaller islands (see Figure 6.17). Persistent cloud cover in this tropical region has hampered acquisition of aerial photographs and Landsat images, but the SIR–A images are of excellent quality. The images were enlarged to 1:250,000 scale and interpreted in four steps:

1. Produce a base map by tracing shorelines, drainage patterns, and the sparse cultural features (roads, cities, and airports) from the images. The contrasting signatures of water and vegetation made this a straightforward task.

2. Define terrain categories that are recognizable throughout the region and map their distribution.

3. Map geologic structures such as faults, folds, and lineaments.

4. Evaluate the completed map by comparing it with existing maps.

The remainder of this section describes steps 2, 3, and 4 and summarizes a report by Sabins (1983a).

Terrain Categories

Much of Indonesia is densely forested, but the forest conforms to the terrain because the top of the tree canopy is parallel with the underlying topography. The topography is controlled by geologic structure and by the erosional characteristics of the underlying rocks. The six terrain categories shown in Figure 6.40 were recognized on the basis of their expression on the SIR–A images. These categories are not restricted to images of Indonesia but are recognizable on radar images of forested regions throughout the world.

Carbonate Terrain (Figure 6.40A) In humid environments, solution and collapse of carbonate rocks produces karst topography, which is readily recognized by the distinctive pitted surface. In Indonesia, carbonate terrain generally occurs as uplands surrounded by lowlands eroded from less resistant rocks. Because of the relatively coarse spatial resolution of SIR–A images, small patches of karst terrain may not be recognizable. Also, faulting and stream erosion may obscure the expression of karst topography.

Clastic Terrain (Figure 6.40B) Terrain formed on clastic sedimentary rocks, primarily sandstone and shale, is recognized by the stratification that forms asymmetric ridges, called *cuestas* and *hogbacks,* where the rocks are dipping, as seen in this example. Flat-lying clastic rocks form mesas, terraces, and associated erosional scarps. The lack of karst topography generally distinguishes clastic terrain from carbonate terrain in humid regions.

Volcanic Terrain (Figure 6.40C) Young volcanic rocks form irregular flows associated with cinder cones or eroded volcanic necks. Because of erosion and deformation, older volcanic terrains lack these distinctive features and so cannot be recognized on the images without additional information.

Alluvial and Coastal Terrain (Figure 6.40D) This category is characterized by low relief, a uniform bright signature of heavily vegetated floodplains, and numerous estuaries and meandering streams. Despite the generally featureless nature of this terrain, careful interpretation of the images often reveals subtle lineaments and drainage anomalies that may be expressions of geologic structure.

Melange Terrain (Figure 6.40E) *Melange* refers to rocks formed in subduction zones as a mixture of clastic sediments and oceanic crustal and mantle rocks. Lenticular rock fragments of a wide range of sizes, up to kilometers in length, are enclosed in a matrix of clay. Erosion of these rocks produces an irregular, rounded terrain with unsystematic drainage patterns. At the scale of the radar images, neither stratification nor individual rock fragments are detectable.

Metamorphic Terrain (Figure 6.40F) Sedimentary rocks that have been metamorphosed to slate, quartzite, and schist occur in portions of Irian Jaya. The original stratification is no longer recognizable. The strongly dissected metamorphic terrain has high relief and angular ridges that distinguish it from the lower relief and rounded appearance of melange terrain. Foliation trends are not discernible in Indonesian metamorphic terrain at the scale of SIR–A images.

A. SEASAT IMAGE WITH LOOK DIRECTION TOWARD NORTHWEST.

B. SEASAT IMAGE WITH LOOK DIRECTION TOWARD NORTHEAST.

C. AERIAL PHOTOGRAPH.

FIGURE **6.38** Land-use and land-cover signatures on Seasat images acquired with different look directions. New Orleans, Louisiana. Images courtesy J. P. Ford, Jet Propulsion Laboratory.

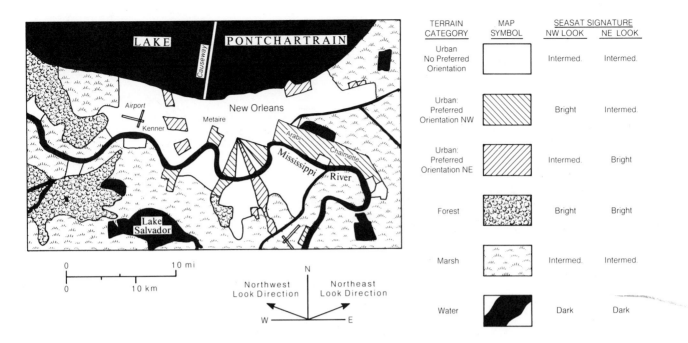

TERRAIN CATEGORY	MAP SYMBOL	SEASAT SIGNATURE	
		NW LOOK	NE LOOK
Urban No Preferred Orientation		Intermed.	Intermed.
Urban: Preferred Orientation NW		Bright	Intermed.
Urban: Preferred Orientation NE		Intermed.	Bright
Forest		Bright	Bright
Marsh		Intermed.	Intermed.
Water		Dark	Dark

FIGURE 6.39 Map of land-use and land-cover categories for the Seasat images of New Orleans, Louisiana.

Undifferentiated Bedrock (Not Illustrated) There are areas where several terrain types are juxtaposed in such a complex manner that individual categories are not recognizable at the scale and resolution of SIR–A images. The category "undifferentiated bedrock" is used for these areas.

Regional Interpretation

The second step in the interpretation process was to recognize the terrain categories and map their distribution on regional SIR–A images.

The image in Figure 6.41A is located in the northwest part of the Vogelkop region at the west end of Irian Jaya. The image was acquired with the radar look direction toward the northeast, which causes shadows to extend from the lower margin toward the upper margin of the image. This shadow orientation may cause topographic inversion for some viewers, which may be corrected by rotating the image 180° so the shadows extend toward the lower margin of the image.

The geologic interpretation map (Figure 6.41B) shows the distribution of the five terrain categories that occur in this SIR–A image. The central part of the area, called the Kemum block, consists of metamorphic rocks with typical rugged terrain. Toward the south (lower right corner of Figure 6.41A,B) the metamorphic rocks are overlain by clastic strata, which are overlain by car-

bonate rocks forming a broad expanse of karst terrain. This vertical sequence of rocks is shown at the southeast end (right side) of the cross section (Figure 6.41C). In the northwest part of the image (left side), the Tamrau Mountains consist of a variety of rocks and are mapped as undifferentiated bedrock. Belts of clastic strata occur on the north and south sides of the Tamrau Mountains.

After mapping the terrain types, the next step was to interpret geologic structure. Strikes and dips are inferred from the attitudes of hogbacks and cuestas formed by stratified rocks. Lineaments are mapped with a dot pattern and belong to two major types. The most common type is formed by straight streams and aligned segments of streams. These lineaments are especially abundant in the metamorphic terrain of the Kemum block. The second type of lineament consists of straight escarpments that are expressed in the image as linear highlights or shadows depending on their orientation relative to the radar look direction. Additional structural information is gained from the outcrop pattern of rock types, which may indicate the plunge of folds and the sense of offset along faults.

In the fourth and final step, the completed SIR–A interpretation map at a scale of 1:250,000 (Figure 6.41B) was evaluated by comparing it with recently published geologic maps of the area at the same scale. Distribution of the terrain types conforms to the patterns in the published maps, and the attitudes of the beds are in agree-

A. CARBONATE TERRAIN.

B. CLASTIC TERRAIN.

C. VOLCANIC TERRAIN.

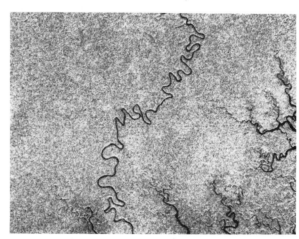

D. ALLUVIAL AND COASTAL TERRAIN.

E. MELANGE TERRAIN.

F. METAMORPHIC TERRAIN.

FIGURE 6.40 SIR–A images of typical Indonesian terrain categories. Areas are approximately 28 km wide. Look direction is from upper to lower margin of each image. From Sabins (1983a, Figure 4).

A. SIR–A IMAGE. LOOK DIRECTION TOWARD UPPER MARGIN OF IMAGE.

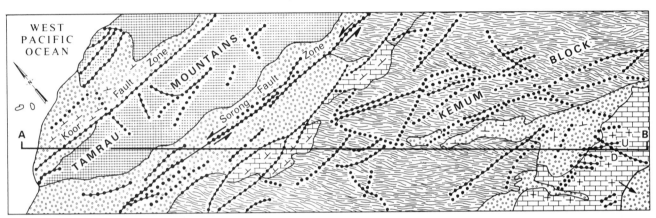

B. GEOLOGIC INTERPRETATION MAP.

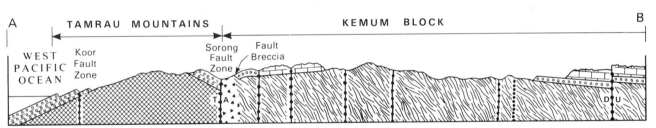

C. CROSS SECTION ALONG LINE A–B.

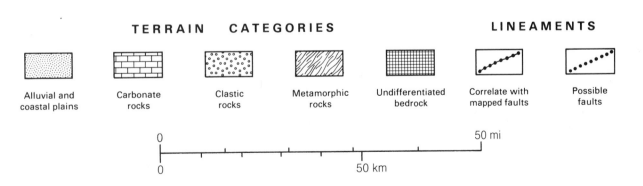

TERRAIN CATEGORIES

Alluvial and coastal plains

Carbonate rocks

Clastic rocks

Metamorphic rocks

Undifferentiated bedrock

LINEAMENTS

Correlate with mapped faults

Possible faults

FIGURE 6.41 SIR–A image and interpretation in the northwest Vogelkop region, Irian Jaya, Indonesia. From Sabins (1983a, Figures 10 and 11).

ment. The radar lineaments were compared with mapped faults; where the two correspond, a solid line was added to the dot pattern of the lineament (Figure 6.41B). Many lineaments correlate with mapped faults; others may represent faults or extensions of faults that were not recognized in the field because of dense forest cover and inaccessible terrain. For example, one prominent lineament in Figure 6.41B correlates with faults at each end; the lineament may indicate a long continuous fault that was not mapped in the field. The Tamrau Mountains are bounded on the north and south by two major west-trending lineaments that are formed by prominent linear valleys and stream segments. The northern lineament correlates with the Koor fault zone (Figure 6.41B). The lineament on the south side of the Tamrau Mountains extends for 80 km across the image and marks the trace of the Sorong fault. This fault is a major tectonic feature in the Vogelkop region and is described later in the chapter in the section "Strike-Slip Faults."

The comparison of regional SIR–A maps with published geologic maps confirms the validity of the radar interpretations. This example from the Vogelkop region is a limited subset of the SIR–A interpretation maps by Sabins (1983a), which cover portions of the mainland of Irian Jaya and of Kalimantan. Published geologic maps are virtually nonexistent in Kalimantan, so the SIR–A maps provide the most detailed and recent information. Additional Indonesian images acquired by SIR–B in 1984 will be interpreted and published.

Detailed Interpretation

The small-scale SIR–A image and interpretation map in the Vogelkop region show the regional use of these data but lack the details of larger-scale versions of the image. To illustrate representative geologic features in the region, selected portions of the Indonesian images were enlarged to a uniform scale (Figure 6.42). Figure 6.43 shows the geologic interpretation for each image.

Strike and Dip (Figure 6.42A) Strike and dip information can be interpreted in carbonate and clastic terrains and in some volcanic terrains where flow surfaces are well expressed. The radar signature of dipping layers depends on the relationship between the dip direction and the radar look direction. In the image (Figure 6.42A) and the map (Figure 6.43A), the strata are dipping generally toward the top of the image, which is toward the radar antenna. The dip slopes therefore have bright signatures and the antidip scarps, which are in the radar shadow, have dark signatures. Where beds dip away from the radar antenna, the antidip scarps are bright and

the dip slopes are dark as seen in the flanks of the folds in Figure 6.42C.

Thrust Faults (Figure 6.42B) Thrust faults are difficult to interpret from radar and other remote sensing images for the following reasons:

1. The planes of thrust faults are commonly parallel or nearly parallel with bedding planes of associated strata. Therefore, most thrust faults do not cause the discordant geometric relationships that are associated with many normal and strike-slip faults.

2. Thrust faults are commonly recognized by anomalous rock relationships such as older beds over younger beds, repetition of beds, and omission of beds. These relationships are difficult to recognize on images without the aid of field data.

Despite these difficulties, thrust faults were interpreted from the SIR–A image of the Paniai Lake region, mainland Irian Jaya, where several imbricate thrust plates occur. Figures 6.42B and 6.43B, show thrust plates that dip gently toward the lower margin of the image and are terminated undip at eroded antidip scarps. The trend of these scarps is locally discordant with the trend of the underlying rocks. At places such as the right margin of Figure 6.42B an upper thrust plate overrides and truncates an underlying plate. Elsewhere in the Paniai region, thrust faults are recognized by the repetition of belts of distinctive karst topography that alternate with belts of clastic terrain.

Folds, Moderately Eroded (Figure 6.42C) Folds may be recognized by attitudes of beds, outcrop patterns, and topographic expression. In this example, of moderately eroded folds, taken from Kalimantan, the strike-and-dip is readily interpreted using the criteria described earlier. Outcrop patterns are also diagnostic in Figure 6.42C, where the youngest beds (carbonate rocks with characteristic karst topography) are preserved as mesas in the troughs of synclines, but are eroded from the crest of the anticline. These erosion patterns commonly result in *topographic reversal*, in which topographic elevations (mesas) correspond to structural depressions (synclines). In the upper right portion of Figures 6.42C and 6.43C, note that the resistant clastic beds at the crest of the anticline have been breached to expose older strata.

Folds, Deeply Eroded (Figure 6.42D) Deep erosion may produce a nearly planar surface, where plunging folds are marked by subdued parallel ridges with arcuate patterns that are formed by outcrops of resistant strata. The noses and axes of these folds are readily mapped (Figure 6.43D). Anticlines cannot be distinguished from synclines, however, because dip attitudes cannot be interpreted from the subdued ridges.

Lineaments (Figure 6.42E) In the SIR–A images of Indonesia, lineaments are expressed as scarps, linear valleys, and aligned valleys. In the upper part of the image (Figure 6.42E) and map (Figure 6.43E), two linear scarps form the boundary between melange terrain and alluvial terrain and may be the expression of faults. In this remote area along the Kindjau River in Kalimantan, however, there are no geologic maps with which to evaluate the SIR–A interpretations. Not all topographic scarps are linear; those with an irregular trend, such as in the upper left part of Figure 6.42E, are not classed as lineaments.

Linear valleys are a lineament category formed by a single straight, or curvilinear channel, of which there are several examples in Figure 6.42E. Two or more separate valleys may be aligned end to end to form a lineament. Many linear valleys and aligned valleys follow faults and fracture zones that are preferentially eroded zones of weakness.

Strike-Slip Faults (Figure 6.42F) The image (Figure 6.42F) and map (Figure 6.43F) cover a segment of the Sorong fault enlarged from the regional SIR–A image in Figure 6.41. The prominent lineament is formed by a linear valley that has the following characteristics of a strike-slip fault: (1) aligned notches; (2) shutter ridges; (3) linear terraces; and (4) offset stream channels, which in this example indicate left-lateral fault displacement. The preservation of these tectonic features in this region of heavy rainfall and rapid erosion indicates that the Sorong fault is active; that is, movement has occurred in the past 10,000 years. Strike-slip faults are also recognizable on images by the lateral offset of geologic units on opposite sides of the fault.

Summary

Despite persistent cloud cover, SIR–A acquired excellent images of portions of Indonesia from which geomorphology, rock types, and geologic structure were interpreted. The following section explains why such interpretations were possible despite the dense vegetation cover. Where geologic maps of Irian Jaya were available for comparison, many SIR–A interpretations were confirmed and others suggested locations for additional fieldwork. In Kalimantan, geologic maps are lacking and thus the SIR–A interpretations provide valuable new information.

RADAR INTERACTION WITH VEGETATED TERRAIN

In aircraft and satellite radar images of forested terrain, one can interpret detailed information about lithology and structure of the bedrock, which may lead to the incorrect conclusion that radar penetrates the forest canopy. The SIR–A images of Indonesia are good examples of this apparent penetration. Both radar theory and practice, however, demonstrate that there is no effective penetration of most vegetation.

L-band radar systems operate at the maximum wavelength for available imaging radars and have the greatest potential for penetrating foliage. However, the rough criterion for SIR–A is 8.4 cm and for Seasat is 5.7 cm (Table 6.4). The relief of most vegetated surfaces greatly exceeds these values, and radar theory states that such surfaces will strongly scatter the incident radar energy. Growing vegetation has a high water content, which increases the dielectric constant and the radar reflectivity, as discussed earlier in this chapter. Thus radar theory predicts that because of the rough surfaces and high moisture content, vegetation will scatter and reflect incident radar energy and little or no penetration will occur.

The theory is amply supported by evidence from radar images. Vegetation generally has very bright signatures, indicating strong reflectance and scattering of incident energy. Several investigators use radar images to identify types of cultivated and natural vegetation on the basis of their backscattering characteristics (Ulaby, Batlivala, and Bane, 1980). Such identification would be impossible if radar penetrated the foliage, because the underlying soil rather than the vegetation would determine the radar signature. The lack of penetration is demonstrated by enlarged SIR–A and Seasat images of agricultural areas in the Great Plains (Figure 6.44) where centerpoint irrigation is practiced. Clusters of bright circles are fields, predominantly wheat, with small dark interstices caused by smooth, bare soil. Differences in brightness of the fields are caused by differences in the growth and harvest status of the crops. The relevant point is that if the L-band radar energy had penetrated the vegetation, each image would have over its entirety the dark signature of the underlying smooth soil and one would find it impossible to recognize crops and variations in crops. Since these images demonstrate the inability of radar to penetrate crops less than 1 m in height, it is even less likely that radar could penetrate a forest canopy, such as in Indonesia.

A. STRIKE AND DIP.

B. THRUST FAULTS.

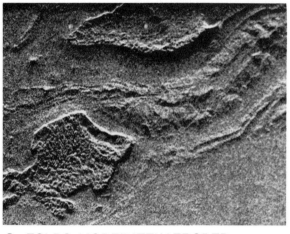

C. FOLDS, MODERATELY ERODED.

D. FOLDS, DEEPLY ERODED.

E. LINEAMENTS.

F. STRIKE–SLIP FAULT.

FIGURE 6.42 Enlarged SIR–A images of typical structural features in Indonesia. Areas are approximately 28 km wide. Look direction is from upper to lower margin of each image. From Sabins (1983a, Figure 5).

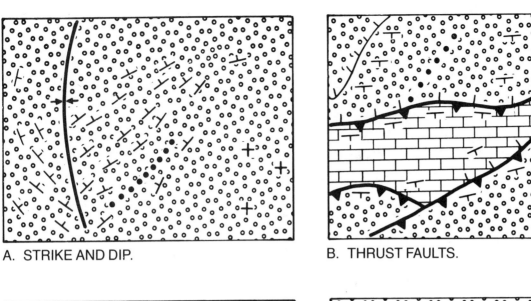

A. STRIKE AND DIP.

B. THRUST FAULTS.

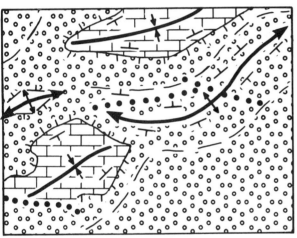

C. FOLDS, MODERATELY ERODED.

D. FOLDS, DEEPLY ERODED.

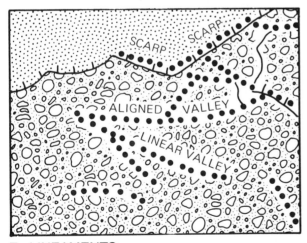

E. LINEAMENTS.

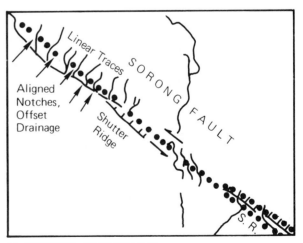

F. STRIKE–SLIP FAULT.

FIGURE 6.43 Interpretation maps of structural features in Indonesia. Symbols are explained in Figure 6.42. From Sabins (1983a, Figure 6).

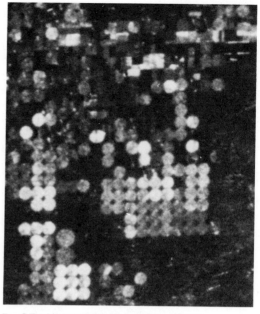

A. SIR–A IMAGE IN NORTHWEST TEXAS, NOVEMBER 1981.

B. SEASAT IMAGE IN SOUTHWEST KANSAS, SEPTEMBER 1978.

FIGURE **6.44** Radar signatures of agricultural fields on enlarged satellite images. Areas are 8 km wide.

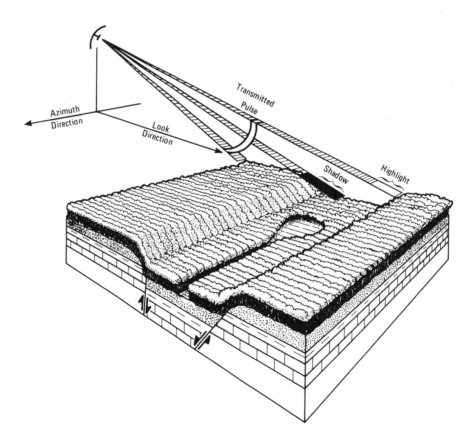

FIGURE **6.45** Block diagram of forested terrain showing radar expression of faults and associated lineaments.

Having established that radar does not penetrate vegetation, we must then understand why bedrock features are so well expressed in images of forested terrain. Figure 6.45 shows two faults and their associated topographic scarps in bedrock covered by forest. The trees are of relatively uniform height and the top surface of the canopy is controlled by the bedrock surface. The radar system produces an image of the canopy (Figure 6.46) that is enhanced by two mechanisms:

1. The inclined illumination produces highlights and shadows that emphasize lineaments, as shown in Figures 6.45 and 6.46. Shadows and highlights also enhance other structural features, such as folds and dipping strata.

2. Because of the relatively large ground resolution cells (10 to 15 m for aircraft radar; 25 m for Seasat; and 38 m for SIR–A), individual trees are not resolved, which improves the subtle topographic expression of geologic features with dimensions of hundreds and thousands of meters. In other words, the radar's large resolution cell acts like a filter to remove the high-frequency spatial detail of vegetation ''noise,'' thereby enhancing the lower frequency geologic ''signals.''

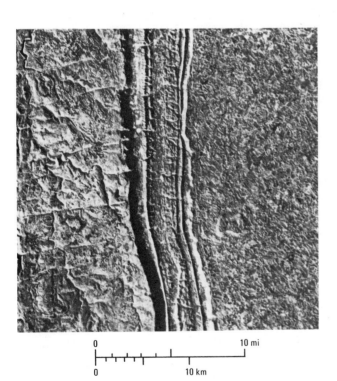

FIGURE 6.46 Radar image of forested terrain showing expression of linear scarps bounding a valley. Look direction is toward the east. This X-band, synthetic-aperture, aircraft image covers part of the Cumberland Plateau, Georgia.

In addition to these factors, radar is not affected by the cloud cover that is associated with most forested regions, especially in the tropics. For these reasons, radar is an excellent system for mapping in forested areas.

RADAR INTERACTION WITH SAND-COVERED TERRAIN

Over the years, many aircraft and Seasat images have been acquired of sand-covered desert areas, principally in North America. Sheets of sand and fine gravel have dark signatures because of their smooth surfaces, and sand dunes have highlights and shadows. The radar signatures are clearly determined by the surface roughness and landforms. Because of this experience, interpreters have been astounded by SIR–A images of the Selima sand sheet of the eastern Sahara in Egypt and the Sudan, for reasons described below.

Selima Sand Sheet, Eastern Sahara

Figure 6.47A is a Landsat MSS image of a portion of the Selima sand sheet that shows a featureless surface of windblown sand. Based on this signature in a visible spectral band, interpreters expected the sand sheet to have a uniform dark signature on SIR–A images. Instead the SIR–A image (Figure 6.47B) show details of ancient drainage patterns eroded into the bedrock that underlies the sand sheet. The SIR–A radar energy has clearly penetrated the sand sheet and returned from the bedrock surface. Scientists from the U.S. Geological Survey, University of Arizona, Egyptian Geological Survey, and Jet Propulsion Laboratory visited the region several times; their investigations are reported in McCauley and others (1982). The sand sheet is so devoid of features that portable satellite navigation systems were used in field vehicles to locate specific sites on the SIR–A images. A number of pits were excavated and established that

1. the bright areas in the SIR–A images are the rough, eroded surface of the Nubian Sandstone.

2. the dark signatures indicate extinct drainage channels filled in with pebbly alluvium.

These relationships are shown in the interpretation map (Figure 6.48). The diagrammatic cross section (Figure 6.49) shows that some of the transmitted radar energy is reflected by the smooth sand surface. However, much of the energy penetrates the sand and is refracted to a steeper depression angle as it enters this denser medium. The energy that encounters the rough bedrock surface is strongly backscattered, and a relatively large

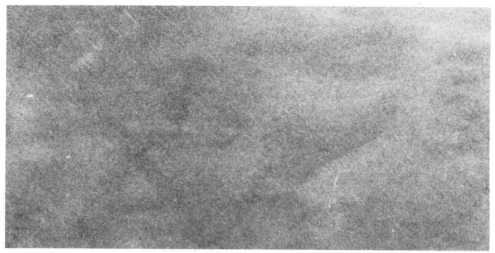

A. LANDSAT MSS BAND–5 IMAGE.

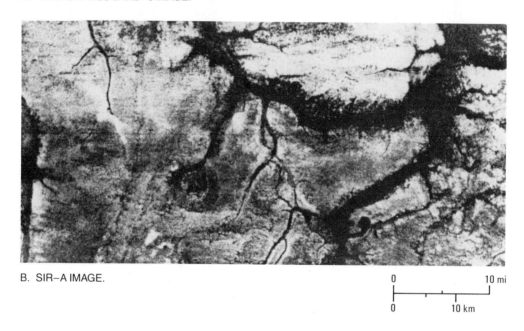

B. SIR–A IMAGE.

0   10 mi

0   10 km

FIGURE 6.47 Selima sand sheet, northwest Sudan. From McCauley and others (1982). Courtesy G. G. Schaber, U.S. Geological Survey.

proportion returns to the antenna to produce a bright signature. The transmitted energy that encounters the smooth surface of a channel is specularly reflected (Figure 6.49) and no energy is returned to the antenna. The lack of any return energy causes the dark signature of the channels.

The radar penetration of the Selima sand sheet is due to the hyperarid environment, where rainfall occurs at intervals of 30 to 50 years and the sun evaporates any absorbed moisture. Laboratory measurements (McCauley and others, 1982, Table 2) showed that the dry sand and other surface materials have very low values of dielectric constant (3.1 to 3.7). These values, together with other measurements, indicate a potential radar penetration, or *skin depth,* ranging from 1 to 6 m in the sand sheet, which agrees with excavations in the field.

Radar images of other desert regions have been examined for evidence of surface penetration, but only one additional example, from an arid valley in the Mojave Desert, California, has been reported (Blom, Crippen, and Elachi, 1984). The general lack of radar penetration of other deserts is due to the presence of subsurface moisture, since most deserts have rainfall every few years. The water soaks into the sand and later evaporates from a thin surface layer that insulates the deeper sand. In many deserts the sand has appreciable

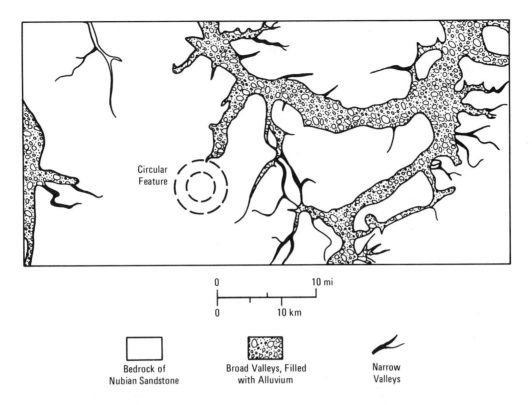

FIGURE 6.48 Interpretation map of SIR–A image of Selima sand sheet, northwest Sudan.

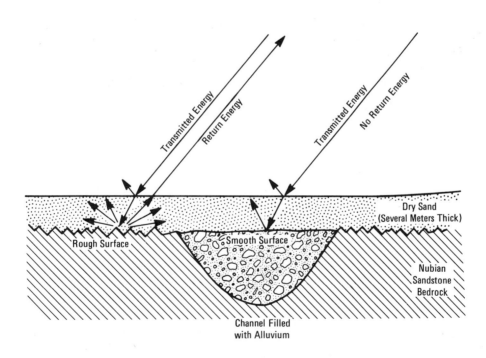

FIGURE 6.49 Diagrammatic cross section showing interaction between SIR–A radar energy and hyperarid terrain in northwest Sudan. From information provided by G. G. Schaber, U.S. Geological Survey.

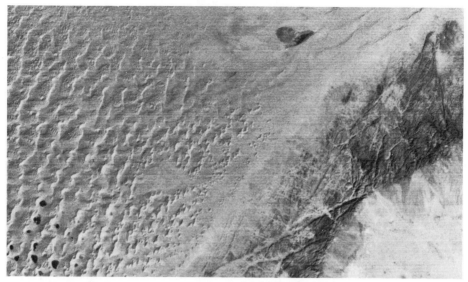

A. LANDSAT MSS BAND–5 IMAGE ACQUIRED OCTOBER 7, 1973.

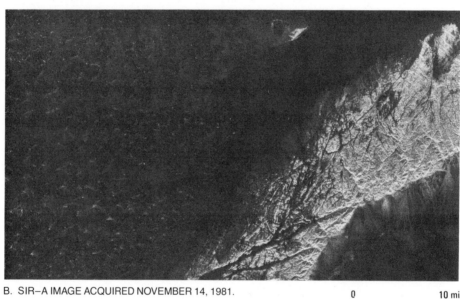

B. SIR–A IMAGE ACQUIRED NOVEMBER 14, 1981.

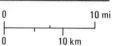

FIGURE 6.50 Landsat and SIR–A images of the Badain Jaran Desert and the Yabrai Shan Mountains, inner Mongolia, China. From Ford, Cimino, and Elachi (1983, Figure 19). Courtesy J. P. Ford, Jet Propulsion Laboratory.

moisture at depths of less than a meter. The Badain Jaran Desert illustrates the radar response of typical deserts.

Badain Jaran Desert, Inner Mongolia

Figure 6.50 illustrates Landsat and SIR–A images of the Badain Jaran Desert and the Yabrai Shan Mountains. As shown in the geologic map and cross section (Figure 6.51), the mountains form an upland with a steep scarp on the southeast that is bordered by alluvial fans. The upland slopes gently northwestward and is overlapped by a belt of sheet sand that grades into an extensive field of coalescing barchan sand dunes. In the southwest portion of the Landsat image (lower left corner of Figure 6.50A), dark oval signatures indicate a number of small interdune lakes. A similar dark signature marks a large lake and marsh in the top center of the image. In the

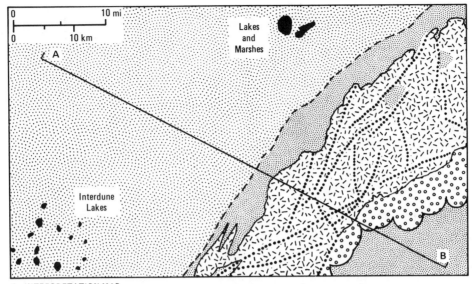

A. INTERPRETATION MAP.

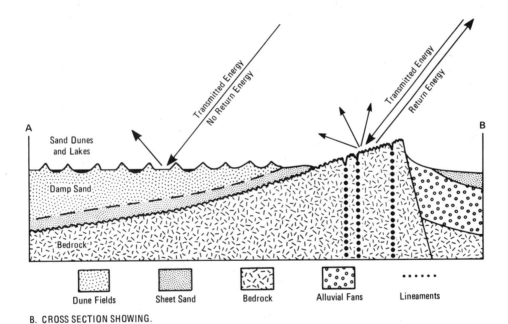

Dune Fields   Sheet Sand   Bedrock   Alluvial Fans   Lineaments

B. CROSS SECTION SHOWING.

FIGURE 6.51 Map and cross section for a SIR–A image showing the Badain Jaran Desert and the Yabrai Shan Mountains, inner Mongolia, China.

SIR–A image (Figure 6.50B), the smooth sand and water have dark signatures. The steep slopes of dunes that face the radar look direction have bright signatures. Very bright signatures around the large lake and marsh are caused by vegetation. Bedrock is dark in the Landsat image because of desert varnish and is bright in the SIR–A image because of the rough surface. The bedrock is cut by many linear valleys filled with sand, which have dark signatures in the radar images.

The key area to evaluate for penetration by SIR–A is the southeast margin of the sheet sand which laps onto the bedrock (Figure 6.51A). In the Landsat image the margin of the sheet sand is clearly visible. If this sand were not penetrated by radar, the contact between dark sand and bright bedrock in the SIR–A image should occur at the same position as in the Landsat image. If the radar had penetrated the sand, the apparent contact between sand and bedrock would have shifted toward

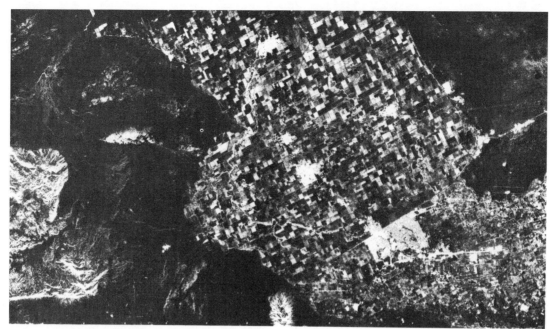

A. SIR–A IMAGE ACQUIRED NOVEMBER 1981.

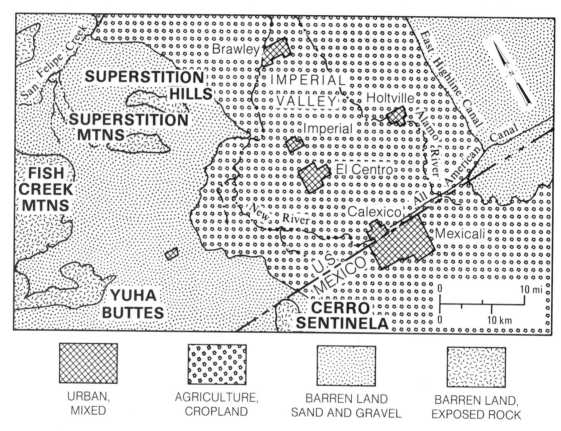

B. MAP SHOWING LAND–USE AND LAND–COVER CATEGORIES.

URBAN, MIXED

AGRICULTURE, CROPLAND

BARREN LAND SAND AND GRAVEL

BARREN LAND, EXPOSED ROCK

FIGURE 6.52 Land use and land cover of the Imperial Valley, California, interpreted from a SIR–A image. From Sabins (1983b, Figure D-6).

the northwest in the SIR–A image. Comparison of the two images and the map shows the location of the sand-bedrock contact to be nearly identical in both images; therefore no radar penetration by SIR–A has occurred. Figure 6.51B shows that the incident energy is back-scattered by exposed bedrock to produce bright signatures but is specularly reflected by the smooth sand to produce dark signatures.

Moisture in the sand has prevented radar penetration.

LAND USE AND LAND COVER, IMPERIAL VALLEY, CALIFORNIA

Land use and land cover may be interpreted from radar images, as shown by the SIR–A image of part of the Imperial Valley in southern California and Mexico (Figure 6.52A). At the intermediate scale of this image, level-II categories of land use and land cover from Table 10.2 are interpreted on the basis of radar signature and pattern (Figure 6.52B).

The Imperial Valley is a broad, flat depression with elevations below sea level; its fertile soil and warm climate support year-round irrigated agriculture. The rectangular pattern in California reflects the township and range system of land ownership in the United States. Mexico does not use this system, so the pattern is irregular. On both sides of the border the agriculture belongs to the category "cropland and pasture" (210) and consists predominantly of cotton, sugar beets, alfalfa, and vegetables, all with bright signatures. Fallow fields have dark signatures. Towns, which are readily recognized by their bright signatures, are assigned to the category "mixed" (180). At greater magnifications of the image, street patterns are recognizable. The agriculture area is bordered on the east and west by desert of the category "sand and gravel other than beaches" (730). There is no evidence of radar penetration of the sand in this area, which receives 10 cm or more of rain annually. San Felipe Creek and other ephemeral streams in the desert have bright signatures caused by gravel and native vegetation along the stream courses. The category "exposed rock" (740) is represented by the mountains and hills of bedrock that project through the sand and gravel in the western part of the image. These bedrock outcrops are identified by their bright signature and rugged topography.

This SIR–A image with 38-m resolution and intermediate scale has been interpreted at level II. Aircraft radar images of higher resolution and larger scale may be interpreted at level III.

FUTURE SATELLITE RADAR SYSTEMS

The success of Seasat, SIR–A, and SIR–B has encouraged the development of several future satellite-based radar systems. Table 6.8 lists and summarizes the characteristics of these future missions. The Space Shuttle will carry another SIR–B, along with SIR–C and the SAR Facility. A major disadvantage is the short duration of Space Shuttle missions. ERS-1 (European Space Agency), Radarsat (Canada), and ERS-1 (Japan) will be deployed on unmanned satellites with long durations in orbit. The manned NASA Space Station will also have a radar capability, but the system characteristics have not yet been defined.

COMMENTS

Because radar is an active system that supplies its own illumination at long wavelengths, images can be acquired at night and through cloud cover. The illumination direction can be oriented to enhance particular linear trends. Interaction between materials and radar energy

TABLE 6.8 Future satellite radar systems

System and agency	Planned launch	Wavelength band	Comments
SIR–B reflight, NASA	1987	L	Repeat of SIR–B mission on the Space Shuttle.
SIR–C, NASA	1988	C and L	System carried on the Space Shuttle.
SAR Facility, European Space Agency	1988	X and C	System carried on the Space Shuttle.
ERS-1, European Space Agency	1988	C	Part of Earth Resources Satellite 1. Designed to monitor ocean conditions.
Radarsat, Canadian Center for Remote Sensing	1990	C	Designed to monitor sea-ice conditions.
ERS-1, National Space Development Agency of Japan	1991	L	Similar to Seasat. Carried on Japanese ERS-1 satellite.

is determined by the wavelength, depression angle, and polarization of the radar beam and by the dielectric properties, surface roughness, and orientation of the material. Rough surfaces produce stronger radar backscatter, which is recorded as brighter tones on the images. The oblique angle of radar illumination causes highlights and shadows, which can enhance the appearance of geologic features such as lineaments, faults, and fracture patterns.

On many images of heavily forested terrain, geologic features are clearly expressed, which leads to the mistaken assumption that radar energy penetrates the tree canopy. This is not the case, however, for vegetation strongly scatters radar energy. The relatively large ground resolution cell of radar does not resolve individual trees, thus enhancing the appearance of geologic features that are measured in hundreds and thousands of meters. In hyperarid environments, radar energy can penetrate sand to a depth of several meters and produce images of buried bedrock surfaces.

Beginning with Seasat in 1978 and followed by the Space Shuttle imaging radar missions in 1981 and 1984, radar images of limited portions of the earth have been acquired from satellites. Beginning in the late 1980s, several planned satellite radar missions will increase this source of information.

QUESTIONS

1. A typical C-band radar system operates at a frequency of 5 GHz. What is the wavelength of this radar?

2. Calculate the range resolution for a radar system with a pulse length of 0.2 μsec and a depression angle of 30°.

3. For a real-aperture, X-band system with an antenna length of 300 cm, calculate the azimuth resolution at a slant-range distance of 15 km.

4. What is the principal difference between real-aperture and synthetic-aperture radar systems?

5. Summarize the differences between images acquired by real-aperture systems and by synthetic-aperture systems.

6. Summarize the differences (aside from wavelength) between radar stereo images and stereo pairs of aerial photographs.

7. Define and discuss the optimum depression angle (relatively steep or relatively shallow) for imaging terrain (a) with low relief, such as coastal plains; and (b) high relief, such as mountain chains.

8. What is the principal difference between images acquired by SIR–A and by Seasat?

9. Prepare a version of Table 6.5 for aircraft systems at Ka-, X-, and L-band wavelengths, all with a depression angle of 30°.

10. Design an experiment to evaluate quantitatively the extent to which Shuttle imaging radar penetrates soil and vegetation. You have access to an electronics laboratory, and a future SIR mission orbit will cross your field test sites.

11. Predict the dielectric constant for Yuma sand, Vernon clay, and Miller clay at 20 percent and at 40 percent volumetric water content.

12. On Seasat images of the Guymon, Oklahoma, test site, predict the relative tone (brightness) of milo fields with soil moistures of (a) 5 percent and (b) 30 percent.

13. What property of different rock types in Indonesia enables them to be distinguished in SIR–A images?

REFERENCES

Barr, D. J., 1969, Use of side-looking airborne radar (SLAR) imagery for engineering studies: U.S. Army Engineer Topographic Laboratories, Technical Report 46-TR, Fort Belvoir, Va.

Blanchard, B. J., A. J. Blanchard, S. Theis, W. D. Rosenthal, and C. L. Jones, 1981, Seasat SAR response from water resources parameter: U.S. Department of Commerce/NOAA Contract 78-4332, Final Report 3891.

Blom, R. G., R. E. Crippen, and C. Elachi, 1984, Detection of subsurface features in Seasat radar images of Means Valley, Mojave Desert, California: Geology, v. 12, p. 346–349.

Bryan, M. L., 1979, The effect of radar azimuth angle on cultural data: Photogrammetric Engineering and Remote Sensing, v. 45, p. 1097–1107.

Cimino, J. B., and C. Elachi, eds., 1982, Shuttle Imaging Radar-A (SIR–A) experiment: Jet Propulsion Laboratory Publication 8277, Pasadena, Calif.

Craib, K. B., 1972, Synthetic aperture SLAR systems and their application for regional resources analysis *in* Sahrokhi, F. ed., Remote sensing of earth resources, University of Tennessee Space Institute: v. 1, p. 152–178, Tullahoma, Tenn.

Donovan, N., D. Evans, and D. Held, eds., 1985, NASA/JPL aircraft SAR workshop proceedings: Jet Propulsion Laboratory Publication 85-39, Pasadena, Calif.

Fischer, W. A., ed., 1975, History of remote sensing *in* Reeves, R. G., ed., Manual of remote sensing, first edition: ch. 2, p. 27–50, American Society of Photogrammetry, Falls Church, Va.

Ford, J. P., R. G. Blom, M. L. Bryan, M. I. Daily, T. H. Dixon, C. Elachi, and E. C. Xenos, 1980, Seasat views North America, the Caribbean, and western Europe with imaging radar: Jet Propulsion Laboratory Publication 80-67, Pasadena, Calif.

Ford, J. P., J. B. Cimino, and C. Elachi, 1983, Space Shuttle Columbia views the world with imaging radar—the SIR–A experiment: Jet Propulsion Laboratory Publication 82-95, Pasadena, Calif.

Hoffer, R. M., P. W. Mueller, and D. F. Lozano-Garcia, 1985, Multiple incidence angle Shuttle Imaging Radar data for discriminating forest cover: American Society of Photogrammetry and Remote Sensing, Technical Papers, p. 476–485, Falls Church, Va.

Hunt, C. B., and D. R. Mabey, 1966, Stratigraphy and structure of Death Valley, California: U.S. Geological Survey Professional Paper 494-A.

Jensen, H., L. C. Graham, L. J. Porcello, and E. N. Leith, 1977, Side-looking airborne radar: Scientific American, v. 237, p. 84–95.

McCauley, J. F., and others, 1982, Subsurface valleys and geoarcheology of the eastern Sahara revealed by Shuttle radar: Science, v. 318, p. 1004–1020.

Moore, R. K., 1983, Radar fundamentals and scatterometers in Colwell, R. N., ed., Manual of remote sensing, second edition: ch. 9, p. 369–427, American Society of Photogrammetry, Falls Church, Va.

Moore, R. K., L. J. Chastant, L. Porcello, and J. Stevenson, 1983, Imaging radar systems in Colwell, R. N., ed., Manual of remote sensing, second edition: ch. 10, p. 429–474, American Society of Photogrammetry, Falls Church, Va.

Peake, W. H., and T. L. Oliver, 1971, The response of terrestrial surfaces at microwave frequencies: Ohio State University Electroscience Laboratory, 2440-7, Technical Report AFAL-TR-70-301, Columbus, Ohio.

Sabins, F. F., 1973, Engineering geology applications of remote sensing in Geology, Seismicity, and Environmental Impact: Association of Engineering Geologists Special Publication, p. 141–155, Los Angeles, Calif.

Sabins, F. F., 1983a, Geologic interpretation of Space Shuttle radar images of Indonesia: American Association of Petroleum Geologists Bulletin, v. 67, p. 2076–2099.

Sabins, F. F., 1983b, Remote sensing laboratory manual, second edition: Remote Sensing Enterprises, La Habra, Calif.

Sabins, F. F., 1984, Geologic mapping of Death Valley from thematic mapper, thermal infrared, and radar images: Proceedings Third Thematic Conference, Remote Sensing for Exploration Geology, p. 139–152, Environmental Research Institute of Michigan, Ann Arbor, Mich.

Sabins, F. F., R. Blom, and C. Elachi, 1980, Seasat radar images of San Andreas fault, California: American Association of Petroleum Geologists Bulletin, v. 64, p. 619–628.

Schaber, G. G., G. L. Berlin, and W. E. Brown, 1976, Variations in surface roughness within Death Valley, California—geologic evaluation of 25-cm wavelength radar images: Geological Society of America Bulletin, v. 87, p. 29–41.

Schaber, G. G., G. L. Berlin, and D. J. Pitrone, 1975, Selection of remote sensing techniques—surface roughness information from 3-cm wavelength SLAR images: American Society of Photogrammetry, Proceedings of 42d Annual Meeting, p. 103–117, Washington, D.C.

Ulaby, F. T., A. Aslam, and M. C. Dobson, 1982, Effects of vegetation cover on radar sensitivity to soil moisture: IEEE Transactions on Geoscience and Remote Sensing, v. GE-20, p. 476–481.

Ulaby, F. T., P. T. Batlivala, and J. E. Bane, 1980, Crop identification with L-band radar: Photogrammetric Engineering and Remote Sensing, v. 46, p. 101–105.

Ulaby, F. T., B. Brisco, and M. C. Dobson, 1983, Improved spatial mapping of rainfall events with spaceborne SAR imagery: IEEE Transactions on Geoscience and Remote Sensing, v. GE-21, p. 118–121.

Wang, J. R., and T. J. Schmugge, 1980, An empirical model for the complex dielectric permittivity of soils as a function of water content: IEEE Transactions on Geoscience and Remote Sensing, v. GE-18, p. 288–295.

ADDITIONAL READING

Bryan, M. L., 1979, Bibliography of geologic studies using imaging radar: Jet Propulsion Laboratory Publication 79-53, Pasadena, Calif.

Elachi, C., 1982, Radar images of the Earth from space: Scientific American, v. 247, p. 54.

Elachi, C., and others, 1982, Shuttle imaging radar experiment: Science, v. 218, p. 996–1003.

Ford, J. P., 1980, Seasat orbital radar imagery for geologic mapping—Tennessee, Kentucky, Virginia: American Association of Petroleum Geologists Bulletin, v. 64, p. 2064–2094.

Jet Propulsion Laboratory, 1980, Radar geology—an assessment: Jet Propulsion Laboratory Publication 80-61, Pasadena, Calif.

Long, M. W., 1975, Radar reflectivity of land and sea: D. C. Heath and Co., Lexington, Mass.

MacDonald, H. C., 1969, Geologic evaluation of radar imagery from Darien Province, Panama: Modern Geology, v. 1, p. 1–63.

Skolnik, M. I., ed., 1970, Radar handbook: McGraw-Hill Book Co., N.Y.

Wing, R. S., 1971, Structural analysis from radar imagery: Modern Geology, v. 2, p. 1–21.

7

Digital Image Processing

Many types of remote sensing images are routinely recorded in digital form and then processed by computers to produce images for interpreters to study. The simplest form of *digital image processing* employs a microprocessor that converts the digital data tape into a film image with minimal corrections and calibrations. At the other extreme, large mainframe computers are employed for sophisticated interactive manipulation of the data to produce images in which specific information has been extracted and highlighted.

A number of books have been published on the subject of digital image processing; a representative selection is listed at the end of this chapter. Indeed, more books are available that deal with image processing than with image interpretation, which is ironic because processing is really just a preparatory step for the all-important activity of interpretation.

Digital processing did not originate with remote sensing and is not restricted to these data. Many image-processing techniques were developed in the medical field to process X-ray images and images from sophisticated body-scanning devices. For remote sensing, the initial impetus was the program of unmanned planetary satellites in the 1960s that *telemetered,* or transmitted, images to ground receiving stations. The low quality of the images required the development of processing techniques to make the images useful. Another impetus was

the Landsat program, which began in 1972 and provided repeated worldwide coverage in digital format. A third impetus is the continued development of faster and more powerful computers, peripheral equipment, and software that are suitable for image processing.

This chapter describes and illustrates the major categories of image processing. The publications cited at the end of the chapter describe in detail the methods and the mathematical transformations. The processes are illustrated with Landsat examples because these are the most familiar and are readily available to readers; the digital processes, however, are equally applicable to all forms of digital image data.

IMAGE STRUCTURE

One can think of any image as consisting of tiny, equal areas, or *picture elements,* arranged in regular rows and columns. The position of any picture element, or *pixel,* is determined on an *xy* coordinate system; in the case of Landsat images, the origin is at the upper left corner of the image. Each pixel also has a numerical value, called a *digital number* (DN), that records the intensity of electromagnetic energy measured for the ground resolution cell represented by that pixel. Digital numbers range from zero to some higher number on a gray scale.

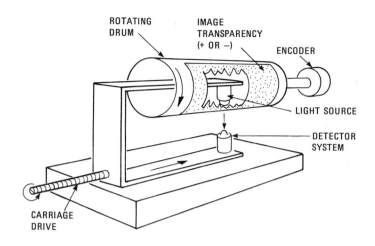

ROTATING DRUM

IMAGE TRANSPARENCY (+ OR −)

ENCODER

LIGHT SOURCE

DETECTOR SYSTEM

CARRIAGE DRIVE

FIGURE 7.1 System for digitizing an image. From Bryant (1974, Figure 2).

The image may be described in strictly numerical terms on a three-coordinate system with x and y locating each pixel and z giving the DN, which is displayed as a grayscale intensity value. Images may be originally recorded in this digital format, as in the case of Landsat. An image recorded initially on photographic film may be converted into digital format by a process known as digitization.

Digitization Procedure

Digitization is the process of converting an image recorded on photographic film (radar, thermal IR, or aerial photographs) into an ordered array of pixels. Maps and other information may also be digitized. The digitized information is recorded on magnetic tape for subsequent computer processing. Digitizing systems belong to several categories: drum, flatbed, flying spot, video camera, and linear array.

A typical *drum digitizing system* (Figure 7.1) consists of a rotating cylinder with a fixed shaft and a movable carriage that holds a light source and a detector similar to a photographic light meter. The positive or negative transparency is placed over the opening in the drum, and the carriage is positioned at a corner of the film. As the drum rotates, the detector measures intensity variations of the transmitted light caused by variations in film density. Each revolution of the drum records one scan line of data across the film. At the end of each revolution the encoder signals the drive mechanism, which advances the carriage by the width of one scan line to begin the next revolution.

Width of the scan line is determined by the optical aperture of the detector, which typically is a square opening measuring 50 μm on a side. The analog electrical signal of the detector, which varies with film density changes, is sampled at 50 μm intervals along the scan line, digitized, and recorded as a DN value on magnetic tape. Each DN is recorded as a series of *bits,* which form an ordered sequence of ones and zeroes. Each bit represents an exponent of the base 2. An eight-bit series is commonly used to represent 256 values on the gray scale ($2^8 = 256$ levels of brightness), with 0 for black and 255 for white. A group of eight bits is called a *byte.* The film image is thus converted into a regular array of pixels that are referenced by scan line number (y coordinate) and pixel count along each line (x coordinate). A typical 23-by-23-cm (9-by-9-in.) aerial photograph digitized with a 50-μm sampling interval thus converts into 21 million pixels.

The intensity of the digitizer light source is calibrated at the beginning of each scan line to compensate for any fluctuations of the light source. These drum digitizers are fast, efficient, and relatively inexpensive. Color images may be digitized into the values of the component primary colors on three passes through the system using blue, green, and red filters over the detector. A digital record is produced for each primary color, or wavelength band.

In *flatbed digitizers,* the film is mounted on a flat holder that moves in the x and y directions between a fixed light source and a detector. These devices are usually slower and more expensive than drum digitizers, but they are also more precise. Some systems employ reflected light in order to digitize opaque paper prints. Another digitizing system is the *flying-spot scanner,* which is essentially a modified television camera; it electronically scans the original image in a raster pattern. The resulting voltage signal for each line is digitized and recorded. *Linear-array digitizers* are similar to the along-track scanners described in Chapter 1. An optical system projects the image of the original photograph onto the focal plane of the system. A mechanical transport system moves a linear array of detectors across the focal plane to record the original photograph in digital form.

The digitized image is recorded on magnetic tape that can be read into a computer for various processing operations. The processed data are then displayed as an image on a viewing device or plotted onto film as described in the following section.

Image-Generation Procedure

Hard-copy images are generated by *film writers* (Figure 7.2), which operate in reverse fashion to digitizers. Recording film is mounted on a drum. With each rotation, a scan line is exposed on the film by a light source, the intensity of which is modulated by the DNs of the pixels. On completion of each scan line, the carriage advances the light source to commence the next line. The exposed film is developed to produce a transparency from which one can make contact prints and enlargements. Some plotters produce only black-and-white film. To produce color images, the red, green, and blue components are plotted as separate films. These films are photographically combined into color images using methods described in Chapter 2. Other plotters record data directly onto color film; three passes are made, each with a red, green, or blue filter over the light source.

Flying-spot plotters produce an image with a beam that sweeps in a raster pattern across the recording film. This system is also employed to display images on a television screen, which enables the interpreter to view the processed image in real time rather than to wait for the processed data to be plotted onto film, developed, and printed.

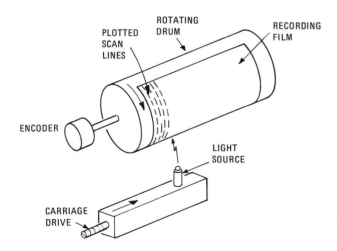

FIGURE 7.2 System for plotting an image from digital data. From Bryant (1974, Figure 3).

Landsat Images

Data acquired by Landsat MSS and TM systems are recorded on *computer-compatible tapes* (CCTs), which can be read and processed by computers. Spectral bands and other characteristics of the Landsat systems were described in Chapter 4. Figure 7.3 describes the format of the digital data that make up MSS and TM images. Both systems employ an oscillating mirror that sweeps scan lines that are 185 km in length across the terrain and oriented at right angles to the satellite orbit path. The scanning is a continuous process; the data are subdivided into scenes that cover 185 km along the orbit direction for MSS and 170 km for TM. As described below, the two systems also differ in other characteristics that are important for image processing.

Landsat MSS Data The ground resolution cell of MSS is a 79-by-79-m square. An image consists of 2340 scan lines, each of which is 185 km long in the scan direction and 79 m wide in the orbit direction (Figure 7.3A). During each east-bound sweep of the scanner mirror, reflected solar energy is separated into the four spectral bands and focused onto detectors to generate a response that varies in amplitude proportionally with the intensity of the reflected energy. A segment of an MSS scan line showing variation in terrain reflectance is illustrated in Figure 7.4A together with the 79-m dimension of the ground resolution cell. The curve showing reflectance recorded as a function of distance along the scan line is called an *analog display* because the curve is a graphic analogy of the property being measured. *Digital displays* record information as a series of numbers. Analog data are converted into digital data by sampling the analog display at regular intervals and recording the analog value at each sample point in digital form. The optimum interval between sample points is one that records significant features of the analog signal with the minimum number of samples. For MSS the optimum sample interval was determined to be 57 m. Each sample is converted into a DN that is transmitted to ground receiving stations. Thus each scan line consists of 3240 pixels each covering an area measuring 57 by 79 m. For each pixel, four digital numbers are recorded for reflectance, one each in bands 4, 5, 6, and 7. Spatial resolution of the image is ultimately determined by the 79-by-79-m dimensions of the ground resolution cell; the smaller sample interval does not alter this relationship. Brightness values for bands 4, 5, and 6 are recorded on CCTs using a seven-bit scale ($2^7 = 128$ values: 0 to 127); band 7 is recorded on a six-bit scale ($2^6 = 64$ values: 0 to 63). Most processing systems employ eight-bit scales; bands 4, 5, and 6 are multiplied by 2 and band 7 is multiplied by 4 for conversion to eight-bit scales.

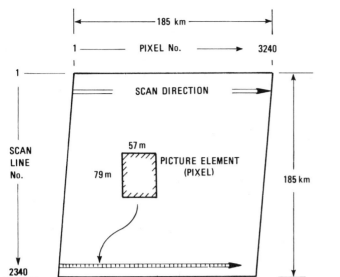

2340 scan lines × 3240 pixels = 7.6 × 10⁶ pixels per band
7.6 × 10⁶ pixels × 4 bands = 30.4 × 10⁶ pixels per scene

A. MULTISPECTRAL SCANNER.

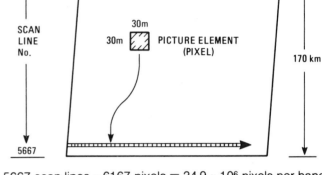

5667 scan lines × 6167 pixels = 34.9 × 10⁶ pixels per band
34.9 × 10⁶ pixels × 7 bands = 244.3 × 10⁶ pixels per scene

B. THEMATIC MAPPER.

FIGURE 7.3 Arrangement of scan lines and pixels in Landsat MSS and TM images.

Each sweep of the MSS scan mirror reflects light onto four arrays of six detectors, one array for each spectral band. Thus each mirror sweep records six scan lines for each band (Figure 7.5). The advantage of this design over one using a single detector is that the number of mirror sweeps and the velocity of the mirror scan are reduced by a factor of 6, which results in a longer dwell time and a higher signal-to-noise ratio. A disadvantage is that the response of one detector may differ from that of the other five, causing every sixth scan line to be defective. Digital methods for correcting these image defects are discussed later in this chapter. As mentioned previously, each MSS band consists of 2340 scan lines, each with 3240 pixels, for a total of 7.6 × 10⁶ pixels or 30.4 × 10⁶ pixels for the four bands (Figure 7.3A). Holkenbrink (1978) describes details of the MSS digital format.

Landsat TM Data The ground resolution cell of TM is a 30-by-30-m square. An image consists of 5965 scan lines, each of which is 185 km long in the scan direction and 30 m wide in the orbit direction. The analog signal is sampled at 30-m intervals to produce 30-by-30-m pixels (Figure 7.4B). Each scan line consists of 6167 pixels, and each band consists of 34.9 × 10⁶ pixels (Figure 7.3B). The seven bands have a total of 244.3 × 10⁶ pixels, which is more than eight times the data that is contained in an MSS image. Each of the six TM visible

and reflected IR bands employs an array of 16 detectors, and unlike MSS, recording occurs during both sweep directions; together these factors allow adequate dwell time for the TM's relatively small ground resolution cells. Because of the lower energy flux radiated in the thermal IR region, band 6 (10.5 to 12.5 μm) is recorded by only four detectors, each with a 120-by-120-m ground resolution cell. The cells are resampled into 30-by-30-m pixels to be consistent with the other bands. TM data are recorded on an eight-bit scale, which provides a greater dynamic range than the lower bit scales of MSS. Bernstein and others (1984) have analyzed the performance of the TM system; NASA (1983) gives details of the CCT format for TM data.

Structure of Landsat Digital Images The structure of a TM digital image is illustrated in Figure 7.6, which is a greatly enlarged portion of band 4 from the Thermopolis, Wyoming, subscene (Figure 4.20D). Figure 7.6A shows an array of 66 by 66 pixels, and Figure 7.6B is a map that shows the location for the printout of DNs in Figure 7.6C. The correlation of the image gray scale to the eight-bit DN scale is shown in Figure 7.6D, where a DN of 0 is displayed as black and 255 as white. Referring again to the printout, note the low values associated with the Wind River and the high values of the adjacent vegetated flood plains. These values are con-

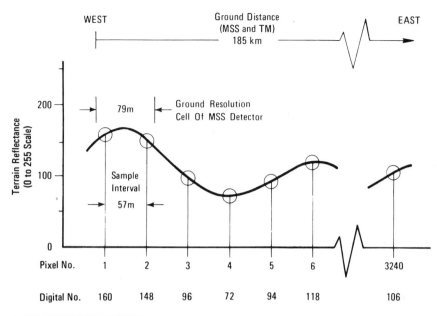

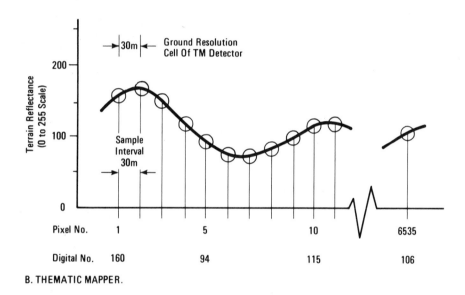

FIGURE 7.4 Plot of terrain reflectance along a Landsat scan line. Sample intervals and DNs are shown for MSS detectors and for TM detectors.

sistent with the strong absorption by water and strong reflection by vegetation of the IR energy recorded by TM band 4. In the river, note that the lowest digital numbers (DN = 8 to 17) correspond to the center of the stream; pixels along the stream margin have higher values (DN = 20 to 80). The 30-by-30-m ground resolution cells in the center of the river are wholly occupied by water, whereas marginal cells are occupied partly by water and partly by vegetation resulting in a DN intermediate between water and vegetation. These mixed

pixels may also be observed in the gray-scale image (Figure 7.6A). A useful method of displaying statistics of an image is the *histogram* (Figure 7.6D), in which the vertical scale records the number of pixels associated with each DN, shown on horizontal scale. This discussion deals with TM data, but MSS data have a similar format.

Availability of Landsat Digital Data Landsat CCTs of the world are available from EOSAT at a cost of $660

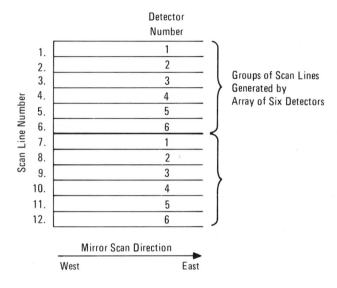

FIGURE 7.5 Detail of the 6-detector array for a spectral band of Landsat MSS data. A band of TM data is recorded with a 16-detector array.

Mirror Scan Direction

West　　　　East

for an MSS scene and $3300 for a TM scene. Many of the Landsat distribution centers listed in Chapter 4 can provide CCTs for their area of coverage.

IMAGE-PROCESSING OVERVIEW

Image-processing methods may be grouped into three functional categories; these are defined below together with lists of typical processing routines.

1. *Image restoration* compensates for data errors, noise, and geometric distortions introduced during the scanning, recording, and playback operations.
    a. Restoring periodic line dropouts
    b. Restoring periodic line striping
    c. Filtering of random noise
    d. Correcting for atmospheric scattering
    e. Correcting geometric distortions
2. *Image enhancement* alters the visual impact that the image has on the interpreter in a fashion that improves the information content.
    a. Contrast enhancement
    b. Intensity, hue, and saturation transformations
    c. Density slicing
    d. Edge enhancement
    e. Making digital mosaics
    f. Producing synthetic stereo images

3. *Information extraction* utilizes the decision-making capability of the computer to recognize and classify pixels on the basis of their digital signatures.
    a. Producing principal-component images
    b. Producing ratio images
    c. Multispectral classification
    d. Producing change-detection images

These routines are described and illustrated in the following sections. The routines are illustrated with Landsat examples, but the techniques are equally applicable to other digital-image data sets. A number of additional routines are described in the various reference publications.

IMAGE RESTORATION

Restoration processes are designed to recognize and compensate for errors, noise, and geometric distortion introduced into the data during the scanning, transmission, and recording processes. The objective is to make the image resemble the original scene. Image restoration is relatively simple because the pixels from each band are processed separately.

Restoring Periodic Line Dropouts

On some MSS images, data from one of the six detectors are missing because of a recording problem. On the CCT every sixth scan line is a string of zeros that plots as a black line on the image (Figure 7.7A). These are called *periodic line dropouts*. The first step in the restoration process is to calculate the average DN value per scan line for the entire scene. The average DN value for each scan line is then compared with this scene average. Any scan line deviating from the average by more than a designated threshold value is identified as defective. The next step is to replace the defective lines. For each pixel in a defective line, an average DN is calculated using DNs for the corresponding pixel on the preceding and succeeding scan lines. The average DN is substituted for the defective pixel, as shown on the printout of Figure 7.7B. The resulting image is a major improvement, although every sixth scan line consists of artificial data. This restoration program is equally effective for random line dropouts that do not follow a systematic pattern.

Restoring Periodic Line Striping

For each spectral band, the detectors (6 for MSS, 16 for TM) were carefully calibrated and matched before

Pixel Number

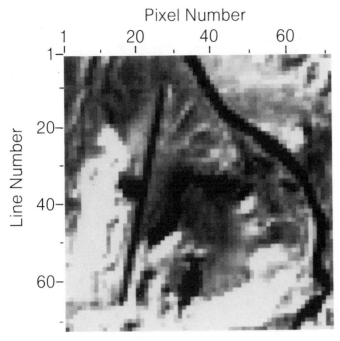

A. GRAY–SCALE IMAGE.

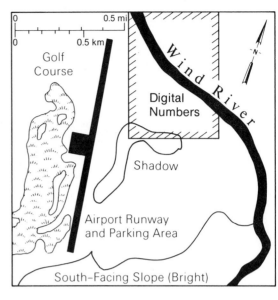

B. LOCATION MAP.

Pixel Number

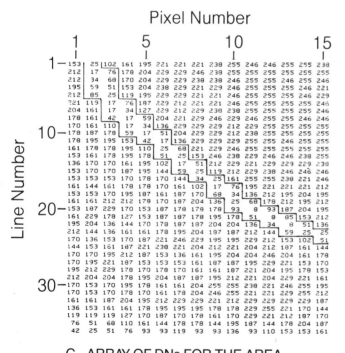

C. ARRAY OF DNs FOR THE AREA
   SHOWN IN THE LOCATION MAP.

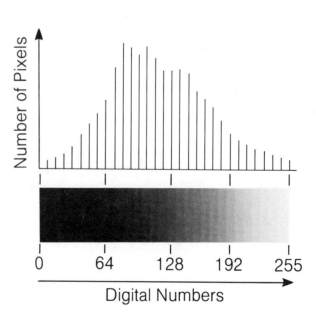

D. HISTOGRAM AND GRAY SCALE.

FIGURE 7.6 Digital structure of a Landsat TM image. The area is a portion of the Thermopolis, Wyoming, subscene of the TM band-4 image illustrated in Figure 4.20.

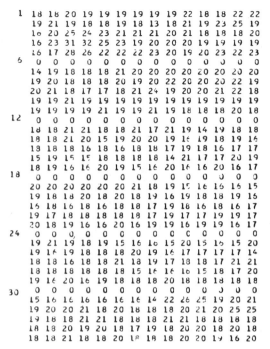

```
 1  18 18 20 19 19 19 19 19 19 22 18 18 22 22
    19 21 19 18 18 19 18 13 18 21 19 23 25 19
    16 20 25 24 23 21 21 21 20 21 18 18 18 20
    16 23 31 32 25 23 19 20 20 20 19 19 19 19
    16 17 28 26 22 22 22 23 20 19 20 23 22 23
 6   0  0  0  0  0  0  0  0  0  0  0  0  0  0
    14 19 18 18 18 21 20 20 20 20 20 20 20 20
    19 20 18 18 18 20 19 20 22 20 20 20 22 19
    20 21 18 17 17 18 21 24 19 20 20 21 22 18
    19 19 21 19 19 19 19 19 19 19 19 19 19 19
    19 19 19 19 21 19 19 21 19 18 18 18 20 18
12   0  0  0  0  0  0  0  0  0  0  0  0  0  0
    18 18 21 21 18 18 21 17 21 19 14 19 18 18
    18 18 21 20 15 19 20 20 19 16 19 18 19 16
    13 18 18 16 18 16 18 18 17 19 18 16 17 17
    15 19 15 15 18 18 18 18 14 21 17 17 20 19
    18 19 16 16 20 19 15 16 20 16 16 20 16 17
18   0  0  0  0  0  0  0  0  0  0  0  0  0  0
    20 20 20 20 20 20 21 18 19 15 16 16 16 15
    19 18 18 20 18 20 18 19 16 19 18 18 19 15
    16 18 16 18 16 18 18 17 19 18 16 18 16 17
    19 17 18 18 18 18 18 17 19 17 17 19 19 17
    20 18 19 16 16 20 16 19 19 16 19 19 16 17
24   0  0  0  0  0  0  0  0  0  0  0  0  0  0
    19 21 19 18 19 15 16 16 15 20 15 16 15 20
    19 16 19 18 18 18 20 19 16 17 17 17 17 14
    18 18 16 18 18 21 18 19 17 18 18 17 21 21
    18 18 18 18 18 18 15 16 16 16 15 18 17 20
    19 16 20 16 19 18 18 20 18 18 18 18 18 18
30   0  0  0  0  0  0  0  0  0  0  0  0  0  0
    15 16 16 16 16 16 16 14 22 26 25 19 20 21
    19 20 20 21 18 20 18 18 20 20 21 20 25 25
    19 18 18 21 21 18 18 18 21 21 18 18 18 18
    18 18 20 19 20 18 17 19 18 20 20 18 20 18
    18 18 21 18 18 20 18 18 18 20 20 19 16 20
```

A. BEFORE DIGITAL RESTORATION.

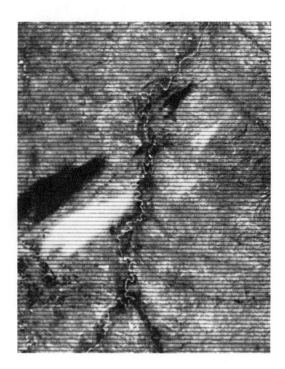

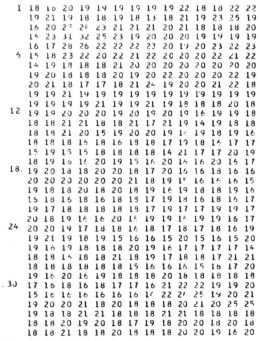

```
 1  18 18 20 19 19 19 19 19 19 22 18 18 22 22
    19 21 19 18 18 19 18 13 18 21 19 23 25 19
    16 20 25 24 23 21 21 21 20 21 18 18 18 20
    16 23 31 32 25 23 19 20 20 20 19 19 19 19
    16 17 28 26 22 22 22 23 20 19 20 23 22 23
 6  15 18 23 22 20 22 21 22 20 20 20 22 21 22
    14 19 18 18 18 21 20 20 20 20 20 20 20 20
    19 20 18 18 18 20 19 20 22 20 20 20 22 19
    20 21 18 17 17 18 21 24 19 20 20 21 22 18
    19 19 21 19 19 19 19 19 19 19 19 19 19 19
    19 19 19 19 21 19 19 21 19 18 18 18 20 18
12  19 19 20 20 20 19 20 19 20 19 16 19 19 18
    18 18 21 21 18 18 21 17 21 19 14 19 18 18
    18 18 21 20 15 19 20 20 19 16 19 18 19 16
    18 18 18 15 18 16 19 18 17 19 18 16 17 17
    15 19 15 15 18 18 18 18 14 21 17 17 20 19
    18 19 16 16 20 19 15 16 20 16 16 20 16 17
18  19 20 18 18 20 20 18 17 20 16 16 18 16 16
    20 20 20 20 20 20 21 18 19 15 16 16 16 15
    19 18 18 20 18 20 18 19 16 19 18 18 19 16
    15 18 16 18 16 18 19 17 19 18 16 18 16 17
    19 17 18 18 18 18 17 19 17 17 19 19 17
    20 18 19 16 16 20 16 19 19 16 19 19 16 17
24  20 20 19 17 18 18 16 18 17 18 17 18 16 19
    19 21 19 18 19 15 16 16 15 20 15 16 15 20
    19 16 19 18 18 18 20 19 16 17 17 17 17 14
    18 18 16 18 18 21 18 19 17 18 18 17 21 21
    18 18 18 18 18 18 15 16 16 16 15 18 17 20
    19 16 20 16 19 18 18 20 18 18 18 18 18 18
30  17 16 18 16 18 17 17 16 21 22 22 19 19 20
    15 16 16 16 16 16 16 14 22 26 25 19 20 21
    19 20 20 21 18 20 18 18 20 20 21 20 25 25
    19 18 18 21 21 18 18 18 21 21 18 18 18 18
    18 18 20 19 20 18 17 19 18 20 20 18 20 18
    18 18 21 18 18 20 18 18 18 20 20 19 16 20
```

B. AFTER DIGITAL RESTORATION.

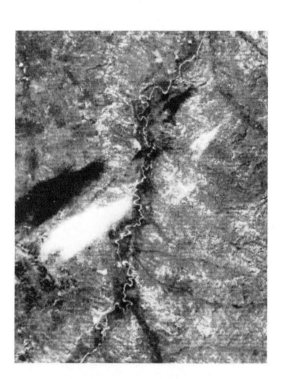

FIGURE 7.7 Restoration of periodic line dropouts on an MSS image of an area in southeast Montana. The arrays of DNs represent a portion of each image. Image processed at Jet Propulsion Laboratory, California Institute of Technology.

the Landsat was launched. With time, however, the response of some detectors may drift to higher or lower levels; as a result every scan line recorded by that detector is brighter or darker than the other lines. The general term for this defect is *periodic line striping*. The defect is called sixth-line striping for MSS where every sixth line has a brightness offset. Figure 7.8A illustrates an MSS image in which the digital numbers recorded by detector number 6 are twice that of the other detectors, causing every sixth scan line to be twice as bright as the scene average. Valid data are present in the defective lines, but must be corrected to match the overall scene. One restoration method is to plot six histograms for the DNs recorded by each detector and compare these with a histogram for the entire scene. For each detector the mean and standard deviation are adjusted to match values for the entire scene. In Figure 7.8B, DNs recorded by detector number 6 were multiplied by a factor of 0.5 to produce the corrected values from which the restored image is plotted.

Another restoration method plots a histogram of DNs for each of the six detectors. Deviations in mean and median values for the histograms are used to recognize and determine corrections for detector differences.

Filtering of Random Noise

The periodic line dropouts and striping are forms of nonrandom noise that may be recognized and restored by simple means. Random noise, on the other hand, requires more sophisticated restoration method. Random noise typically occurs as individual pixels with DNs that are much higher or lower than the surrounding pixels. In the image these pixels produce bright and dark spots that mar the image. These spots also interfere with information extraction procedures such as classification. Random-noise pixels may be removed by digital filters.

Figure 7.9A shows an array of Landsat pixels most of which have DN values ranging from 40 to 60. There are, however, two pixels with DN values of 0 and 90, which would produce dark and bright spots in the image. These spots may be eliminated by a moving average filter in the following steps:

1. Design a *filter kernel,* which is a two-dimensional array of pixels with an odd number of pixels in both the *x* and *y* dimension. The odd-number requirement means that the kernel has a central pixel, which will be modified by the filter operation. In Figure 7.9A, a three-by-three kernel is shown by the outlined box.

2. Calculate the average of the nine pixels in the kernel; for the initial location, the average is 43.

3. If the central pixel deviates from the average DN of the kernel by more than a threshold value (in this example, 30), replace the central pixel by the average value. In Figure 7.9B the original central pixel with a value of 0 has been replaced by 43.

4. Move the kernel to the right by one column of pixels and repeat the operation. The new average is 41. The new central pixel (original value = 40), is within the threshold limit of the new average and remains unchanged. In the third position of the kernel, the central pixel (DN = 90) is replaced by a value of 53.

5. When the right margin of the kernel reaches the right margin of the pixel array, the kernel returns to the left margin, drops down one row of pixels, and the operation continues until the entire image has been subjected to the moving average filter.

This filter concept is also used in many other image processing applications.

Correcting for Atmospheric Scattering

As discussed in Chapter 2, the atmosphere selectively scatters the shorter wavelengths of light. For Landsat MSS images, band 4 (0.5 to 0.6 μm) has the highest component of scattered light and band 7 (0.8 to 1.1 μm) has the least. Atmospheric scattering produces haze, which results in low image contrast. The contrast ratio of an image is improved by correcting for this effect. Figure 7.10 shows two techniques for determining the correction factor for different MSS bands. Both techniques are based on the fact that band 7 is essentially free of atmospheric effects. This can be verified by examining the DNs corresponding to bodies of clear water and to shadows. Both the water and the shadows have values of either 0 or 1 on band 7. The first restoration technique (Figure 7.10A) employs an area within the image that has shadows caused by irregular topography. For each pixel the DN in band 7 is plotted against the DN in band 4, and a straight line is fitted through the plot, using a least-squares technique. If there were no haze in band 4, the line would pass through the origin; because there is haze, the intercept is offset along the band 4 axis, as shown in Figure 7.10A. Haze has an additive effect on scene brightness. To correct the haze effect on band 4, the value of the intercept offset is subtracted from the DN of each band-4 pixel for the entire image. Bands 5 and 6 are also plotted against band 7, and the procedure is repeated.

The second restoration technique also requires that the image have some dense shadows or other areas of

```
 1  22 17 20 18 23 26 26 26 26 25 22 27 24 24
    24 24 25 24 26 30 30 26 24 28 24 25 19 19
    23 21 22 32 31 28 27 22 24 23 23 22 23 20
    13 16 24 32 27 23 24 22 18 23 22 23 23 19
    16 21 30 31 25 23 24 23 23 24 23 23 22 22
 6  18 20 36 40 26 24 26 22 20 26 24 22 28 22
    16 19 33 31 27 25 25 23 20 23 24 29 25 26
    19 16 20 24 27 27 25 22 25 25 23 31 25 30
    17 18 21 20 22 23 24 23 22 22 23 23 28 23
    23 19 18 20 16 19 24 24 23 23 23 23 24 24
    23 24 18 20 18 18 23 27 24 23 22 22 23 23
12  44 44 40 40 44 44 44 44 48 44 40 44 36 44
    21 21 21 21 24 21 20 27 20 22 19 19 19 22
    24 20 18 23 25 22 24 23 23 24 23 19 20 19
    22 22 23 23 22 23 19 23 20 17 23 23 20
    19 17 23 23 17 20 18 24 19 17 19 17 19 18
    18 18 19 17 19 18 20 19 20 18 17 18 16 18
18  36 36 36 36 32 40 32 40 36 44 36 36 36 40
    20 18 17 17 20 18 17 17 20 18 17 19 18 18
    25 24 18 20 20 19 18 25 20 18 18 20 18 17
    23 21 23 23 23 22 23 20 18 18 18 19 17
    18 18 24 18 19 20 18 17 17 19 17 19 18 19
    18 18 17 20 20 18 20 19 19 20 17 19 19 17
24  44 36 40 32 36 44 36 44 36 44 36 44 44 36
    22 18 18 20 18 20 19 20 18 17 19 18 19 17
    20 12 20 21 19 20 20 20 19 19 20 18 14 15
    23 24 23 23 16 19 17 19 18 21 18 17 18 18
    23 18 19 18 24 18 19 20 17 16 18 18 19 13
    18 20 18 17 23 22 22 23 18 20 17 17 25 23
30  36 36 44 36 40 32 36 32 36 32 32 40 44 44
    18 18 20 19 19 19 19 19 22 18 18 22 22
    19 21 19 18 18 19 18 13 18 21 19 23 25 19
    16 20 25 24 23 21 21 21 20 21 18 18 18 20
    16 23 31 32 25 23 19 20 20 20 19 19 19 19
    16 17 28 26 22 22 22 23 20 19 20 23 22 23
```

A. BEFORE DIGITAL RESTORATION.

```
 1  22 17 20 18 23 26 26 26 26 25 22 27 24 24
    24 24 25 24 26 30 30 26 24 28 24 25 19 19
    23 21 22 32 31 28 27 22 24 23 23 22 23 20
    13 16 24 32 27 23 24 22 18 23 22 23 23 19
    16 21 30 31 25 23 24 23 23 24 23 23 22 22
 6  18 20 36 40 26 24 26 22 20 26 24 22 28 22
    16 19 33 31 27 25 25 23 20 23 24 29 25 26
    19 16 20 24 27 27 25 22 25 25 23 31 25 30
    17 18 21 20 22 23 24 23 22 22 23 23 28 23
    23 19 18 20 16 19 24 24 23 23 23 23 24 24
    23 24 18 20 18 18 23 27 24 23 22 22 23 23
12  22 22 20 20 22 22 22 22 24 22 20 22 18 22
    21 21 21 21 24 21 20 27 20 22 19 19 19 22
    24 20 18 23 26 22 24 23 23 24 23 19 20 19
    22 22 23 23 23 22 23 19 23 20 17 23 23 20
    19 17 23 23 17 20 18 24 19 17 19 17 19 18
    18 18 19 17 19 18 20 19 20 18 17 18 16 18
18  18 18 18 18 18 16 20 16 20 18 22 18 18 18 20
    20 18 17 17 20 18 17 17 20 18 17 19 18 18
    25 24 18 20 20 19 18 25 20 18 18 20 18 17
    23 21 23 23 23 22 23 20 18 18 18 19 17
    18 18 24 18 19 20 18 17 17 19 17 19 18 19
    18 18 17 20 20 18 20 19 19 20 17 19 19 17
24  22 18 20 16 19 22 18 22 18 22 18 22 22 18
    22 18 18 20 18 20 19 20 18 17 19 18 19 17
    20 12 20 21 19 20 20 20 19 19 20 18 14 15
    23 24 23 23 15 19 17 19 18 21 18 17 18 18
    23 18 19 18 24 18 19 20 17 16 18 18 19 13
    18 20 18 17 23 22 22 23 18 20 17 17 25 23
30  18 18 22 18 20 16 18 16 18 16 16 20 22 22
    18 18 20 19 19 19 19 19 22 18 18 22 22
    19 21 19 18 18 19 18 13 18 21 19 23 25 19
    16 20 25 24 23 21 21 21 20 21 18 18 18 20
    16 23 31 32 25 23 19 20 20 20 19 19 19 19
    16 17 28 26 22 22 22 23 20 19 20 23 22 23
```

B. AFTER DIGITAL RESTORATION.

FIGURE 7.8 Restoration of periodic striping on an MSS image of an area in southeast Montana. The arrays of DNs represent only a portion of the image. Image processed at Jet Propulsion Laboratory, California Institute of Technology.

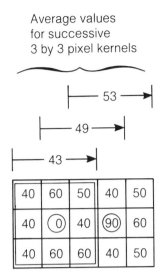

Average values
for successive
3 by 3 pixel kernels

A. ORIGINAL DATA SET
SHOWING OUTLINE OF
FILTER KERNEL (3 BY
3 PIXELS).

B. FILTERED DATA SET
WITH NOISE PIXELS
REPLACED BY
AVERAGE VALUES,
SHOWN IN CIRCLES.

FIGURE 7.9 Filtering to remove random-noise pixels.

zero reflectance on band 7. The histogram of band 7 in Figure 7.10B shows some values of 0. The histogram of band 4 lacks zero values, and the peak is offset toward higher DNs because of the additive effect of the haze. Band 4 also shows a characteristic abrupt increase in the number of pixels on the low end of the DN scale (left side of the histogram). The lack of DNs below this threshold is attributed to illumination from light scattered into the sensor by the atmosphere. For band 4 this threshold typically occurs at a DN of 11; this value is subtracted from all the band 4 pixels to correct for haze. For band 5 the correction is generally a DN of 7 and for band 6 a DN of 3. There can be a problem with this correction technique if the Landsat scene lacks dense shadows and if there are no band-7 pixels with DNs of 0.

Correcting Geometric Distortions

During the scanning process, a number of systematic and nonsystematic geometric distortions are introduced into MSS and TM image data. These distortions are corrected during production of the master images, which are remarkably orthogonal. The corrections may not be included in the CCTs, however, and geometric corrections must be applied before plotting images.

Nonsystematic Distortions *Nonsystematic distortions* (Figure 7.11A) are not constant because they result from variations in the spacecraft attitude, velocity, and altitude and therefore are not predictable. These distortions must be evaluated from Landsat tracking data or ground-control information. Variations in spacecraft velocity cause distortion in the along-track direction only and are known functions of velocity that can be obtained from tracking data.

The amount of earth rotation during the 28 sec required to scan an image results in distortion in the scan direction that is a function of spacecraft latitude and orbit. In the correction process, successive groups of scan lines (6 for MSS, 16 for TM) are offset toward the west to compensate for earth rotation, resulting in the parallelogram outline of the image.

Variations in attitude (roll, pitch, and yaw) and altitude of the spacecraft cause nonsystematic distortions that must be determined for each image in order to be corrected. The correction process employs geographic features on the image, called *ground control points* (GCPs), whose positions are known. Intersections of major streams, highways, and airport runways are typical GCPs. Differences between actual GCP locations and their positions in the image are used to determine the geometric transformations required to restore the image. The original pixels are resampled to match the correct geometric coordinates. The various resampling methods are described by Rifman (1973), Goetz and others (1975), and Bernstein and Ferneyhough (1975).

Systematic Distortions Geometric distortions whose effects are constant and can be predicted in advance are called *systematic distortions*. Scan skew, cross-track distortion, and variations in scanner mirror velocity belong to this category (Figure 7.11B). *Cross-track distortion* results from sampling of data along the MSS scan line at regular time intervals that correspond to spatial intervals of 57 m. The length of the ground interval is actually proportional to the tangent of the scan angle and therefore is greater at either margin of the scan line. The resulting marginal compression of Landsat images is identical to that described for airborne thermal IR scanner images (Chapter 5); however, distortion is min-

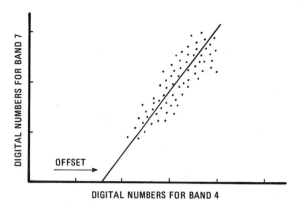

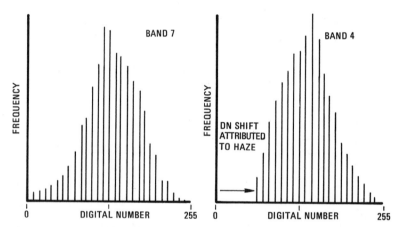

A. PLOT OF BAND 7 VERSUS BAND 4 FOR AN AREA WITHIN THE IMAGE THAT HAS SHADOWS. OFFSET OF THE LINE OF LEAST—SQUARES FIT ALONG THE BAND 4 AXIS IS ATTRIBUTED TO ATMOSPHERIC SCATTERING IN THAT BAND.

B. HISTOGRAMS FOR BANDS 7 AND 4. THE LACK OF LOW DN'S ON BAND 4 IS CAUSED BY ILLUMINATION FROM LIGHT SCATTERED BY THE ATMOSPHERE (HAZE).

FIGURE 7.10 Methods for determining atmospheric corrections on individual Landsat MSS bands. From Chavez (1975, Figures 2 and 3).

imal in Landsat images because the scan angle is only 5.8° on either side of vertical in contrast to the 45° to 60° angles of airborne cross-track scanners.

Tests before Landsat was launched determined that the velocity of the MSS mirror would not be constant from start to finish of each scan line, resulting in minor systematic distortion along each scan line. The known mirror velocity variations may be used to correct for this effect. Similar corrections are employed for TM images.

Scan skew is caused by the forward motion of the spacecraft during the time required for each mirror sweep. The ground swath scanned is not normal to the ground track but is slightly skewed, producing distortion across the scan line. The known velocity of the satellite is used to restore this geometric distortion.

IMAGE ENHANCEMENT

Enhancement is the modification of an image to alter its impact on the viewer. Generally enhancement distorts the original digital values; therefore enhancement is not done until the restoration processes are completed.

Contrast Enhancement

Chapter 1 demonstrated the strong influence of contrast ratio on resolving power and detection capability of images. Techniques for improving image contrast are among the most widely used enhancement processes.

The sensitivity range of Landsat TM and MSS detectors was designed to record a wide range of terrain brightness from black basalt plateaus to white sea ice

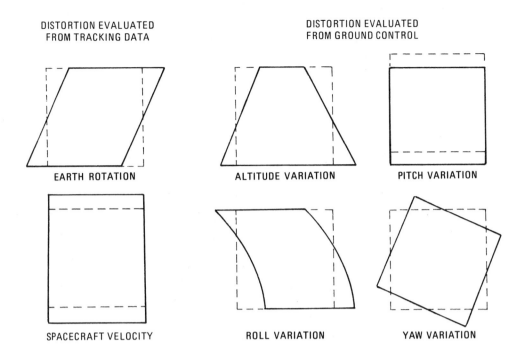

A. NONSYSTEMATIC DISTORTIONS. DASHED LINES
INDICATE SHAPE OF DISTORTED IMAGE; SOLID LINES
INDICATE SHAPE OF RESTORED IMAGE.

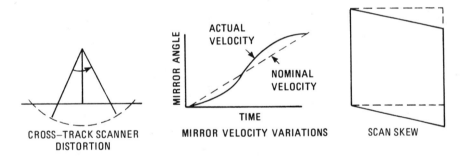

B. SYSTEMATIC DISTORTIONS.

FIGURE 7.11 Geometric distortions of Landsat images. From Bernstein and Ferneyhough (1975, Figure 3).

under a wide range of lighting conditions. Few individual scenes have a brightness range that utilizes the full sensitivity range of the MSS and TM detectors. To produce an image with the optimum contrast ratio, it is important to utilize the entire brightness range of the display medium, which is generally film. Figure 7.12A is a portion of an MSS band-4 image of the Andes along the Chile-Bolivia border that was produced directly from the CCT data with no modification of original DNs. The accompanying histogram shows the number of pixels that correspond to each DN. The central 92 percent of the his-

togram has a range of DNs from 49 to 106, which utilizes only 23 percent of the available brightness range [(106 − 49)/256 = 22.3%]. This limited range of brightness values accounts for the low contrast ratio of the original image. Three of the most useful methods of contrast enhancement are described in the following sections.

Linear Contrast Stretch The simplest contrast enhancement is called a *linear contrast stretch* (Figure 7.12B). A DN value in the low end of the original histogram is assigned to extreme black, and a value at the

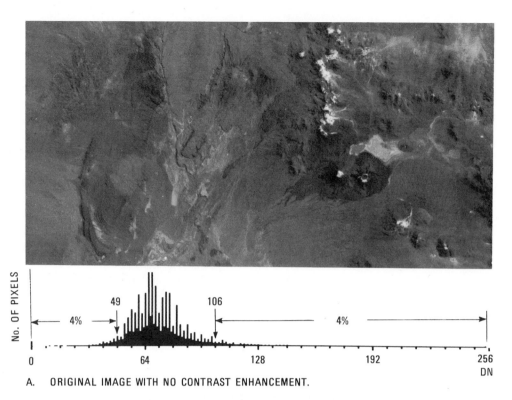

A. ORIGINAL IMAGE WITH NO CONTRAST ENHANCEMENT.

B. LINEAR CONTRAST STRETCH WITH LOWER AND UPPER FOUR
PERCENT OF PIXELS SATURATED TO BLACK AND WHITE RESPECTIVELY.

FIGURE 7.12 Contrast-enhancement methods. Portion of Landsat MSS band-4
image of an area in the northern Andes, Chile and Bolivia. From Soha and others
(1976, Figures 4 and 5). Courtesy Jet Propulsion Laboratory, California Institute
of Technology.

C.   UNIFORM DISTRIBUTION STRETCH.

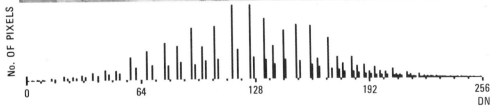

D.   GAUSSIAN STRETCH.

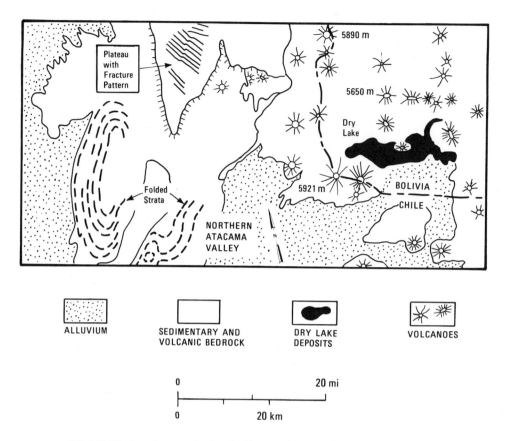

FIGURE 7.13 Location map for Landsat image of an area in the northern Andes, Chile and Bolivia.

high end is assigned to extreme white. In this example the lower 4 percent of the pixels (DN < 49) are assigned to black, or DN = 0, and the upper 4 percent (DN > 106) are assigned to white, or DN = 255. The remaining pixel values are distributed linearly between these extremes, as shown in the enhanced image and histogram of Figure 7.12B. The map of Figure 7.13 is useful in locating features for comparison on the images. The improved contrast ratio of the image with linear contrast stretch is particularly evident in the enhancement of individual rock units in the area of folded strata west of the Atacama Valley. Northwest-trending fractures occur in the plateau in the north-central portion of the area (Figure 7.13). The fractures are visible on the stretched image but are obscure on the original image. Most of the Landsat images in this book have been linearly stretched. For color images, the individual bands were stretched before being combined in color.

The linear contrast stretch greatly improves the contrast of most of the original brightness values, but there is a loss of contrast at the extreme high and low end of DN values. In the northeast portion of the original image

(Figure 7.12A), the lower limits of snow caps on volcanoes are clearly defined. On the stretched image (Figure 7.12B), however, the white tone includes both the snow and the alluvium lower on the flanks of the mountains. In the small dry lake north of the border, patterns that are visible on the original image are absent on the stretched image. Brightness differences within the dry lake and between the snow and alluvium were in the range of DNs greater than 106. On the stretched image, all these DNs are now white, as shown on the histogram (Figure 7.12B) by the spike at a DN of 255. A similar effect occurs at the low end of the DN range. On the original image, some individual dark lava flows are differentiated in the vicinity of the 5921-m volcano (see Figure 7.13 for location), but all of the flows are black on the stretched image because all DNs less than 49 are now black. In comparison to the overall contrast improvement in Figure 7.12B, these contrast losses at the brightness extremes are acceptable unless one is specifically interested in these elements of the scene.

Because of the flexibility of digital methods, an investigator could, for example, cause all DNs less than

106 to become black (0) and then linearly stretch the remaining high DNs greater than 105 through a range from 1 through 255. This extreme stretch would enhance the contrast differences within the bright pixels at the expense of the remainder of the scene. For the original Andes image (Figure 7.12A), 4 and 96 percent happened to be the optimum dark and bright limits; for the other MSS bands of this scene and for other images, different limits may be optimum and can be determined by inspection of the original histograms.

Nonlinear Contrast Stretch   Nonlinear contrast enhancement is also possible. Figure 7.12C illustrates a *uniform distribution stretch* (or histogram equalization) in which the original histogram has been redistributed to produce a uniform population density of pixels along the horizontal DN axis. This stretch applies the greatest contrast enhancement to the most populated range of brightness values in the original image. In Figure 7.12C the middle range of brightness values are preferentially stretched, which results in maximum contrast in the alluvial deposits around the flanks of the mountains. The northwest-trending fracture pattern is also more strongly enhanced by this stretch. As both histogram and image show, the uniform distribution stretch strongly saturates brightness values at the sparsely populated light and dark *tails* of the original histogram. The resulting loss of contrast in the light and dark ranges is similar to that in the linear contrast stretch but not as severe.

A *Gaussian stretch* (Figure 7.12D) is a nonlinear stretch that enhances contrast within the tails of the histogram. This stretch fits the original histogram to a normal distribution curve between the 0 and 255 limits, which improves contrast in the light and dark ranges of the image. The different lava flows are distinguished, and some details within the dry lake are emphasized. This enhancement occurs at the expense of contrast in the middle gray range; the fracture pattern and some of the folds are suppressed in this image.

An important step in the process of contrast enhancement is for the user to inspect the original histogram and determine the elements of the scene that are of greatest interest. The user then chooses the optimum stretch for his needs. Experienced operators of image processing systems bypass the histogram examination stage and adjust the brightness and contrast of images that are displayed on a CRT. For some scenes a variety of stretched images are required to display fully the original data. It also bears repeating that contrast enhancement should not be done until other processing is completed, because the stretching distorts the original values of the pixels.

Intensity, Hue, and Saturation Transformations

Chapter 2 described the additive system of primary colors (red, green, and blue, or RGB system). An alternate approach to color is the *intensity, hue, and saturation system* (IHS), which is useful because it presents colors more nearly as the human observer perceives them. The IHS system is based on the color sphere (Figure 7.14) in which the vertical axis represents intensity, the radius is saturation, and the circumference is hue. The *intensity* (*I*) axis represents brightness variations and ranges from black (0) to white (255); no color is associated with this axis. *Hue* (*H*) represents the dominant wavelength of color. Hue values commence with 0 at the midpoint of red tones and increase counterclockwise around the circumference of the sphere to conclude with 255 adjacent to 0. *Saturation* (*S*) represents the purity of color and ranges from 0 at the center of the color sphere to 255 at the circumference. A saturation of 0 represents a completely impure color, in which all wavelengths are equally represented and which the eye will perceive a shade of gray that ranges from white to black depending on intensity. Intermediate values of saturation represent pastel shades, whereas high values represent purer and more intense colors. The range from 0 to 255 is used here for consistency with

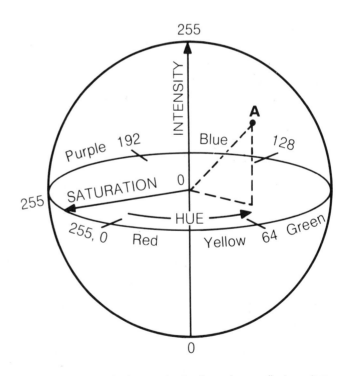

FIGURE 7.14 Intensity, hue, and saturation color coordinate system. The color at point A has the following values: *I* = 195, *H* = 75, *S* = 135.

the Landsat eight-bit scale; any range of values (0 to 100, for example) could be used as IHS coordinates. Buchanan (1979) describes the IHS system in detail.

When any three spectral bands of MSS or TM data are combined in the RGB system, the resulting color images typically lack saturation, even though the bands have been contrast-stretched. Plate 10A is an RGB normal color image prepared from TM bands 3, 2, and 1 of the Thermopolis, Wyoming, subscene that was described in Chapter 4. The individual bands were linearly stretched, but the color image has the pastel appearance that is typical of many Landsat images. The undersaturation is due to the high degree of correlation between spectral bands. High reflectance values in the green band, for example, are accompanied by high values in the blue and red bands, so pure colors are not produced. To correct this problem, a method of enhancing saturation was developed that consists of the following steps:

1. Transform any three bands of data from the RGB system into the IHS system in which the three component images represent intensity, hue, and saturation. (The equations for making this transformation are described shortly.) This IHS transformation was applied to TM bands 1, 2, and 3 of the Thermopolis subscene to produce the intensity, hue, and saturation images illustrated in Figure 7.15. The intensity image (Figure 7.15A) is dominated by albedo and topography. Sunlit slopes have high intensity values (bright tones), and shadowed areas have low values (dark tones). The airport runway and water in the Wind River have low values (see Figure 4.21 for locations) Vegetation has intermediate intensity values, as do most of the rocks. In the hue image (Figure 7.15B), the low DNs used to identify red hues in the sphere (Figure 7.14) cause the Chugwater redbeds to have conspicuous dark tones. Vegetation has intermediate to light gray values assigned to the green hue. The lack of blue hues is shown by the absence of very bright tones. The original saturation image (Figure 7.15C) is very dark because of the lack of saturation in the original TM data. Only the shadows and rivers are bright, indicating high saturation for these features.

2. Apply a linear contrast stretch to the original saturation image. Note the overall increased brightness of the enhanced image (Figure 7.15D). Also note the improved discrimination between terrain types.

3. Transform the intensity, hue, and enhanced saturation images from the IHS system back into three images of the RGB system. These enhanced RGB images were used to prepare the new color composite image of Plate 10B, which is a significant improvement over the original version in Plate 10A. In the IHS version, note the wide range of colors and improved discrimination between colors. Some bright green fields are distinguishable in vegetated areas that were obscure in the original. The wider range of color tones helps separate rock units.

The following description of the IHS transformation is a summary of the work of R. Hayden (Short and Stuart, 1982). Figure 7.16 shows graphically the relationship between the RGB and IHS systems. Numerical values may be extracted from this diagram for expressing either system in terms of the other. In Figure 7.16 the circle represents a horizontal section through the equatorial plane of the IHS sphere (Figure 7.14), with the intensity axis passing vertically through the plane of the diagram. The corners of the equilateral triangle are located at the position of the red, green, and blue hues. Hue changes in a counterclockwise direction around the triangle, from red ($H = 0$), to green ($H = 1$), to blue ($H = 2$), and again to red ($H = 3$). Values of saturation are 0 at the center of the triangle and increase to a maximum of 1 at the corners. Any perceived color is described by a unique set of IHS values; in the RGB system, however, different combinations of additive primaries can produce the same color. The IHS values can be derived from RGB values through the transformation equations

$$I = R + G + B \qquad (7.1)$$

$$H = \frac{G - B}{I - 3B} \qquad (7.2)$$

$$S = \frac{I - 3B}{I} \qquad (7.3)$$

for the interval $0 < H < 1$, extended to $1 < H < 3$. After enhancing the saturation image, the IHS values are converted back into RGB images by inverse equations.

The IHS transformation and its inverse are also useful for combining images of different types. For example, a digital radar image could be geometrically registered to the Thermopolis TM data. After the TM bands are transformed into IHS values, the radar image data may be substituted for the intensity image. The new combination (radar, hue, and enhanced saturation) can then be transformed back into an RGB image that incorporates both radar and TM data.

Density Slicing

Density slicing converts the continuous gray tone of an image into a series of density intervals, or slices, each corresponding to a specified digital range. Digital

A. INTENSITY IMAGE.

B. HUE IMAGE.

C. SATURATION IMAGE, ORIGINAL.

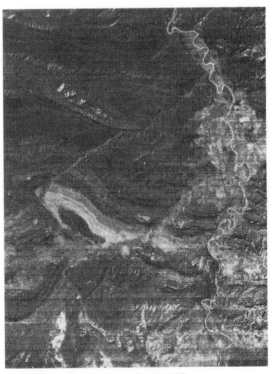

D. SATURATION IMAGE, STRETCHED.

FIGURE 7.15   Intensity, hue, and saturation transformation images of Landsat TM bands 1, 2, and 3 of the Thermopolis, Wyoming, subscene.

slices may be displayed as separate colors or as areas bounded by contour lines. This technique emphasizes subtle gray-scale differences that may be imperceptible to the viewer. In Chapter 5 digital density slicing was illustrated in the description of calibrated image displays.

Edge Enhancement

Most interpreters are concerned with recognizing linear features in images. Geologists map faults, joints, and lineaments. Geographers map manmade linear features such as highways and canals. Some linear features occur as narrow lines against a background of contrasting brightness; others are the linear contact between adjacent areas of different brightness. In all cases, linear features are formed by edges. Some edges are marked by pronounced differences in brightness and are readily recognized. More typically, however, edges are marked by subtle brightness differences that may be difficult to recognize. Contrast enhancement may emphasize brightness differences associated with some linear features. This procedure, however, is not specific for linear features because all elements of the scene are enhanced equally, not just the linear elements. Digital filters have been developed specifically to enhance edges in images and fall into two categories: directional and nondirectional.

Nondirectional Filters Laplacian filters are *nondirectional filters* because they enhance linear features having almost any orientation in an image. The exception applies to linear features oriented parallel with the direction of filter movement; these features are not enhanced. A typical *Laplacian filter* (Figure 7.17A) is a kernel with a high central value, 0 at each corner, and −1 at the center of each edge. The Laplacian kernel is placed over a three-by-three array of original pixels (Figure 7.17A), and each pixel is multiplied by the corresponding value in the kernel. The nine resulting values (four of which are 0 and four are negative numbers) are summed. The resulting value for the filter kernel is combined with the central pixel of the three-by-three data array, and this new number replaces the original DN of the central pixel. For example, the Laplacian kernel is placed over the array of three-by-three original pixels indicated by the box in Figure 7.17A. When the multiplication and summation are performed, the value for the filter kernel is 5. The central pixel in the original array (DN = 40) is combined with the filter value to produce the new value of 45, which is used in the filtered data set (Figure 7.17B). The kernel moves one column of pixels to the right, and the process repeats until the kernel reaches the right margin of the pixel array. The kernel then shifts back to the left margin, drops down one line of pixels, and continues the same process. The result is the data set

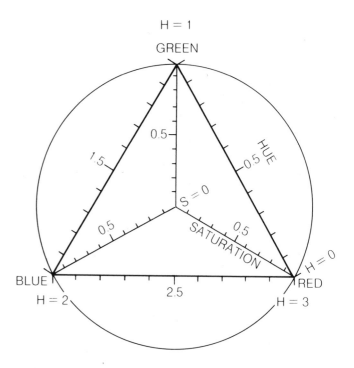

FIGURE 7.16 Diagram showing the relationship between the RGB and IHS systems.

of filtered pixels shown in Figure 7.17B. In this filtered data set, the outermost column and line of pixels are blank because they cannot form central pixels in an array.

The effect of the edge-enhancement operation can be evaluated by comparing profiles A–B of the original and the filtered data (Figure 7.17C,D). The original data set has a regional background value of 40 that is intersected by a darker, north-trending lineament that is three pixels wide and has DN values of 35. The contrast ratio, between the lineament and background, as calculated from Equation 1.3 is 40/35, or 1.14. In the enhanced profile, the contrast ratio is 45/30, or 1.50, which is a major enhancement. In addition the image of the original lineament, which was three pixels wide, has been widened to five pixels in the filtered version, which further enhances its appearance.

In the eastern portion of the original data set (Figure 7.17A), the background values (40) change to values of 45 along a north-trending linear contact. In an image, this contact would be a subtle tonal lineament between brighter and darker surfaces. The original contrast ratio (45/40 = 1.13) is enhanced by 27 percent in the filtered image (50/35 = 1.43). After the value of the filter kernel has been calculated and prior to combining it with the original central data pixel, the calculated value may be multiplied by a weighting factor. The factor weight may

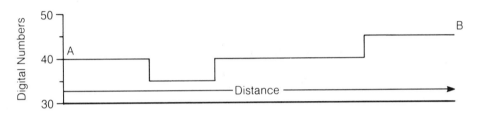

A. ORIGINAL DATA SET AND LAPLACIAN FILTER KERNEL.

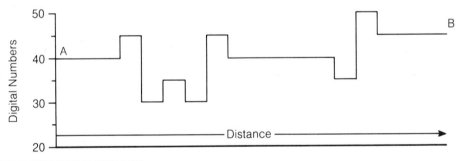

B. FILTERED DATA SET.

C. PROFILE OF THE ORIGINAL DATA.

D. PROFILE OF THE FILTERED DATA.

FIGURE 7.17 Nondirectional edge enhancement using a Laplacian filter.

be less than 1 or greater than 1 in order to diminish or to accentuate the effect of the filter. The weighted filter value is then combined with the central original pixel to produce the enhanced image.

Figure 7.18A is a computer-generated synthetic image with a uniform background (DN = 127). The left portion of the image is crossed by a darker lineament (DN = 107). The central portion is crossed by a brighter lineament (DN = 147). In the right portion of the image, a tonal lineament is formed where the background becomes somewhat brighter (DN = 137). These three linear features have subtle expressions in the original im-

age. The dark lineament and the tonal lineament of the synthetic image are similar to the features in the digital pixel arrays of Figure 7.17A. The Laplacian filter of Figure 7.17A was applied to the synthetic image (Figure 7.18A), and the calculated kernel value was multiplied by a factor of 2.0 to produce the enhanced image in Figure 7.18B; a factor of 5.0 was used to produce Figure 7.18C; a factor of 10.0 was used for Figure 7.18D, which saturates the brightest values to a DN of 255 and the darkest to a DN of 0. The filter has significantly improved the expression of the original lineaments in Figure 7.18A.

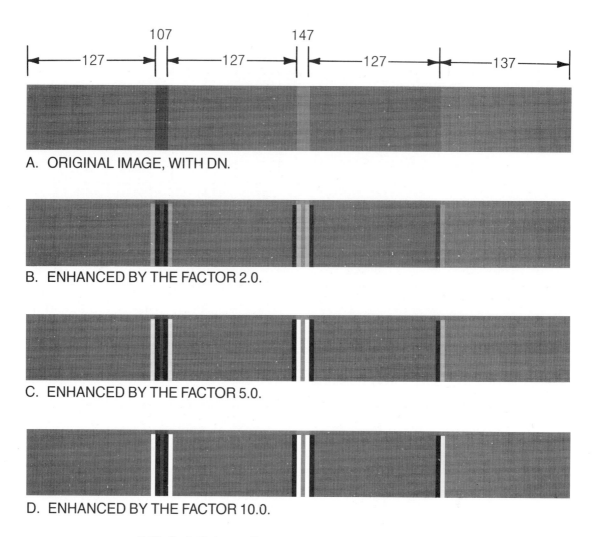

A. ORIGINAL IMAGE, WITH DN.

B. ENHANCED BY THE FACTOR 2.0.

C. ENHANCED BY THE FACTOR 5.0.

D. ENHANCED BY THE FACTOR 10.0.

FIGURE 7.18 Synthetic images illustrating nondirectional edge enhancement with a Laplacian filter and different weighting factors.

Having described the design and operation of a filter for nondirectional edge enhancement, the next step is to apply the filter to a real image. Figure 7.19A is a portion of a Landsat MSS image of an area in Chile. As shown by the map (Figure 7.19F), the central portion of the image is a plateau capped by a layer of resistant rocks that is cut by two sets of fractures. The set that is more abundant and better expressed in the original image trends northwest. A second, less abundant set with a weaker expression trends north. The nondirectional Laplacian filter kernel of Figure 7.17A was applied to the original Landsat image with a factor of 0.8, and the enhanced image is shown in Figure 7.19B. In comparing the original image and the enhanced image, the following features are significant:

1. In the northern part of the plateau, the northwest-trending fractures visible in the original image are

strongly enhanced with light and dark borders, as demonstrated earlier in Figure 7.18.

2. In the southern part of the plateau, northwest-trending fractures that are almost indiscernible in the original image are clearly visible in the enhanced version.

3. The north-trending fractures are somewhat more visible in the enhanced image but are largely obscured by the northwest-trending features.

4. Topographic features such as drainage channels, ridges, and scarps are sharply defined in the enhanced image.

Directional Filters *Directional filters* are used to enhance specific linear trends in an image. Equation 5 and Figure 1 of Haralick (1984) describe the concept of directional filters. A typical filter (Figure 7.20A) consists of two kernels, each of which is an array of three-by-

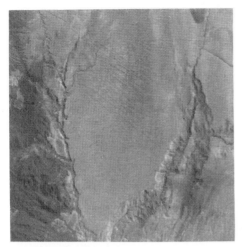

A. ORIGINAL IMAGE.

B. NONDIRECTIONAL ENHANCEMENT
(FACTOR 0.8).

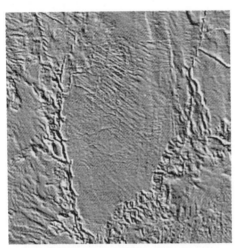

C. DIRECTIONAL ENHANCEMENT OF
NORTHWEST–TRENDING LINEAMENTS.

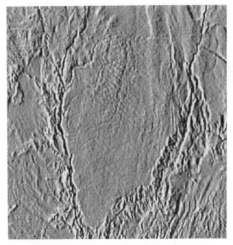

D. DIRECTIONAL ENHANCEMENT OF
NORTH–TRENDING LINEAMENTS.

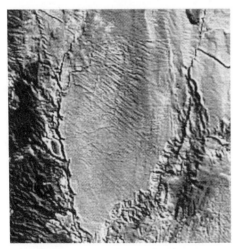

E. ORIGINAL IMAGE COMBINED WITH
IMAGE C.

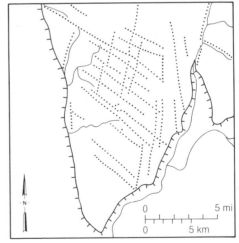

F. INTERPRETATION MAP.

FIGURE 7.19 Nondirectional and directional edge enhancement compared. Landsat
MSS band-5 image in the Altiplano region, Chile.

$$\text{Cos } A \cdot \begin{array}{|c|c|c|} \hline -1 & 0 & 1 \\ \hline -1 & 0 & 1 \\ \hline -1 & 0 & 1 \\ \hline \end{array} + \text{Sin } A \cdot \begin{array}{|c|c|c|} \hline 1 & 1 & 1 \\ \hline 0 & 0 & 0 \\ \hline -1 & -1 & -1 \\ \hline \end{array} = \text{Filter Value, Where}$$

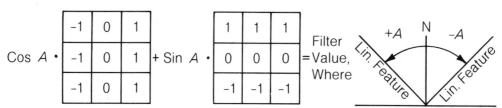

A. DIRECTIONAL FILTER KERNEL.

40	40	40	40	40	40
40	40	40	40	40	35
40	40	40	(40)	35	35
40	40	40	35	35	35
40	40	35	35	35	35
40	35	35	35	35	35

A • … • B

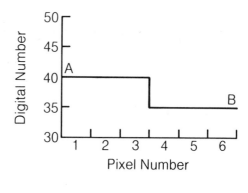

B. ORIGINAL DATA. EDGE DIRECTION IS N45°E(−).

0	0	0	0	-8	-14
0	0	0	-8	-14	-14
0	0	-8	(-14)	-14	-8
0	-8	-14	-14	-8	0
-8	-14	-14	-8	0	0
-14	-14	-8	0	0	0

A • … • B

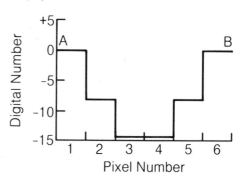

C. KERNEL VALUES.

40	40	40	40	32	26
40	40	40	32	26	21
40	40	32	(26)	21	27
40	32	26	21	27	35
32	26	21	27	35	35
26	21	27	35	35	35

A • … • B

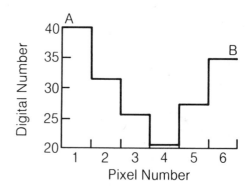

D. SUM OF ORIGINAL DATA AND KERNEL VALUES (IMAGE C).

FIGURE 7.20 Edge enhancement using a directional filter.

three pixels. The left kernel is multiplied by the cos A, where A is the angle, relative to north, of the linear direction to be enhanced. The right kernel is multiplied by sin A. Angles in the northeast quadrant are considered negative; angles in the northwest quadrant are positive. The filter can be demonstrated by applying it to a data set (Figure 7.20B) in which a brighter area (DN = 40) is separated from a darker area (DN = 35) along a tonal lineament that trends northeast ($A = -45°$). The profile A–B shows the brightness difference (DN = 5) across the lineament.

The filter is demonstrated by applying it to the array of nine pixels shown by the box over the original data set (Figure 7.20B). The sequence of operations is as follows:

1. Place the right filter kernel over the array of original pixels and multiply each pixel by the corresponding filter value. When these nine resulting values are summed, the filter result is 10.

2. Determine the sine of the angle (sin $-45° = -0.71$) and multiply this by the filter result (-0.71×10) to give a filter value of -7.

3. Place the left filter kernel over the pixel array and repeat the process; the a resulting value is -7.

4. Sum the the two filtered values (-14); this value replaces the central pixel in the array of the original data. When steps 1 through 4 are applied to the entire pixel array, the resulting filtered values are those shown in the array and profile of Figure 7.20C.

5. The filtered values for each pixel are then combined with the original value of the pixel to produce the array and profile of Figure 7.20D.

The contrast ratio of the lineament in the original data set (40/35 = 1.14) is increased in the final data set to (40/21 = 1.90), which is a contrast enhancement of 67 percent. As shown in profile A–B, the original lineament occurs across an interface only one pixel wide; the enhanced lineament has an image four pixels wide.

The directional filter of Figure 7.20 was applied to the Landsat image of Figure 7.19A. In the first case the angle of the features to be enhanced was specified as N55°W, (angle $A = +55°$) the direction of the northwest-trending fractures (Figure 7.19F). The resulting kernel values (Figure 7.20C) are shown in the image of Figure 7.19C. An alternate way to express this operation is to say that the filter has passed through the data in a direction normal to the specified lineament direction. In this case the filter direction is S35°W. In addition to enhancing features oriented normal to this direction of movement, the filter also enhances linear features that trend obliquely to the direction of filter movement. As a result, many additional edges of diverse orientations are enhanced in Figure 7.19C. The filter values of Figure 7.19C are combined with the original data (Figure 7.19A) to produce the enhanced image of Figure 7.19E. The directionality of the filter may also be demonstrated by passing the filter in a direction parallel with a linear trend. The filter values in Figure 7.19D were produced by using an orientation of N35°E for angle A. The effect is to move the filter in a direction N55°W, which is parallel with the direction of the major fractures. As a result these fractures appear subdued in Figure 7.19D, whereas north- and northeast-trending features are strongly enhanced.

Making Digital Mosaics

Mosaics of Landsat images may be prepared by matching and splicing together individual images, as described in Chapter 4. Differences in contrast and tone between adjacent images cause the checkerboard pattern that is common on many mosaics. This problem can be largely eliminated by preparing mosaics directly from the digital CCTs, as described by Bernstein and Ferneyhough, 1975.

Adjacent images are geometrically registered to each other by recognizing ground control points (GCPs) in the regions of overlap. Pixels are then geometrically adjusted to match the desired map projection. The next step is to eliminate from the digital file the duplicate pixels within the areas of overlap. Optimum contrast stretching is then applied to all the pixels, producing a uniform appearance throughout the mosaic. Figure 7.21A is a digital mosaic prepared in this manner from eight MSS band-7 images covering the portions of southeast Montana and adjacent Wyoming shown on the location map (Figure 7.21B). Note the excellent geometric agreement between adjacent images. The eastern strip was imaged on July 30 and the western strip on July 31, 1973. The only tonal mismatches in the mosaic result from changes in cloud pattern and illumination levels on the two days. Bernstein and Ferneyhough (1975) repeated the registration procedure for bands 4 and 5 of the eight images and composited them with band 7 to produce an IR color mosaic of the area (not shown).

Producing Synthetic Stereo Images

Ground-control points may be used to register Landsat pixel arrays to other digitized data sets, such as topographic maps. This registration causes an elevation value to be associated with each Landsat pixel. With this information the computer can then displace each

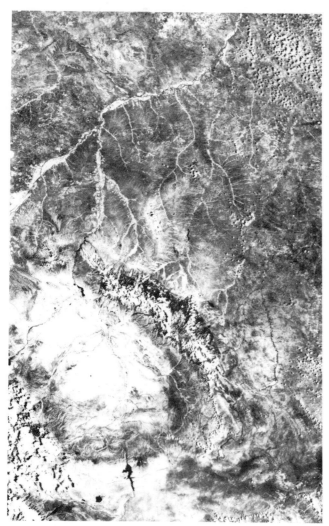

A. MOSAIC OF EIGHT IMAGES (BAND 7) ACQUIRED
JULY 30 AND 31, 1973.

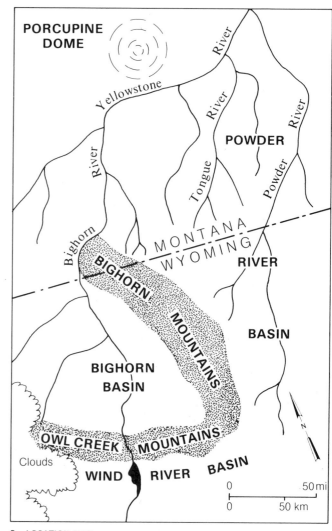

B. LOCATION MAP.

FIGURE 7.21 Digitally composited and plotted mosaic of eight Landsat MSS band-7
images of southeast Montana and northeast Wyoming. From Bernstein and
Ferneyhough (1975). Courtesy R. Bernstein, IBM.

pixel in a scan line relative to the central pixel of that
scan line. Pixels to the west of the central pixel are
displaced westward by an amount that is determined by
the elevation of the pixel and by its distance from the
central pixel. The same procedure determines eastward
displacement of pixels east of the central pixel. The
resulting image simulates the parallax of an aerial photograph. The principal point is then shifted, and a second
image is generated with the parallax characteristics of
the overlapping image of a stereo pair.

This form of enhancement is illustrated in Figure 7.22,
which covers the Gunnison River area in west-central
Colorado. Elevation data from a topographic map were
digitized with 79-by-79-m pixels that were registered to

the MSS image using GCPs. The process described above
was used to produce the pair of synthetic stereo images
in Figure 7.22A. View the images with a stereoscope to
appreciate the three-dimensional effect and vertical exaggeration. Compare the stereo model with the map
(Figure 7.22B) to evaluate the topographic features of
the area.

A synthetic stereo model is superior to a model from
side-lapping portions of adjacent Landsat images because (1) the vertical exaggeration can be increased and
(2) the entire image may be viewed stereoscopically.
Two disadvantages of computer images are that they are
expensive and that a digitized topographic map must be
available for elevation control.

A. SYNTHETIC STEREO PAIR.

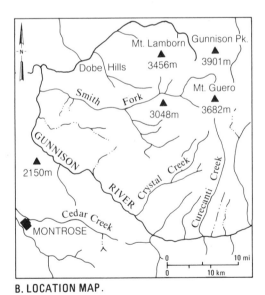

B. LOCATION MAP.

FIGURE 7.22 Synthetic stereo image and topographic map of the Gunnison River area, west-central Colorado, plotted from Landsat MSS band-5 data. From Batson, Edwards, and Eliason (1976, Plate 1).

Simpson and others (1980) employed a version of this method to combine Landsat images with *aeromagnetic maps* (in which the contour values represent the magnetic properties of rocks). The digitized magnetic data are registered to the Landsat band so that each Landsat pixel has an associated magnetic value. The stereoscopic transformation is then used to produce a stereo pair of Landsat images in which magnetic values determine the vertical relief in the stereo model. Elevated areas represent high magnetic values and depressions represent low magnetic values. The stereo model enables the interpreter to compare directly the magnetic and visual signatures of the rocks.

INFORMATION EXTRACTION

Image restoration and enhancement processes utilize computers to provide corrected and improved images for study by human interpreters. The computer makes no decisions in these procedures. However, processes that identify and extract information do utilize the computer's decision-making capability to identify and extract specific pieces of information. A human operator must instruct the computer and must evaluate the significance of the extracted information.

Principal-Component Images

For any pixel in a multispectral image, the DN values are commonly highly correlated from band to band. This correlation is illustrated schematically in Figure 7.23, which plots digital numbers for pixels in TM bands 1 and 2. The elongate distribution pattern of the data points indicates that as brightness in band 1 increases, brightness in band 2 also increases. A three-dimensional plot (not illustrated) of three TM bands, such as 1, 2, and 3, would show the data points in an elongate ellipsoid, indicating correlation of the three bands. This correlation means that if the reflectance of a pixel in one band (TM band 2, for example) is known, one can predict the reflectance in adjacent bands (TM bands 1 and 3). The correlation also means that there is much redundancy in a multispectral data set. If this redundancy could be reduced, the amount of data required to describe a multispectral image could be compressed.

The *principal-components transformation,* originally known as the Karhunen-Loéve transformation (Loéve, 1955), is used to compress multispectral data sets by calculating a new coordinate system. For the two bands of data in Figure 7.23, the transformation defines a new axis (y_1) oriented in the long dimension of the distribution and a second axis (y_2) perpendicular to y_1. The mathematical operation makes a linear combination of

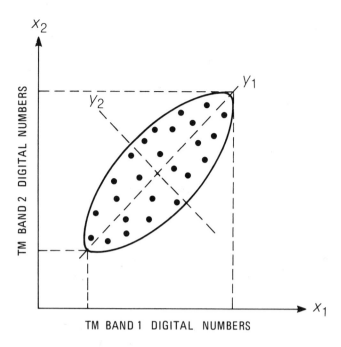

FIGURE 7.23 Scatter plot of Landsat TM band 1 (x_1 axis) and band 2 (x_1 axis) showing correlation between these bands. The principal-components transformation was used to generate a new coordinate system (y_1, y_2). After Swain and Davis (1978, Figure 7.9).

pixel values in the original coordinate system that results in pixel values in the new coordinate system:

$$y_1 = \alpha_{11}x_1 + \alpha_{12}x_2 \qquad (7.4)$$

$$y_2 = \alpha_{21}x_1 + \alpha_{22}x_2 \qquad (7.5)$$

where (x_1, x_2) are the pixel coordinates in the original system; (y_1, y_2) are the coordinates in the new system; and α_{11}, α_{12}, α_{21}, and α_{22} are constants. In Figure 7.23 note that the range of pixel values for y_1 is greater than the ranges for either of the original coordinates, x_1 or x_2 and that the range of values for y_2 is relatively small.

The same principal-components transformation may be carried out for multispectral data sets consisting of any number of bands. Additional coordinate directions are defined sequentially. Each new coordinate is oriented perpendicular to all the previously defined directions and in the direction of the remaining maximum density of pixel data points. For each pixel, new DNs are determined relative to each of the new coordinate axes. A set of DN values is determined for each pixel relative to the first principal component. These DNs are then used to generate an image of the first principal component. The same procedure is used to produce images for the remaining principal components. The preceding description of the principal-components transformation is summarized from Swain and Davis (1978); additional information is given in Moik (1980).

A principal-components transformation was performed on the three visible and three reflected IR bands of TM data for the Thermopolis, Wyoming, subscene (Figure 4.20). Each pixel was assigned six new DNs for the first through the sixth principal-component coordinate axes. Figure 7.24 illustrates the six PC images, which have been enhanced with a linear contrast stretch. As noted earlier, each successive principal component accounts for a progressively smaller proportion of the variation of the original multispectral data set. These percentages of variation are indicated for each PC image in Figure 7.24 and are plotted graphically in Figure 7.25. The first three PC images contain 97 percent of the variation of the original six TM bands, which is a significant compression of data. The PC image 1 (Figure 7.24A) is dominated by topography, expressed as highlights and shadows, that is highly correlated in all six of the original TM bands. PC image 2 (Figure 7.24B) is dominated by differences in albedo that also correlate from band to band because pixels that are bright in one TM band tend to be bright in adjacent bands. The least correlated data are noise, such as line striping and dropouts, which occur in different detectors and different bands. In the example of Figure 7.24, noise dominates PC images 4, 5, and 6, which together account for only 2.6 percent of the original variation. It is interesting to note, however, that PC image 6 (Figure 7.24F) displays parallel arcuate dark bands in the outcrop belt of the Chugwater Formation that are clearly related to lithologic variations in this unit.

Any three principal-component images can be combined to create a color image by assigning the data that make up each image to a separate primary color. Plate 10C was produced by combining PC images from Figure 7.24 in the following fashion: PC image 2 = red, PC image 3 = green, PC image 4 = blue. PC image 1 was not used in order to minimize topographic effects. As a result, the color PC image displays a great deal of spectral variation in the vegetation and rocks, although the three images constitute only 10.7 percent of the variation of the original data set.

In summary, the principal-components transformation has several advantages: (1) most of the variance in a multispectral data set is compressed into one or two PC images; (2) noise may be relegated to the less-correlated PC images; and (3) spectral differences between materials may be more apparent in PC images than in individual bands.

Ratio Images

Ratio images are prepared by dividing the DN in one band by the corresponding DN in another band for each pixel, stretching the resulting value, and plotting the new values as an image. Figure 7.26 illustrates some ratio images prepared from the visible and reflected TM bands of the Thermopolis subscene (Figure 4.20). A total of 15 ratio images plus an equal number of inverse ratios (reciprocals of the first 15 ratios) may be prepared from these six original bands. In a ratio image the black and white extremes of the gray scale represent pixels with the greatest difference in reflectivity between the two spectral bands. The darkest signatures are areas where the denominator of the ratio is greater than the numerator. Conversely the numerator is greater than the denominator for the brightest signatures. Where denominator and numerator are the same, there is no difference between the two bands.

For example, the spectral reflectance curve (Figure 4.14) shows that the maximum reflectance of vegetation occurs in TM band 4 (reflected IR) and that reflectance is considerably lower in band 2 (green). Figure 7.26C shows the ratio image 4/2, the image that results when the DNs in band 4 are divided by the DNs in band 2. The brightest signatures in this image correlate with the cultivated fields along the Wind River and Owl Creek. For location of features in this and other discussions of the Thermopolis subscene, see the description and map of the area in Chapter 4. Red materials, such as the Chugwater Formation with its high content of iron oxide, have their maximum reflectance in band 3. Thus, in the ratio image 3/1 (red/blue) of Figure 7.26A, the Chugwater outcrops have very bright signatures.

Like PC images, any three ratio images may be combined to produce a color image by assigning each image to a separate primary color. In Plate 10D the ratio images 3/1, 5/7, and 3/5 are combined as red, green, and blue respectively. This image should be compared with the normal color and IR color images of Plate 5C,D and with the geologic map of Figure 4.21. The color variations of the ratio color image express more geologic information and have greater contrast between units than do the conventional color images. Ratio images emphasize differences in slopes of spectral reflectance curves between the two bands of the ratio. In the visible and reflected IR regions, the major spectral differences of materials are expressed in the slopes of the curves; therefore individual ratio images and ratio color images enable one to extract reflectance variations. A disadvantage of ratio images is that they suppress differences in albedo; materials that have different albedos but similar slopes of their spectral curves may be indistinguishable in ratio images.

Ratio images also minimize differences in illumination conditions, thus suppressing the expression of topography. In Figure 7.27 a red siltstone bed crops out on both the sunlit and shadowed sides of a ridge. In the individual Landsat TM bands 1 and 3 the DNs of the siltstone are lower in the shadowed area than in the sunlit outcrop, which makes it difficult to follow the siltstone bed around the hill. Values of the ratio image 3/1, however, are identical in the shadowed and sunlit areas, as shown by the chart in Figure 7.27; thus the siltstone has similar signatures throughout the ratio image. This suppression of topography can be seen by comparing the ratio images of the Thermopolis subscene (Figure 7.26) with images of the individual TM bands (Figure 4.20).

In addition to ratios of individual bands, a number of other ratios may be computed. An individual band may be divided by the average for all the bands, resulting in normalized ratios. Another combination is to divide the difference between two bands by their sum: for example, (band 4 − band 5)/(band 4 + band 5). Ratios of this type are used to process data from environmental satellites, as described in Chapter 9.

Multispectral Classification

For each pixel in a Landsat MSS or TM image, the spectral brightness is recorded for four or seven different wavelength bands respectively. A pixel may be characterized by its *spectral signature,* which is determined by the relative reflectance in the different wavelength bands. *Multispectral classification* is an information-extraction process that analyzes these spectral signatures and then assigns pixels to categories based on similar signatures.

Procedure Multispectral classification is illustrated diagrammatically with a Landsat MSS image of southern California that covers the Salton Sea, Imperial Valley, and adjacent mountains and deserts (Figure 7.28A). Reflectance data from the four MSS bands are shown in Figure 7.29A for representative pixels of water, agriculture, desert, and mountains. The data points are plotted at the center of the spectral range of each MSS band. In Figure 7.29B the reflectance ranges of bands 4, 5, and 6 form the axes of a three-dimensional coordinate system. The solid dots are the loci of the four pixels in Figure 7.29A. Plotting additional pixels of the different terrain types produces three-dimensional clusters or ellipsoids. The surface of the ellipsoid forms a *decision boundary,* which encloses all pixels for that terrain category. The volume inside the decision boundary is called the *decision space.* Classification programs differ in their

A. PC IMAGE 1 (88.4%).

B. PC IMAGE 2 (6.6%).

C. PC IMAGE 3 (2.4%).

D. PC IMAGE 4 (1.7%).

FIGURE 7.24 PC images of the Thermopolis, Wyoming, subscene. PC images were generated for the six visible and reflected IR bands of the Landsat TM image. Percentage of variance represented by each PC image is shown.

E. PC IMAGE 5 (0.5%).

F. PC IMAGE 6 (0.3%).

FIGURE 7.24   (CONTINUED)

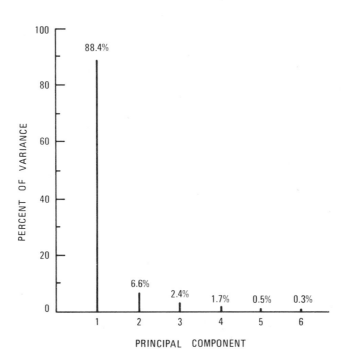

FIGURE 7.25   Percentage of variance represented by the six PC images of the Thermopolis subscene from Landsat TM data.

A. RATIO IMAGE 3/1.

B. RATIO IMAGE 3/5.

C. RATIO IMAGE 4/2.

D. RATIO IMAGE 5/7.

FIGURE 7.26 Ratio images of Landsat TM bands for the Thermopolis, Wyoming, subscene. The images have been stretched to enhance contrast.

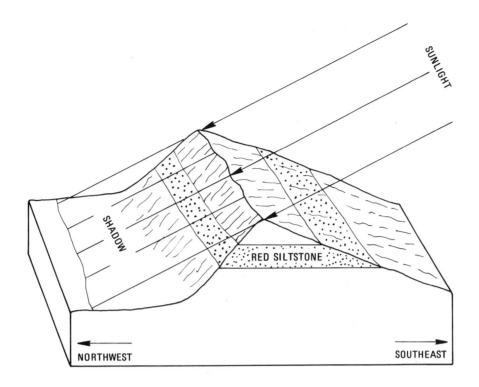

ILLUMINATION	TM BAND 3	TM BAND 1	RATIO 3/1
Sunlight	94	42	2.24
Shadow	76	34	2.23

SILTSTONE REFLECTANCE

FIGURE 7.27 Suppression of illumination differences on a ratio image.

criteria for defining the decision boundaries. In many programs the analyst is able to modify the boundaries to achieve optimum results. For the sake of simplicity the cluster diagram (Figure 7.29B) is shown with only three axes. In actual practice the computer employs a separate axis for each spectral band of data: four for MSS and six or seven for TM.

Once the boundaries for each cluster, or *spectral class,* are defined, the computer retrieves the spectral values for each pixel and determines its position in the classification space. Should the pixel fall within one of the clusters, it is classified accordingly. Pixels that do not fall within a cluster are considered unclassified. In practice the computer calculates the mathematical probability that a pixel belongs to a class; if the probability

exceeds a designated threshold (represented spatially by the decision boundary), the pixel is assigned to that class. Applying this method to the original data of the Salton Sea and the Imperial Valley scene produces the classification map of Figure 7.28B. Note that the blank areas (unclassified category) occur at the boundaries between classes where the pixels include more than one terrain type.

There are two major approaches to multispectral classification:

1. *Supervised classification.* The analyst defines on the image a small area, called a *training site,* which is representative of each terrain category, or class. Spectral values for each pixel in a training site are

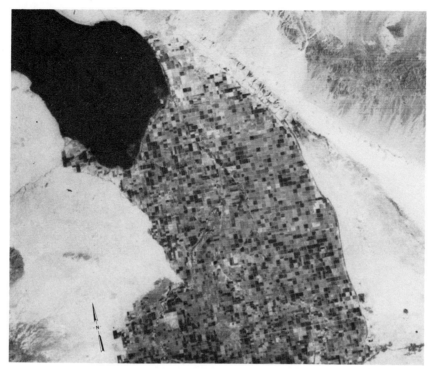

A. LANDSAT BAND–5 IMAGE.

B. CLASSIFICATION MAP. A = AGRICULTURE, D = DESERT,
M = MOUNTAINS, W = WATER, BLANK = UNCLASSIFIED.

FIGURE 7.28 Multispectral classification of Landsat MSS data for the Salton Sea and the Imperial Valley, California.

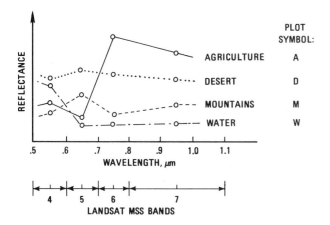

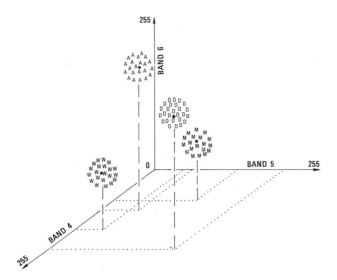

A. SPECTRAL REFLECTANCE
CURVES FOR MAJOR TERRAIN TYPES.

B. THREE–DIMENSIONAL CLUSTER DIAGRAM FOR
CLASSIFICATION.

FIGURE 7.29 Spectral reflectance curves and cluster diagram for Landsat MSS data of the Salton Sea and the Imperial Valley.

used to define the decision space for that class. After the clusters for each training site are defined, the computer then classifies all the remaining pixels in the scene.

2. *Unsupervised classification.* The computer separates the pixels into classes with no direction from the analyst.

The two classification methods can be compared by applying them to the TM data for the Thermopolis, Wyoming, subscene. As background for the following dis-

cussion, the reader should review the images, map, and description of this subscene in Chapter 4, together with the normal color and IR color images in Plate 5C,D.

Supervised Classification The first step in the supervised classification is to select training sites for each of the terrain categories. In Figure 7.30 the training sites are indicated by black rectangles. Some categories are represented by more than one training site in order to cover the full range of reflectance characteristics. Figure 7.31 shows TM reflectance spectra for the terrain categories. Note the wider range of spectral variation in the reflected IR bands (4, 5, and 7) than in the visible bands (1, 2, and 3). In the first attempt at supervised classification, training sites were selected for each of the geologic formations, but the results showed that individual formations of the same rock type had similar or overlapping spectral characteristics and were assigned to the same class. For example, many of the formations are sandstones, which proved to be spectrally similar despite their different geologic ages. The final classification identified rocks as sandstone, shale, and redbeds. The resulting supervised-classification map is shown in Plate 11A, in which the colors represent the six major terrain classes and black indicates unclassified pixels. The legend identifies the terrain class represented by each color and the percentage of the map occupied by each class as calculated by the classification program. The color classification map may be evaluated by comparing it with Figure 7.30, which is based on geologic maps of the Thermopolis area. Plate 11A shows the major terrain categories accurately. Some narrow belts of the shale class are shown interbedded with the sandstone class; this is correct because several sandstone formations, such as the Frontier Formation, include shale beds. The proportion of shale in the classification map is probably understated for the following reason. Shale weathers to very fine particles that are easily removed by wind and water. Sandstone, however, weathers to relatively coarse fragments of sand and gravel that may be transported downslope to partially conceal underlying shale beds. This relationship is shown where the course of dry streams that cross the outcrop of Cody Shale are classified as sandstone because of the transported sand and sandstone gravel in the stream beds.

Unsupervised Classification The unsupervised classification results are shown in the map and legend of Plate 11B. This particular program automatically defined 16 classes for the scene. After the 16 unsupervised classes were calculated, each class was displayed for inspection by the analyst, who then combined various categories into the 8 classes shown in Plate 11B. Distribution patterns of the sandstone and shale classes are similar to

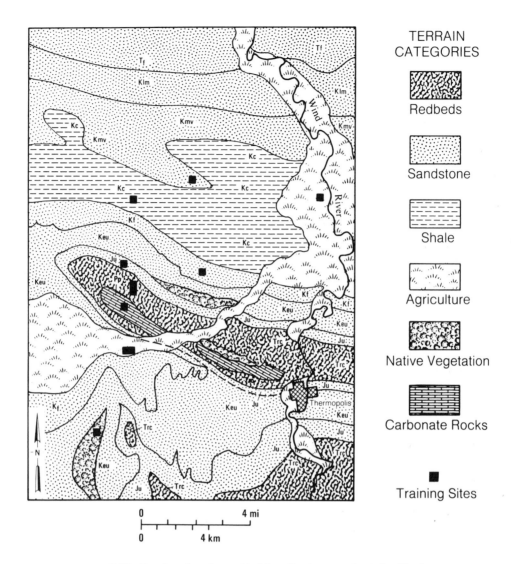

TERRAIN
CATEGORIES

Redbeds

Sandstone

Shale

Agriculture

Native Vegetation

Carbonate Rocks

Training Sites

0   4 mi

0   4 km

FIGURE 7.30 Terrain categories and training sites for supervised classification of the Thermopolis, Wyoming, subscene. Formation symbols are explained in Figure 4.21.

the patterns in the supervised map. The computer recognized two spectral subdivisions of the Chugwater Formation: a lower member shown in red and an upper member shown in orange (Plate 11B). The lower member is confined to the Chugwater outcrop belt, but the upper member is recognized beyond the Chugwater outcrops, where it apparently represents patches of reddish soil. The unsupervised classification recognized native vegetation and 3 distinct classes of agriculture.

Neither classification map recognized outcrops of carbonate rocks of the Phosphoria Formation, shown in the map of Figure 7.30. Aside from the cover of native vegetation, the weathered carbonate outcrops are classified as sandstone in both maps. Apparently the reflectance spectrum of the carbonate rocks in the TM bands is not

distinguishable from that of sandstone. Planned field measurements using a portable reflectance spectrometer will evaluate these properties.

Change-Detection Images

Change-detection images provide information about seasonal or other changes. The information is extracted by comparing two or more images of an area that were acquired at different times. The first step is to register the images using corresponding ground-control points. Following registration, the digital numbers of one image are subtracted from those of an image acquired earlier or later. The resulting values for each pixel will be positive, negative, or zero; the latter indicates no change.

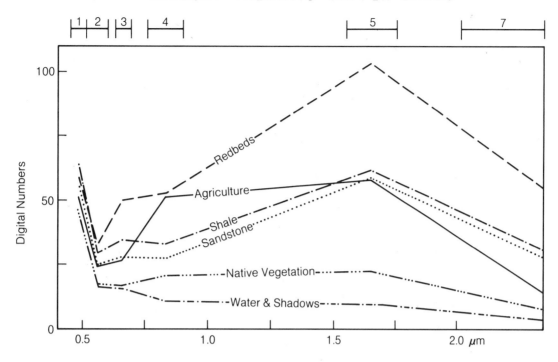

FIGURE 7.31 Reflectance spectra of training sites used in the supervised classification of the Thermopolis subscene.

The next step is to plot these values as an image in which a neutral gray tone represents zero. Black and white tones represent the maximum negative and positive differences respectively. Contrast stretching is employed to emphasize the differences.

The change-detection process is illustrated with Landsat MSS band-5 images of the Goose Lake area of Saskatchewan, Canada (Figure 7.32). The DN of each pixel in image B is subtracted from the DN of the corresponding registered pixel in image A. The resulting values are linearly stretched and displayed as the change-detection image (Figure 7.32C). The location map (Figure 7.32D) aids in understanding signatures in the difference image. Neutral gray tones representing areas of little change are concentrated in the northwest and southeast and correspond to forested terrain. Forest terrain has a similar dark signature on both original images. Some patches within the ephemeral Goose dry lake have similar light signatures on images A and B, resulting in a neutral gray tone on the different image. The clouds and shadows present only on image B produce dark and light tones respectively on the difference image. The agricultural practice of seasonally alternating between cultivated and fallow fields is clearly shown by the light and dark tones on the difference image. On the original images, the fields with light tones have crops or stubble and the fields with dark tones are bare earth.

Change-detection processing is also useful for producing difference images for other remote sensing data, such as between nighttime and daytime thermal IR images, as illustrated in Chapter 5.

SYSTEMS AND STRATEGY FOR IMAGE PROCESSING

Preceding sections described the major routines for processing digital image data; the present section describes the systems and strategy for implementing the processing routines. Image-processing systems consist of hardware and software. The hardware components are typically located in two separate rooms (Figure 7.33). Tape drives, disk drives, computers, and processing units are located in a central computer facility where they are maintained and serviced by computer operators. The color monitor, control terminal, and cursor control are located in an image-processing laboratory and are operated by image analysts (Figure 7.34). The host computer and tape drive may be shared among multiple users; the remaining hardware in Figure 7.33 is dedicated solely

A. SEPTEMBER 7, 1973, IMAGE.

B. JUNE 27, 1973, IMAGE.

BLACK   GRAY   WHITE
◄——————— 0 ———————►
(A < B)   (A > B)

C. DIFFERENCE IMAGE
(IMAGE A MINUS IMAGE B).

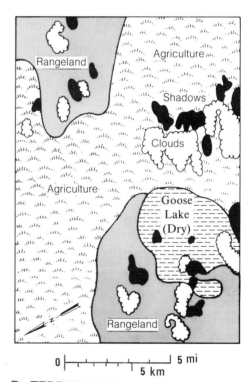

D. TERRAIN MAP.

FIGURE 7.32 Change-detection image computed from seasonal Landsat MSS band-5 images in Saskatchewan, Canada. From Rifman and others (1975, Figures 2-14, 2-15, and 2-17). Images courtesy TRW, Incorporated.

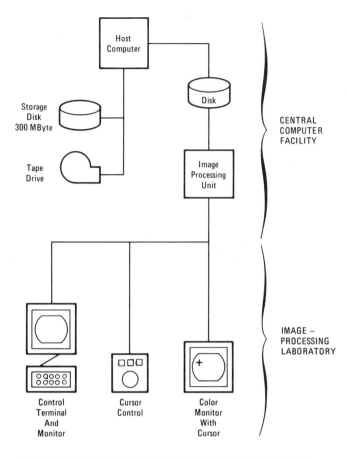

FIGURE 7.33 Diagram of an interactive digital image-processing system.

to image processing. Software for the various interactive processing routines runs in the image processing unit. Operation of the system can be described by summarizing a typical image-processing session.

Processing Session

The image analyst begins an interactive session by typing appropriate commands on the keyboard of the control terminal to actuate the system. The black-and-white screen of the control terminal displays the status of processing throughout the session. An operator in the computer facility loads the appropriate Landsat CCT onto the tape drive, which reads and transfers the data to the 300-Mbyte disk. The disk can store up to 10 Landsat MSS scenes or a single TM scene. During this data transfer phase, many systems automatically reformat the data to correct for systematic geometric distortion and to resample MSS pixels into a square (79-by-79-m) format. The analyst is now ready to view the image on the color display unit, which is typically a television screen that displays 512 horizontal scan lines, each with 512 pixels. A resampled MSS image band is 2340 lines long by 3240 pixels wide, and a TM image is 5667 lines long by 6167 pixels wide (Figure 7.3). To display the entire area covered by these images on the 512-by-512 color monitor, the data are subsampled: only every fifth line and fifth pixel of MSS data are shown, and for TM only every thirteenth line and pixel are shown. Despite

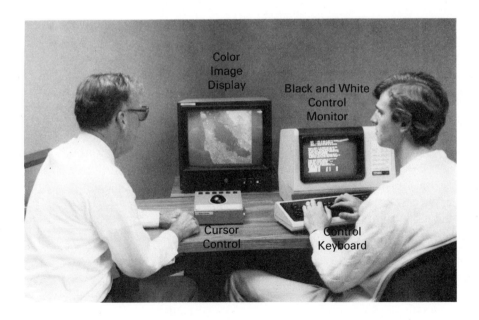

FIGURE 7.34 Interactive system in an image-processing laboratory.

the reduction of data, sufficient detail is visible on the screen to select a subscene for processing. The analyst has the option of displaying any combination of three bands on any combination of screen colors. Generally the green and red bands and a reflected IR band of data are displayed on the blue, green, and red phosphors of the screen to produce an IR color image. Images are traditionally oriented with north at the top of the screen.

The analyst selects a subscene by using the *cursor,* which is a bright cross that is moved across the screen with the aid of the cursor control. The *cursor control* is a track ball that the analyst manipulates to shift the cursor on the screen. For each position of the cursor, its location in the image is displayed on the screen by numbers showing the corresponding scan line and pixel. The analyst positions the cursor at the northwest corner of the desired 512-by-512 subscene and types a command that identifies this locality. The analyst then positions the cursor at the southeast corner of the subscene and enters the command. The system then selects this subscene from the storage disk, transfers it to the local disk in the image processing unit, and displays the 512-by-512 array of pixels on the color monitor. With this display the analyst is ready to begin a session of interactive image processing. The analyst specifies the desired digital processing, views the results on the monitor in near real time, and makes further adjustments. The analyst continues to interact with the data via the cursor and keyboard until obtaining the desired results.

A typical interactive processing session can involve combinations of restoration, enhancement, and information extraction. The first step is to examine the 512-by-512 display for defects such as line dropouts and noise that are corrected. Next the image is enhanced using a contrast stretch routine. The analyst examines each of the three spectral bands separately, together with its histogram, and produces an optimum enhanced image for each band. The bands are then viewed together in color, and any contrast adjustments are made. At this point, the image may be transformed into intensity, hue, and saturation components. The saturation is stretched, and the IHS image is then transformed back into RGB components and viewed on the monitor. When the results are satisfactory for the 512-by-512 subscene, the various restoration and enhancement programs and parameters are then applied to the entire Landsat scene in the batch-processing mode. In *batch processing* the analyst does not interact with the data during the session and sees no results until processing is completed. For this reason, batch jobs may be run during the night shift of the computer facility when they will not compete for computer resources with interactive users on the day shift. In some systems, a separate large mainframe computer, such as a Cray supercomputer, is employed for batch processing. At the conclusion of a batch-processing session, the computer creates a file of processed pixels. The tape drive (Figure 7.33) copies the processed pixels onto a plot tape, which is used with an image-plotting system (Figure 7.2) to produce the photographic images.

Once an enhanced image is generated, the analyst may proceed to extract information by creating ratio images using the original data for the 512-by-512 subscene, not the enhanced data. Different ratio images are interactively combined in different colors to produce the optimum combination for recognizing desired features. At this point, a color ratio composite image for the entire scene may be produced in the batch mode. The same processing strategy is used to perform supervised multispectral classifications.

Image-Processing Systems

The author has conducted an informal survey of most of the commercial image-processing systems that are available at a wide range of prices and capabilities. The survey included the price of software and hardware. Price does not include the host computer, tape drive, and storage disks because many facilities already have this equipment, which can be shared for image processing. The systems belong to two major categories: (1) those costing less than $50,000, listed in Table 7.1; and (2) those costing more than $50,000, listed in Table 7.2.

The less expensive systems are hosted on personal computers or minicomputers. Because of their small memories, most of these systems can process only small subscenes of Landsat images but are useful for teaching purposes. The more expensive systems are hosted on larger computers and can process entire Landsat scenes in the batch-processing mode. These systems can be connected to even more powerful computers for faster processing.

Some potential users have monitors, terminals, and processors but lack the software for image processing. Software for various purposes, including image processing, that was developed under contract to the U.S. government is available to the public at modest prices. This software is available through the Computer Software Management and Information Center (COSMIC) at the University of Georgia, Athens, GA 30602. COSMIC will provide a copy of their catalog, which has a section describing available software for image processing.

Some users may require processed images but do not wish to invest in a processing system. There are a number of facilities that for a fee can process images to the user's specifications (Table 7.3).

TABLE 7.1 Smaller interactive digital image-processing systems

Vendor	System	Price range, $1000s	Host computer
ERDAS, Inc. 430 Tenth St., NW Atlanta, GA 30318	ERDAS PC	15 to 20	IBM PC
Gould, Inc. 1870 Lundy Avenue San Jose, CA 95131	FD5000	14 to 26	LSI 11/23
International Imaging Systems 1500 Buckeye Drive Milpitas, CA 95035	I²S 75	45	PDP
Spectral Data Corp. P.O. Box 11356 Hauppauge, NY 11788	VIP	7	Apple, IBM, PC, Epson, Compaq, Kaypro
Terra-Mar 2113 Landings Drive Mountain View, CA 94043	Microimage I/A	25	Compaq Desk Pro

TABLE 7.2 Larger interactive digital image-processing systems

Vendor	System	Price range, $1000s*	Host computer
Dipix, Inc. 10220 Old Columbia Road Columbia, MD 21046	ARIES III	150 to 250	VAX, PDP-11, SEL 32
ESL, Inc. 495 Java Drive Sunnyvale, CA 94088	IDIMS	350	VAX
Global Imaging 201 Lomas Santa Fe Solana Beach, CA 92075	GS9000	95	HP 9000
Gould, Inc. 1870 Lundy Avenue San Jose, CA 95131	IP 8400 IP 8500	50 55	VAX, PDP-11
IBM Corp. P.O. Box 1369 Houston, TX 77251	7350 HLIPS	123 to 173	IBM 370, 4300
International Imaging Systems 1500 Buckeye Drive Milpitas, CA 95035	I²S 6500	50	VAX, MASSCOMP, HP 3000
MacDonald Dettweiler and Associates 3751 Shell Road Richmond, British Columbia Canada V6X 229	Meridian	100 to 200	VAX

*Price is for image-processing hardware and software only. In addition to this equipment, the user must provide support equipment, including host computer, tape drive, disk storage, and a device to produce hard copy.

TABLE 7.3 Organizations providing image-processing services

Earth Satellite Corp.
7222 47th Street
Chevy Chase
Washington, DC 20815

Environmental Research Institute of Michigan
P.O. Box 618
Ann Arbor, MI 48107

Geospectra Corp.
P.O. Box 1387
Ann Arbor, MI 48106

Aero Service Corp.
8100 West Park Drive
Houston, TX 77063

COMMENTS

Digital image processing has been demonstrated in this chapter using examples of Landsat images that are available in digital form. It is emphasized, however, that any image can be converted into a digital format and processed in similar fashion. The three major functional categories of image processing are

1. *image restoration* to compensate for data errors, noise, and geometric distortions introduced during the scanning, recording, and playback operations.

2. *image enhancement* to alter the visual impact that the image has on the interpreter, in a fashion that improves the information content.

3. *information extraction* to utilize the decision-making capability of the computer to recognize and classify pixels on the basis of their digital signatures.

In all of these operations the user should be aware of the tradeoffs involved, as demonstrated in the discussion of contrast stretching.

A common query is whether the benefits of image processing are commensurate with the cost. This is a difficult question that can only be answered by the context of the user's needs. If digital filtering, for example, reveals a previously unrecognized fracture pattern that in turn leads to the discovery of major ore deposits, the cost benefits are obvious. On the other hand, it is difficult to state the cost benefits of improving the accuracy of geologic and other maps through digital processing of remote sensing data. However, it should also be noted that technical advances in software and hardware are steadily increasing the volume and complexity of the processing that can be performed, often at a reduced unit cost.

QUESTIONS

1. Many users advocate higher spatial resolution (smaller ground resolution cells) for imaging systems without considering the operational consequences. Assume that the Landsat TM ground resolution cell is reduced to 10 by 10 m. Refer to Figure 7.3B and calculate the following:

   a. Number of pixels per band
   b. Number of pixels per scene (seven bands)
   c. Number of computer disk drives (300-Mbyte capacity) required to store the data.

2. Refer to the digital number array of the Thermopolis subscene (Figure 7.6C). For line number 15 construct a plot of terrain reflectance as a function of pixel number. Use same horizontal and vertical scales as in Figure 7.4B.

3. For Figure 7.9A replace 0 with 10 and 90 with 80. Calculate the filtered values to replace noise pixels.

4. For your particular application (forestry, geography, geology, oceanography, and so forth) of Landsat MSS images, which of the contrast-enhancement methods in Figure 7.12 are optimum? Explain the reasoning for your selections and any tradeoffs that might occur.

5. For Figure 7.17A replace all the 35s with 47s and all the 45s with 38s. Produce the following enhanced results:

   a. Filtered data set
   b. Profile of the revised original data
   c. Profile of your filtered data

6. For Figure 7.20B replace all the 35s with 45s. Assume a new edge direction *A* of N50°E. Use the directional filter kernel (Figure 7.20A) and this new set of original data to calculate the following:

   a. Plot of kernel values and profile A–B plot.
   b. Plot of sum of your original data plus your kernel values. Plot profile A–B for these data.

REFERENCES

Batson, R. M., K. Edwards, and E. M. Eliason, 1976, Synthetic stereo and Landsat pictures: Photogrammetric Engineering, v. 42, p. 1279–1284.

Bernstein, R., and D. G. Ferneyhough, 1975, Digital image processing: Photogrammetric Engineering, v. 41, p. 1465–1476.

Bernstein, R., J. B. Lottspiech, H. J. Myers, H. G. Kolsky, and R. D. Lee's, 1984, Analysis and processing of Landsat

4 sensor data using advanced image processing techniques and technologies: IEEE Transactions on Geoscience and Remote Sensing, v. GE-22, p. 192–221.

Bryant, M., 1974, Digital image processing: Optronics International Publication 146, Chelmsford, Mass.

Buchanan, M. D., 1979, Effective utilization of color in multidimensional data presentation: Proceedings of the Society of Photo-Optical Engineers, v. 199, p. 9–19.

Chavez, P. S., 1975, Atmospheric, solar, and MTF corrections for ERTS digital imagery: American Society of Photogrammetry, Proceedings of Annual Meeting in Phoenix, Ariz.

Goetz, A. F. H., and others, 1975, Application of ERTS images and image processing to regional geologic problems and geologic mapping in northern Arizona: Jet Propulsion Laboratory Technical Report 32-1597, Pasadena, Calif.

Haralick, R. M., 1984, Digital step edges from zero crossing of second directional filters: IEEE Transactions on Pattern Analysis and Machine Intelligence, v. PAMI-6, p. 58–68.

Holkenbrink, 1978, Manual on characteristics of Landsat computer-compatible tapes produced by the EROS Data Center digital image processing system: U.S. Geological Survey, EROS Data Center, Sioux Falls, S.D.

Loéve, M., 1955, Probability theory: D. van Nostrand Company, Princeton, N.J.

Moik, H., 1980, Digital processing of remotely sensed images: NASA SP no. 431, Washington, D.C.

NASA, 1983, Thematic mapper computer compatible tape: NASA Goddard Space Flight Center Document LSD-ICD-105, Greenbelt, Md.

Rifman, S. S., 1973, Digital rectification of ERTS multispectral imagery: Symposium on Significant Results Obtained from ERTS-1, NASA SP-327, p. 1131–1142.

Rifman, S. S., and others, 1975, Experimental study of application of digital image processing techniques to Landsat data: TRW Systems Group Report 26232-6004-TU-00 for NASA Goddard Space Flight Center Contract NAS 5-20085, Greenbelt, Md.

Short, N. M., and L. M. Stuart, 1982, The Heat Capacity Mapping Mission (HCMM) anthology: NASA SP 465, U.S. Government Printing Office, Washington, D.C.

Simpson, C. J., J. F. Huntington, J. Teishman, and A. A. Green, 1980, A study of the Pine Creek geosyncline using integrated Landsat and aeromagnetic data in Ferguson, J., and A. B. Goleby, eds., International Uranium Symposium on the Pine Creek Geosyncline: International Atomic Energy Commission Proceedings, Vienna, Austria.

Soha, J. M., A. R. Gillespie, M. J. Abrams, and D. P. Madura, 1976, Computer techniques for geological applications: Caltech/JPL Conference on Image Processing Technology, Data Sources and Software for Commercial and Scientific Applications, Jet Propulsion Laboratory SP 43-30, p. 4.1–4.21, Pasadena, Calif.

Swain, P. H., and S. M. Davis, 1978, Remote sensing—the quantitative approach: McGraw-Hill Book Co., N.Y.

ADDITIONAL READING

Andrews, H. C., and B. R. Hunt, 1977, Digital image restoration: Prentice-Hall, Englewood Cliffs, N.J.

Castleman, K. R., 1977, Digital image processing: Prentice-Hall, Englewood Cliffs, N.J.

Gonzalez, R. C., and P. Wintz, 1977, Digital image processing: Addison-Wesley Publishing Co., Reading, Mass.

Pratt, W. K., 1977, Digital image processing: John Wiley & Sons, N.Y.

Rosenfeld, A., and A. C. Kak, 1982, Digital picture processing, second edition: Academic Press, Orlando, Fla.

8

Resource Exploration

This chapter deals with nonrenewable resources, specifically minerals and fossil fuels. Other resources such as vegetation, water, and soil are discussed in Chapter 9 and elsewhere. Remote sensing methods have great promise as techniques for both reconnaissance and detailed exploration of nonrenewable resources. The following sections give theory, techniques, and examples of these applications.

MINERAL EXPLORATION

Remote sensing has proven valuable for mineral exploration in at least four ways:

1. Mapping regional lineaments along which groups of mining districts may occur

2. Mapping local fracture patterns that may control individual ore deposits

3. Detecting hydrothermally altered rocks associated with ore deposits

4. Providing basic geologic data

Regional Lineaments and Ore Deposits of Nevada

Prospectors and mining geologists have long realized that, in many mineral provinces, mining districts occur along linear trends that range from tens to hundreds of kilometers in length. These trends are referred to as *mineralized belts* or *zones,* and many deposits have been found by exploring along the projections of such trends. The state of Nevada is rich in historic and active mining districts of great wealth. In the late 1800s, rich gold and silver deposits (Virginia City and Goldfield) were discovered in the western part of the state. Later, porphyry copper deposits such as the Ruth, Ely, Eureka, and Yerington deposits were discovered. Exploration has continued, resulting in the more recent discovery of large deposits of gold at Carlin, Alligator Ridge, Cortez, and elsewhere.

Lineament Interpretation   It was long recognized that Nevada mining districts were not randomly distributed but tended to occur in linear zones or belts. The availability of Landsat images has enabled geologists to evaluate the relationship between mineral deposits and linear structural features. Rowan and Wetlaufer (1975) of the U.S. Geological Survey interpreted a mosaic of Landsat MSS images of Nevada (Figure 8.1); they recognized 367 lineaments, 80 percent of which correlated with previously mapped faults. Fifty-seven percent of the lineaments are formed by the linear contact between the bedrock of mountain ranges and the alluvium of adjacent valleys. These lineaments are the expression of basin-and-range faults that dominate the structure of Nevada. Other lineaments are formed by linear ridges, aligned ridges, and tonal boundaries.

Seven lineaments of regional extent and importance are shown in Figure 8.2A. These lineaments transect the topography of the Basin and Range province and are several hundred kilometers in length. The Walker Lane, Las Vegas, Midas Trench, and Oregon-Nevada lineaments have previously been documented as major crustal features. The following description of the major lineaments is summarized from Rowan and Wetlaufer (1975).

Walker Lane lineament This zone of right-lateral transcurrent faulting extends southeast from Pyramid Lake to south-central Nevada, where it merges with the lineament called the Las Vegas shear zone. The Walker Lane lineament is expressed by a distinct northwest-trending discontinuity on the mosaic (Figure 8.1). On either side of the lineament, mountain ranges have the regional northeast trend; within the zone, however, the orientation is abruptly changed to northwest trends. Right-lateral, strike-slip movement is thought to be the dominant sense of displacement. A number of mining districts occur along the lineament, from Virginia City in the northwest to Tonopah and Goldfield in the southeast.

Las Vegas shear zone The extension of the Walker Lane lineament into southern Nevada has long been known as the Las Vegas shear zone. Prior to Landsat it was recognized from regional topographic maps that the mountain ranges on either side of the zone are bent in a pattern suggesting drag along a right-lateral strike-slip fault. In the Landsat mosaic, this drag effect is clearly seen in the mountain ranges northwest of Las Vegas. Few ore deposits occur along the Las Vegas shear zone, possibly because the sedimentary bedrock in southern Nevada is less suitable for ore formation than the volcanic and plutonic rocks elsewhere in the state.

Midas Trench lineament This feature extends northeast from Lake Tahoe for 460 km to the northern border of Nevada. The lineament is named for the old mining camp of Midas, where it forms a prominent linear depression. Tuscarora, Golconda, Winnemucca, and the Eagle-Picher mining districts also occur along the lineament. Elsewhere the Midas Trench is expressed by linear escarpments and aligned stream segments. Recent lateral movements along faults of the lineament zone are indicated by offset stream channels.

Oregon-Nevada lineament This lineament extends 750 km from central Nevada to central Oregon and consists of aligned tonal and textural boundaries caused by closely spaced faults.

A linear belt of lava flows and flow domes of late Miocene age mark the Nevada portion of the lineament. An aeromagnetic map shows a prominent linear belt of high values in this portion of the lineament that is attributed to dikes that provided magma for the volcanic features (Stewart, Walker, and Kleinhampl, 1975).

Rye Patch lineament This 250-km-long feature trends northwest and is located midway between the Walker Lane and Oregon-Nevada lineaments. The Rye Patch lineament, named for the Rye Patch reservoir, has a diffuse appearance on the Landsat mosaic where it is mapped by tonal and textural changes. Mountain ranges are terminated or offset by the lineament, which coincides with the existence of regional deep-seated fracture zones proposed by earlier workers.

East-northeast lineament This feature is actually a pair of parallel lineaments that terminate or disrupt the north-trending ranges that they intersect. The lineaments are marked by aligned streams, canyons and tonal boundaries. Both lineaments are associated with mapped faults with right-lateral strike-slip displacement.

Ruby Mountains lineament This relatively poorly documented lineament has a length of 230 km. The southern part coincides with the normal fault that forms the east boundary of the Ruby Mountains. The northern part of the lineament is marked by a series of smaller faults.

Analysis of Lineaments The maps in Figure 8.2 were prepared by Rowan and Wetlaufer (1975) to evaluate the relationship between the lineaments and ore deposits. Nevada mining districts, ranked by dollar value of production, are plotted in Figure 8.2B. Dollar values are based on prices at time of production. During late 1800s and early 1900s, when much of the ore was mined, the price of gold was $20 per ounce and silver was less than $1 per ounce. One linear belt of mining districts coincides with the northeast-trending Midas Trench lineament. The districts in the southwest part of the state occur in a broad belt parallel with the northwest-trending Walker Lane lineament.

To aid in evaluating the relationship of mining districts to lineaments, Rowan and Wetlaufer laid a grid over Figure 8.2B and counted the number of districts in each grid square. These data were contoured to produce the map of Figure 8.2C. This map emphasizes the concentration of districts along the Midas Trench lineament. The high concentration is interrupted by an area of low mining density at the intersection with the Oregon-Nevada lineament. The lack of mining districts along the Oregon-Nevada lineament may be caused by the extensive cover of Tertiary volcanic rocks that masks any underlying deposits. The greatest density of ore deposits along the Midas Trench occurs at the intersection with the Rye Patch lineament. In south-central Nevada the East-Northeast lineament system coincides with two of the three east-trending belts of high mining density on Figure 8.2C. Ore deposits along the Walker Lane

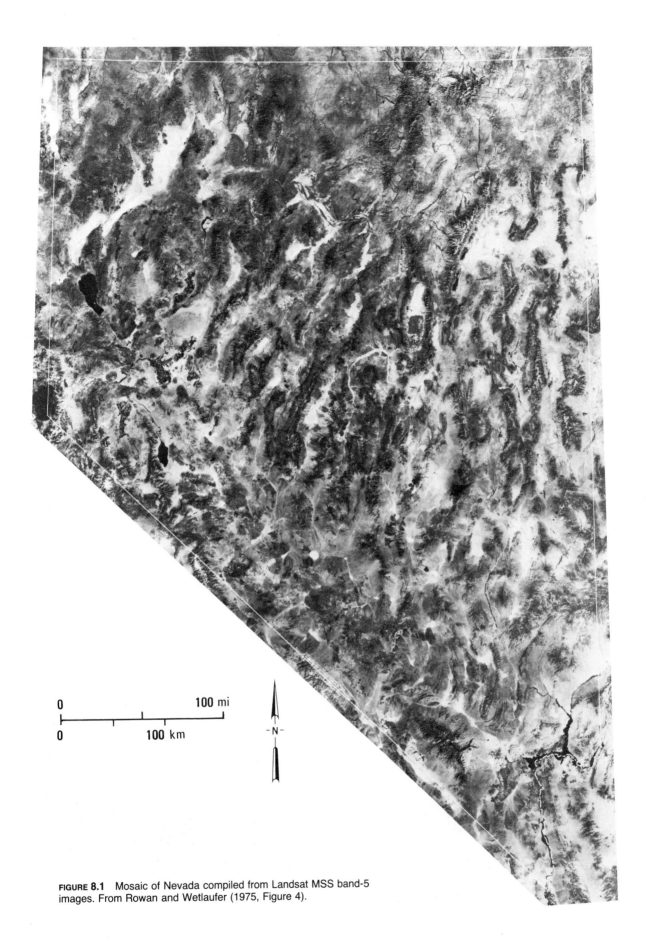

0   100 mi

0   100 km

-N-

FIGURE 8.1   Mosaic of Nevada compiled from Landsat MSS band-5 images. From Rowan and Wetlaufer (1975, Figure 4).

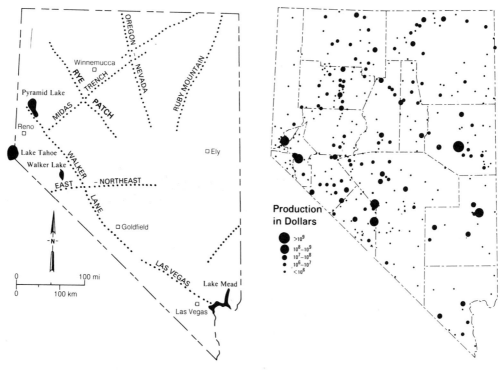

A. MAJOR LINEAMENTS INTERPRETED
FROM A LANDSAT MOSAIC.

B. MINING DISTRICTS.

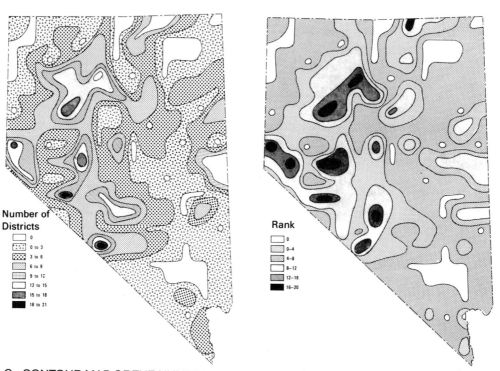

C. CONTOUR MAP OF THE NUMBER OF
MINING DISTRICTS.

D. CONTOUR MAP OF THE DISTRIBUTION
OF MINING DISTRICTS WEIGHTED
ACCORDING TO DOLLAR VALUE.

FIGURE 8.2 Landsat lineaments and mining districts of Nevada. Maps from Rowan
and Wetlaufer (1975); mining data from Horton (1964).

lineament are generally concentrated at the intersections with east-trending lineaments.

Figure 8.2D was prepared by gridding and contouring the weighted dollar value of production for the districts. These value trends closely resemble the trends of mining density. The Midas Trench and Walker Lane lineaments are marked by aligned concentrations of mining districts.

Local Fractures and Ore Deposits of Central Colorado

Within a mineral province, areas with numerous fracture intersections are good prospecting targets because fractures are conduits for ore-forming solutions. Local fracture patterns are mappable on enlarged Landsat images, especially those acquired at low sun angles and those that have been digitally filtered to enhance fractures.

Relationships between Landsat fracture patterns and ore deposits are illustrated in the example from central Colorado, which is summarized from the work of Nicolais (1974). A winter image (Figure 8.3) was used for interpretation because the snow cover and low sun elevation enhance the expression of fractures. On the interpretation map (Figure 8.4), fractures and circular features are plotted together with location of major mining districts. The Landsat interpretation reduced the original 33,500 km^2 image to 10 target areas, each 165 km^2 in area. These areas were selected because they show concentrations of fracture intersections or intersections of fractures and circular features. Five of these target areas coincide with, or are directly adjacent to, major mining districts. The other five target areas may be sites of undiscovered ore deposits.

Mapping Hydrothermally Altered Rocks

Many ore bodies are deposited by hot aqueous fluids, called *hydrothermal solutions,* that invade the host rock, or *country rock.* During formation of the ore minerals, these solutions also interact chemically with the country rock to alter the mineral composition for considerable distances beyond the site of ore deposition. The hydrothermally altered country rocks contain distinctive assemblages of secondary, or alteration, minerals that replace the original rock constituents. Alteration minerals commonly occur in distinct sequences, or *zones of hydrothermal alteration,* relative to the ore body. These zones are caused by changes in temperature, pressure, and chemistry of the hydrothermal solution at progressively greater distances from the ore body. At the time of ore deposition, the zones of altered country rock may not extend to the surface of the ground. Later uplift and erosion expose successively deeper alteration zones and eventually the ore body itself.

Not all alteration is associated with ore bodies, and not all ore bodies are marked by alteration zones, but these zones are valuable indicators of possible deposits. Fieldwork, laboratory analysis of rock samples, and interpretation of aerial photographs have long been used to explore for hydrothermal alteration zones.

In regions where bedrock is exposed, multispectral remote sensing is useful for recognizing altered rocks because their reflectance spectra differ from those of the country rock. An instructive example of remote sensing of alteration zones is provided by the gold and silver vein deposits of Goldfield, Nevada.

Goldfield, Nevada

The Goldfield district in southwest Nevada (Figure 8.2A) was noted for the richness of its ore. Over 4 million troy ounces (130,000 kg) of gold with silver and copper were produced, largely in the boom period between 1903 and 1910. During peak production the town had a population of 15,000 but today is largely a ghost town.

Geology and Hydrothermal Alteration The geology and hydrothermal alteration have been thoroughly mapped and analyzed by the U.S. Geological Survey (Ashley, 1974, 1979). Volcanism began in the Oligocene epoch with eruption of rhyolite and quartz latite flows and the formation of a small caldera and ring-fracture system. Hydrothermal alteration and ore deposition occurred during a second period of volcanism in the early Miocene epoch when the dacite and andesite flows that host the ore deposits were extruded. Heating associated with volcanic activity at depth caused convective circulation of hot, acidic, hydrothermal solutions through the rocks. Fluid movement was concentrated in the fractures and faults of the ring-fracture system. Following ore deposition, the area was covered by younger volcanic flows. Later doming and erosion have exposed the older volcanic center with altered rocks and ore deposits.

In the generalized map (Figure 8.5), the hydrothermally altered rocks are cross-hatched and the unaltered country rocks are blank. The map also identifies alluvial deposits and post-ore (formed after ore deposition) volcanic rocks. Approximately 40 km^2 of the area are underlain by altered rocks, but less than 2 km^2 of the altered area are underlain by mineral deposits, shown in black. The irregular oval band of alteration was controlled by the zone of ring fractures, which have a linear extension toward the east. The central patch of alteration shown in Figure 8.5 was controlled by closely spaced faults and fractures.

Figure 8.6 is a vertical section through an ore-bearing vein and the associated altered rocks. Alteration is most intense where hydrothermal solution entered the dacite

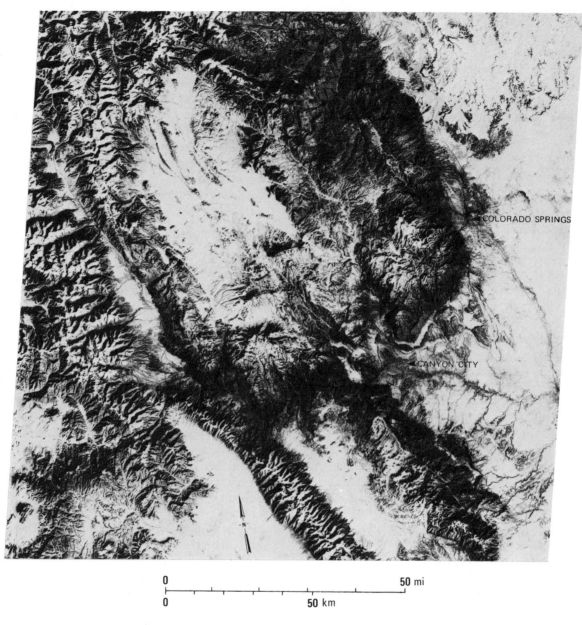

0   50 mi

0   50 km

FIGURE 8.3 Landsat MSS band-7 image of central Colorado acquired January 11, 1973, at a sun elevation of 23°.

and andesite country rocks through a fault or fracture, and gold was deposited to form a vein. Intensity of alteration decreases laterally away from the vein. Characteristics of the various alteration zones are summarized below.

Silicic zone Predominantly quartz replaces the ground mass of host rock; subordinate amounts of alunite and kaolinite replace feldspar phenocrysts. Fresh rock of this zone, which is gray and resembles chert, is resistant to erosion and weathers to ridges with conspicuous dark coatings of desert varnish. Contact with adjacent argillic zones is sharp. All ore deposits occur in veins of the silicic zone, but not all veins contain ore.

Argillic zone Alteration minerals are predominantly clay. The argillic zone is divided into three subzones (Figure 8.6) based on the predominant clay species. Disseminated grains of pyrite (iron sulfide) are present that weather to iron oxides. The argillic rocks generally have a bleached appearance, but

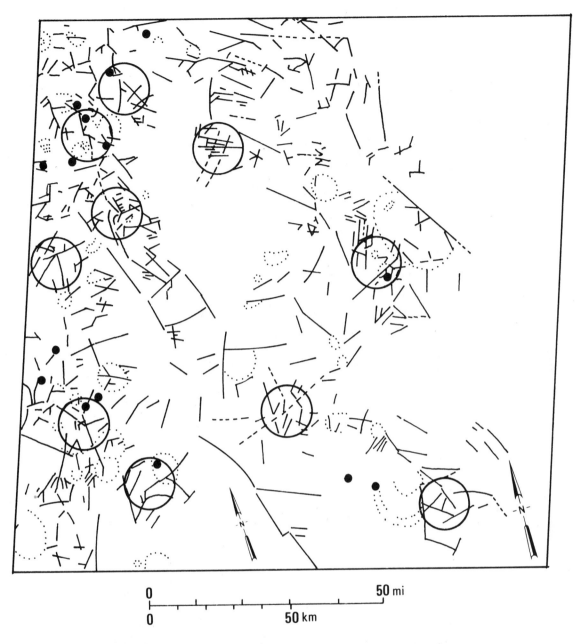

```
0                                          50 mi
├────┼────┼────┼────┼────┼────┼────┤
0                                  50 km
```

FIGURE 8.4 Interpretation of Landsat image of central Colorado. Solid lines are distinct lineaments, dashed lines are possible lineaments, and dotted lines are curvilinear features. Large circles are target areas selected for exploration. Solid dots indicate major mining districts. From Nicolais (1974, Figure 3).

the secondary iron oxide minerals (limonite and goethite) impart local patches of red, yellow, and brown to the outcrops. No ore deposits occur in the argillic zone, but the presence of these altered rocks may be a clue to the occurrence of veins.

Alunite-kaolinite subzone Relatively narrow and locally absent. In addition to alunite and kaolinite, some quartz is present.

Illite-kaolinite subzone Marked by the occurrence of illite.

Montmorillorite subzone Montmorillonite is the dominant clay

mineral in this subzone, which has a pale yellow color due to jarosite, an iron sulfate mineral.

Propylitic zone These rocks represent regional alteration of lower intensity than the argillic and silicic zones. Chlorite, calcite, and antigorite are typical minerals in this zone and impart a green or purple color to the rocks. Propylitic alteration is absent at numerous localities in Goldfield, where the argillic zone is in sharp contact with unaltered rocks.

Country rock Dacite and andesite. These gray volcanic rocks are hard and resistant to erosion. As shown by the blank

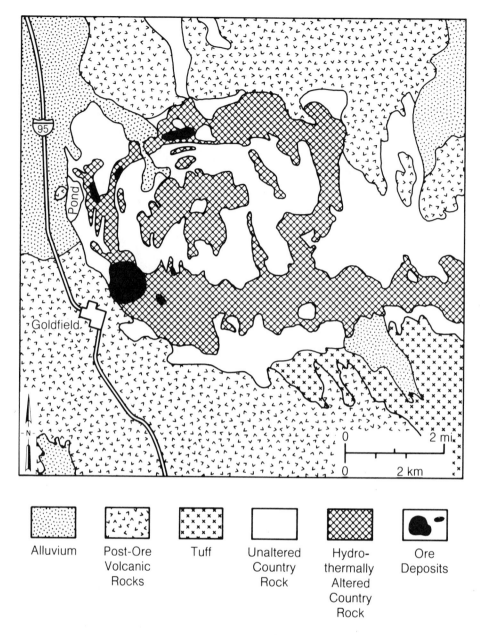

Alluvium   Post-Ore   Tuff   Unaltered   Hydro-   Ore
          Volcanic   Country   thermally   Deposits
          Rocks   Rock   Altered
                                            Country
                                            Rock

FIGURE 8.5   Map showing hydrothermal alteration and geology of the Goldfield district, Nevada. After Ashley (1979, Figures 1 and 8).

area in the map (Figure 8.5), the unaltered rocks surround the inner and outer margins of the circular belt of altered rocks.

In the field the orderly sequence of subzones shown in Figure 8.6 rarely occurs because the veins are so closely spaced that the subzones coalesce and overlap to form the altered outcrops shown in the map (Figure 8.5).

Clay and iron minerals of the altered rocks have distinctive spectral characteristics that are recognizable in multispectral images such as Landsat thematic mapper. Reflectance spectra of quartz, alunite, and the clay minerals are shown in Figure 8.7A. Spectral bands of Landsat MSS and TM are shown in Figure 8.7C. The spectrum of quartz has no distinctive feature, but the spectra of the clay minerals (kaolinite, illite, and montmorillonite) and alunite have distinctive absorption features at wavelengths of approximately 2.2 μm that coincide with band 7 of TM. Spectra of the iron oxide minerals limonite, hematite, and goethite are shown in Figure 8.7B.

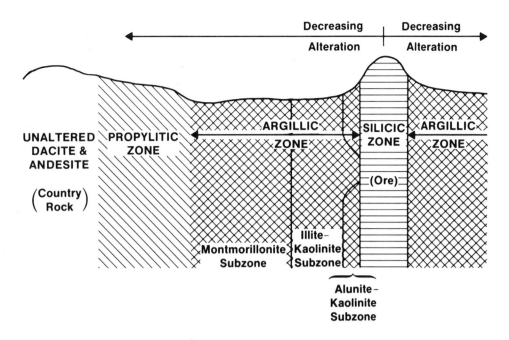

Decreasing | Decreasing
Alteration | Alteration

UNALTERED
DACITE &
ANDESITE

(Country Rock)

PROPYLITIC ZONE

ARGILLIC ZONE

SILICIC ZONE

ARGILLIC ZONE

(Ore)

Montmorillonite Subzone

Illite–Kaolinite Subzone

Alunite–Kaolinite Subzone

FIGURE 8.6 Model cross section of zones of hydrothermal alteration at Goldfield, Nevada. Not to scale. Silicic zone has maximum width of a few meters. Argillic zone is several tens of meters wide. After Ashley (1974) and Harvey and Vitaliano (1964).

Reflectance spectra of altered and unaltered rocks were measured in the field at Goldfield using a portable spectrometer (Rowan, Goetz, and Ashley, 1979). In the average rock spectra (Figure 8.7C), the two gaps at 1.4 μm and 1.9 μm are due to absorption by water vapor in the atmosphere, as explained in Chapter 1.

This background information on geology and spectral properties of rocks at Goldfield sets the stage for using remote sensing methods to recognize the hydrothermally altered rocks as a guide for ore exploration.

Earlier Remote Sensing Investigations An early investigation by Rowan and others (1974) employed Landsat MSS data. As shown in Figure 8.7C these detectors are restricted to wavelengths of less than 1.1 μm and do not record the diagnostic absorption features of clay and alunite at 2.2 μm. The iron oxide staining commonly associated with altered rocks is recognizable in the visible region, however, and may be detected in MSS data. The red color of iron oxide causes high reflectance in MSS band 5 and low reflectance in band 4. The ratio of band 4 divided by band 5 (Chapter 7) has a low value for reddish rocks because the higher digital numbers of band 5 are in the denominator. Rowan and others (1974, Figure 17) prepared a color ratio image by projecting the ratio 4/5 image with blue light, the ratio 5/6 image with yellow light, and the ratio 6/7 image with magenta light. In the resulting color image, green tones correlate with limonitic areas of hydrothermally altered rocks at Goldfield and the other mining districts in the area. The green tones also correlate with outcrops of ferruginous sandstone and siltstone and with some plutonic rocks.

Digitally Processed TM Images Investigations at Goldfield and elsewhere in the mid-1970s pointed out some shortcomings of Landsat MSS spectral bands for mineral exploration. At that time, NASA had designed the Landsat TM scanner with six spectral bands: bands 1, 2, and 3 in the visible region, band 4 centered at 0.85 μm, band 5 centered at 1.6 μm, and band 6 in the thermal region. No coverage was planned for the critical band at 2.2 μm. Representatives from the remote sensing user community for exploration (Geosat Committee, U.S. Geological Survey, NASA, and JPL) argued successfully for adding band 7, centered at 2.2 μm, to the TM scanner.

Landsat 4 acquired an excellent TM image of Goldfield on October 4, 1984. The data were digitally processed at Chevron Oil Field Research Company, using methods described in Chapter 7. The normal color image of the Goldfield subscene (Plate 12A) was enhanced with a linear contrast stretch and an IHS transformation. The subscene covers the area shown in Figure 8.5. A yellow patch directly northeast of the town of Goldfield is caused by the mine dumps and disturbed ground of the main mineralized area. A white patch 3 km north of Goldfield

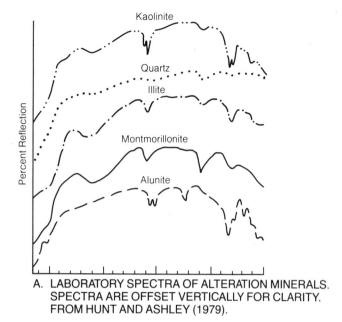

A. LABORATORY SPECTRA OF ALTERATION MINERALS. SPECTRA ARE OFFSET VERTICALLY FOR CLARITY. FROM HUNT AND ASHLEY (1979).

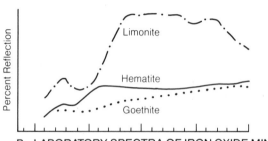

B. LABORATORY SPECTRA OF IRON OXIDE MINERALS. SPECTRA ARE OFFSET VERTICALLY. FROM HUNT, SALISBURY, AND LENHOF (1971).

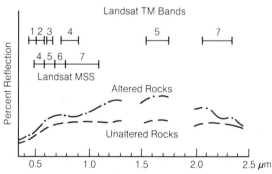

C. FIELD SPECTRA AVERAGED FOR ALTERED AND UNALTERED ROCKS AT GOLDFIELD, NEVADA. FROM ROWAN, GOETZ, AND ASHLEY (1977).

FIGURE 8.7 Reflectance spectra of hydrothermally altered rocks and minerals.

is the dry tailings pond of the abandoned Columbia Mill, where gold was separated from the altered host rock. The tailings pond is a useful reference standard since it contains a concentration of alteration minerals. The dark rocks in the margins of the image are volcanic flows that are younger than the ore deposits and altered rocks. Distinctive light blue signatures in the southeast portion are outcrops of volcanic tuff. Some of the hydrothermally altered rocks have a tan signature in the normal color image (Plate 12A), but other rocks have similar colors.

Neither the normal color TM image nor the IR color image (not illustrated) are diagnostic for recognizing the hydrothermally altered rocks. The spectra of altered and unaltered rocks at Goldfield (Figure 8.7C) suggest a means of distinguishing between these rocks. Absorption caused by alunite and clay minerals results in low reflectance at 2.2 μm, which corresponds to TM band 7. Altered rocks have high reflectance at 1.6 μm, which corresponds to TM band 5. Enhanced images of these bands are shown in Figure 8.8A,B where the altered rocks have brighter signatures in band 5 than in band 7. A ratio 5/7 image (Figure 8.8C) has bright signatures for altered rocks because the lower reflectance values of band 7 are in the denominator, which results in higher ratio values of approximately 1.5.

The unaltered rocks, however, have nearly equal reflectance values in bands 5 and 7 (Figure 8.7C) resulting in a ratio 5/7 of approximately 1.0. In the ratio 5/7 image (Figure 8.8C), the high values of altered rocks have brighter signatures than unaltered rocks. A color density slice was applied to the ratio image to convert gray-scale variations into color (Plate 12C). Red and yellow colors were assigned to the highest ratio values; green and blue were assigned to the lowest values. Red and yellow hues closely match the distribution of altered rocks in the map of Figure 8.5.

Spectra of the weathered iron minerals (Figure 8.7B) have weak reflectance in the blue region (TM band 1) and strong reflectance in the red region (TM band 3). The ratio 3/1 has high values for iron-stained areas, resulting in bright tones on the image (Figure 8.8D). A color density slice of this ratio (Plate 12D) assigns red and yellow colors to the highest ratio values. These colors delineate the iron-stained portions of the altered rocks.

Another way to use ratio images is to combine three selected ratios in color as shown in Plate 12B, where ratios 5/7, 3/1, and 3/5 are combined in red, green, and blue respectively. The orange and yellow tones delineate the outer and inner areas of altered rocks in a fashion similar to that of the density slice of the ratio 5/7 image (Plate 12C). An advantage of the color ratio image is

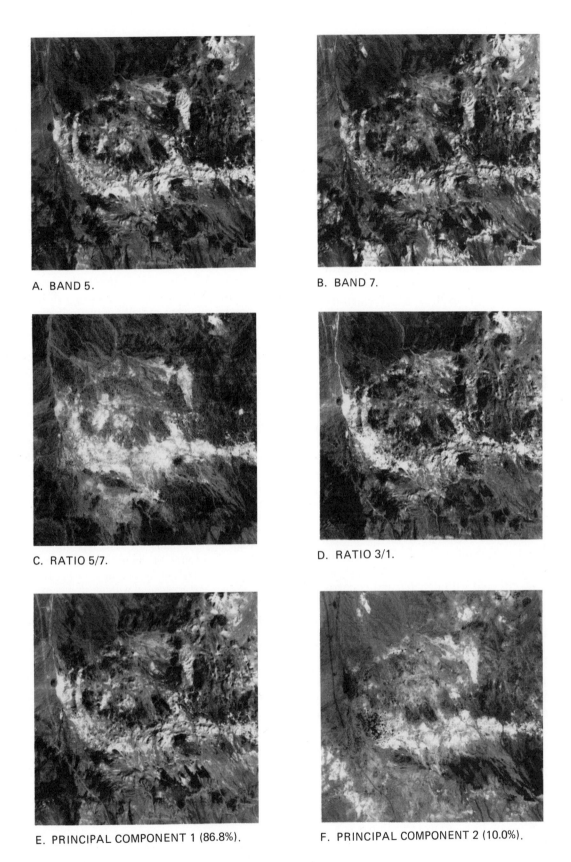

A. BAND 5.

B. BAND 7.

C. RATIO 5/7.

D. RATIO 3/1.

E. PRINCIPAL COMPONENT 1 (86.8%).

F. PRINCIPAL COMPONENT 2 (10.0%).

FIGURE 8.8 Digitally processed Landsat TM images of the Goldfield, Nevada, subscene.

that it displays distribution patterns of both iron-staining and hydrothermal clays. A disadvantage is that the individual patterns of iron-staining and clay minerals may overlap and obscure each other in the color ratio image.

Principal component (PC) images were prepared for the six TM visible and reflected IR bands of the Goldfield subscene. In PC 1 (Figure 8.8E) the hydrothermally altered rocks and the tailings pond have bright signatures that distinguish them from the other materials. In PC 2 (Figure 8.8F) the altered rocks are bright, but so are outcrops of unaltered volcanic rocks southwest of Goldfield town site. The tailings pond and the mine dumps north of Goldfield are dark in PC 2. A statistical analysis of the PC images show that TM band 5 is the most heavily weighted component of PC 1 and PC 2. Altered rocks have their maximum reflectance in TM band 5 (Figure 8.7C), which explains their bright signatures in PC 1 and PC 2.

An unsupervised multispectral classification applied to the six visible and reflected IR bands of the subscene resulted in 12 classes. These classes were aggregated into the 6 classes shown in the map and explanation of Plate 12E,F. Two types of altered rocks were classified. The class shown in red ("altered rocks, A") is confined to the altered rocks and to the tailings pond but does not indicate the full extent of alteration. The class shown in orange ("altered rocks, B") includes all of the remaining altered rocks as well as some rocks outside the alteration zone. Basalt (blue), volcanic tuff (purple), and unaltered rocks (green) are reasonably portrayed. Alluvium (yellow) is considerably more extensive in the classification map (Plate 12E) than in the geologic map (Figure 8.5). Field checking and comparison with the normal color image (Plate 12A) shows that much of the bedrock is thinly covered with detritus and is correctly classed as alluvium by the computer. The field geologist, however, is able to infer and map the lithology of the underlying bedrock.

In summary, the hydrothermally altered rocks at Goldfield are successfully distinguished by three types of digitally processed TM data: ratio images, principal-component images, and unsupervised multispectral classification maps. Much of this success is due to the availability of TM band 7, but the high spatial resolution and spectral sensitivity of TM also contribute.

Porphyry Copper Deposits

Much of the world's copper is mined from porphyry deposits, which occur in a different geologic environment from the gold deposits of the Goldfield type. Porphyry deposits are named for the *porphyritic* texture of the granitic host rock, in which large feldspar crystals are surrounded by a fine-grained matrix of quartz and other minerals. Granite porphyry occurs as plugs (or *stocks*) up to several kilometers in diameter that intruded the older country rock and reached to within several kilometers of the surface. Intensive fracturing of the porphyry and country rock occurred during the emplacement and cooling of the stock. The heat of the magma body caused convective circulation of hydrothermal fluids through the fracture system that resulted in alteration of the porphyry stock and adjacent country rock.

Hydrothermal Alteration The model of alteration zones shown in Figure 8.9 was developed from the study of many deposits in southwest United States (Lowell and Guilbert, 1970) and is applicable elsewhere.

The most intense alteration occurs in the core of the stock and diminishes radially outward in a series of zones described below.

Potassic zone Most intensely altered rocks in the core of the stock. Characteristic minerals are quartz, sericite, biotite, and potassium feldspar.

Phyllic zone Quartz, sericite, and pyrite are common.

Ore zone Disseminated grains of chalcopyrite, molybdenite, pyrite, and other metal sulfides. Much of the ore occurs in a cylindrical shell near the gradational boundary between the potassic and phyllic zones. Copper typically constitutes only a few tenths of a percent of the ore, but the large volume of ore is suitable for open pit mining. Where the ore zone is exposed by erosion, pyrite may oxidize to form a red to brown limonitic crust called a *gossan*. Gossans can be useful indicators of underlying mineral deposits, although not all gossans are associated with ore deposits.

Argillic zone Quartz, kaolinite, and montmorillonite are characteristic minerals of the argillic zone in porphyry deposits, just as they are associated with the argillic zone at Goldfield and elsewhere.

Propylitic zone Epidote, calcite, and chlorite occur in these weakly altered rocks. Propylitic alteration may be of broad extent and have little significance for ore exploration.

Few, if any, porphyry ore deposits and alteration patterns have the symmetry and completeness of the model in Figure 8.9. Structural deformation, erosion, and deposition commonly conceal large portions of the system. Nevertheless, recognition of small patches of altered rock on remote sensing images may be a valuable clue to a potential ore deposit. Porphyry copper deposits are one category of nonrenewable resource that was investigated by the NASA/Geosat test case project described in the following section.

NASA/Geosat Test Case Project—Copper Mines
The Geosat Committee is a nonprofit organization that promotes the use and development of remote sensing

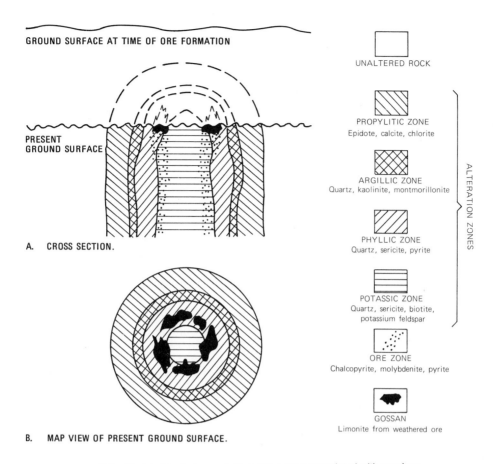

GROUND SURFACE AT TIME OF ORE FORMATION

PRESENT
GROUND SURFACE

A.   CROSS SECTION.

B.   MAP VIEW OF PRESENT GROUND SURFACE.

UNALTERED ROCK

PROPYLITIC ZONE
Epidote, calcite, chlorite

ARGILLIC ZONE
Quartz, kaolinite, montmorillonite

PHYLLIC ZONE
Quartz, sericite, pyrite

POTASSIC ZONE
Quartz, sericite, biotite,
potassium feldspar

ORE ZONE
Chalcopyrite, molybdenite, pyrite

GOSSAN
Limonite from weathered ore

ALTERATION ZONES

FIGURE 8.9 Model of hydrothermal alteration zones associated with porphyry copper deposits. From Lowell and Guilbert (1970, Figure 3).

by industry. The committee consists of, and is supported by, approximately 100 companies involved in various aspects of remote sensing. In 1977, NASA and Geosat jointly agreed to evaluate the use of remote sensing data in exploration for porphyry copper, uranium, and oil and gas. Test sites were selected for these three categories. NASA then acquired a variety of remote sensing data for the sites, which were digitally processed at JPL. Representatives of 38 Geosat member companies plus JPL personnel formed teams to interpret the image data, collect and analyze samples, and prepare a report (Abrams, Conel, and Lang, 1985). Settle (1985) prepared an executive summary of the project.

The image-processing and interpretation phases of the project were essentially completed before Landsat TM images became available in late 1982. To compensate for the lack of TM images, NASA acquired images with an airborne *thematic mapper simulator* (TMS). The eight TMS bands compare with Landsat TM bands as follows:

TMS band	TM band	Spectral range, μm
1	1	0.45 to 0.52
2	2	0.52 to 0.60
3	3	0.63 to 0.69
4	4	0.76 to 0.90
5	No band	1.00 to 1.30
6	5	1.55 to 1.75
7	7	2.08 to 2.35
8	6	10.40 to 12.50

Aircraft images were also acquired with a *modular multispectral scanner* (M²S) in the 0.3-to-1.1-μm range. Spatial resolution of TMS and M²S images ranges from 10 to 20 m.

The Geosat Committee chose three porphyry copper sites: the Silver Bell, Helvetia, and Safford deposits, all located in southern Arizona. The alteration zones shown in Figure 8.9 crop out in the Silver Bell district, where TMS data were digitally processed to produce a color ratio composite image (TMS ratio 3/2 = green, ratio 4/5 = blue, ratio 6/7 = red). The phyllic and potassic alteration zones have a distinct yellow-orange hue; the adjacent argillic and propylitic zones have yellowish-green and yellowish-brown hues (Abrams and Brown, 1985, Figure 4-32). A supervised classification map of TMS data also defined the outcrops of altered rocks (Abrams and Brown, 1985, Figure 4-41).

These results at Silver Bell using TMS data are similar to those described earlier for Goldfield using TM data. This is surprising because the geology and ore deposits of the two districts are completely different. The comparable remote sensing results are explained by the following similarities:

1. Hydrothermal alteration produced similar suites of secondary minerals (quartz, clays, and iron oxide) at both districts. Alunite occurs at Goldfield, but is absent at Silver Bell.

2. Spectral ranges of the TMS bands used at Silver Bell are comparable to the TM bands used at Goldfield.

3. The digital-processing methods were similar, although Silver Bell data were processed at JPL and Goldfield data were processed at Chevron Oil Field Research Company.

The similarity in results for the two mining districts indicates the potential value of remote sensing and digital image processing for mineral exploration.

Basic Data for Geologic Mapping

In addition to locating specific mineral target areas of fracture intersections or rock alteration, remote sensing provides data for preparing and improving geologic maps, which are the fundamental tool for exploration. Geologic maps, even at reconnaissance scales, are not available for large areas of the earth. For example, approximately two-thirds of southern Africa lacks published geologic maps at scales of 1:500,000 or larger; this can be improved by use of Landsat images. A previously unknown major fault was discovered on a Landsat image of southern South-west Africa and the Cape Province of South Africa by Viljoen and others (1975, Figure 3), who named it the Tantalite Valley fault zone. The fault zone appears to have right-lateral strike-slip displacement and has been mapped for 450 km along the strike. A number of large mafic intrusives have been emplaced along the Tantalite Valley fault zone and are recognized on Landsat images. On a Landsat color mosaic of the northwest Cape Province of South Africa, Viljoen and others (1975, Figures 11 and 12) mapped a pronounced structural discontinuity, called the Brakbos fault zone, that separates the Kaapvaal craton on the east from the Bushmanland Metamorphic Complex on the west. The contact between these structural provinces is obscure in the field and had previously been drawn approximately 30 km to the east of the Brakbos fault zone, which is also defined on gravity maps. The use of seasonal Landsat images for mapping rock types was illustrated for the Transvaal Province in Chapter 4.

In the Nabesna quadrangle of east-central Alaska, Albert (1975) combined lineament analysis and digital processing of Landsat MSS data to evaluate known and potential mineral deposits. A preliminary analysis indicates that 56 percent of the known mineral deposits occur within 1.6 km of Landsat lineaments. Color anomalies on the enhanced images coincide with 72 percent of the known mineral occurrences. Of the remaining color anomalies, some coincide with areas of known rock alteration and others constitute potential exploration targets.

URANIUM EXPLORATION

Most uranium deposits in the United States occur in nonmarine fluvial sandstone and conglomerate beds of the Colorado Plateau, Rocky Mountain basins, and southern Texas. The solutions that deposited the uranium also altered the host rocks and caused local color anomalies on the outcrops. Aerial photographs have been used extensively in exploring for these areas of altered rock. Digital processing of Landsat CCTs has the potential to recognize subtle alteration effects that may not be obvious to the eye or on aerial photographs. A brief description of typical sedimentary uranium deposits will aid in understanding the following examples of Landsat applications.

Origin of Sedimentary Uranium Deposits

The model shown in Figure 8.10 for the formation of sedimentary uranium deposits is widely accepted, although there is debate about the origin of the uranium and the chemistry of the transporting solutions. Granite, granitic detritus, and silicic volcanic ash and flows contain disseminated uranium in concentrations up to 10 parts per million. Some of this uranium is leached from the source rocks by oxygen-rich groundwater that then migrates into porous sandstone and conglomerate beds carrying the uranium in solution. Within these sedimentary rocks, the migrating water encounters reducing conditions caused by the presence of organic material, nat-

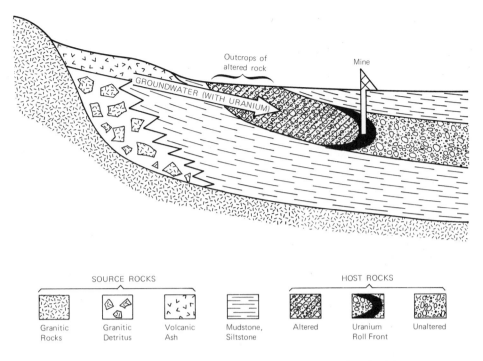

SOURCE ROCKS

Granitic Rocks

Granitic Detritus

Volcanic Ash

Mudstone, Siltstone

HOST ROCKS

Altered

Uranium Roll Front

Unaltered

FIGURE 8.10 Model for the formation of sedimentary uranium deposits.

ural gas, or hydrogen sulfide and pyrite. The change from oxidizing to reducing conditions causes the uranium to precipitate as oxide minerals, primarily uraninite, that coat sand grains and fill pore spaces in the host rock. The ore deposits typically contain from 0.1 to 0.5 percent U_3O_8 and occur as tabular layers or as arcuate bodies called *roll fronts* (Figure 8.10). Later uplift and erosion may expose the ore to secondary oxidation.

Outcrops of altered host rock are clues to possible ore deposits below the surface. The unaltered host rocks are typically drab in color and contain organic carbon and pyrite. Oxidation by the migrating solutions destroys the carbon and converts the dark pyrite to iron oxide minerals that impart characteristic yellow, red, and brown colors to the altered rocks.

The following sections describe remote sensing applications to uranium exploration in Arizona and Texas.

Cameron Uranium District, Arizona

In the Cameron district of north-central Arizona, uranium host rocks are conglomerates, sandstones, and siltstones interbedded with mudstone and limestone of the Chinle Formation of Upper Triassic age (Figure 8.11). The original pyrite, calcite, and aluminous mineral components of the host rocks were altered by the mineralizing solutions to limonite, alunite, gypsum, and jarosite. The resulting light brownish yellow color contrasts with

the typical purple to gray color of unmineralized parts of the Chinle Formation. Areas of altered rock form elongate halos up to 400 m long surrounding the ore deposits. The alteration colors are valuable guides to the ore deposits but are not uniquely related to ore deposits for the following reasons: (1) the normal color of some unmineralized parts of the Chinle Formation resembles that of the altered zones; (2) uranium may have been remobilized and removed after the alteration occurred; and (3) alteration may have occurred without any deposition of ore. This geologic description is summarized from the work of Spirakis and Condit (1975) of the U.S. Geological Survey, who also reported the following Landsat interpretation.

Plate 13A shows a digitally processed subscene of a Landsat MSS image of the Cameron district. Grabens, faults, collapse structures, volcanic cones, basalt flows, and sedimentary rock formations are recognizable. On this color image the light gray and light brown altered rocks cannot be distinguished from the surrounding unaltered rocks of the Chinle Formation, which have similar colors. To enhance the appearance of altered rocks, a color ratio image (Plate 13B) was prepared by combining the ratio images 4/7, 6/4, and 7/4 in blue, green, and red, respectively. The color ratio image is dominated by yellow and brown tones, but there are a few conspicuous blue patches located along the outcrop of the Chinle Formation (Figure 8.11). These blue patches

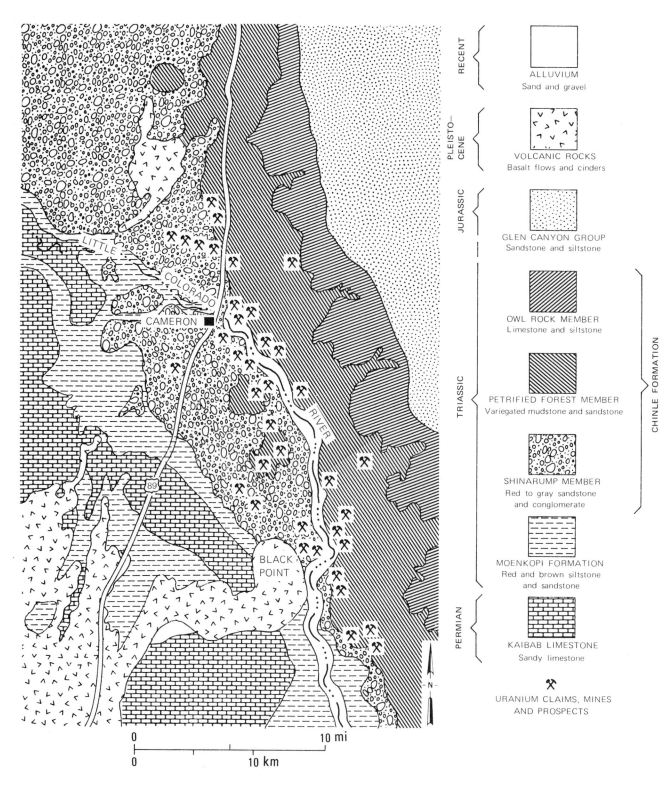

FIGURE 8.11 Geologic map of the Cameron uranium district, north-central Arizona. From Chenoweth and Magleby (1971).

correlate with areas of altered rock as shown by an aerial reconnaissance of the region (Spirakis and Condit, 1975). The uranium mines and claims shown in the geologic map occur within or adjacent to the altered outcrops indicated by the blue signatures in the color ratio image.

The color ratio image is not a perfect exploration method because some unaltered rocks may have reflectance properties similar to those of altered host rocks. In the southeast portion of the Cameron subscene, for example, some outcrops of unaltered Moenkopi Formation west of the Colorado River have blue signatures on the color ratio image. These areas can be eliminated as exploration targets because they are not associated with known host rocks.

Freer–Three Rivers Uranium District, Texas

In this typical south Texas uranium district, the host rocks are channels filled with sandstone or conglomerate in the Catahoula Tuff (Miocene age). One important exploration method is to map the occurrence of outcrops of the channels filled with potential host rocks. Geologic mapping in this area of low relief is hampered by lack of outcrops, nondistinctive rock types, a partial cover of younger gravel, and restricted land access. Digital processing of Landsat images of the district has not been successful. The U.S. Geological Survey conducted airborne scanner surveys in an attempt to map the sandstone channel deposits (Offield, 1976).

Figure 8.12A shows a daytime cross-track scanner image acquired in the visible band. The road network and boundaries of agricultural fields are shown by reflectance differences. Some known occurrences of sandstone channels are indicated by arrows, but these rocks are not recognizable in the daytime visible image. The nighttime thermal IR image (Figure 8.12B) was acquired following a week of heavy rain. The high moisture content greatly reduced thermal contrasts between different rock types. Despite these poor conditions for image acquisition, the sandstone channels have distinct bright (warm) signatures that contrast with the dark (cool) signatures of the clay-rich tuff units. These nighttime thermal IR signatures are consistent with the densities and thermal characteristics described in Chapter 5. Sandstone has a higher density, which results in higher thermal inertia and a warmer nighttime radiant temperature relative to tuff, which has lower density, lower thermal inertia, and cooler nighttime temperature. The warm signature of the sandstone host rock is caused by the thermal properties of the rock, not by radiogenic heat (heat from decay of radioactive elements). Calculations show that radiogenic heat produced by typical sedimentary uranium deposits is insufficient to produce a detectable thermal anomaly (Kappelmeyer and Haenel, 1974, p. 170).

NASA/Geosat Test Case Project— Uranium Mines

The Lisbon Valley, Utah, and Copper Mountain, Wyoming, areas were selected as uranium test sites for the NASA/Geosat project. At Lisbon Valley, uranium occurs in the Chinle Formation (Triassic age), which is poorly exposed at the surface. The overlying Wingate Sandstone (Triassic age) is widely exposed at Lisbon Valley and has a characteristic red color caused by iron-oxide minerals. Where it overlies uranium deposits in the Chinle Formation, however, the Wingate Sandstone is white because the iron oxides are removed (bleached). Mining geologists have attributed the bleaching of the Wingate Sandstone to the reducing conditions that precipitated uranium in the Chinle Formation. Under reducing conditions, ferric iron is converted to ferrous iron, which is soluble and may be removed by groundwater. Thus the present-day white outcrops of Wingate Sandstone record the subsurface geochemical conditions that precipitated uranium in the Chinle Formation. Conel and Alley (1985) prepared and interpreted principal-component images from TMS data. These images distinguished the white from the red portion of the Wingate Sandstone and separated a number of geologic formations. This ability to recognize anomalous color patterns may be useful in uranium exploration.

FUTURE MINERAL EXPLORATION METHODS

Landsat MSS was a new tool for mineral exploration; Landsat TM was a major improvement with its higher spatial resolution and additional spectral bands. The Goldfield example demonstrated the importance of TM band 5 centered at 1.6 μm and band 7 centered at 2.2 μm for mapping areas of hydrothermal alteration. TM bands have relatively broad spectral ranges, however, and are not capable of discriminating the various zones and subzones of the alteration models. The subzones of the Goldfield model (Figure 8.6) are characterized by the presence of specific alteration minerals. Within the atmospheric window from 2.0 to 2.5 μm, kaolinite, montmorillonite, illite, and alunite have distinctive absorption features in their reflectance spectra as shown in Figure 8.13. All four minerals have an absorption minimum near 2.2 μm, but the exact location of the minimum is different for each mineral. There are additional spectral features that can be used to identify specific minerals. For example, kaolinite has a "shoulder"

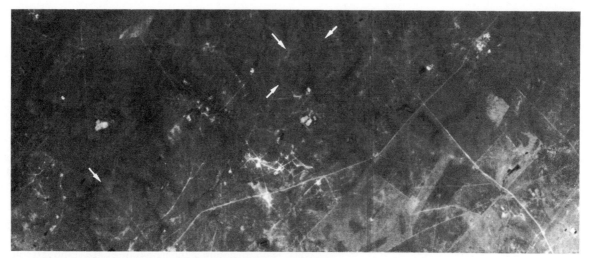

A. IMAGE ACQUIRED AT MIDDAY IN THE VISIBLE
   SPECTRAL REGION. ARROWS MARK CHANNEL–FILL
   CONGLOMERATES WITH TOPOGRAPHIC EXPRESSION.

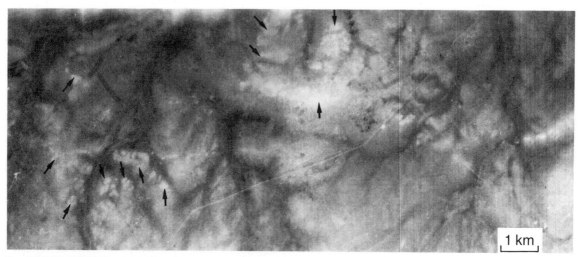

1 km

B. PREDAWN THERMAL IR IMAGE (8 TO 14 μm).
   ARROWS MARK CHANNEL–FILL CONGLOMERATES THAT
   ARE WARMER (BRIGHTER TONE) THAN THE
   SURROUNDING CATAHOULA TUFF.

FIGURE 8.12 Freer–Three Rivers uranium district, southern Texas. Airborne
scanner images acquired November 1974. From Offield (1976, Figure 4). Courtesy
T. W. Offield, U.S. Geological Survey.

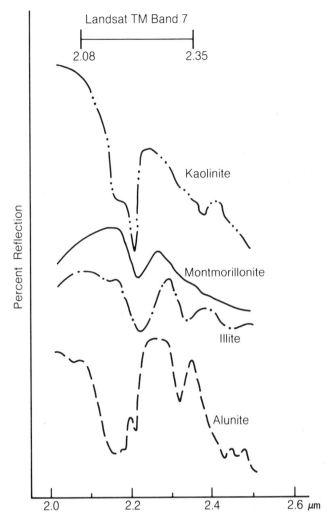

FIGURE 8.13 Laboratory spectra of alteration minerals in the 2.0-to-2.5-μm atmospheric window. Spectra are offset vertically for clarity.

tered. Figure 8.14 shows one of the AIS image bands of the Cuprite district with several ground resolution cells indicated. For each of these cells, a reflectance spectrum was plotted and displayed in Figure 8.14 together with laboratory reference spectra of alunite and kaolinite. The AIS spectra from the northern part of the district are similar to the alunite reference spectrum; those from southern part of the district are similar to the kaolinite reference. The distribution of alteration minerals determined remotely by AIS agree with the mineral patterns mapped at Cuprite. This example demonstrates the exploration potential of multispectral scanner images with high spectral resolution. AIS is currently an experimental system, but this technology will soon become operational.

MINERAL EXPLORATION IN COVERED TERRAIN

The examples of mineral exploration in this book, except for the Freer–Three Rivers deposit, are in arid to semiarid terrain with extensive exposures of bedrock and little vegetation cover. Remote sensing images from these and similar areas can be analyzed for evidence of alteration zones that are surface expressions of mineral deposits. In most of the temperate and humid climate zones of the world, however, mineral deposits are commonly

in the absorption feature at 2.18 μm, and alunite has a secondary absorption minimum at 2.21 μm. The broad spectral range of Landsat TM band 7 (Figure 8.13) is incapable of distinguishing these spectral features that characterize the different alteration minerals.

The airborne imaging spectrometer (AIS), described in Chapter 2, has the capability of recording detailed spectral features and thereby the potential for identifying specific alteration minerals. AIS acquires multiple images at wavelength intervals of 0.01 μm. In the important region from 2.00 to 2.30 μm, AIS acquires 30 spectral readings for each ground resolution cell that can be displayed as a reflectance spectrum.

Jet Propulsion Laboratory has acquired AIS image data over the Cuprite district, 20 km south of Goldfield, where the volcanic rocks have been hydrothermally al-

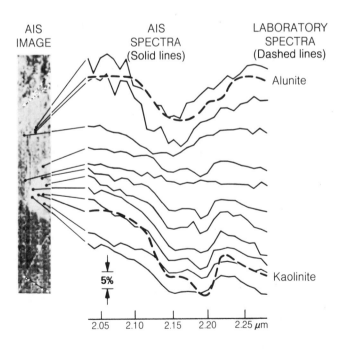

FIGURE 8.14 Reflectance spectra derived from AIS data of the Cuprite district, Nevada. From Goetz (1984, Figure 14). Courtesy A. F. H. Goetz, Jet Propulsion Laboratory.

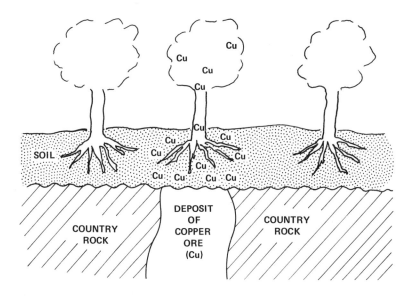

FIGURE 8.15 Copper enrichment of vegetation and soil overlying a concealed copper deposit.

concealed beneath a cover of soil and vegetation. The composition of residual soil reflects the composition of the underlying bedrock from which the soil was derived by weathering processes.

Geochemical exploration techniques involve collecting water and soil samples and analyzing their metal content. Areas with high metal concentrations are then tested by core drilling. Figure 8.15 illustrates the copper enrichment of soil overlying a copper deposit in the bedrock. Vegetation growing in mineralized soil may have higher metal content in its tissue than vegetation in normal, or background, soil, as the figure shows. This concentration of metals in vegetation is the basis for biogeochemical and geobotanical prospecting methods. However, these techniques for testing soil and analyzing vegetation are not applicable in areas where the soil has been transported rather than formed in place. Alluvial and glacial soils are examples of transported soils.

Biogeochemical and Geobotanical Exploration

Biogeochemical exploration consists of collecting vegetation samples that are analyzed chemically for high metal concentrations that may indicate a concealed ore deposit.

Geobotanical exploration searches for unusual vegetation conditions that may be caused by high metal concentrations in the soil. Sampling and chemical anal-

yses of vegetation are not required, but skill and experience are needed to recognize the more subtle vegetation anomalies. Remote sensing techniques are being investigated as possible geobotanical exploration methods. The major geobotanical criteria for recognizing concealed ore deposits are:

1. *Lack of vegetation.* This may be caused by concentrations of metals in the soil that are toxic to plants. These areas are sometimes called *copper barrens* where they are caused by high concentrations of that metal. Areas that lack vegetation may be seen on aerial photographs. These barren areas may result from causes other than mineralization, however.

2. *Indicator plants.* These are species that grow preferentially on outcrops and soils enriched in certain elements. Cannon (1971) prepared an extensive list of indicator plants. For example, in the Katanga region of southern Zaire, a small blue-flowered mint, *Acrocephalus robertii*, is restricted entirely to copper-bearing rock outcrops.

3. *Physiological changes.* High metal concentrations in the soil may cause abnormal size, shape, and spectral reflectance characteristics of leaves, flowers, fruit, or entire plants. A relationship between spectral reflectance properties of plants and the metal content of their soils could form the basis for remote sensing of mineral deposits in vegetated terrain.

Remote Sensing for Minerals in Vegetated Terrain

Chlorosis, or yellowing of leaves, is an example of a spectral change visible to the eye. Chlorosis results from an upset of the iron metabolism of plants that may be caused by an excess concentration of copper, zinc, manganese, or other elements. Relatively high metal concentrations are required to produce chlorosis, which has been observed in plants growing near mineral deposits. Chlorosis is not a reliable prospecting criterion, however, for these reasons: (1) Many areas of known mineralized soil support apparently healthy plants with no visible toxic symptoms. (2) Chlorosis may result from conditions unrelated to mineral deposits, such as soil salinity.

The large, low-grade, copper-molybdenum deposit at Catheart Mountain, Maine, has been used as a geobotanical remote sensing test site (Yost and Wenderoth, 1971). Field spectrometers measured reflectance of trees growing in normal soil and in mineralized soil overlying the deposit (Figure 8.16). Red spruce and balsam fir growing in the mineralized soil had higher metal concentrations than trees of the same species in unmineralized soil. In the reflected IR spectral region, the mineralized balsam firs have a higher reflectance than the normal trees, whereas mineralized red spruce have a lower reflectance than the normal trees (Figure 8.16). In the green spectral region, the mineralized trees of both species have a higher reflectance.

In the years following the 1970 study, a number of additional investigations have been made and were summarized by Labovitz and others (1983, Figure 1). With some exceptions, vegetation reflectance in the green and red bands generally increased with increasing metal concentration in the soil. In the reflected IR region, however, there is less agreement; some studies show an increase in vegetation reflectance while others show a decrease. Labovitz and others (1983, p. 759) also noted that the geobotanical model of Figure 8.15 is not universally true. In Virginia they found that the leaves of oak trees growing in metal-rich soil may have a lower metal content than leaves from trees in normal soil.

Geophysical Environmental Research has taken a different approach to geobotanical remote sensing. The company operates a nonimaging airborne system that acquires detailed reflectance spectra. The spectra in Figure 8.17 were acquired from conifers growing over a mineralized area and in an adjacent nonmineralized area. In the green band (0.5 to 0.6 μm) reflectance is higher for trees in the mineralized area, which is consistent with other studies. Beginning at a wavelength of about 0.7 μm, vegetation spectra have a steep upward slope

to the high reflectance values in the IR region. In Figure 8.17, this steep slope is shifted slightly toward shorter wavelengths for the conifers growing in the mineralized area. This shift, called the *blue shift,* has been noted in vegetation over several mineralized areas (Collins and others, 1983) and has exploration potential.

In summary, remote sensing of mineral deposits in vegetated terrain is in the research and development stage. The relationships between mineral deposits, soil chemistry, reflectance properties of vegetation, and remote sensing systems are complex, but they are being investigated at several research centers.

OIL EXPLORATION

The search for oil in unexplored onshore areas, such as portions of Africa and China, normally follows a pattern.

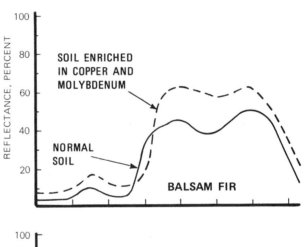

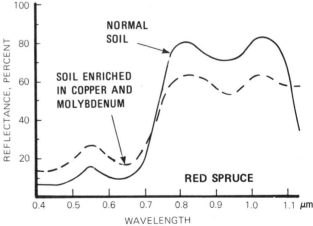

FIGURE 8.16   Reflectance spectra of balsam fir and red spruce growing in normal soil and in soil enriched in copper and molybdenum. Spectra were recorded with a field spectrometer by Yost and Wenderoth (1971, Figures 5 and 6).

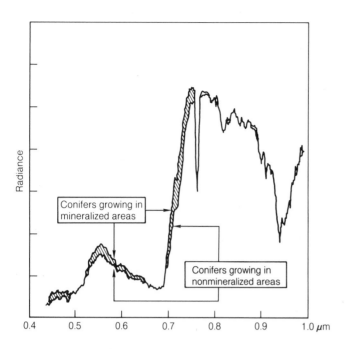

FIGURE 8.17 Airborne reflectance spectra of conifers in the Cotter Basin, Lewis and Clark County, Montana. Note the blue shift for conifers in the mineralized area. From Collins and others (1983, Figure 4B).

It begins with regional reconnaissance and is succeeded by progressively more detailed (and expensive) steps that culminate in drilling a wildcat well. *Wildcat wells* are exploratory tests in previously undrilled areas, whereas *development wells* are drilled to produce oil from a previously discovered field. A typical exploration program proceeds as follows:

1. *Regional remote sensing reconnaissance.* Small-scale Landsat mosaics covering hundreds of thousands of square kilometers are especially useful in this phase. The objective is to locate *sedimentary basins*, which are areas underlain by thick sequences of sedimentary rocks. These basins are essential for the formation of oil fields.

2. *Reconnaissance geophysical surveys. Aerial magnetic* surveys are made to produce maps that record the intensity of the earth's magnetic field. Sedimentary basins have lower magnetic intensities than do areas underlain by nonsedimentary rocks such as granite and volcanic rocks. The aerial magnetic maps thus can confirm the presence of sedimentary basins. Surface *gravity surveys* are made that record the intensity of the earth's gravity field. Sedimentary rocks have a lower specific gravity than nonsedimentary rocks; hence sedimentary basins are shown by lower values on the gravity maps. Gravity and magnetic maps may also show regional structural features.

3. *Detailed remote sensing interpretation.* Individual digitally processed Landsat images are studied to identify and map geologic structures, such as anticlines and faults, that may form oil traps. Promising structures may be mapped in detail using stereo pairs of aerial photographs. Radar images are used in regions of poor weather where it is difficult to acquire good photographs and Landsat images. At this stage, geologists go into the field to check the interpretation and collect samples of the exposed rocks.

4. *Seismic surveys.* Explosives or mechanical devices are used to transmit waves of sonic energy into the subsurface, where they are reflected by geologic structures. The reflected waves are recorded at the surface and processed to produce *seismic maps and cross sections* that show details of subsurface geologic structure.

5. *Drilling.* Wildcat wells are drilled to test the subsurface targets defined by the preceding steps. Because of the inevitable uncertainties of predicting geologic conditions thousands of meters below the surface, the success rate in 1983 of wildcat wells drilled in the United States was only 17.5 percent.

Each of the three following examples illustrates a particular aspect of remote sensing for oil exploration. The Kenya project was Chevron's first major utilization of Landsat images. No oil was discovered, but Landsat was shown to be a reliable source of regional geologic information. As a result, Chevron geologists then interpreted Landsat images of southern Sudan to locate a major sedimentary basin. The Sudan project, a classic example of the systematic exploration process described above, resulted in the discovery of several oil fields. Kenya and the Sudan are frontier exploration areas where relatively few wells have been drilled. The final example in northwest Colorado is a mature area that has been thoroughly explored and drilled. Thus it is a good training site for learning how to recognize oil-producing structures on Landsat images.

Kenya

Chevron Overseas Petroleum acquired an exploration license in eastern Kenya and completed a photogeologic and field study in 1972. Landsat MSS images of the area became available in 1973 and were compiled into the mosaic shown in Figure 8.18 that was interpreted by Miller (1975). The drainage patterns on the interpretation map (Figure 8.19) provide valuable geographic reference in this region of generalized base maps. Much geologic information is present on the Landsat images, despite the relatively featureless nature of the terrain.

Lineaments are a major feature in Kenya, as shown in the interpretation map and in the image covering the northwest part of the Chevron license area (Figure 8.20A). The lineaments and other features are more apparent on color composite images (not illustrated) than on the black-and-white versions shown here. The Lagh Bogal lineament that trends northwest across Figure 8.20A is particularly significant because it marks the northeast boundary of a sedimentary basin that was confirmed by later geophysical surveys. The major lineaments extend beyond the limits of the mosaic. Large volcanoes occur along the extensions of several lineaments beyond the Chevron license area.

Young volcanic flows form the black pattern in the west-central part of Figure 8.20A. In the southeast part, large arcuate tonal bands may represent depositional patterns in clastic beds. A dark lobe extends southeast from the northwest part of the image and marks the outcrop of crystalline basement rocks with discernible foliation trends.

The image in Figure 8.20B covers the southwest part of the license area, where dark basement rocks, with north-trending foliation patterns, crop out in the southwest part of the image. East of the basement outcrops, and probably in fault contact, are sands and clays of Pliocene age that form a triangular light-colored outcrop through which flows the Tana River with its associated vegetation. North of the river, the Pliocene strata are capped by a dark layer of duricrust. In the southeast part of Figure 8.20B, a thin wedge of strata with a gray tone occurs between the duricrust and the underlying light-toned strata. The eastward expansion of this wedge represents basinward thickening of the sedimentary section that was first observed on the Landsat image.

Figure 8.21 summarizes the results of gravity, magnetic, and seismic surveys. The Lagh Bogal lineament coincides in part with a subsurface fault, independently interpreted from the geophysical surveys, that forms the northeast boundary of the basin. Geophysical data indicate that the west border of the basin is not a single major fault zone but a combination of faulting and tilting. The northeast and northwest trends of the image lineaments in the northeast part of the Chevron license area are parallel with the gravity and magnetic positive trends shown in Figure 8.21. Based largely on the Landsat interpretations, Chevron acquired a second exploration license area adjoining the original area on the northwest (Figure 8.21). Several wildcat wells were drilled to test geophysical prospects but were dry because subsurface conditions were not conducive to hydrocarbon generation. However, the geophysical surveys and drilling confirmed the accuracy of the geologic interpretations of Landsat images, which encouraged Chevron to use remote sensing to explore farther north in the Sudan.

The Sudan

The Sudan, the largest nation in Africa, is about equal in area to the United States east of the Mississippi River. The Sudan is incompletely mapped, both geographically and geologically. Little oil exploration had been done prior to 1975. Using the experience gained in Kenya, Chevron compiled a mosaic of Landsat MSS images at a scale of 1:1,000,000 for the southern part of the Sudan. Figure 8.22 shows a reduced-scale version of part of the mosaic. The White Nile River flows northward from the highlands of Uganda into the vast Sudd Swamp, which occupies much of the eastern portion of the mosaic (Figure 8.22). The river emerges from the swamp and flows northward toward Khartoum.

The mosaic was interpreted by J. B. Miller of Chevron, who recognized the presence of a previously unknown sedimentary basin in the vicinity of the present-day Sudd Swamp. Miller also analyzed the drainage patterns and noted that while there were numerous local bends and meanders, the major streams were relatively straight at the regional scale of the Landsat mosaic. These stream lineaments were interpreted as the expression of faults that formed the boundaries of the major basin and smaller subbasins. Based on this regional Landsat interpretation, Chevron negotiated with the Sudanese government to obtain exploration rights to the area outlined in the mosaic (Figure 8.22), which included the inferred sedimentary basin. Figure 8.23A shows the location of the original concession, which covered an area of 5.1×10^5 km^2. For comparison, note that the state of California covers an area of less than 4.1×10^5 km^2.

The next step in the campaign was to have the Landsat MSS images covering the concession area digitally processed, enlarged, and registered to ground-control points to produce a base map for the area that was used throughout the campaign. Additional geologic information was interpreted from the individual enhanced color images to guide subsequent exploration.

Chevron then made aerial magnetic surveys that confirmed the existence of the sedimentary basin. A gravity survey added details to the evolving regional picture. The next step was to conduct seismic surveys, which were difficult and expensive in this area of few roads and towns. Equipment and supplies were moved by aircraft to airstrips built for that purpose. The seismic surveys defined a number of potential oil traps, and wildcat drilling began in 1977. The first five wells were dry, but in 1979 the first oil was discovered in Sudan at Abu Gabra (Figure 8.23B). Subsequent drilling discovered the Heglig, Unity, and Melut fields. Full extent of the fields is not known at this early stage of development, but reserves exceed several hundred million barrels. A

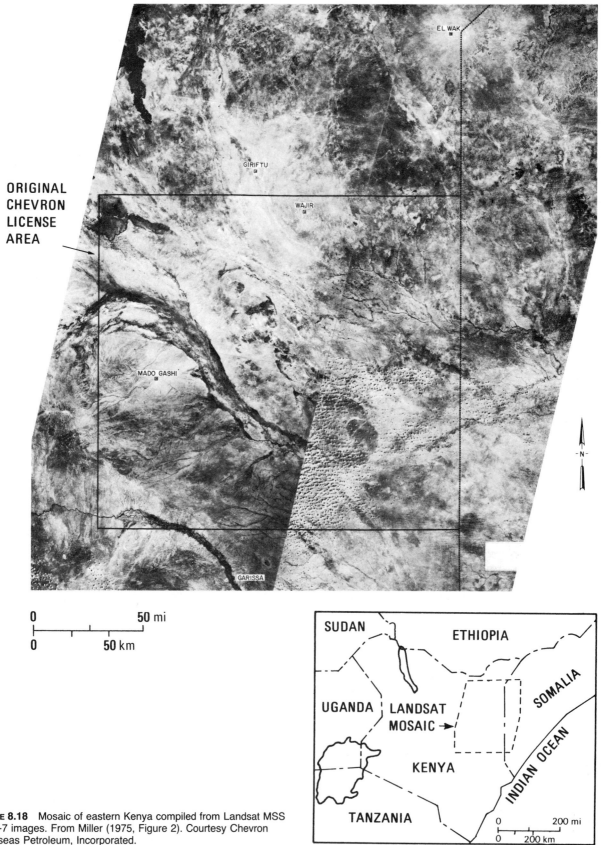

ORIGINAL
CHEVRON
LICENSE
AREA

EL WAK

GIRIFTU

WAJIR

MADO GASHI

GARISSA

0   50 mi

0   50 km

SUDAN   ETHIOPIA

UGANDA   LANDSAT   SOMALIA
          MOSAIC →

               KENYA   INDIAN OCEAN

TANZANIA

0   200 mi

0   200 km

FIGURE 8.18   Mosaic of eastern Kenya compiled from Landsat MSS
band-7 images. From Miller (1975, Figure 2). Courtesy Chevron
Overseas Petroleum, Incorporated.

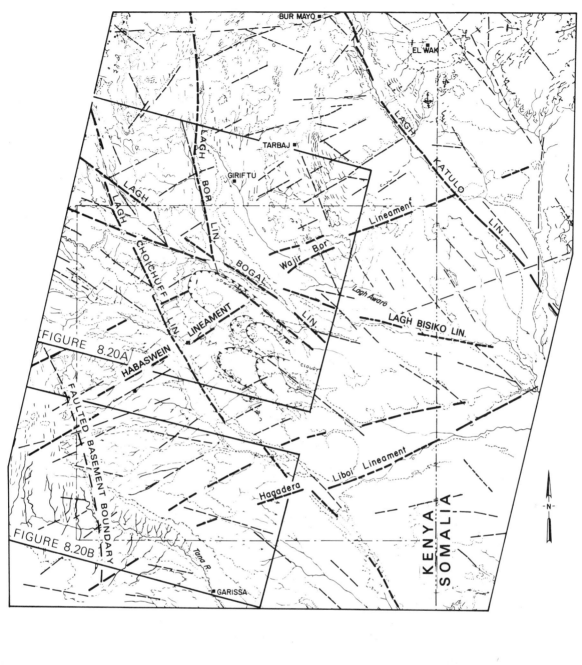

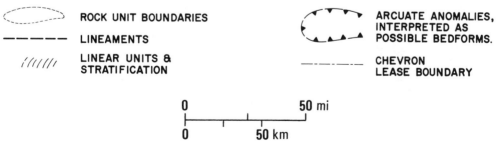

ROCK UNIT BOUNDARIES

LINEAMENTS

LINEAR UNITS &
STRATIFICATION

ARCUATE ANOMALIES,
INTERPRETED AS
POSSIBLE BEDFORMS.

CHEVRON
LEASE BOUNDARY

0   50 mi

0   50 km

FIGURE 8.19 Interpretation of Landsat mosaic of eastern Kenya. From Miller
(1975, Figure 4). Courtesy Chevron Overseas Petroleum, Incorporated.

A. NORTHWEST PART OF THE CHEVRON LICENSE AREA.

B. SOUTHWEST PART OF THE CHEVRON
   LICENSE AREA.

FIGURE 8.20 Landsat MSS band-5 images that cover parts of the Chevron
exploration license area in Kenya. See Figure 8.19 for locations. From Miller
(1975). Courtesy Chevron Overseas Petroleum, Incorporated.

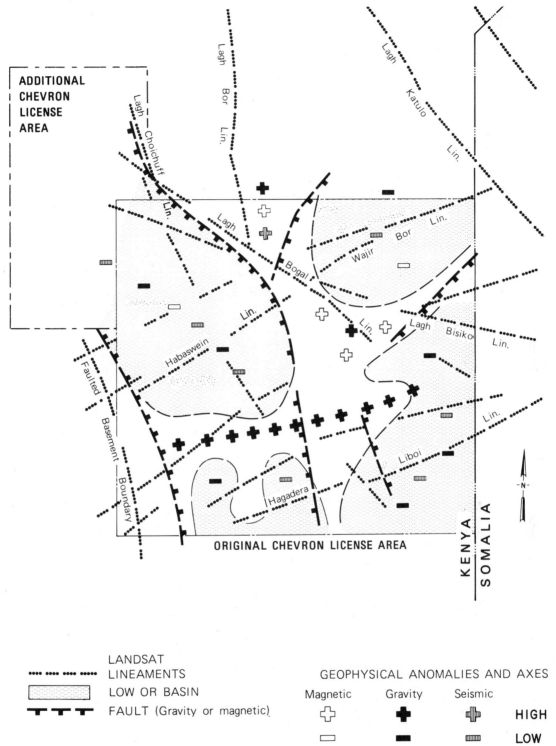

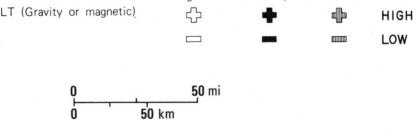

FIGURE 8.21 Comparison of Landsat features with geophysical trends. From Miller (1975, Figure 6). Courtesy Chevron Overseas Petroleum, Incorporated.

FIGURE 8.22 Mosaic of Landsat MSS band-5 images of southern Sudan showing the outline of the original Chevron exploration concession area.

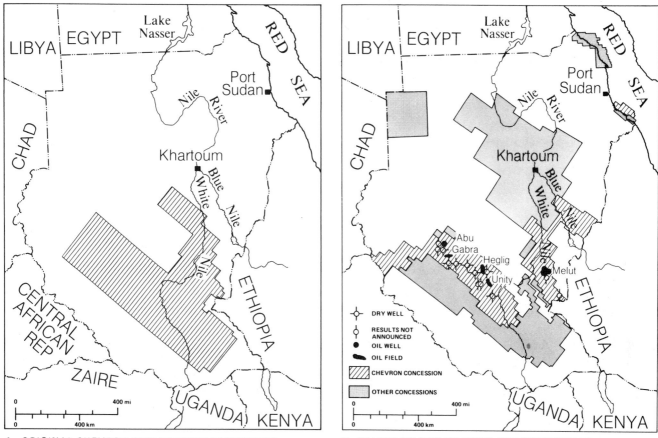

A. ORIGINAL CHEVRON EXPLORATION CONCESSION GRANTED IN 1974.

B. STATUS OF EXPLORATION AND CONCESSIONS IN 1982.

FIGURE 8.23 Maps showing status of oil exploration in the Sudan.

pipeline is planned to Port Sudan on the Red Sea, where a shipping terminal and refinery will be constructed.

The maps in Figure 8.23 show major changes between 1974 and 1982. The original Chevron exploration area has diminished in size because the agreement requires the company to relinquish portions of the concession at stated times. Rights to much of the relinquished areas were acquired by competing companies, which also acquired other large concession areas. The progress of the Sudan project has been reported by Miller and Vandenakker (1977), Vandenakker and Ryan (1983), and Schull (1984).

Landsat data were also employed in the operational phase of the project. Southern Sudan has a dry season and a wet season with drastically different terrain con-

ditions. Plate 13C is a subscene of an MSS image in the Sudd Swamp acquired during the dry season. The sinuous red band is a stream channel with vegetation, and associated small lakes which have dark signatures. The dark terrain on either side of the channel is grassland that was burned by the local people to produce better forage during the ensuing wet season. Plate 13D is the same subscene imaged during the wet season and digitally processed in the following manner. Field crews in the area noted localities of major terrain categories and communicated this information to Chevron staff personnel in San Francisco. The field localities were used as training sites to produce the supervised classification map of Plate 13D. The map colors and corresponding terrain categories are listed below.

Color	Terrain category
Dark blue	Open water
Light blue	Shallow water with vegetation
Red	Papyrus and water hyacinth
Orange	Wet grass with standing water
Green	Bullrushes
Black	Bullrushes with standing water
Dark yellow	Dry grass
Light yellow	Upland areas, driest areas

These terrain classification maps were valuable to the seismic crews for planning operations. For example, swamp buggies were used in marsh areas, but where papyrus plants occurred, the stalks jammed the drive mechanisms. Fieldworkers were able to avoid driving through areas with papyrus plants by consulting the classification map (Plate 13D), which shows these areas in red.

Northwest Colorado

In contrast to Kenya and the Sudan, northwest Colorado is a mature exploration area where the surface structures have been drilled and one can directly evaluate the relationship of Landsat features to oil and gas fields. The winter MSS image of Figure 8.24 demonstrates the advantages of low sun elevation and light snow cover for structural mapping. The major geologic features and the local structures associated with oil and gas fields are shown in the map of Figure 8.25. The White River and Uinta Mountain uplifts and the Piceance Creek and Sand Wash basins are well expressed. The Piceance Creek basin contains major oil shale reserves of the Green River Formation (Eocene age) that crops out around the basin margins. The Grand Hogback monocline separating the White River uplift from the Piceance Creek basin is especially prominent on the image.

The Rangely anticline, in the southwest part of the map (Figure 8.25), is a major Chevron oil field that is outlined by strike ridges of resistant Cretaceous sandstones surrounding the eroded Mancos Shale outcrops in the core. The asymmetry of the anticline is indicated on the image by the gentle dip slopes on the north flank and the steeper slopes on the south. The Blue Mountain anticline to the north is equally well expressed, but is nonproductive. Moffat, Iles, and Thornburgh are small oil fields trapped at anticlinal closures that are outlined by resistant sandstone ridges on the image. The Danforth Hills and Wilson Creek oil fields are also anticlinal

structures but are marked on the image by a change to very fine texture. The Piceance Creek gas field is formed by an anticline that is clearly indicated by the radial drainage pattern and by the streams that "wrap around" the structure.

Radar Images for Oil Exploration

The essentially all-weather capability and the ability to enhance geologic structure in forested terrain have made side-looking airborne radar useful for oil exploration, especially in tropical regions. Wing and Mueller (1975) of Continental Oil Company described their structural reconnaissance mapping in Irian Jaya, Indonesia, using SLAR images. Magnier, Oki, and Kaartidiputra (1975) published a SLAR mosaic of the Mahakam delta on the east coast of Kalimantan, Indonesia. The anticlinal trends of the onshore oil fields are clearly visible on the mosaic despite the dense vegetation cover. In the 1980s the Space Shuttle acquired radar images of Indonesia and other oil-producing tropical areas (Chapter 6). One can readily interpret lithologic terrains and geologic structures from these images. When more coverage is available, satellite radar images will be a valuable exploration resource.

Oil Exploration Research

Landsat and radar images may be analyzed for surface evidence of geologic structures such as folds and faults that may form petroleum traps at depth. Some oil fields are marked at the surface by direct indications of the underlying hydrocarbons. Surface seeps of oil and gas that have leaked from subsurface traps are a well-known example. The original Drake well in Pennsylvania was located on the basis of oil seeps. As recently as the early 1900s, oil fields were located in California on the basis of oil seeps. The recognition of oil and gas seeps is called *direct detection,* and much research has been done on this subject.

Escaping oil and gas may interact with surface rocks, soil, and vegetation to produce anomalous conditions that may be clues to the underlying hydrocarbon deposit. Classic examples of rock alteration occur at the Cement and Velma oil fields in southern Oklahoma. Sandstone outcrops in the area are typically red, but over the fields they are tan and gray. Gypsum ($CaSO_4 \cdot nH_2O$) is locally replaced by calcite ($CaCO_3$) over the fields. The color change from red to gray has long been attributed to escaping hydrocarbons that have chemically reduced the red iron oxide in the sandstone to a nonred iron compound. Donovan (1974) recognized another surface alteration effect: the secondary calcite and dolomite ($Ca_{1/2}Mg_{1/2}CO_3$) in the surface rocks have

A. TIMS COLOR COMPOSITE IMAGE. FROM
   KAHLE (1983, COVER). COURTESY A. B. KAHLE,
   JET PROPULSION LABORATORY.

B. IR COLOR IMAGE FROM LANDSAT TM.

PLATE 9   Thermal IR multispectral scanner (TIMS) image and Landsat image in the
Panamint Mountains and Death Valley, California. Each image covers a width of 12
km.

A. NORMAL COLOR IMAGE BEFORE IHS
    TRANSFORMATION.

B. NORMAL COLOR IMAGE AFTER IHS
    TRANSFORMATION.

C. PRINCIPAL–COMPONENT COLOR IMAGE.
    PC IMAGE 2 = RED, PC IMAGE 3 = GREEN,
    AND PC IMAGE 4 = BLUE.

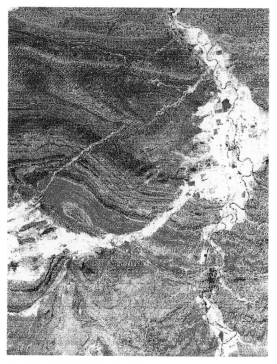

D. RATIO COLOR IMAGE.
    3/1 = RED, 5/7 = GREEN, 3/5 = BLUE.

PLATE 10   Digital enhancement and information extraction of Landsat TM data for
the Thermopolis, Wyoming, subscene. Each image covers a width of 20 km.

SYMBOL	CLASS	PERCENT
	Redbeds	8.4
	Sandstone	48.3
	Shale	18.9
	Agriculture	16.2
	Native vegetation	5.2
	Water and shadows	1.9
	Unclassified	1.1

A. SUPERVISED–CLASSIFICATION MAP AND EXPLANATION.

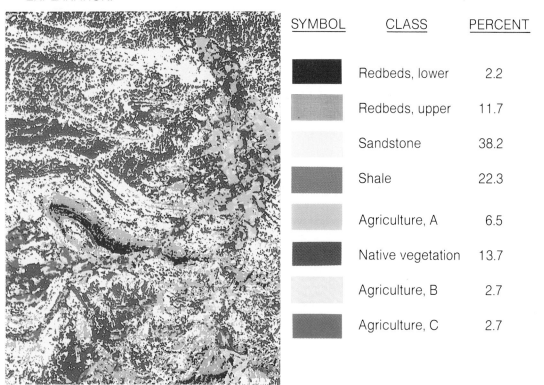

SYMBOL	CLASS	PERCENT
	Redbeds, lower	2.2
	Redbeds, upper	11.7
	Sandstone	38.2
	Shale	22.3
	Agriculture, A	6.5
	Native vegetation	13.7
	Agriculture, B	2.7
	Agriculture, C	2.7

B. UNSUPERVISED–CLASSIFICATION MAP AND EXPLANATION.

PLATE 11   Multispectral classification maps of Landsat TM data for the Thermopolis, Wyoming, subscene. Each image covers a width of 20 km.

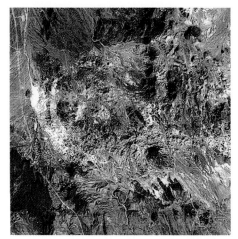

A. NORMAL COLOR IMAGE, ENHANCED.

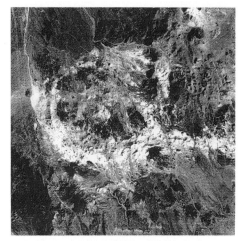

B. COLOR RATIO IMAGE. 5/7 = RED, 3/1 = GREEN, 3/5 = BLUE.

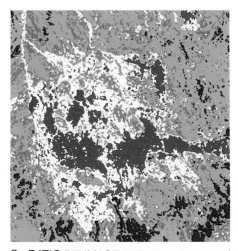

C. RATIO 5/7 IMAGE WITH DENSITY SLICE.

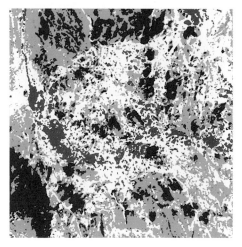

D. RATIO 3/1 IMAGE WITH DENSITY SLICE.

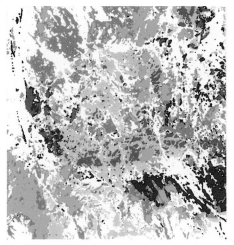

E. UNSUPERVISED–CLASSIFICATION MAP.

F. CLASSIFICATION EXPLANATION.

SYMBOL	CLASS	PERCENT
	Alluvium	39.2
	Basalt	14.0
	Tuff	6.6
	Altered rocks, A	5.3
	Altered rocks, B	18.3
	Unaltered rocks	16.6

PLATE 12 Digitally processed Landsat TM images of the Goldfield, Nevada, subscene. Each image covers a width of 15 km.

A. COLOR COMPOSITE IMAGE, CAMERON URANIUM DISTRICT, ARIZONA. AREA COVERS A WIDTH OF 3.5 km.

B. COLOR RATIO IMAGE, CAMERON URANIUM DISTRICT, ARIZONA.

C. COLOR COMPOSITE IMAGE, SUDD SWAMP, THE SUDAN. AREA COVERS A WIDTH OF 45 km.

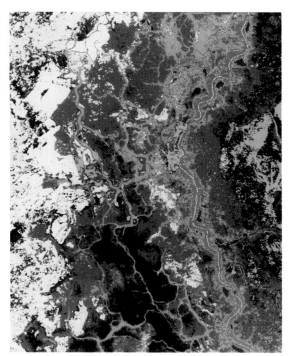

D. SUPERVISED–CLASSIFICATION MAP, SUDD SWAMP, THE SUDAN.

PLATE 13 Digitally processed Landsat MSS images for resource exploration.
Plates A and B are from Spirakis and Condit (1975, Figures 4, 6). Plates C and D
are from Vandenakker and Ryan (1983).

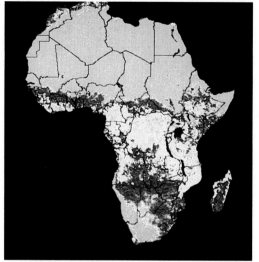

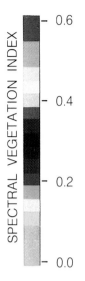

SPECTRAL VEGETATION INDEX

- 0.6

- 0.4

- 0.2

- 0.0

A. SPECTRAL-VEGETATION-INDEX
MAP OF AFRICA, APRIL 12
TO MAY 2, 1982.

B. VEGETATION CLASSIFICATION
MAP OF AFRICA. COLORS ARE
EXPLAINED IN CHAPTER 9.

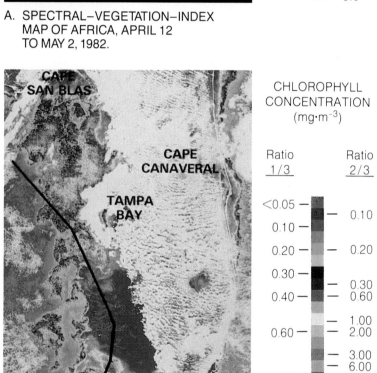

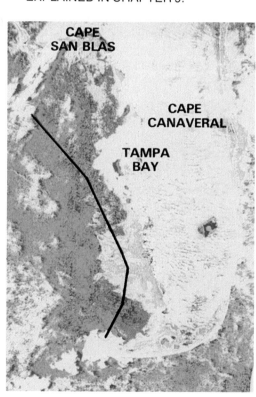

CHLOROPHYLL
CONCENTRATION
$(mg \cdot m^{-3})$

Ratio 1/3		Ratio 2/3
<0.05		0.10
0.10		0.20
0.20		
0.30		0.30
0.40		0.60
		1.00
0.60		2.00
		3.00
		6.00
1.00		10.00
>1.00		>10.00

Ship Track _____

C. CZCS RATIO 1/3 IMAGE,
GULF OF MEXICO AND FLORIDA.
AREA COVERS A WIDTH OF 600 km.

D. CZCS RATIO 2/3 IMAGE,
GULF OF MEXICO AND FLORIDA.

PLATE 14 Digitally processed images from NOAA environmental satellites. Plates
A and B are from Tucker, Townshend, and Goff (1985, Figures 1 and 5). Plates C
and D are from Gordon and others (1980, Figure 2).

A. MSS IR COLOR IMAGE, THE OXNARD
PLAIN, CALIFORNIA. AREA COVERS
A WIDTH OF 14 km.

B. SUPERVISED–CLASSIFICATION MAP.
COLORS ARE EXPLAINED IN CHAPTER 10.

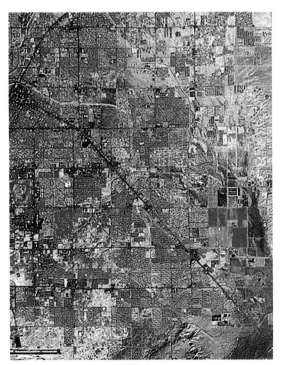

C. TM IR COLOR IMAGE, LAS VEGAS,
NEVADA. AREA COVERS A WIDTH OF 10 km.

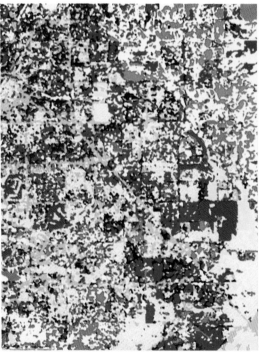

D. UNSUPERVISED–CLASSIFICATION MAP.
COLORS ARE EXPLAINED IN CHAPTER 10.

PLATE 15 Land use and land cover interpreted from digitally classified Landsat
images. Plates A and B are from Estes and others (1979, Figures 2 and 3). Plates
C and D were digitally processed at Chevron Oil Field Research Company.

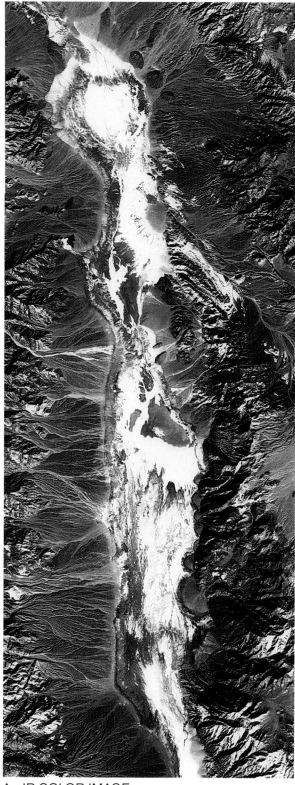

A. IR COLOR IMAGE.

B. RATIO 5/1, DENSITY SLICED.

PLATE **16** Landsat TM images of Death Valley, California. Each image covers a width of 30 km.

unusual carbon isotopic values. These values indicate that hydrocarbons leaking from the reservoir were oxidized and the carbon incorporated into the secondary carbonate minerals of the surface rocks. Similar carbon isotopic values occur over the Davenport oil field in central Oklahoma (Donovan, Friedman, and Gleason, 1974).

Several attempts have been made to identify on Landsat MSS and TM images any spectral signatures of the color and mineralogic alteration patterns at the Cement oil field. Such signatures could then be used to explore for other fields. Various digital processes have been applied to Landsat data of the Cement field, but no successes have been reported. A field investigation found that the color changes at the Cement and Velma fields are only visible at limited exposures in road cuts and stream beds. Most of the area is covered with soil and agriculture that obscures the alteration effects.

Everett and Petzel (1973) interpreted Landsat MSS images of the Anadarko Basin in the Texas Panhandle and western Oklahoma. They reported ''hazy'' anomalies over a number of oil and gas fields, that were said to appear as if image detail had been smudged or erased. They are recognizable only on certain Landsat images and are not visible on aerial photographs. However, the anomalies are not artifacts of the image reproduction process. No conclusive explanation has been given for them, but it has been suggested that they are caused by human activities. These anomalies were of interest to other investigators, but no one has reported similar features. The Anadarko Basin investigation was summarized by Short (1975), who noted the lack of agreement about causes of the hazy anomalies.

NASA/Geosat Test Case Project—Oil and Gas Fields

The NASA/Geosat project selected three oil and gas fields for study: the Coyanosa field in west Texas, the Lost River field in West Virginia, and the Patrick Draw field in southwest Wyoming. Lang, Nicolais, and Hopkins (1985) interpreted TMS images and Landsat MSS images of the Coyanosa test site. They found no evidence of surface alteration caused by hydrocarbon seepage, such as the bleached rocks that occur at the Cement and Velma fields in Oklahoma. Some significant structural features were expressed in the images, however.

The Lost River field is particularly interesting because it is located in forested terrain of the Appalachian Mountains and provided the opportunity to investigate possible vegetation anomalies associated with a gas field. Lang, Curtis, and Kovacs (1985) prepared a supervised classification map from TMS data that shows distribution of plant types. Concentrations of maple trees occurred at localities normally occupied by oak and hickory trees. A soil gas survey identified unusually high concentrations of methane and ethane that coincide with the maple anomaly. It is postulated that the gas concentration causes anaerobic soil conditions that inhibit growth of oak and hickory trees but does not inhibit maples.

At the Patrick Draw field, the vegetation cover is predominantly sagebrush. Lang, Alderman, and Sabins (1985) described a ratio color image of TMS data that indicated an anomalous area of sagebrush at the west margin of the field. Ground investigation found a few square kilometers of blighted sagebrush with stunted growth and small leaves. A soil gas survey showed high concentrations of gas that could be responsible for this blighted sagebrush. The blighted vegetation, however, does not extend over a significant proportion of the field, and soil gas concentrations can occur with no associated vegetation anomaly.

GEOTHERMAL ENERGY

Geothermal energy is obtained from subsurface reservoirs of steam or hot water that are shallow enough to be drilled and exploited economically. Most of the steam and hot water is used to generate electricity, but some is used directly for heating, as in Iceland. The following conditions are necessary for a *geothermal reservoir:*

1. A large, high-temperature heat source must be present at relatively shallow depth. Intrusive masses of young igneous rock are the usual heat source, and most geothermal areas are associated with surface or subsurface igneous rocks of Cenozoic age.

2. Porous and permeable reservoir rocks filled with steam or hot water must occur near the heat source. A variety of rocks can serve as geothermal reservoirs. In the Imperial Valley, California, poorly consolidated sandstones of late Tertiary age are the reservoir rocks. At the Geysers in northern California, the reservoir occurs in highly fractured sedimentary and volcanic rocks.

3. A natural recharge system must replenish the reservoir with water as steam or hot water is produced.

4. An impermeable zone above the reservoir is necessary to prevent the escape of steam and hot water. Convective flow to the surface would dissipate the heat of an unconfined reservoir. Heat losses due to thermal conduction through the rocks are relatively minor because of the low thermal conductivity of rocks.

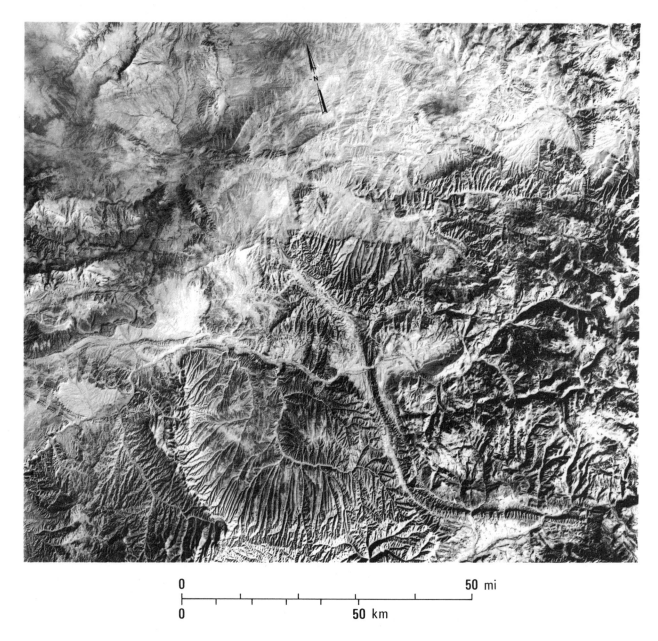

0   50 mi

0   50 km

FIGURE 8.24 Landsat MSS band-5 winter image of northwest Colorado.

Some geothermal reservoirs have no visible or thermal expression at the surface and are not detectable by remote sensing methods. Many geothermal reservoirs, however, have surface thermal expressions ranging in intensity from a minor increase in ground temperature to the presence of hot springs and geysers. Hot springs and geysers commonly occur along faults and fractures that allow hot water to escape from the reservoir. Thermal springs and zones of hydrothermal alteration associated with a geothermal area in Mexico have been interpreted from thermal IR images (Valle and others,

1970). At the Geysers area, hot springs and fumaroles were also detected on thermal IR images and there was some local evidence of higher ground temperatures (Moxham, 1969). However, there was little evidence of a regional surface temperature anomaly on the images of the Geysers area.

Iceland Geothermal Reconnaissance

Iceland is the site of frequent volcanic eruptions, including six since 1946, and associated geothermal activ-

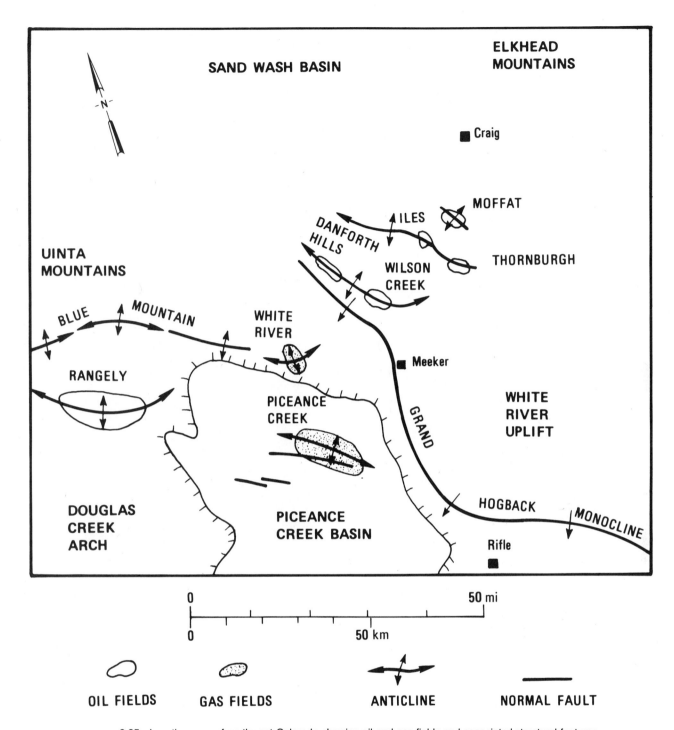

FIGURE 8.25 Location map of northwest Colorado showing oil and gas fields and associated structural features.

ity. Space heating and hot water for the capital city of Reykjavik have long been supplied from geothermal sources. The 17 high-temperature geothermal areas are concentrated along the zones of active rifting and volcanism. Vatnajökull is an ice cap approximately 100 km in diameter that overlies two known high-temperature geothermal areas.

The aerial photograph and thermal IR image of Figure 8.26 show the Kverkfjöll geothermal area, which is located on the northern edge of the Vatnajökull ice cap. As shown on the interpretation map of Figure 8.27, the geothermal area is located between the Kverkjökull outlet glacier on the east and a re-entrant of bedrock on the west. The thermal IR image (Figure 8.26B) is not

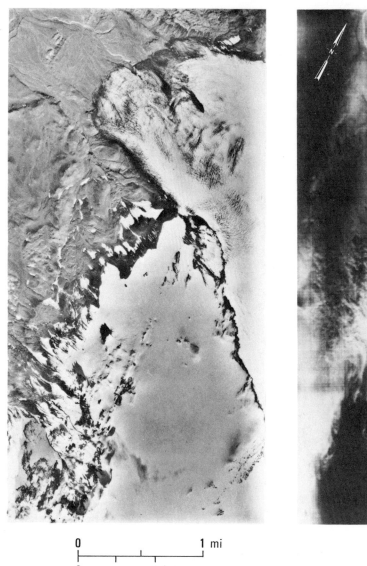 

A. AERIAL PHOTOGRAPH ACQUIRED
   AUGUST 24, 1960, BY THE U.S. AIR
   FORCE.

B. NIGHTTIME THERMAL IR (1.0 TO 5.5 μm)
   IMAGE (NOT RECTILINEARIZED).
   ACQUIRED AUGUST 22,1966, BY THE
   U.S. AIR FORCE, CAMBRIDGE
   RESEARCH LABORATORIES.

FIGURE 8.26 Kverkfjöll geothermal area and Kverkjökull outlet glacier, Iceland.
From Friedman and others (1969, Figures 10 and 11). Courtesy R. S. Williams,
U.S. Geological Survey.

rectilinearized, which accounts for the geometric compression at the east and west margins. The ice and bedrock have relatively cool signatures (dark tones). The geothermal features, the meltwater, and the bedrock ridges that confine the outlet glacier have warm signatures. The north-trending geothermal feature is at least 2 km long and includes two separate hot areas at

the northeast end. Near the south margin of the image, warm signatures mark concentric crevasses and an ice cauldron subsidence feature that are caused by subsurface melting of the glacier (Figure 8.27). The warm stream emerging from the snout of the outlet glacier is meltwater that has flowed along the base of the glacier from a subglacial geothermal source (R. S. Williams, Jr., per-

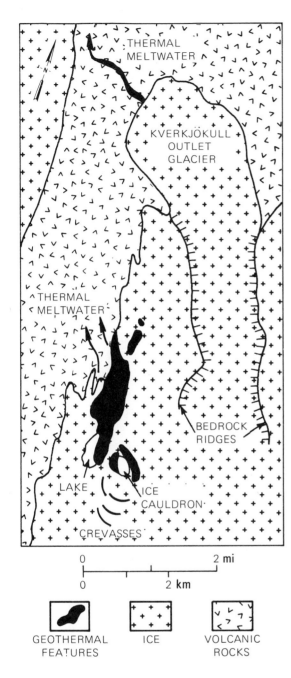

FIGURE 8.27 Interpretation map of the thermal IR image of the Kverkfjöll geothermal area. From Friedman and others (1969, Figure 11).

sonàl communication). The topographic expressions of some geothermal features are detectable on the aerial photograph of Figure 8.26A, but are less pronounced than on the IR image.

Geothermal vents and hot springs have been detected on thermal IR images in such diverse localities as Japan, Italy, Ethiopia, and the United States. Remote sensing is especially useful in remote areas where the surface expression of geothermal activity has not been located by conventional means.

Low-Intensity Geothermal Anomalies

Thermal IR images are ideal for detecting fumaroles, steaming ground, and hot springs associated with very active geothermal areas such as Iceland and Yellowstone National Park. Elsewhere, however, there are so-called *blind geothermal areas* that lack surface thermal or alteration expression. Intermediate in character between the active areas and the blind areas is a broad category of geothermal areas where the surface temperature is only slightly higher than the surrounding areas. These areas with low-intensity surface temperature anomalies are difficult to detect on IR images or on airborne radiometer profiles. Watson (1975) pointed out that natural variations of geology and topography can readily overwhelm surface geothermal anomalies of several hundred *heat-flow units* (HFU) (1 HFU $= 1 \times 10^{-6}$ cal $\cdot$ cm^{-1} $\cdot$ sec^{-1}). Mathematical modeling was used to evaluate the relative effect of various factors on the surface radiant temperature. This theoretical analysis suggests that both thermal and reflectance images should be acquired at least three times during the diurnal cycle (Watson, 1975, p. 136). Comparison of the images may reveal subtle anomalies. An 8-to-14-μm image in the Raft River area of Idaho revealed a weak thermal anomaly that was confirmed by ground measurements.

OTHER ENERGY SOURCES

The location and distribution of large reserves of coal, oil shale, and tar sands are already known in the United States and Canada. Therefore exploration is relatively inactive and there is little application of remote sensing. Landsat images are potentially useful during mining of coal. The status of strip mining and land reclamation may be monitored by digitally processing data acquired during the repetitive cycles of Landsat. Roof falls are hazards in underground coal mines. In Indiana it was demonstrated that areas of intense fracturing on Landsat images coincided with areas of roof falls in the mines and could be used to predict the hazards (Wier and others, 1973).

COMMENTS

Remote sensing has proven a valuable aid in exploring for mineral and energy resources. Many ore deposits are localized along regional and local fracture patterns that provided conduits along which ore-forming solutions penetrated host rocks. Landsat images are used to map these fracture patterns. Ore-forming hydrothermal solutions alter the host rocks to distinctive assemblages of secondary minerals. Iron sulfide minerals weather to iron oxide minerals with reddish colors that are recognizable on Landsat MSS ratio 4/5 images and Landsat TM ratio 3/1 images. Clay minerals and alunite formed in hydrothermally altered rocks have distinctive absorption features at wavelengths between 2.08 and 2.35 μm that are recorded by band 7 of Landsat TM. The TM ratio 5/7 image has distinctive bright signatures associated with hydrothermally altered rocks. In the future, multispectral scanners with high spectral resolution may identify specific alteration minerals.

Detection of hydrothermally altered rocks is not possible in vegetated areas, so this environment requires other remote sensing methods. Reflectance spectra of foliage growing over mineralized areas may differ from spectra of foliage in adjacent nonmineralized areas. The spectral differences, however, are variable for different plant species. More research and development is needed for remote detection of mineral deposits in vegetated terrain.

For oil exploration, remote sensing is especially valuable in poorly mapped regions. Interpretation of Landsat images of southern Sudan revealed the presence of a previously unrecognized sedimentary basin in which several oil fields were subsequently discovered. Landsat images are also useful as base maps and for planning field operations. In forested areas of perennial cloud cover, radar images acquired from aircraft and satellites make it possible for geologists to map lithologic terrain and geologic structures that may be exploration clues for oil fields.

Hot springs and other surface indications of geothermal areas are recognizable in thermal IR images, but low-intensity thermal anomalies are generally obscured by variations in surface temperature.

QUESTIONS

1. You are employed by an international mineral exploration company that plans to explore for hydrothermal gold deposits in the southern third of the Andes Mountains of South America. Your assignment is to plan the remote sensing phase of the exploration campaign ranging from regional reconnaissance to definition of individual prospects. You can utilize the image-acquisition systems and image-processing systems described in this book. Prepare the remote sensing exploration campaign, with reasons and justification for each step.

2. As part of your Chile project you need to identify remotely the alteration minerals kaolinite, montmorillonite, illite, and alunite. Your company has an airborne scanner that can digitally record five bands of data in the 2.0-to-2.5-μm region. Each band has a spectral range of 0.05 μm, such as 2.10 to 2.15 μm. Use Figure 8.13 to select the five optimum bands for identifying the alteration minerals. List your reasons for selecting each band. Describe how you would digitally process the resulting airborne data to produce maps showing distribution of the different minerals.

3. The mineral industry is depressed, and you are now employed by an international oil exploration company. Your company plans to evaluate the petroleum potential of the western portion of the People's Republic of China in preparation for negotiating an exploration concession. Because of limited accessibility, deadlines, and competitor pressure, concession areas must be selected solely on the basis of remote sensing evaluations. Prepare such a plan for your management, giving reasons for each step.

4. Assume that western China is completely covered by Landsat TM images, SIR images, and LFC overlapping black-and-white photographs. List the advantages and disadvantages of each of these kinds of satellite images for your project.

REFERENCES

Abrams, M. J., and D. Brown, 1985, Silver Bell, Arizona, porphyry copper test site: The Joint NASA/Geosat test case study, section 4, American Association of Petroleum Geologists, Tulsa, Okla.

Abrams, M. J., J. E. Conel, and H. R. Lang, 1985, The joint NASA/Geosat test case study: American Association of Petroleum Geologists, Tulsa, Okla.

Albert, N. R. D., 1975, Interpretation of Earth Resource Technology Satellite imagery of the Nabesna quadrangle, Alaska: U.S. Geological Survey Miscellaneous Field Map MP 655J.

Ashley, R. P., 1974, Goldfield mining district: Nevada Bureau of Mines and Geology Report 19, p. 49–66.

Ashley, R. P., 1979, Relation between volcanism and ore deposition at Goldfield, Nevada: Nevada Bureau of Mines and Geology Report 33, p. 77–86.

Cannon, H. L., 1971, The use of plant indicators in groundwater surveys, geologic mapping, and mineral prospecting: Taxon, v. 20, p. 227–256.

Chenoweth, W. L., and D. N. Magleby, 1971, Mine location map, Cameron uranium area, Coconino County, Arizona: U.S. Atomic Energy Commission Preliminary Map 20.

Collins, W., S. H. Chang, G. Raines, F. Canney, and R. P. Ashley, 1983, Airborne biogeophysical mapping of hidden mineral deposits: Economic Geology, v. 78, p. 737–749.

Conel, J. E., and R. E. Alley, 1985, Lisbon Valley, Utah,

uranium test site report: The joint NASA/Geosat test case study, section 8, American Association of Petroleum Geologists, Tulsa, Okla.

Donovan, T. J., 1974, Petroleum microseepage at Cement field, Oklahoma—evidence and mechanism: Bulletin of the American Association of Petroleum Geologists, v. 58, p. 429–446.

Donovan, T. J., I. Friedman, and J. D. Gleason, 1974, Recognition of petroleum-bearing traps by unusual isotopic compositions of carbonate-cemented surface rocks: Geology, v. 2, p. 351–354.

Everett, J. R., and G. Petzel, 1973, An evaluation of the suitability of ERTS data for the purposes of petroleum exploration: Third Earth Resources Technology Satellite Symposium, NASA SP-356, p. 50–61.

Friedman, J. D., R. S. Williams, G. Pálmason, and C. D. Miller, 1969, Infrared surveys in Iceland: U.S. Geological Survey Professional Paper 650-C, p. C89–C105.

Goetz, A. F. H., 1984, High spectral resolution remote sensing of the land: Proceedings of the International Society for Optical Engineering, v. 475, p. 56–68.

Harvey, R. D., and C. J. Vitaliano, 1964, Wall-rock alteration in the Goldfield District, Nevada: Journal of Geology, v. 72, p. 564–579.

Horton, R. C., 1964, An outline of the mining history of Nevada, 1924–1964: Nevada Bureau of Mines Report 7, pt. 2.

Hunt, G. R., and R. P. Ashley, 1978, Spectra of altered rocks in the visible and near infrared: Economic Geology, v. 74, p. 1613–1629.

Hunt, G. R., J. W. Salisbury, and C. J. Lenhof, 1971, Visible and near-infrared spectra of minerals and rocks—III, oxides and hydroxides: Modern Geology, v. 2, p. 195–205.

Kappelmeyer, O., and R. Haenel, 1974, Geothermics with special reference to application: Geoexploration Monographs, series 1, no. 4, Gebrüder Borntraeger, Berlin.

Labovitz, M. L., E. J. Masuoka, R. Bell, A. W. Segrist, and R. F. Nelson, 1983, The application of remote sensing to geobotanical exploration for metal sulfides—results from the 1980 field season at Mineral, Virginia: Economic Geology, v. 78, p. 750–760.

Lang, H. R., W. H. Alderman, and F. F. Sabins, 1985, Patrick Draw, Wyoming, petroleum test case report: The NASA/Geosat test case project, section 11, American Association of Petroleum Geologists, Tulsa, Okla.

Lang, H. R., J. B. Curtis, and J. S. Kovacs, 1985, Lost River, West Virginia, petroleum test site: The NASA/Geosat test case project, section 12, American Association of Petroleum Geologists, Tulsa, Okla.

Lang, H. R., S. M. Nicolais, and H. R. Hopkins, 1985, Coyanosa, Texas, petroleum test site: The NASA/Geosat test case project, section 13, American Association of Petroleum Geologists, Tulsa, Okla.

Lowell, J. D., and J. M. Guilbert, 1970, Lateral and vertical alteration-mineralization zoning in porphyry ore deposits: Economic Geology, v. 65, p. 373–408.

Magnier, P., T. Oki, and L. W. Kaartidiputra, 1975, The Mahakam delta: World Petroleum Congress Proceedings, v. 2, p. 239–250.

Miller, J. B., 1975, Landsat images as applied to petroleum exploration in Kenya: NASA Earth Resources Survey Symposium, NASA TM X-58168, v. 1-B, p. 605–624.

Miller, J. B., and J. Vandenakker, 1977, Sudan interior exploration project—planimetry and geology: American Association of Petroleum Geologists, Third Pecora Conference, Sioux Falls, S.D.

Moxham, R. M., 1969, Aerial infrared surveys at the Geysers geothermal steam field, California: U.S. Geological Survey Professional Paper 630-C, p. C106–C122.

Nicolais, S. M., 1974, Mineral exploration with ERTS imagery: Third ERTS-1 Symposium, NASA SP-351, v. 1, p. 785–796.

Offield, T. W., 1976, Remote sensing in uranium exploration: Exploration of uranium ore deposits, International Atomic Energy Proceedings, p. 731–744, Vienna, Austria.

Rowan, L. C., and P. H. Wetlaufer, 1975, Iron-absorption band analysis for the discrimination of iron-rich zones: U.S. Geological Survey, Type III Final Report, Contract S-70243-AG.

Rowan, L. C., A. F. H. Goetz, and R. P. Ashley, 1977, Discrimination of hydrothermally altered and unaltered rocks in the visible and near infrared: Geophysics, v. 42, p. 522–535.

Rowan, L. C., P. H. Wetlaufer, A. F. H. Goetz, F. C. Billingsley, and J. H. Stewart, 1974, Discrimination of rock types and detection of hydrothermally altered areas in southcentral Nevada by the use of computer-enhanced ERTS images: U.S. Geological Survey Professional Paper 883.

Schull, T. J., 1984, Oil exploration in nonmarine rift basins of interior Sudan (abstract): American Association of Petroleum Geologists, v. 68, p. 526.

Settle, M., 1985. The joint NASA/Geosat Test Case Project, executive summary: American Association of Petroleum Geologists, Tulsa, Okla.

Short, N. M., 1975, Exploration for fossil and nuclear fuels from orbital altitudes: Remote sensing energy related studies, p. 189–232, Hemisphere Publishing Corporation, Washington, D.C.

Spirakis, C. S., and C. D. Condit, 1975, Preliminary Report on the use of Landsat-1 (ERTS-1) reflectance data in locating alteration zones associated with uranium mineralization near Cameron, Arizona: U.S. Geological Survey Open File Report 75–416.

Stewart, J. H., G. W. Walker, and F. J. Kleinhampl, 1975, Oregon-Nevada lineament: Geology, v. 3, p. 251–268.

Valle, R. G., J. D. Friedman, S. J. Gawarecki, and C. J. Banwell, 1970, Photogeologic and thermal infrared reconnaissance surveys of the Los Negritos–Ixtlan de Los Hervores geothermal area, Michoacan, Mexico: Geothermics Special Issue 2, p. 381–398.

Vandenakker, J., and J. Ryan, 1983, Landsat applications for geophysical field operations (abstract): Geophysics, v. 48, p. 475.

Viljoen, R. P., M. J. Viljoen, J. Grootenboer, and T. G. Longshaw, 1975, ERTS-1 imagery—an appraisal of applications in geology and mineral exploration: Minerals Science and Engineering, v. 7, p. 132–168.

Watson, K., 1975, Geologic applications of thermal infrared images: Proceedings of the IEEE, n. 501, p. 128–137.

Wier, C. E., F. J. Wobber, O. R. Russell, R. V. Amato, and

T. V. Leshendok, 1973, Relationship of roof falls in underground coal mines to fractures mapped on ERTS-1 imagery: Third ERTS-1 Symposium, NASA SP-351, p. 825–843.

Wing, R. S., and J. C. Mueller, 1975, SLAR reconnaissance, Mimika-Eilanden Basin, southern trough of Irian Jaya: NASA Earth Resources Survey Symposium, NASA TM X-58168, v. 1-B, p. 599–604.

Yost, E., and S. Wenderoth, 1971, The reflectance spectra of mineralized trees: Proceedings of Seventh International Symposium on Remote Sensing of Environment, University of Michigan, v. 1, p. 269–284, Ann Arbor, Mich.

ADDITIONAL READING

Abrams, M. J., D. Brown, L. Lepley, and R. Sadowski, 1983, Remote sensing for porphyry copper deposits in southern Arizona: Economic Geology, v. 78, p. 591–604.

Goetz, A. F. H., B. N. Rock, and L. C. Rowan, 1983, Remote sensing for exploration—an overview: Economic Geology, v. 78, p. 573–590. (This issue includes a number of articles on remote sensing exploration.)

Offield, T. W., E. A. Abbott, A. R. Gillespie, and S. O. Laguercio, 1977, Structural mapping on enhanced Landsat images of southern Brazil—Tectonic control of mineralization and speculations on metallogeny: Geophysics, v. 42, p. 482–500.

Sabins, F. F., 1979, Oil occurrence and plate tectonics as viewed on Landsat images: Proceedings of the Tenth World Petroleum Congress, p. 105–109, Bucharest, Romania.

Schmidt, R., B. B. Clark, and R. Bernstein, 1975, A search for sulfide-bearing areas using Landsat-1 data and digital image-processing techniques: NASA Earth Resources Survey Symposium, NASA TM X-58168, v. 1-B, p. 1013–1027.

9

Environmental Applications

Remote sensing is a valuable source of environmental information about the atmosphere, continents, and oceans. All wavelength regions of the electromagnetic spectrum from UV through microwave are useful. In addition to the general-purpose imaging systems such as aerial photography and Landsat, a number of satellites have been deployed for specific environmental applications.

ENVIRONMENTAL SATELLITES

Beginning in 1960, the United States has put into orbit more than 40 unmanned satellites for environmental and meteorologic monitoring, in addition to the Landsat, HCMM, and other systems described in previous chapters. The early satellites carried miniature television cameras that produced low-resolution, visible-band images of cloud patterns for meteorologic use. There has been both a steady improvement in spatial resolution and an expansion of the spectral range into both the reflected IR and thermal IR bands. Because of these improvements the satellite data are now used for oceanographic, hydrologic, and vegetation applications in addition to meteorology. Three major systems—GOES, AVHRR, and CZCS—are summarized in the following sections. Details of these systems are given by Epstein

and others (1984). Predecessors of the currently operational systems are described by Allison and Schnapf (1983).

Data collected by the environmental satellites are available from

NOAA Satellite Services Division
Room 100
World Weather Building
Washington, D.C. 20233

This facility can provide catalogs, price lists, and ordering information.

Geostationary Operational Environmental Satellite (GOES)

Geostationary satellites travel at the angular velocity at which the earth rotates; as a result they remain above the same point on earth at all times. The two active geostationary satellites operated by NOAA are essentially "parked" above the equator at an altitude of 35,000 km and a velocity of 11,000 km · h^{-1}. GOES East is positioned above the equator at 75° west longitude, and GOES West sits at 135° west longitude. Together they provide continuous coverage of North America, South America, and adjacent oceans.

FIGURE 9.1 GOES East visible-band image (0.55 to 0.70 μm) of North and South America with an 8-km spatial resolution. Courtesy NOAA Satellite Services Division.

The scanner imaging system aboard GOES acquires one band of visible data (0.55 to 0.70 μm) with a 1-km spatial resolution. Images of the full hemisphere, such as Figure 9.1, are produced at a spatial resolution of 8 km to lower both the number of pixels and the amount of data processing. At this small scale the advantages of higher spatial resolution would not be apparent. One band of thermal IR data (10.50 to 12.60 μm) is acquired at a spatial resolution of 8 km. The image in Figure 9.2 was acquired simultaneously with the visible-band image of Figure 9.1. In GOES IR images, the temperature

signatures are reversed from those of other IR images in this book (Landsat TM band 6, HCMM, and aircraft images). Bright tones record relatively cool radiant temperatures and dark tones are relatively warm ones. This arrangement causes clouds, which are colder than water and land, to have familiar bright signatures. In the daytime IR image of Figure 9.2, land areas are warm (darker signatures) relative to oceans (brighter signatures) for reasons given in Chapter 5.

The clouds in Figure 9.2 show a range of thermal IR signatures due to different temperatures caused by dif-

FIGURE 9.2 GOES East thermal IR band image (10.5 to 12.6 μm) of North and South America with an 8-km spatial resolution. Dark tones record warm radiant temperatures and bright tones record cool temperatures. Courtesy NOAA Satellite Services Division.

ferent cloud heights. The swirl of gray clouds adjacent to the west coast of South America is noticeably warmer than the nearby linear belts of white clouds. These differences are not discernible in the visible image of Figure 9.1. Another advantage of using the IR band is that images may be acquired both day and night.

Images from GOES East and West satellites are acquired every 30 minutes, 24 hours daily, but a faster schedule may be used during periods of severe weather. Data are available in a range of digital and photographic products. Data are also distributed over telephone cir-

cuits to seven National Weather Service Satellite Service Units, where additional communication facilities provide further distribution to subscribers over a "GOES Tap" system. In addition to NOAA facilities, 140 federal users and 75 industry and university users receive the image data.

Epstein and others (1984) summarized the following NOAA applications of GOES data:

1. Cloud-motion wind vectors are automatically computed for low-level winds and are manually processed

A. VISIBLE–BAND IMAGE OF HURRICANE ALICIA APPROACHING GALVESTON, TEXAS, AUGUST 17, 1983.

B. ENHANCED THERMAL IR IMAGE OF A SEVERE WEATHER OUTBREAK IN WESTERN KANSAS AND SOUTHEAST OKLAHOMA, APRIL 2, 1982.

FIGURE 9.3 Enlarged GOES images of storms. From Epstein, Callicott, Cotter, and Yates (1984, Figures 3 and 5). Courtesy W. M. Callicott, NOAA Satellite Services Division.

for mid- and high-level winds. Over 1200 vectors are generated each day.

2. Precipitation resulting from convective storm systems and hurricanes is estimated.

3. Freeze warnings are provided by defining surface-temperature patterns associated with rapid radiative cooling in citrus and vegetable growing areas. The nearly real-time, continuous monitoring allows farmers to delay the start of preventative measures until a freeze is imminent, thus saving thousands of dollars.

4. Snow-cover analyses are used to predict the snowmelt runoff potential in river basins across the nation. Analyses are derived from both the meteorological satellites and from Landsat images.

5. Hurricane classification is performed using a semiobjective method of classifying the stage of development and wind intensity from cloud patterns. Figure 9.3A shows Hurricane Alicia in August 1983 when its center was about to make landfall at the Texas-Louisiana border.

6. Messages describing weather systems and features interpreted from satellite images are prepared and distributed over the National Weather Service communications links at regular intervals daily.

7. The 30-minute image sequences are combined to form animated cloud images. The animation reveals the dynamics associated with the cloud development, providing the analyst with an added dimension for interpreting the data.

8. Severe convective storm development is monitored by combining cloud-top temperatures with cloud motions at different levels to describe and classify the events in terms of severity and to anticipate where severe weather will eventually develop. Figure 9.3B is an enhanced GOES infrared image showing a severe weather outbreak over the midwestern states from which tornadoes developed in April 1982. The enhancement scheme reverses the gray scale at key temperatures to accentuate important features and results in the coldest clouds with the deepest convection having black signatures.

9. Detection of fog and the rate of dissipation are determined to support shipping, fishing, and aviation interests.

NOAA Polar-Orbiting Satellites and the Advanced Very High Resolution Radiometer

In addition to its geostationary satellites, NOAA also operates the current generation of polar-orbiting environmental satellites. The two operating satellites are in sun-synchronous orbits at an altitude of 850 km. One crosses the equator at 7:30 a.m. and 7:30 p.m. local sun time and the other at 2 a.m. and 2 p.m. Each satellite orbits the earth 14 times daily and acquires complete global coverage every 24 hours. Data are recorded on tape recorders on the satellite and transmitted to the two receiving stations at Fairbanks, Alaska, and Wallops Island, Virginia.

The primary imaging system is the *advanced very high resolution radiometer* (AVHRR), a cross-track multispectral scanner that acquires images with an image-swath width of 2700 km and a ground resolution cell of 1.1 by 1.1 km. At this spatial resolution, however, only 15 percent of the data can be stored onboard. For complete coverage the data are resampled to form 4-by-4-km pixels. Table 9.1 lists the spectral bands of AVHRR.

Southern Louisiana Figure 9.4 is a subscene processed by W. A. Miller of EDC that illustrates the three commonly used bands of AVHRR images. Band 1, which is in the red region (Figure 9.4A), is approximately equivalent to band 3 of Landsat TM. In the land areas, variation in gray tones relate to differences in vegetation. The cities of Baton Rouge and New Orleans have distinct dark signatures. In the water areas, brighter

tones correlate with higher concentrations of suspended silt and mud. The Mississippi River and its discharge into the Gulf of Mexico are notably bright. Plumes of sediment-laden water discharged into Atchafalaya Bay and Lake Pontchartrain form conspicuous bright patches. Clouds are bright.

AVHRR band 2 (Figure 9.4B), which is in the reflected IR region, is approximately equivalent to TM band 4. Water bodies are clearly distinguished from land by their very dark signatures, which are caused by absorption of these wavelengths. The thin clouds present in the band-1 image are readily penetrated by reflected IR energy and are not visible in band 2.

Band 4 (Figure 9.4C) is a daytime thermal IR image equivalent to TM band 6. Bright signatures represent relatively cool, and dark signatures relatively warm, radiant temperatures. In this early spring season, water in the Gulf of Mexico is warm (dark tones) relative to the fresh water discharged at the Mississippi Delta and Atchafalaya Bay (bright tones). Lake Pontchartrain and adjacent lakes are also cool. The stringers of clouds are cool.

Africa Spectral ranges of AVHRR bands 1 and 2 were positioned to record significant vegetation properties. As shown by the vegetation reflectance curve in Figure 9.5, band 1 (B_1) records the chlorophyll absorption of red wavelengths. Band 2 (B_2) records the strong reflectance of IR wavelengths by the cell structure of leaves. The ratio B_2/B_1 is one index of vegetation. Another is the *spectral vegetation index,* a normalized relationship defined as

$$\text{Spectral vegetation index} = \frac{B_2 - B_1}{B_2 + B_1} \qquad (9.1)$$

This ratio is more useful because it largely eliminates reflectance variations due to differences in solar elevation (Tucker, Townshend, and Goff, 1985). The values for B_1 and B_2 are the average values for the reflectance curves at those wavelength intervals. For the vegetation spectrum in Figure 9.5, the spectral vegetation index is calculated as 0.41. For the dry soil spectrum, the index is only 0.30. Various proportions of soil and vegetation in a ground resolution cell of an AVHRR image will result in intermediate values. Also, different types of vegetation and soil may have different index values from those in Figure 9.5. The spectral vegetation index is also called the *normalized difference vegetation index.*

AVHRR images are well suited for studying vegetation distribution and seasonal changes on a continent-wide scale for the following reasons:

TABLE 9.1 Characteristics of the advanced very high resolution radiometer (AVHRR)

Band	Wavelength, μm	Remarks
1	0.55 to 0.68	Red—for daytime clouds and vegetation
2	0.73 to 1.10	Reflected IR—for shorelines and vegetation
3	3.55 to 3.93	Thermal IR—for hot targets such as fires and volcanoes
4	10.50 to 11.50	Thermal IR—for sea temperature and for daytime and nighttime clouds
5	11.50 to 12.50	Thermal IR—recorded only on the NOAA 7 satellite

A. BAND 1 (0.58 TO 0.68 μm).

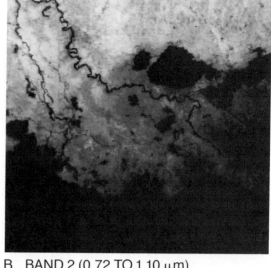

B. BAND 2 (0.72 TO 1.10 μm).

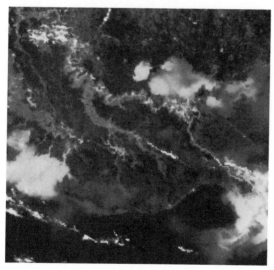

C. BAND 4 (10.5 TO 11.50 μm).

D. LOCATION MAP.

FIGURE 9.4 AVHRR images of the southern Louisiana subscene acquired March 25, 1984. Courtesy W. A. Miller, EROS Data Center.

1. The 2700-km image-swath width of AVHRR covers a continent such as Africa with a few images, whereas 1100 Landsat MSS or TM images are required.

2. The daily repetition of AVHRR provides a wide selection of images for seasonal changes and for cloud-free coverage. Landsat TM operates on a 16-day repetition cycle.

3. The 4-km pixels of AVHRR are adequate for regional studies, whereas the 79-m or 30-m pixels of MSS or TM result in far too much data for economical processing.

For these reasons, Tucker, Townshend, and Goff (1985) used AVHRR data to analyze vegetation cover of Africa in the following manner. NOAA maps band-1 and band-2 data into arrays of 1024 by 1024 grid cells for the Northern and Southern Hemispheres each day. Approximate dimensions of the grid cells are 15 km at the equator and 30 km at the poles. Each week, NOAA

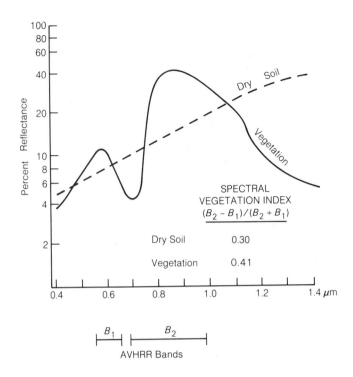

FIGURE 9.5 Reflectance spectra of vegetation and dry soil. Spectral vegetation index was calculated from AVHRR bands 1 and 2 for soil and vegetation.

principal-component images for a supervised classification. Training sites were areas of known vegetation types based on their field experience. The resulting classification map (Plate 14B) has the following color code:

Tan	Desert and semidesert
Light green	Semiarid wooded grassland and brushland
Purple	Sudanian woodland and grassland
Dark blue	Interspersed tropical forest and grassland
Red	Tropical rain forest and montane forest
Dark green	Woodland, especially Miombo woodland
Light blue	Brushland and thicket
Yellow	Wooded grassland and deciduous brushland

generates a composite data set by selecting the highest band-1 and band-2 values for each grid cell from the seven daily images. For Africa the spectral vegetation index (Equation 9.1) was calculated at NASA Goddard Space Flight Center for each of the weekly data sets from April 1982 to November 1983. For each 3-week period the highest index value for each grid cell was recorded; together these index values produced a single image for the period. The data were then density-sliced and displayed in color. Plate 14A is a spectral-vegetation-index map for April 12 to May 2, 1982. As shown by the accompanying color scale, the highest values of spectral vegetation index (shown in blue, white, and red tones) are concentrated in the equatorial belt of Africa. Low values (yellow and buff tones) occur in the arid terrain of northern and extreme southern Africa. Small areas of higher index values occur along the Mediterranean coast and the Nile River and delta. A total of eight spectral-vegetation-index maps were prepared, each representing a 3-week period. Significant seasonal changes are apparent in these images.

Tucker, Townshend, and Goff (1985) used the individual spectral-vegetation-index maps to prepare a vegetation classification map in the following manner. They applied a principal-components transformation (Chapter 7) to the eight index images and used the first and second

Comparison of the AVHRR classification map with published vegetation maps shows considerable agreement, although some errors occur. The extent of woodland (dark green in Plate 14B) in eastern Africa is probably overestimated. The amount of rain forest (red) in Gabon and the Republic of Cameroon may be somewhat underestimated because of the persistent clouds and haze in this area. Despite these problems the combination of AVHRR data and digital image processing has produced useful maps in an expeditious fashion.

Justice and others (1985) used AVHRR data and the methods described above to produce global maps of seasonal changes in vegetation.

Coastal Zone Color Scanner

Polar-orbiting NASA meteorologic satellite Nimbus 7, launched October 23, 1978, carried the *coastal zone color scanner* (CZCS), which is a cross-track multispectral scanner. The six bands listed in Table 9.2 operate in the visible, reflected IR, and thermal IR regions. At the spacecraft altitude of 955 km, the ground resolution cell at nadir is 825 by 825 m. Image-swath width of the images is 1600 km. Sunlight reflected from ripples on the sea surface (called *sunglint*) may be a problem. To avoid *sunglint* the scanner may be tilted up to 20° ahead of or behind the spacecraft. Figure 9.6 illustrates the visible and reflected IR CZCS images of a subscene covering Florida and the adjacent Gulf of Mexico and Atlantic Ocean. These images are 800 km wide and show half of the full 1600-km image-swath width. Spectral bandwidth of each visible band (bands 1 through 4) is only 0.02 μm, which is needed to record spectral properties of chlorophyll in seawater, as the following section describes.

TABLE 9.2 Characteristics of the coastal zone color scanner (CZCS)

Band	Wavelength, μm	Characteristics
1	0.43 to 0.45	Blue
2	0.51 to 0.53	Green
3	0.54 to 0.56	Green
4	0.66 to 0.68	Red
5	0.70 to 0.80	Reflected IR
6	10.50 to 12.50	Thermal IR

Spectral Properties of Chlorophyll The color of ocean water is more precisely called the *upwelled spectral radiance,* which is the sunlight that is reradiated from just beneath the sea surface. Near coastlines and the mouths of rivers, the radiance is largely determined by the amount of suspended clay and silt. For most of the ocean surface, however, variations in suspended organic constituents cause the variations in ocean color. The most important constituent is phytoplankton, microscopic plants that form the lowest link in the ocean food chain. Plankton contain chlorophyll, which strongly absorbs blue and red light. Figure 9.7 shows radiance spectra in the visible region for four types of seawater in which chlorophyll content ranges from 0.09 to 60.40 mg $\cdot$ m^{-3}. Chlorophyll content increases by approximately an order of magnitude for each of the four water samples. In the blue spectral region, radiance ranges from a maximum of 1.0 μW $\cdot$ cm^{-2} $\cdot$ srad^{-1} $\cdot$ nm^{-1} for the lowest chlorophyll content to a minimum of 0.001 for the highest chlorophyll content. A steradian (srad) is the solid angle subtended at the center of a sphere by an area equal to the radius squared on the surface of the sphere. Radiance in the green region is less influenced by chlorophyll absorption and ranges from 1.0 to 0.15 μW $\cdot$ cm^{-2} $\cdot$ srad^{-1} $\cdot$ nm^{-1}. These radiance differences are seen in the color of seawater, which is deep blue in the open ocean where concentration of nutrients and phytoplankton is low. Water in coastal zones and zones of upwelling, however, is rich in nutrients that support phytoplankton. The resulting absorption of blue light by chlorophyll causes green hues in the water.

Southern Florida The CZCS digital data of the southern Florida subscene were used to map chlorophyll distribution in the Gulf of Mexico. The spectral radiance curves in Figure 9.7 indicate that CZCS bands 1, 2 and 3 are most suitable for interpreting chlorophyll concen-

trations. These bands are in the blue and green regions, however, which are subject to intense atmospheric scattering, as described in Chapter 2. The atmospheric effect is seen by comparing the images of these bands with the image of band 5 in the reflected IR region, which is much less subject to atmospheric scattering (Figure 9.6). As shown in Chapter 7, it is possible to correct for the effects of atmospheric scattering in multispectral data. The raw data that formed the images in Figure 9.6 were corrected using algorithms described by Gordon and others (1980), who then produced the CZCS ratio 1/3 and 2/3 images shown in Plate 14C,D. Ratio images have the advantage of partially compensating for the influence of nonchlorophyllic material, such as mud and silt, in the water. Values of these two ratios for various chlorophyll concentrations are shown in the chart in Figure 9.7. Ratio 1/3 is sensitive to chlorophyll variations at low concentrations, but at the two highest concentrations the differences in this ratio are too small (0.01) to be reliable. Ratio 2/3 is more sensitive to chlorophyll variations at high concentrations. The ratio images (Plate 14C,D) have been color-sliced to display variations. Ratio values represented by different colors correlate with chlorophyll concentration according to an algorithm developed by Gordon and others (1980). The color scale indicates chlorophyll concentrations for the two ratio images.

The line in the images of Plate 14C,D is the 500-km-long track of a research vessel that measured chlorophyll concentrations off the west coast of Florida during November 13 and 14, 1978, which spans the time when the CZCS images were acquired. There is general agreement between the ship measurements and the CZCS measurements, although CZCS values are consistently lower. This ability to map phytoplankton distribution from satellites is important for understanding ocean productivity, water-mass boundaries, and circulation patterns.

IMAGING THE SEAFLOOR

Oceans cover 70 percent of the earth, but much of the sea floor is mapped only in a general fashion and some charts are inaccurate. Accurate, updated charts are needed for shallow shelf areas, where deposition, erosion, and growth of coral reefs change bottom topography within a few years after a *bathymetric survey* (which measures ocean depths) has been completed. Knowledge of the geomorphology of oceanic spreading centers, transform faults, and submarine volcanoes is needed to improve our understanding of plate tectonics. The seafloor may be mapped or interpreted from the following remote sensing images: Landsat images, side-scanning sonar, and radar images of the sea surface.

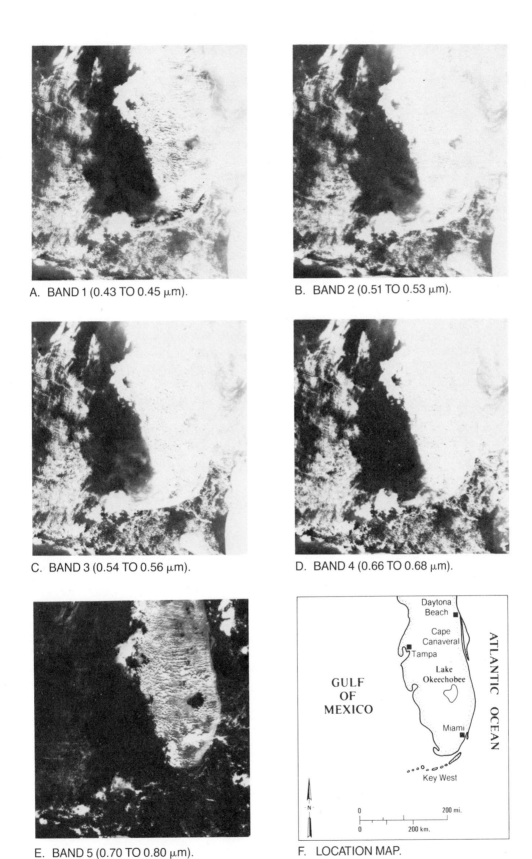

A. BAND 1 (0.43 TO 0.45 μm).

B. BAND 2 (0.51 TO 0.53 μm).

C. BAND 3 (0.54 TO 0.56 μm).

D. BAND 4 (0.66 TO 0.68 μm).

E. BAND 5 (0.70 TO 0.80 μm).

F. LOCATION MAP.

FIGURE 9.6 CZCS individual-band images of Florida subscene. Courtesy D. K. Clark, NOAA Satellite Services Division.

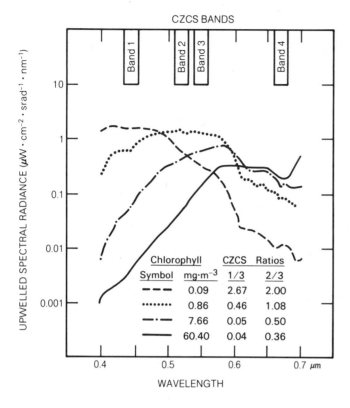

FIGURE 9.7 Spectra for seawater with various concentrations of chlorophyll. Note positions of CZCS visible bands. From Hovis and others (1980, Figure 1).

Landsat Images

Remote sensing of the seafloor from satellites or aircraft is restricted by the fact that water absorbs or reflects most wavelengths of electromagnetic energy. Only visible wavelengths penetrate water, and the depth of penetration is influenced by turbidity of the water. Figure 9.8 shows penetration of different wavelengths through 10 m of different types of water. This figure demonstrates that penetration is essentially restricted to the visible region (0.4 to 0.7 μm). Energy at reflected IR wavelengths (greater than 0.7 μm) is almost totally absorbed by water, as illustrated by black-and-white IR photographs in Chapter 2. Ten meters of clear ocean water transmits almost 50 percent of incident blue and green wavelengths (0.4 to 0.6 μm) but transmits less than 10 percent of red light (0.6 to 0.7 μm). As shown in Figure 9.8, the optical density, or turbidity, gets progressively higher as one moves from ocean to coastal to bay water. The increased turbidity causes a decrease in light transmittance and also shifts the wavelength of maximum transmittance toward longer wavelengths (0.5 to 0.6 μm). For Landsat MSS, band 4 (0.5 to 0.6 μm) has maximum water penetration. Band 5 (0.6 to 0.7 μm) has limited penetration. Band 6 (0.7 to 0.8 μm) has min-

imal penetration. Band 7 (0.8 to 1.1 μm) has no penetration of water.

Figure 9.9 illustrates a subscene of Landsat MSS images and a hydrographic chart of the Sibutu Island group in the Celebes Sea off the northeast coast of Borneo. The central part of the group is a large atoll where the central lagoon has maximum water depths of 8 fathoms (14 m). The east border of the lagoon is formed by Tumindao Island and smaller islands; the west border is the Tumindao Reef, which is submerged. Narrow passages between the islands and through the reef connect the atoll to the ocean. Deep channels separate the atoll from the submerged Meridian Reef on the west and Sibutu Island on the east. The islands are heavily vegetated, as shown by their red signature in Landsat IR color images (not illustrated).

In the band-7 image (Figure 9.9C) the islands have very bright signatures because their vegetation strongly reflects IR wavelengths; water is dark because it absorbs these wavelengths. Boundaries between land and water are precisely shown in this image. In bands 4 and 5, however, water a few fathoms deep and land both have dark signatures and cannot be distinguished. For example, on both these images, South Lagoon, shaped like a bow tie at the south end of the atoll, could readily be mistaken for land. On band 5 the brightest signatures represent the shallowest areas of barely submerged reefs

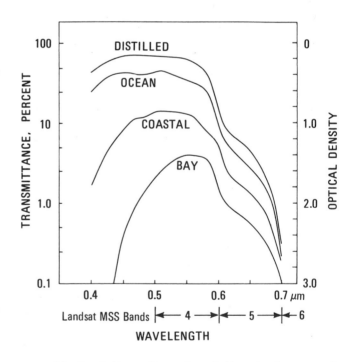

FIGURE 9.8 Spectral transmittance through 10 m of various types of water. From Specht, Needler, and Fritz (1973, Figure 1).

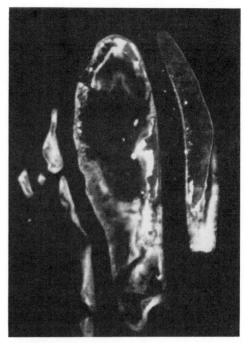

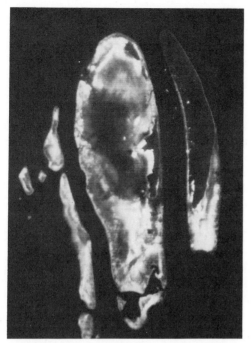

A. BAND 4 (0.5 TO 0.6 μm).

B. BAND 5 (0.6 TO 0.7 μm).

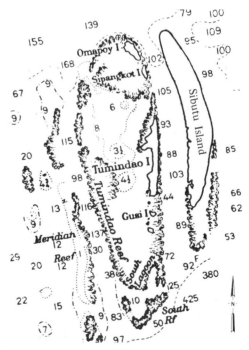

C. BAND 7 (0.8 TO 1.1 μm).

D. HYDROGRAPHIC CHART, DEPTH IN FATHOMS.

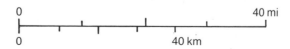

FIGURE **9.9** Landsat MSS bands and hydrographic chart of the Sibutu Island group in the Celebes Sea. Note differences in water penetration of the MSS bands.

and banks of coral sand. The water penetration capability of band 4 (Figure 9.9A) is shown by the extensive bright signatures that represent water depths down to about 3 fathoms (5 m). Intermediate gray signatures correlate with depths of 6 to 7 fathoms (11 to 13 m), such as the central lagoon of the atoll. At greater depths, no energy is reflected from the bottom, which appears dark. The South Lagoon is only 10 fathoms (18 m) deep but produces the same dark signature on band 4 as does the much deeper water in the channels. The 10-to-20-fathom (18-to-36-m) water depths west of the Meridian Reef also produce dark signatures.

Signatures on Landsat images are controlled not only by water depth but also by water clarity, reflectance of the bottom sediment, sunglint from sea surface, and atmospheric conditions. Therefore the depth ranges associated with image signatures in the Sibutu example will not necessarily be the same in other areas. In Figure 9.9 a few small clouds occur over the channel between Sibutu and Tumindao islands and over the eastern part of the Tumindao Lagoon. These may be recognized by their bright signatures on all Landsat bands. This distinguishes the clouds from dry land (dark on bands 4 and 5, bright on band 7) and from shallow reefs (bright on 4 and 5, dark on 7).

The hydrographic chart of the Sibutu Island group appears to be accurate because it closely matches the features on Landsat images. Elsewhere, however, Landsat images have revealed errors of commission or omission on hydrographic charts, many of which are based on surveys made in the 1800s. In the Chagos Archipelago of the Indian Ocean, Landsat images revealed the presence of a previously uncharted reef and showed that a known bank was charted 18 km east of its true position (Hammack, 1977). The U.S. Defense Mapping Agency incorporated these changes on new editions of their published hydrographic charts. Landsat images of the Georgia coast disclosed an island that was not present on existing maps or on aerial photographs acquired several years earlier. The island had formed recently by the accumulation of sand from drift along the shore. In the Red Sea, Chevron's marine seismic surveys in the late 1970s were hampered by uncharted submerged coral heads that snagged cables. Digitally processed Landsat images were used to locate and avoid these hazards.

Normal color aerial photographs have also been used successfully for charting shallow shelf areas. In the early 1970s the Kodak Company manufactured a two-layer color film designed for maximum water penetration, but this product has been discontinued.

To acquire images of the seafloor now, camera and television systems with their own light sources are low-ered on cables or operated from submersible research vessels. These systems acquire high-resolution images, but their coverage is restricted by the limited penetration of light in water.

Side-Scanning Sonar Images

Sound waves are readily transmitted through water and have long been used for detection of submarines and for *fathometers* that measure depth. The general term for this form of active remote sensing is *sound navigation ranging* (sonar). Figure 9.10 illustrates the technique for imaging the seafloor known as *side-scanning sonar,* which is analogous in many respects to SLAR. The system is housed in a torpedo-shaped vehicle, called a *sonar fish,* that is towed near the seafloor from a cable that also provides power and transmits data from the fish to a recording system on the ship. The imaging system uses *transducers,* which are devices that convert electrical energy into sound energy and received sound into electrical energy. As Figure 9.10 shows, tranducers on each side of the fish transmit a narrow pulse of sound at right angles to the ship track. The pulse encounters the sea floor and is reflected back to the transducer, where the received sound generates electrical signals that vary in amplitude proportional to the strength of the received sound. As the ship moves forward, the process continues, generating two strips of imagery separated by a narrow blank strip directly beneath the sonar fish. As in radar terminology, the direction of travel is the azimuth direction and the direction of transmitted pulses is the look direction. The data may be recorded on magnetic tape and played back to produce images. The analogy between sonar and radar systems extends to the geometry of the images; sonar images are also subject to slant-range distortion, which may be corrected during playback. Belderson and others (1972) published a standard reference to side-scanning sonar.

In sonar images, strong sonic returns are recorded as dark signatures and weak returns as bright signatures. Topographic scarps facing the sonic look direction produce strong returns, while surfaces facing or sloping away from the look direction produce weak returns. Smooth surfaces of mud or sand reflect the sonic pulse specularly (Chapter 6); hence these surfaces have weak returns and are recorded with bright signatures. Rough surfaces such as boulder fields and lava scatter much of the incident energy back to the sonar fish and are recorded with dark signatures.

These sonic signatures are illustrated in Figure 9.11, which is a sonar image of the Make It Burn (MIB) submarine volcano, or seamount, on the East Pacific Rise at depths of 2000 m and greater. As shown in the inter-

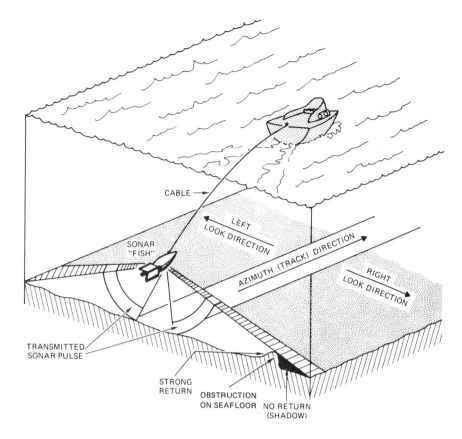

FIGURE 9.10 Side-scanning sonar system.

pretation map (Figure 9.12A), the ship track was toward the west; the upper half of the image was recorded with a look direction toward the right and the lower half was recorded with a look direction toward the left. The narrow blank strip is the area directly beneath the fish that is not imaged. The seamount has sloping flanks, a relatively horizontal summit plateau, and a large caldera surrounded by a scarp that is formed by ring fractures. There are two pit craters and a small cone on the flat floor of the caldera. The floors of the caldera and pit craters are relatively smooth and thus have uniform, bright signatures in the image (Figure 9.11). The summit plateau has a rough, hummocky surface, which is recorded as irregular bright and dark spots. The scarp enclosing the caldera is a steep slope that faces the left and right look directions and is recorded with arcuate dark signatures, as are the walls of the pit craters. The northern and southern flanks of the seamount are in an acoustic shadow zone because the summit plateau blocks the transmitted sonar energy. For this reason the flanks have bright signatures. Smaller features such as gullies, slump blocks, landslides, and volcanic cones are recognizable in the image.

The image in Figure 9.11 was acquired with the Sea MARC I system, which operates at the frequencies of 27 and 30 kHz. The system is towed 200 to 400 m above the bottom and images a swath with maximum total width of 5 km at a spatial resolution of several meters. Another sonar system, Gloria, operates at a 6.5-kHz frequency and is towed near the sea surface to acquire images with total swath width of 60 km and spatial resolution of 45 m. Gloria was used to image the floor of the Pacific Ocean in mosaic fashion to a distance of 200 nautical miles (370 km) off the west coast of the United States. Sea MARC and Gloria are large and complex systems. Smaller sonar systems are commercially available and are used for surveys in shallow water. Chapter 11 illustrates a mosaic of submarine landslides acquired by one of these systems.

Shallow Bathymetric Features on Seasat Images

Radar energy does not penetrate water but is reflected and scattered by the water surface. Seasat images of shallow seas, however, commonly have bright and dark signatures that record the presence of underwater fea-

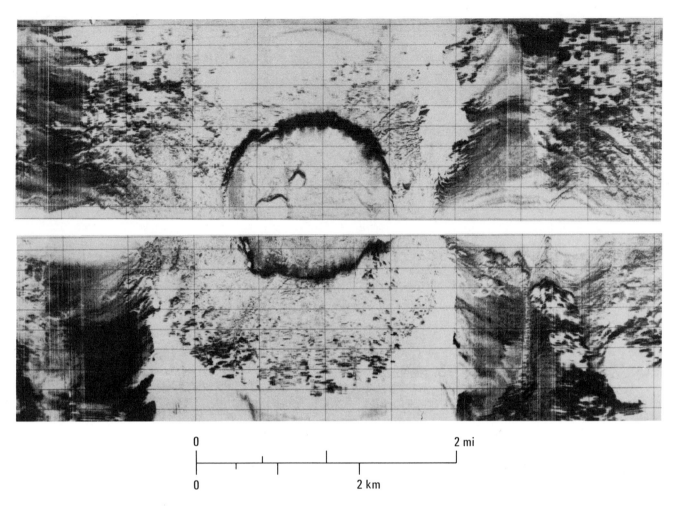

0   2 mi

0   2 km

FIGURE 9.11 Side-scanning sonar image of MIB seamount, East Pacific Rise. Dark signatures record strong sonic returns; bright signatures record weak returns. From Fornari, Ryan, and Fox (1984, Figure 3A). Courtesy D. J. Fornari, Lamont-Doherty Geological Observatory, Columbia University.

tures such as channels and sandbars at depths of several tens of meters. Figure 9.13 is a Seasat image of the English Channel in which the linear bright and dark patterns record the submerged sandbars and depressions at depths down to 30 m. The coincidence between radar patterns and bathymetry is seen by comparing the Seasat image with the corresponding chart (Figure 9.14). This relationship between radar signatures and shallow bathymetry also occurs in Seasat images of the Bahama Banks, Bermuda, the Nantucket Shoals, and Cook Inlet of Alaska. The following explanation of this relationship is summarized from Kasischke and others (1983).

Under normal conditions, the ocean surface is covered by small wind-generated waves called *small-scale waves*. Their wavelengths range from less than 1 cm to 1.5 m. The resulting surface roughness interacts with the depression angle and wavelength of Seasat to produce a typical intermediate gray signature. Local variations in velocity of surface ocean currents, however, can change the size and spacing of the small-scale waves with attendant changes in surface roughness and radar signature. An increase in current velocity stretches the small-scale waves, which decreases their amplitude and reduces surface roughness. Decreased current velocity results in increased roughness. Figure 9.15 is a model cross section of the ocean surface and a shallow sandbar, such as those in the English Channel. A tidal current flows from right to left with relative velocity indicated by the length of the arrows. Over the deeper seafloor (right side), the tidal velocity and small-scale waves are in equilibrium and the resulting ocean surface roughness has an intermediate gray signature in the radar image. As the current approaches the flank and crest of the sandbar, a constant volume of water must flow through

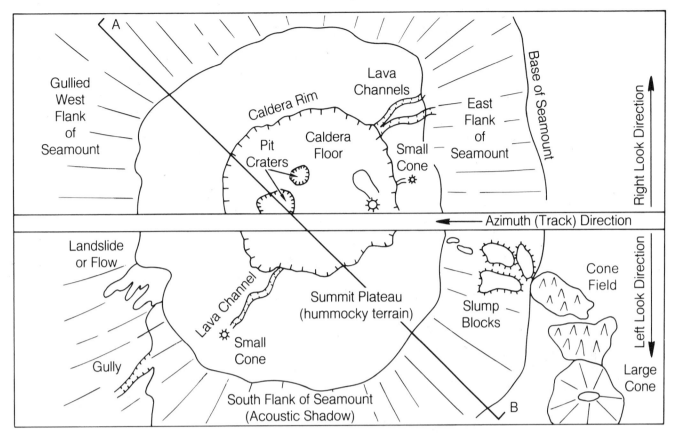

A. INTERPRETATION MAP. AFTER FORNARI, RYAN, AND FOX (1984, FIGURE 3b).

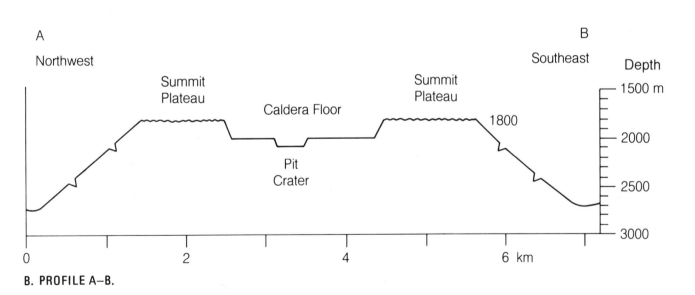

B. PROFILE A–B.

FIGURE 9.12   Interpretation map and profile of MIB seamount, East Pacific Rise.

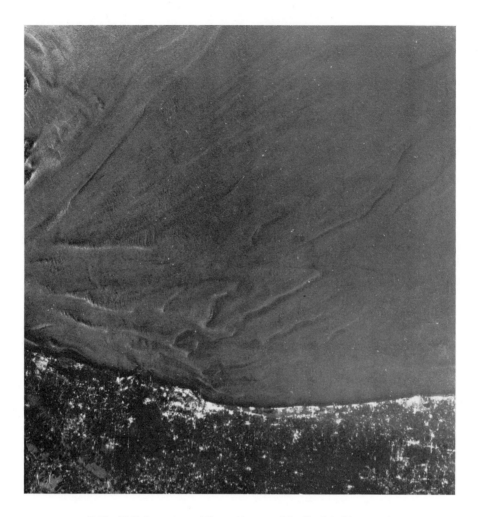

FIGURE 9.13 Digitally processed Seasat image of the English Channel. From Fu and Holt (1982, Figure 31A). Courtesy J. P. Ford, Jet Propulsion Laboratory.

a decreasing cross-sectional area, which causes an increased velocity and decreased roughness of small-scale waves. This strip of smoother water results in a dark band in the Seasat image (Figure 9.13) parallel with the crest of the sandbar. On the down-current flank of the sandbar (left side in Figure 9.15), the velocity decreases and the roughness increases, which results in a bright band in the image. Down-current from the sandbar, the normal pattern of small-scale waves resumes.

The chart of the English Channel (Figure 9.14) has arrows that show direction and velocity of tidal currents at the time the Seasat image was acquired. The tidal current was flowing from east to west, or right to left, which matches conditions of the model. In the Seasat image the dark bands are on the up-current (east) flank of the sandbars and the bright bands are on the down-

current (west) flank. This model for radar expression applies to shallow water (less than 50 m deep) with a tidal current of at least $0.4 \text{ km} \cdot \text{sec}^{-1}$ and a wind velocity of 1.0 to $7.5 \text{ m} \cdot \text{sec}^{-1}$ (Schuchman, 1982).

MAPPING OCEANIC CIRCULATION PATTERNS

Traditional methods for mapping oceanic circulation patterns employ current meters, drift floats, and direct temperature measurements. In addition to being expensive, these methods are hampered by the need for obtaining simultaneous data over a broad expanse of water. These problems are largely overcome by remote sensing systems that can provide nearly instantaneous images of circulation patterns over very large areas. Current

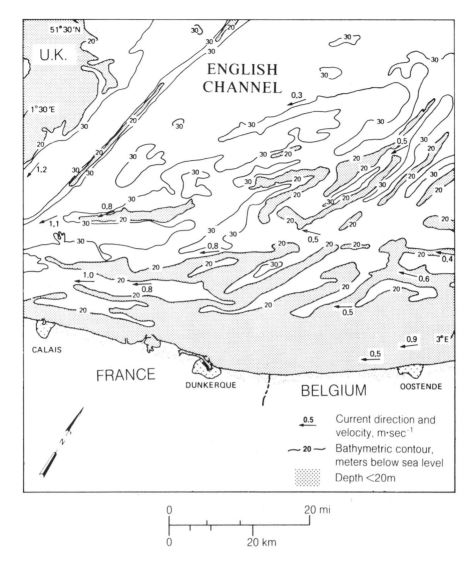

FIGURE 9.14 Bathymetric chart of the English Channel. From Fu and Holt (1982, Figure 31B).

systems are mapped by recognizing some property of water that differs from that of the surrounding water. Remote sensing systems record the following properties of water:

1. *Color due to suspended material such as sediment and plankton.* Images in the visible region acquired by CZCS, Landsat MSS and TM, and other satellite scanners are employed. An example of CZCS images of the Gulf of Mexico was described earlier in this chapter.

2. *Radiant temperature.* The thermal IR bands of GOES, AVHRR, Landsat TM, and others are used. Chapter 5 described a Landsat TM band-6 image of thermal patterns in Lake Ontario.

3. *Surface roughness.* The boundary between faster moving water of a current system and adjacent water is commonly marked by differences in small-scale waves that are readily detected in radar images. Seasat was deployed specifically to record oceanic roughness patterns that indicate circulation patterns, sea state, and bathymetry.

California Current

The small-scale and regional coverage of Landsat images emphasizes major current features and reduces the confusing local details. The repetition cycle of Landsat images enables seasonal changes in circulation patterns to be monitored by mapping the plumes of suspended sediment that serve as tracers for various current sys-

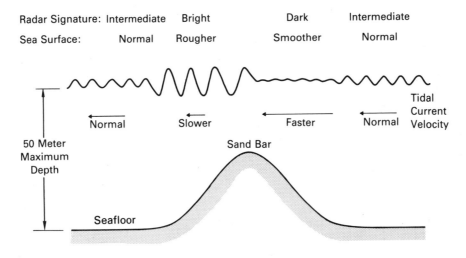

| Radar Signature: | Intermediate | Bright | Dark | Intermediate |
| Sea Surface: | Normal | Rougher | Smoother | Normal |

FIGURE 9.15 Cross-section model showing relationship of sandbar, tidal current, sea-surface roughness, and radar signatures. Modified from Kasischke and others (1983, Figure 4).

tems. Several investigators report that Landsat MSS band 5 is optimum for mapping current patterns because it provides the maximum contrast between clear and turbid water (Maul and Gordon, 1975; Rouse and Coleman, 1976). The interpretation of many Landsat images of the Pacific coast, however, indicates that MSS band-4 images are superior for this region. Reflectance differences between clear and turbid water are determined by the nature of the suspended sediment and other factors. One should examine MSS bands 4 and 5 or TM bands 2 and 3 to select the optimum band for interpretating current patterns in a particular area.

Figure 9.16 shows seasonal changes in near-shore current patterns of the Cape Mendocino region, California. Sediment plumes on both the January and April images originate at the mouths of the Mad, Eel, and Mattole rivers. The California Current flows slowly southward at speeds generally less than $0.25 \text{ m} \cdot \text{sec}^{-1}$ and controls circulation patterns for much of the year, as shown by the southward movement of sediment plumes on the April image (Figure 9.16B).

The Davidson Current is a deep counter-current that flows northward along the California coast. For most of the year it travels at a depth below 200 m, but in the late fall and winter, north winds are weak or absent and the countercurrent appears at the surface, inshore from the main California Current. The January image (Figure 9.16A) shows the northward movement of sediment plumes propelled by the Davidson Current. The effects of the current systems vary from year to year. For example, on a Landsat image (not illustrated) acquired one year after the January 1973 example, the plumes are drifting southward and there is no surface expression of the Davidson Current.

Gulf Stream

Most oceanic current systems are significantly warmer or cooler than the adjacent water and are therefore easily recognized in thermal IR images. Figure 9.17A is an AVHRR image of the Atlantic Ocean off the eastern United States. The Gulf Stream is the distinct band of warmer water (darker signature) that flows northward, roughly parallel with the coast. Interaction between the water bodies is shown by the gyres (circular plumes) of Gulf Stream water that extend into the cooler water. AVHRR thermal IR data are also available as maps (Figure 9.17B) in which contours, rather than gray-scale tones, show temperature variations. Satellite thermal IR images are routinely employed to monitor oceanic phenomena, including upwelling patterns and El Niño currents in the eastern Pacific Ocean which have major impacts on commercial fisheries. The daily acquisition of AVHRR images enables oceanographers to monitor detailed changes in current patterns.

Internal Waves

Waves form at the interface between fluids of different density; the well-known example is surface waves, which form at the interface between water and air. Within water bodies, the *thermocline* is the interface between the surface layer of warmer, less dense water and the underlying layer of colder, denser water. In shallow seas, tidal

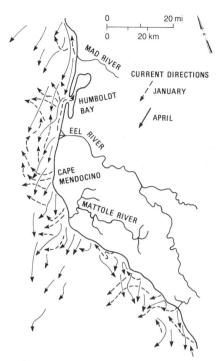

A. JANUARY 6, 1973.

B. APRIL 24, 1973.

C. JANUARY AND APRIL CURRENTS INTERPRETED FROM THE LANDSAT IMAGES. AFTER CARLSON (1976).

FIGURE 9.16 Landsat MSS band-4 images showing seasonal changes in current and sedimentation patterns off Cape Mendocino, California.

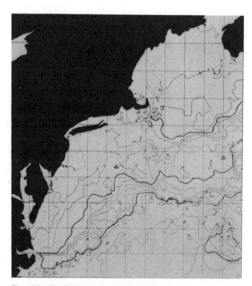

A. THERMAL IR IMAGE ACQUIRED APRIL 29, 1983. DARK TONES ARE WARM; BRIGHT TONES ARE COOL.

B. MAP OF SEA–SURFACE TEMPERATURES FOR DECEMBER 29, 1982.

FIGURE 9.17 AVHRR thermal IR image and temperature map of the Gulf Stream off the east coast of the United States. From Epstein, Callicott, Cotter, and Yates (1984, Figures 6, 7). Courtesy W. M. Callicott, NOAA Satellite Services Division.

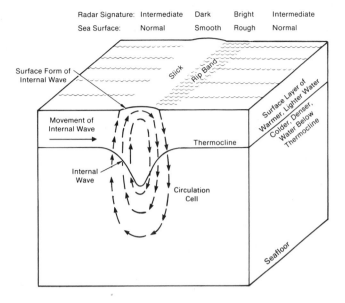

FIGURE 9.18 Model of an internal wave showing the relationships among circulation pattern, surface expression, and radar signatures. Modified from Osborne and Burch (1980, Figure 3).

currents that encounter seafloor irregularities, such as submarine canyons and breaks in slope, may cause waves at the thermocline (Figure 9.18). These waves are commonly called *internal waves* but are also known as *solitons* and *internal gravity waves*. The circulation pattern of an internal wave causes a low linear bulge at the surface that is accompanied by distinctive patterns of small-scale surface waves. The circulation cell rises toward the sea surface at the trailing edge of an internal wave, which decreases the roughness of the small-scale surface waves and results in a linear band of smooth water, called a *slick,* parallel with the crest of the internal wave (Figure 9.18). The circulation cell descends at the leading edge, causing a band of rougher water, called a *rip band.*

Figure 9.19A is an oblique aerial photograph of alternating slicks and rip bands accompanying internal waves in the Gulf of Georgia on the Pacific coast of Canada. The rip bands have bright signatures because of increased sunglint from the rougher water, and the slicks are dark. The aircraft radar image (Figure 9.19B) was acquired 8 minutes after the photograph, which accounts for different wave patterns on the two images. The rip bands have bright radar signatures, and the slicks are dark. The Canadian research vessel *Endeavor* is present in both images; the metal hull and superstructure produce the very bright radar signature.

Internal waves in many regions have been recorded on SIR and Seasat radar images that have provided oceanographers with valuable information. Figure 9.20A

is a Seasat image of the Atlantic Bight that spans the continental slope on the east (right) and the shallow shelf on the west (left). The shelf break trends northeast across the image. Internal waves are generated where tidal currents encounter the shelf break. The waves occur in *packets* of parallel waves, each packet having up to 30 waves. The arcuate wave fronts are convex toward the direction of travel, which is generally shoreward (westward), and they travel at an estimated speed of 0.3 m · sec^{-1}. Crest length is about 50 km, and maximum wavelength within packets is 1.3 km. A few east-trend-

A. OBLIQUE AERIAL PHOTOGRAPH SHOWING THE RESEARCH VESSEL *ENDEAVOR.*

B. AIRCRAFT L−BAND IMAGE (23.5 cm). VERY BRIGHT SPOT IS THE *ENDEAVOR.* SEA CONDITIONS ARE SIMILAR BUT NOT IDENTICAL TO THOSE IN THE AERIAL PHOTOGRAPH.

FIGURE 9.19 Internal waves in the Gulf of Georgia, British Columbia, Canada, July 1978. Courtesy J. F. R. Gower, Canada Institute of Ocean Sciences.

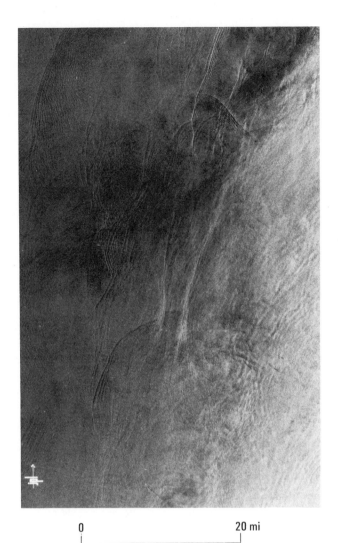

A. SEASAT IMAGE IN MID–ATLANTIC BIGHT, 100 km EAST OF DELAWARE COAST. AUGUST 31, 1978. FROM FU AND HOLT (1982, FIGURE 7).

B. SIR-A IMAGE IN ANDAMAN SEA, 100 km EAST OF ANDAMAN ISLAND. NOVEMBER 14, 1981. FROM FORD, CIMINO, AND ELACHI (1983, FIGURE 52).

FIGURE 9.20   Satellite radar images of internal waves. Courtesy J. P. Ford, Jet Propulsion Laboratory.

ing wave packets may have been generated by currents that interacted with east-trending submarine canyons.

Figure 9.20B is a SIR–A image of a packet of large internal waves in the Andaman Sea with crest lengths of 100 km and wavelengths from 2.5 to 6.5 km. The pattern of rip bands (bright) and slicks (dark) is clearly seen. Osborne and Burch (1980) studied internal waves during offshore oil-drilling operations in the Andaman Sea. Under normal sea conditions the small-scale surface waves were 0.6 m high; the rip bands had wave heights of 1.8 m and traveled at speeds of $2.2 \text{ m} \cdot \text{sec}^{-1}$;

the slicks had ripples approximately 0.1 m high. Additional satellite radar images of internal waves are illustrated in Fu and Holt (1982) and in Ford, Cimino, and Elachi (1983).

Internal waves are also expressed in Landsat images by bright and dark bands that correspond to rip bands and slicks, respectively. Figure 9.21 is a Landsat image with a number of internal wave packets present. The image is of the Atlantic shelf off Cape Cod, Massachusetts, acquired several years before and 500 km to the northeast of the previous Seasat image (Figure 9.20A).

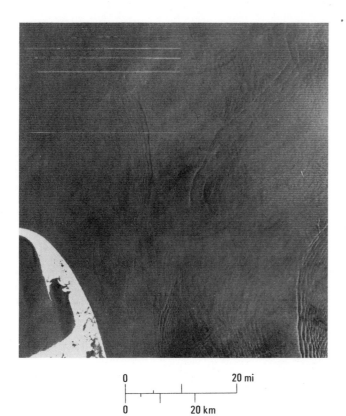

```
0                        20 mi
|--|--|--|----|----|

0              20 km
```

FIGURE 9.21 Landsat MSS band-5 image of the Atlantic shelf off Cape Cod, Massachusetts. Courtesy J. R. Apel, The Johns Hopkins University.

Notice that the patterns and dimensions of the waves in both images are similar. Mariners and oceanographers knew about internal waves for many decades before satellites were launched; satellite images, however, have shown the waves to be far more common and widely distributed than previously suspected.

MONITORING INDUSTRIAL THERMAL PLUMES

Many industrial plants withdraw water from lakes, rivers, and the ocean to cool their processes and then return it to the same water bodies at higher temperatures. The heated water discharges, called *thermal plumes,* may be monitored by airborne thermal IR scanners in the same manner as natural water currents of different temperatures. Nuclear and fossil-fuel electric power plants, refineries, and chemical and steel plants use large volumes of water. Aside from any chemicals or suspended matter, the heated water affects the environment in two ways:

1. Excessively high temperatures may kill organisms or inhibit their growth and reproduction. In some areas, however, the heated discharge water is used for commercial cultivation of lobsters and oysters.

2. The heated water has a lower content of the dissolved oxygen that is essential for aquatic animals and for oxidation of organic wastes.

Environmental legislation has been enacted to regulate thermal discharges. In California coastal waters, for example, the maximum temperature of thermal discharges must not exceed the natural temperature of receiving waters by more than 11°C. At a distance of 300 m from the point of discharge, the surface temperature of the ocean must not increase by more than 2.2°C. Temperature of the discharge water may be lowered by passing the water through cooling towers or by mixing it with cooler water before it is discharged. Thermal IR surveys are an ideal way to monitor the temperature and pattern of the discharge outfalls.

In April 1973 Daedalus Enterprises acquired repeated calibrated IR images of the thermal plume discharged into Montsweag Bay, Maine, from the Maine Yankee nuclear power plant. The images were part of an investigation of the effect of tidal action on shape and distribution of the thermal plume. The image in Figure 9.22A was selected from more extensive coverage to illustrate the maximum extent of the plume at low tide. Tape-recorded data were digitally processed and displayed as an image with eight discrete gray-scale levels. Image data were converted into a thermal contour map (Figure 9.22B) using the quantitative method described in Chapter 5. The branching or merging contour lines occur where the horizontal thermal gradient is so steep that contour lines run together. Both image and map clearly show the location of the plume and the temperature distribution within it. Note, for example, that some parts of the bay are not affected by the plume during low tide. The upstream and downstream extent and temperature level of the plume are precisely shown.

To appreciate the practical value of monitoring thermal plumes using IR surveys, imagine undertaking the following exercise. Design a monitoring system using conventional surface thermometers that will produce a thermal map, with the precision and detail of Figure 9.22B, throughout a tidal cycle. You must use several hundred surface thermometers, precisely positioned and located; you must calibrate them and record from them to an accuracy of 0.5°C; they must all be read at the same times to provide simultaneous data for contouring. Finally, you must deploy and retrieve the thermometers during strong tidal currents. This survey would be im-

A. THERMAL IR IMAGE (8 TO 14 μm).

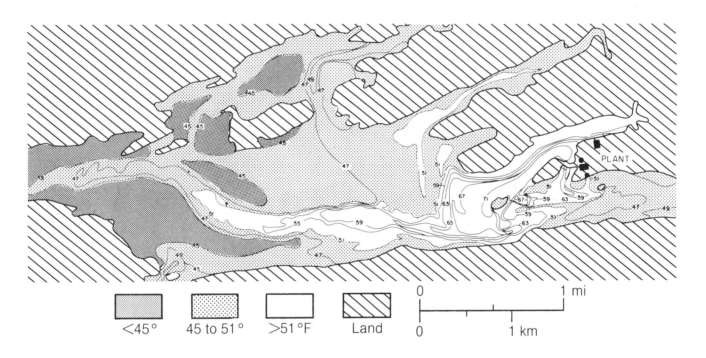

| <45° | 45 to 51° | >51 °F | Land |

B. TEMPERATURE MAP.

FIGURE 9.22 Thermal plume from the Maine Yankee nuclear power plant, Montsweag Bay, Maine. Image was acquired at low tide. Courtesy Maine Yankee Power Company and Daedalus Enterprises, Incorporated.

practical to conduct, but is readily accomplished with several IR scanning flights.

DETECTION OF OIL FILMS

The National Research Council (1985) estimated the amount of oil that enters world oceans annually from various sources (Table 9.3). Note that transportation of oil plus municipal and industrial wastes and runoff account for 81.6 percent of the oil entering the oceans each year. Also note that natural sources (seeps and sediment erosion) contribute five times more oil than do offshore production operations.

The two major applications of remote sensing of oil films are

1. *Law enforcement:* Survey coastal and inland waterways for violations of pollution regulations.

2. *Pollution countermeasures:* Monitor spills to aid cleanup operations and to determine their effectiveness.

The U.S. Environmental Protection Agency (EPA) uses the following nonquantitative classification of oil spills, given in the order of decreasing thickness:

Slick Relatively thick layer with a definite brown or black color

Sheen Thin, silvery layer on the water surface with no black or brown color

Rainbow Very thin iridescent multicolored bands visible on the water surface

Oil commonly mixes with water to form a brown emulsion called *chocolate mousse* because it resembles that dessert. In a typical oil spill, 90 percent of the volume is concentrated in 10 percent of the area, largely in the form of slicks and mousse. For remote sensing interpretation, the two categories of thinner oil spills are commonly lumped together as ''rainbow/sheen'' because they are difficult to distinguish. In this text, *oil film* refers to any detectable oil floating on water with no implication about the thickness. The following sections describe the detection of oil films using images acquired in the UV, visible, thermal IR, and microwave spectral regions.

UV Images

In Figure 9.23, the radiance of a thin layer of crude oil is compared with the radiance of ocean water. The oil film has a noticeably higher radiance in the UV, blue, and reflected IR bands. Intensity of reflectance in the IR band is very low, however. The image in Figure 9.24A was acquired in the daytime with a cross-track aircraft scanner equipped with a UV detector (0.32 to 0.38 μm). The area is the Santa Barbara Channel off the

TABLE 9.3 Estimated annual input of petroleum hydrocarbons into the marine environment

Source	Metric tons per year, millions	Annual input, %
Transportation	1.47	45.3
Municipal and industrial wastes and runoff	1.18	36.3
Atmospheric fallout	0.3	9.2
Natural sources	0.24	7.7
Offshore production	0.05	1.5
Totals	3.24	100.0

Source: From National Research Council (1985, Table 2-22).

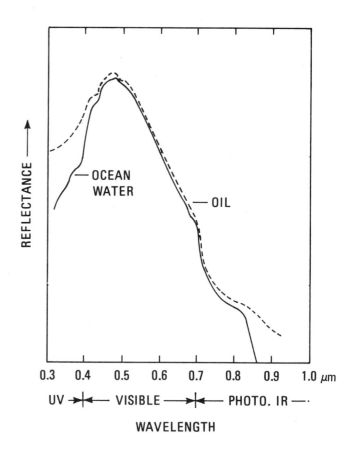

FIGURE 9.23 Spectral radiance of ocean water and of a thin layer of crude oil. From Vizy (1974, Figure 5).

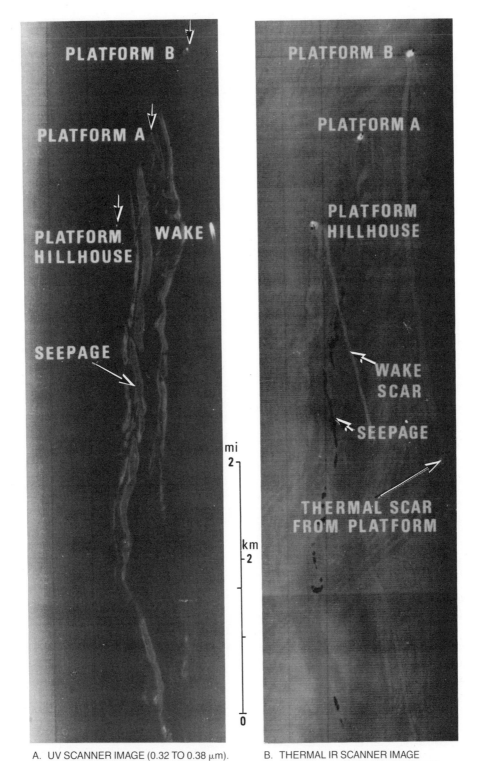

A. UV SCANNER IMAGE (0.32 TO 0.38 μm).
BRIGHT TONES ARE FLUORESCENCE
CAUSED BY OIL FILMS; DARK TONES
ARE CLEAN WATER.

B. THERMAL IR SCANNER IMAGE
(8 TO 13 μm). DARK TONES (COOL)
ARE OIL FILMS; BRIGHT TONES
(WARM) ARE CLEAN WATER.

FIGURE 9.24 Aircraft UV and thermal IR scanner images of natural oil seeps in the
Santa Barbara Channel, California, acquired July 31, 1974. From Maurer and
Edgerton (1976, Figure 9).

coast of southern California, where numerous oil seeps occur on the seafloor and form widespread films on the surface. The oil films have bright signatures in the UV image because sunlight causes oil to fluoresce at UV wavelengths, as shown by the radiance curve for oil in Figure 9.23. The platforms shown in the image are oil production facilities.

UV images are the most sensitive remote sensing method for monitoring oil on water and can detect films as thin as 0.15 mm (Maurer and Edgerton, 1976). Daylight and very clear atmosphere are necessary to acquire UV images. UV energy is strongly scattered by the atmosphere, as discussed in earlier chapters, but the effects are not too severe for images acquired at altitudes below 1000 m. Floating patches of foam and seaweed have bright UV signatures that may be confused with oil. Foam and seaweed can be recognized on images in the visible band acquired simultaneously with the UV images.

Experimental active airborne scanners have been developed to irradiate the water surface with a UV laser that stimulates the oil films to fluoresce. Spectral distribution of the fluorescence is detected and recorded as a trace that can be compared with traces of various oils. Under certain conditions, such a system is capable of recognizing broad classes of oils (Fantasia and Ingrao, 1974). Such active systems can acquire images both day and night.

Visible and Reflected IR Images

In the visible and reflected IR region, the signature of an oil film is determined by two factors:

1. The spectral radiance as shown in Figure 9.23.

2. The surface-tension effect of an oil film, which dampens the small-scale surface waves to create calm water, or slicks. These slicks, like those produced by internal waves, typically have dark signatures in contrast to the bright signatures caused by sunglint from the rougher, clean water.

The daytime images in Figure 9.25 were acquired with an airborne, multispectral, cross-track scanner in the Gulf of Mexico offshore from Galveston, Texas. The extensive oil film was caused by the collision of a tanker and another vessel. The oil occurs as a band extending right to left across the center of the images.

In the green band (Figure 9.25A), the dark signature of the oil film relative to clean water is attributed to the dampening of small-scale waves. In the reflected IR image (Figure 9.25B), the sunglint of clean water produces

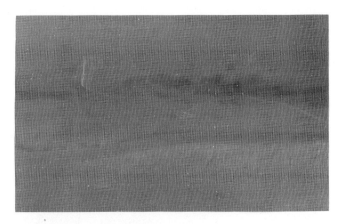

A. GREEN (0.52 to 0.60 μm).

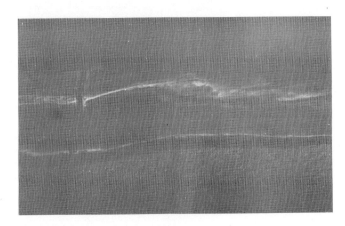

B. REFLECTED IR (1.00 to 1.30 μm).

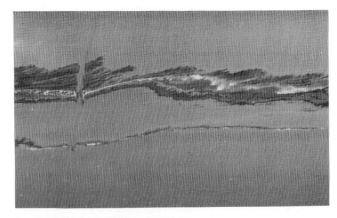

C. THERMAL IR (10.40 to 12.50 μm).

FIGURE 9.25 Aircraft multispectral scanner images of an oil spill in the Gulf of Mexico near Galveston, Texas. Courtesy NASA Johnson Space Center.

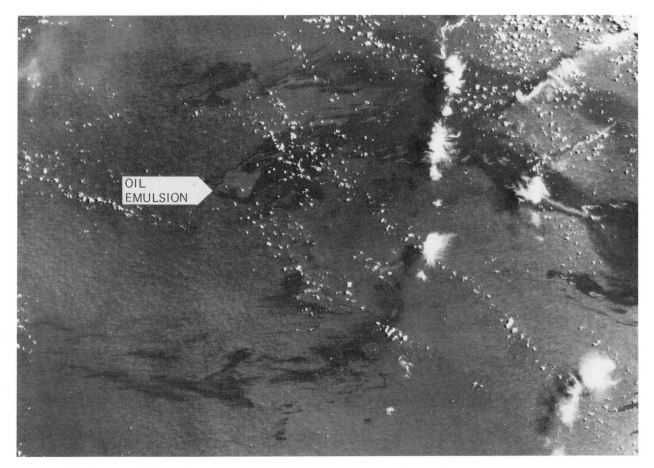

OIL
EMULSION

0   40 mi

0   40 km

FIGURE 9.26 Landsat MSS band-5 image of an oil spill in the Bay of Campeche from a blowout of Ixtoc 1. Image was acquired June 24, 1975 (21 days after blowout). Oil spill is dark relative to the water. Patches of mousse (emulsified oil) are light gray.

a bright band in the lower part of the image where small-scale wave patterns are evident. Most of the slick has a darker signature, and wave patterns are absent. The conspicuous narrow bright streaks are attributed to stringers of emulsified oil and water (mousse).

On June 3, 1979, the Ixtoc 1 offshore well, in the Bay of Campeche, off the Yucatan Peninsula of Mexico, blew out and was not capped until March 24, 1980. It is estimated that a half million metric tons of oil were discharged. Between 30 and 50 percent burned at the wellhead; less than 10 percent was recovered. A number of Landsat MSS and RBV images of the oil spill were analyzed by Deutsch, Vollmers, and Deutsch (1980), who reported that MSS bands 5 and 6 were superior to the single RBV band for recognizing the slick. Figure 9.26

is a Landsat MSS image of the Bay of Campeche northwest of the well. Most of the oil film has a dark signature relative to the bright water that is attributed to oil dampening of the small-scale waves and sunglint. Patches with brighter signatures than the water are thought to be mousse.

Digital processing can enhance the appearance of oil slicks on Landsat images, as demonstrated for the Santa Barbara Channel by Deutsch and Estes (1980).

Thermal IR Images

In the thermal IR images of the Santa Barbara Channel (Figure 9.24B) and the Gulf of Mexico (Figure 9.25C), the oil films have cool radiant temperatures relative to the warmer temperatures of clean water. In these examples oil and water have the same kinetic temperature because the two liquids are in direct contact. Note from Table 5.1, however, that the emissivity of pure water is 0.993, but a thin film of petroleum reduces water's emissivity to 0.972. The radiant temperature of a material is calculated from Equation 5.8 as

$$T_{\text{rad}} = \epsilon^{1/4} T_{\text{kin}}$$

For pure water at a kinetic temperature of 291°K (18°C), the radiant temperature is

$$T_{\text{rad}} = 0.993^{1/4} \times 291°\text{K}$$
$$= 290.5°\text{K, or } 17.5°\text{C}$$

For an oil film at the same kinetic temperature of 291°K, the radiant temperature is

$$T_{\text{rad}} = 0.972^{1/4} \times 291°\text{K}$$
$$= 288.9°\text{K, or } 15.9°\text{C}$$

This difference of 1.6°C in radiant temperature between the oil (15.9°C) and water (17.5°C) is readily measured by thermal IR detectors, which are typically sensitive to temperature differences of 0.1°C.

In the thermal IR image of the Santa Barbara Channel (Figure 9.24B), the patches of oil have cool signatures relative to the surrounding water, as predicted by Equation 5.8. The oil slick in the Gulf of Mexico also has a cool signature in the thermal IR image (Figure 9.25C) that is accompanied by narrow irregular streaks with warm signatures. The warm streaks are caused by mousse, which reradiates absorbed sunlight at thermal IR wavelengths.

The daytime and nighttime capability for acquiring images makes thermal IR imagery a valuable technique for surveillance and for monitoring cleanup activities around the clock. Rain and fog, however, prevent image acquisition. Also, the interpreter must be careful to avoid confusing cool water currents with oil slicks. An experienced interpreter using simultaneously acquired UV and IR images can minimize this problem.

Radar Images

Small-scale waves that cause backscatter on radar images of the sea are dampened by an oil film; this results in an area of low backscatter (dark signature) surrounded by the stronger backscatter (bright signature) from rough, clean water. The Seasat image of the Santa Barbara coast in Chapter 6 (Figure 6.16A) accurately records the oil slicks. On SIR images, slicks are less apparent but can be enhanced by image processing (Estes, Crippen, and Star, 1985). Aircraft images are also effective for recognizing oil films (Maurer and Edgerton, 1976).

Radar images may be acquired day or night under any weather conditions, which is an advantage over other remote sensing systems for monitoring oil spills. As with other images, however, radar images must be interpreted carefully, because dark streaks may be signatures of smooth water that is not caused by oil. Internal waves and shallow bathymetric features are two other possible causes of dark signatures. Experienced interpreters learn to distinguish these features by their patterns and other attributes.

SEA-ICE MAPPING

Increased shipping activity and petroleum exploration in Arctic waters has increased the need for information on sea-ice conditions. Global weather predictions require information about the thermal conditions of the polar seas, which in turn are related to ice abundance. Thus information on ice cover can aid meteorologists. Remote sensing, especially from satellites, is the only practical way to map sea ice on a regional, repetitive basis.

Table 9.4, which lists and defines the most important sea-ice features, is summarized from the more extensive nomenclature of the World Meteorological Organization. Most of these features are illustrated in the images of the following sections.

Landsat Images

The Landsat MSS band-7 image of Dove Bay (Figure 9.27) illustrates many of the features in Table 9.4. The prominent flaw lead separates the pack ice to the east from the fast ice to the west that is attached to the shore. The flaw lead and many other leads in the pack ice are refrozen, as indicated by their dark gray tone; open leads have dark signatures. Fragments of broken floes occur along the southern part of the flaw lead and between large floes of the pack ice. Individual floes have a wide range of sizes and shapes. This pack has a concentration of over 90 percent ice floes. The pack is classified as *consolidated* because the leads are refrozen and there is no open water. *Open* pack ice has approximately equal proportions of ice and water, and the floes are not in contact. A swarm of icebergs in the southwest corner of Figure 9.27 calved from coastal glaciers in the summer of 1972 but were locked in the fjord by winter fast ice before they could enter Dove Bay.

TABLE 9.4 Sea-ice terminology

Feature	Description
Fast ice	Ice that forms adjacent to and remains attached to the shore. May extend seaward for a few meters to several hundred kilometers from the coast.
Floe	Any relatively flat piece of sea ice 20 m or more across. Floes are classified according to size.
Ice concentration	Percentage of total sea surface area that is covered by ice.
Pack ice	General term for any area of sea ice, other than fast ice, regardless of form or forms present. Pack ice is classified by concentration of the floes.
Lead	Any fracture or passageway through sea ice that is navigable by surface vessels. Leads may be open or refrozen. A *flaw lead* separates fast ice from pack ice.
First-year ice	Sea ice of not more than one winter's growth. Thickness ranges from 30 cm to 2 m.
Multiyear ice	Old ice that has survived more than one summer's melt. Compared to first-year ice, floes of multiyear ice are thicker and rougher and have rounder outlines.
Pressure ridge	Wall of broken ice forced up by pressure.
Brash ice	Accumulations of floating ice made up of fragments not more than 2 m across. The wreckage of other forms of ice.
Iceberg	A massive piece of ice extending more than 5 m above sea level that has broken away from a glacier. Icebergs are classified according to shape.

Source: From World Meteorological Organization Publication 259.TP145.

The Landsat orbits provide image coverage as far north and south as 81° latitude, although illumination above 70° latitude is insufficient to acquire images from late October to late March. Convergence of orbits at these high latitudes provides up to three or four consecutive days of coverage of the same area during each 16- or 18-day cycle. These repeated images may be used to measure movement of sea ice. On March 20 and 21, 1973, nearly cloud-free images were acquired of the Davis Strait between Greenland and Baffin Island to the west. Figure 9.28 shows two images of the same area acquired on successive days. The images were registered to each other by the latitude and longitude grid. The eastern one-quarter of both images is covered with brash ice and very small floes. Most of the area is covered by large floes up to 40 km long. On March 20 (Figure 9.28A) there are few open leads, indicated by black signatures, and many of the leads are refrozen, shown by the gray signatures. The open leads are wider and more abundant on March 21 (Figure 9.28B), and some of the larger floes have broken up.

Ice-movement vectors during the 24-hour period are shown by the arrows in Figure 9.28B, which have the same scale as the images. Ice movement was consistently southeastward at an average rate of $0.4 \text{ km} \cdot \text{h}^{-1}$. This rate assumes that the ice traveled a straight path; if it had followed an irregular course, the actual rate of movement would be higher.

Thermal IR Images

Thermal IR systems can acquire images during periods of polar darkness but not when heavy clouds and fog are present. One can estimate relative ice thickness from thermal IR signatures. Figure 9.29 illustrates thermal IR images acquired in November 1982 with an aircraft cross-track scanner of the Arctic Ocean north of Alaska. The bright horizontal line records the aircraft flight path. The irregular bright line is a topographic profile of the surface recorded with a laser altimeter along the flight path; peaks directed toward top of page are topographic highs.

Open water has the highest radiant temperature and brightest signature in the images. Sea ice insulates the relatively warm water beneath it. Larger amounts of radiant energy are transmitted to the surface of thin ice and smaller amounts to the surface of thicker ice. As a result, thin ice appears warmer than thick ice. Figure 9.29A illustrates these relationships between ice thickness and radiant temperatures. The center of the image has an extensive area of open water with a bright signature (warm temperature). The right portion of the image is covered with a thin sheet of first-year ice with a gray signature (intermediate temperature). The left portion is a large rounded floe of thicker, multiyear ice with a dark signature (cool temperature). The open-water area contains both multiyear floes and first-year floes. The multiyear floes are larger and cooler and have rounded outlines. The first-year floes are smaller and warmer with angular outlines. The general relationship between radiant temperature and ice thickness can be altered by other factors, such as variations in emissivity and the presence of water on the ice surface during the summer melting period.

The laser profile shows a relationship between ice type and surface roughness. Patches of open water in the center of Figure 9.29A have smooth profiles; the first-year ice in the right portion has minor irregularities; the multiyear floes have rough surfaces caused by their history of fracturing, thawing, refreezing, and ablation by Arctic storms. The roughness relationships are useful in interpreting radar images.

Figure 9.29B illustrates multiyear floes separated by leads that are largely refrozen and have intermediate temperatures. A few narrow open leads are recognizable

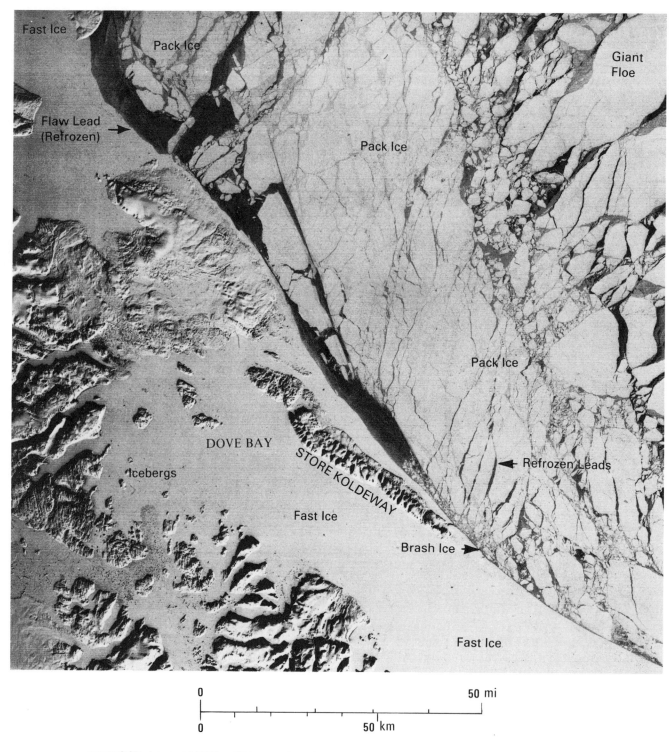

Fast Ice

Pack Ice

Giant Floe

Flaw Lead (Refrozen)

Pack Ice

Pack Ice

DOVE BAY

STORE KOLDEWAY

Icebergs

Refrozen Leads

Fast Ice

Brash Ice

Fast Ice

0   50 mi

0   50 km

FIGURE 9.27 Landsat MSS band-7 image of sea ice in Dove Bay, east coast of Greenland, March 25, 1973.

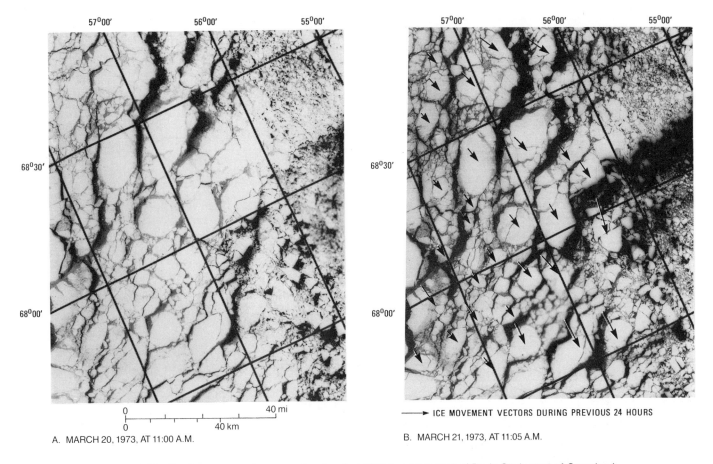

68°30'

68°00'

0   40 mi
0   40 km

A. MARCH 20, 1973, AT 11:00 A.M.

57°00' 56°00' 55°00'

68°30'

68°00'

⟶ ICE MOVEMENT VECTORS DURING PREVIOUS 24 HOURS

B. MARCH 21, 1973, AT 11:05 A.M.

FIGURE 9.28 Sea-ice movement measured on Landsat MSS band-7 images of Davis Strait, west of Greenland.

by their relatively warm temperature. The faint diagonal pattern from lower left to upper right results from wind streaks (Chapter 5). The laser profile drifts toward the top of image because of changing aircraft altitude. The base line of the profile was reset in the right side of the image.

The image in Figure 9.29C has extensive open and refrozen leads. The multiyear floe in the lower right corner of the image is cut by several pressure ridges, one of which is crossed by the profile that records several meters of elevation. The thicker ice of the ridge has an irregular narrow thermal IR signature that is cooler than the surrounding floe. Some ridges are accompanied by parallel fractures with warmer signatures.

Thermal IR images acquired by satellite systems such as AVHRR are used to compile regional maps of ice concentration in polar regions and are especially useful during winter periods. The thermal IR band 6 of Landsat TM with its 120-m spatial resolution has much potential for mapping ice, but few images have been recorded.

Radar Images

The ability of radar to acquire images during both darkness and bad weather makes it an excellent source of information for investigating sea ice. Radar images record surface roughness as a function of radar wavelength and depression angle as defined by the roughness criteria in Chapter 6. Radar signatures of sea-ice features are shown in images of the Beaufort Sea off northern Canada that were acquired with an aircraft, synthetic-aperture, X-band system. Figure 9.30A shows an extensive sheet of first-year ice that encloses some multiyear floes and that is cut by open and refrozen leads and by pressure ridges. Surface roughness profiles (not shown) are similar to those shown in the laser profiles of Figure 9.29. The rough surface of multiyear ice causes strong backscatter and bright radar signatures. Calm water in open leads causes little or no backscatter and has very dark signatures. First-year ice has minor roughness and a dark gray signature. The very narrow, very

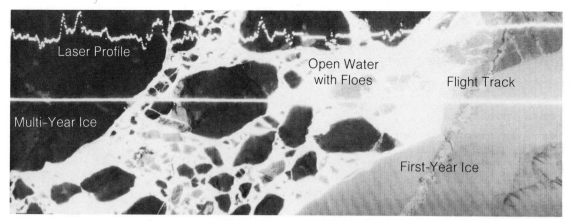

A. FIRST–YEAR AND MULTIYEAR ICE WITH OPEN WATER.

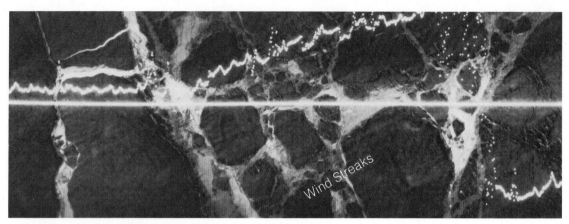

FLOES OF MULTIYEAR ICE SEPARATED BY OPEN AND
REFROZEN LEADS.

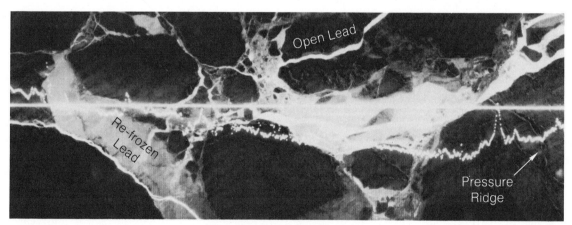

C. MULTIYEAR ICE WITH LEADS AND PRESSURE RIDGES.

FIGURE 9.29 Aircraft thermal IR images (8 to 12 μm) of ice in the Arctic Ocean
north of Alaska, November 1983. Bright signatures are warm; dark signatures are
cool. Cross-track width is 480 m. Courtesy Chevron–USA, Marathon Oil Company,
Shell Development Company, and Sohio Petroleum Company.

A. FIRST–YEAR AND MULTIYEAR ICE WITH OPEN AND REFROZEN LEADS IN THE BEAUFORT
   SEA NORTH OF LIVERPOOL BAY, CANADA. IMAGE WAS ACQUIRED JANUARY 1984.

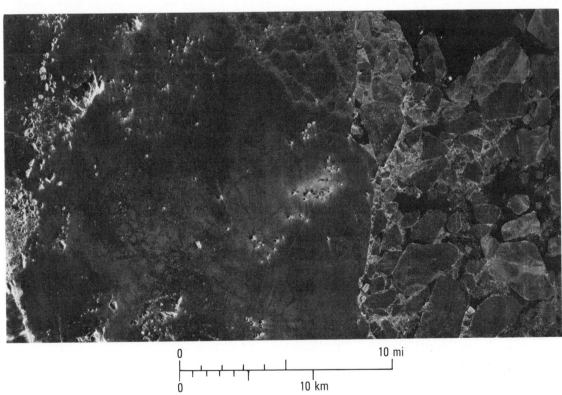

B. FAST ICE AND FLOES OF FIRST–YEAR ICE WITH TABULAR AND IRREGULAR ICEBERGS IN
   BAFFIN BAY NEAR THE WEST COAST OF GREENLAND. IMAGE WAS ACQUIRED JUNE 9, 1984.

FIGURE 9.30   Aircraft synthetic-aperture, X-band radar images (3-cm wavelength) of
sea-ice features. Courtesy M. C. Wride, Intera Technologies Limited.

bright lines crossing the first-year ice are pressure ridges. Broader bands with light gray signatures are leads filled with brash ice.

The area shown in Figure 9.30B is located near the terminus of glaciers in the Arctic islands that produce icebergs similar to those in the Landsat image of Figure 9.27. The left two-thirds of Figure 9.30B shows a sheet of first-year fast ice attached to a small island of bedrock in the northwest. The fast ice encloses a number of icebergs calved from nearby glaciers. Tabular icebergs in the upper left area have flat but rough surfaces; the other icebergs are irregular. The right portion of the image has floes of first-year ice separated by calm open water (dark signature) and by brash ice (bright signature). A few icebergs are included with the floes; three large tabular icebergs occur along the lower right margin of the image, and several irregular icebergs occur in the upper right area. This example demonstrates the value of radar for detecting and tracking icebergs as they enter shipping lanes. Kirby and Lowry (1981) published radar images of bergs and discussed their interpretation.

As stated earlier, younger ice is generally smoother and has darker radar signatures; older ice is rougher and has brighter signatures. During the summer melting season, however, thin sheets of water cover the ice and partially mask the roughness characteristics. Even under these circumstances, experienced interpreters can recognize ice types based on morphology and distribution patterns.

In 1978, Seasat acquired repetitive images of Arctic sea ice within its area of coverage, shown by the index map in Chapter 6. As described in Chapter 6, the L-band Seasat system had a longer wavelength (23.5 cm) and steeper depression angle (70°) than typical aircraft X-band systems. As a result the Seasat smooth and rough criteria (1 and 6 cm respectively) were higher than for the aircraft images in Figure 9.30. Figure 9.31 illustrates two Seasat images acquired at a 3-day interval in October 1978. Fu and Holt (1982) describe these and other images. The northwest corner of Banks Island in the Beaufort Sea occurs in the southeast corner of each image. A sheet of smooth fast ice is attached to the west coast of Banks Island. Smooth ice is distinguished from calm water by the presence of bright pressure ridges. The pack ice consists of floes of multiyear ice, many of which are aggregates of smaller floes that are separated by brash ice. The very rough brash ice has a brighter signature than the floes. A conspicuous floe is Fletcher's ice island, a tabular iceberg 7 by 12 km in size. The bright signature of much of the iceberg is attributed to low corrugated ice ridges and scattered rock debris inherited from its glacial origin on Ellsmere Island. This iceberg, which is also called T-3, was discovered in 1946 and has been tracked since then, remaining in the clockwise circulation pattern of the Beaufort Sea.

Considerable ice movement occurred during the three-day interval in the acquisition of the two Seasat images. The vectors in Figure 9.31B were plotted by connecting positions of individual floes, using the technique applied to repetitive Landsat images. The pack directly north of Banks Island was stable, but on the west and northwest, floes moved southward approximately 20 km at an average rate of $0.3 \text{ km} \cdot \text{h}^{-1}$. In the earlier Seasat image (Figure 9.31A), leads were narrow, but 3 days later the leads were extensive in the moving portion of the pack. Many of the leads in Figure 9.31B have dark signatures indicating smooth, calm water. The gray patches within the leads represent the formation of new ice as the leads began to freeze.

In other Seasat images of the Arctic seas, areas of open water commonly have bright signatures due to small-scale waves generated by wind; these areas should not be mistaken for patches of rough ice. Ketchum (1984) illustrated and described Seasat images of sea ice.

Radar Scatterometer

A *radar scatterometer* is a nonimaging active system for quantitatively measuring the radar backscatter of terrain as a function of the incidence angle. Scatterometer data are useful for characterizing the surface roughness of materials and are particularly useful for identifying types of sea ice. A scatterometer transmits a continuous radar signal directly along the flight path. The 3°-wide beam illuminates terrain both ahead and behind the aircraft, but for simplicity Figure 9.32 shows only the forward portion. The 0° incidence angle is directly beneath the aircraft, and the maximum incidence angle is approximately 60°. At an altitude of 900 m, the ground resolution cell at a 30° incidence angle is a 54-by-54-m square. The return signal is a composite of the backscatter properties of all the terrain features within the cell. The wavelength of this scatterometer system is 2.25 cm (X-band), and both the transmitted and received energy is vertically polarized. As the aircraft advances along the flight path, a ground resolution cell is illuminated initially at a 60° incidence angle and then at successively decreasing angles. The recorded amplitude and frequency of the successive returns are processed with Doppler techniques to obtain the scattering coefficient at each of several incidence angles. Details of scatterometer theory and operation are given in Moore (1983).

Scatterometer data may be displayed either as profiles, or as scattering coefficient curves. Profiles (Figure 9.33B) display the scattering coefficient of the terrain along the flight line for a particular incidence angle. In this example the incidence angles range from 2.5° (almost directly beneath the aircraft) to 52.0°. Scattering coefficient curves (Figure 9.34) display the returns at

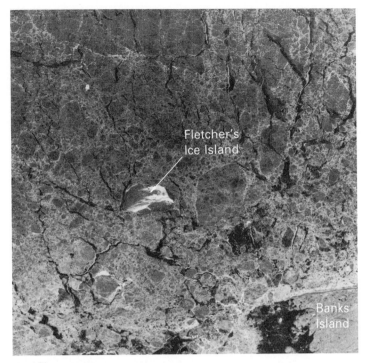

A. OCTOBER 3, 1978.

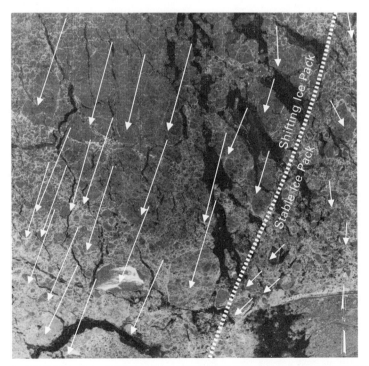

B. OCTOBER 6, 1978. ARROWS SHOW MOVEMENT DURING PRECEDING THREE DAYS.

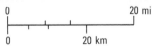

FIGURE 9.31 Seasat radar images (23.5-cm wavelength) of areas in the Beaufort Sea, northwest of Banks Island, Canada, showing ice movement. Images courtesy J. P. Ford, Jet Propulsion Laboratory.

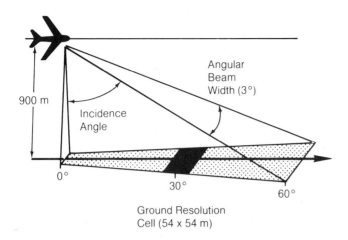

FIGURE 9.32 Geometry of the radar scatterometer system.

different incidence angles for an area on the ground.

Scatterometer profiles and aerial photographs of sea ice off Point Barrow, Alaska, were acquired in 1967 and interpreted by Rouse (1968). The photomosaic (Figure 9.33A) shows the following features:

A and F	Open water
A to C	Smooth first-year ice
B	Narrow open leads
C to F	Slightly ridged first-year ice
C	Major pressure ridge separating the smooth and ridged ice
D	Floe of smooth ice within the ridged ice
B, E, and F	Representative areas of smooth ice, ridged ice, and open water respectively, for which scattering coefficient curves are shown (Figure 9.34)

The scatterometer profiles (Figure 9.33) are plotted at the same horizontal scale as the photomosaic. At low incidence angles (2.5° and 6.7°), open water has a strong return, as explained in Chapter 6. Similarly the smooth ice has stronger returns than ridged ice at these low incidence angles. At the higher incidence angles (25.0° and 52.0°), most of the microwave energy encountering the water and smooth ice is specularly reflected away from the antenna and produces little return. Energy backscattered from the rough ridge ice, however, produces a relatively strong and characteristically spiked

profile at these incidence angles. Note that the floe of smooth ice at D can be recognized on the 52.0° profile by its reduced return within the ridged ice.

Passive Microwave Images

Energy at microwave wavelengths radiates from surfaces, just as at thermal IR wavelengths. The radiant microwave energy may be recorded by scanner systems with detectors operating in the microwave region to produce images of microwave brightness temperature. Energy flux is very low at these long wavelengths, as predicted from Wien's displacement law (Equation 5.1); therefore ground resolution cells must be large, on the order of kilometers, to record sufficient energy. Microwave brightness temperature of a surface is the product of kinetic temperature and emissivity at the wavelength band of the detector. Emissivity values are related to physical characteristics of the ice. In the wavelength region from 1 to 3 cm, using horizontal polarization, the following microwave brightness temperatures have been reported: open ocean (100° to 140°K), multiyear ice (180° to 210°K), and first-year ice (220°K).

Passive microwave imaging systems have been deployed on several environmental satellites. This technology has been described and illustrated in a report by NASA (1984).

COMMENTS

For many environmental applications, remote sensing is the best means of acquiring basic information, particularly on a regional scale and repetitive schedule. Environmental satellites have been launched specifically for these purposes. Digital-processing methods are becoming increasingly important for extracting and interpreting desired information from the extensive databases. The vegetation analysis of the African continent using digitally processed repetitive AVHRR satellite data is one example.

Many aspects of the oceans that were once poorly understood (such as circulation, sea state, productivity, sea-ice distribution, and bathymetry) are becoming better known through remote sensing techniques. Pollution in the form of thermal plumes and oil spills may be monitored.

QUESTIONS

1. Prepare a satellite comparison chart in the following manner. In a vertical column on left margin, list the following systems:

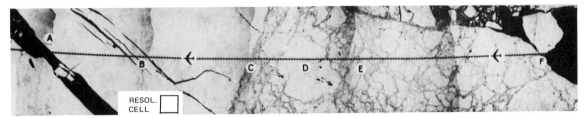

A. PHOTOMOSAIC SHOWING SCATTEROMETER FLIGHT LINE.

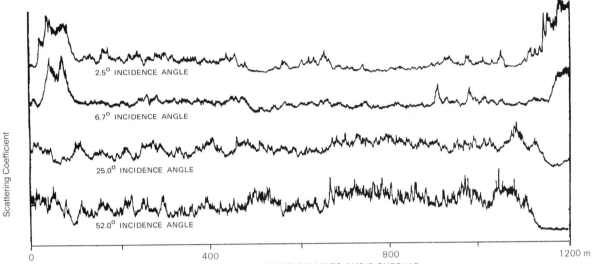

B. SCATTEROMETER PROFILES. PROFILES ARE OFFSET VERTICALLY TO AVOID OVERLAP.

FIGURE 9.33 Scatterometer profiles of sea ice off Point Barrow, Alaska. From Rouse (1968, Figure 10).

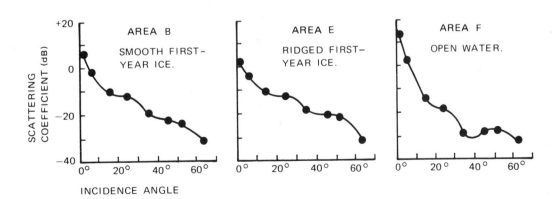

FIGURE 9.34 Scattering coefficient curves of various types of sea ice at localities off Point Barrow, Alaska. Lettered areas refer to regions shown in Figure 9.33A. From Rouse (1968, Figure 10).

Landsat MSS
Landsat TM
GOES
AVHRR
CZCS

Across the top of the chart, list the following characteristics as headings:

Spectral range, μm
Number of spectral bands
Ground resolution cell, m
Image-swath width, km
Repetition rate, days

Complete the chart by listing the characteristics for each satellite.

2. From this chart, select the optimum satellite system for your particular application(s). Give reasons for your selection.

3. Calculate the AVHRR spectral vegetation index for materials with following reflectance characteristics:

	Percent reflectance		Spectral vegetation index
	Band 1	Band 2	
Damp soil	5	11	———
Evergreen foliage	6	15	———

4. Describe how you would employ CZCS images and AVHRR thermal IR images to increase the catches of commercial fisheries in the oceans. (*Hint:* Upwelling occurs where deeper water comes to the surface carrying dissolved nutrients. Under the right conditions, upwelling will support plankton growth, which is the first link in the oceanic food chain.)

5. Describe the various conditions under which you would *not* rely on the following images for bathymetric information in a shallow sea.

Landsat images
Seasat images

6. List the various properties of water currents that enable one to detect them on images. List the imaging system or systems that record each property.

7. Your agency is responsible for monitoring the environment of San Francisco Bay, which has daily tidal fluctuations and inflows from several natural streams. Along the shore are numerous fossil-fuel power plants, chemical plants, oil refineries, and sewage plants. Fog and rain are dense at times. Describe the remote sensing system you would recommend to monitor thermal and chemical pollution in this situation.

8. For monitoring types, distribution, and movement of sea ice in the Arctic Ocean, list the advantages and disadvantages of following images:

Landsat TM band 4
Landsat TM band 6
Seasat

REFERENCES

Allison, L. J., and A. Schnapf, 1983, Meteorological satellites *in* Colwell, R. N., ed., Manual of remote sensing, second edition: ch. 14, p. 651–679, American Society for Photogrammetry and Remote Sensing, Falls Church, Va.

Belderson, R. H., N. H. Kenyon, A. H. Stride, and A. R. Stubbs, 1972, Sonographs of the sea floor, a picture atlas: Elsevier Publishing Co., N.Y.

Carlson, P. R., 1976, Mapping surface current flow in turbid nearshore waters of the northeast Pacific *in* Williams, R. S., and W. D. Carter, eds., ERTS-1, a new window on our planet: U.S. Geological Survey Professional Paper 929, p. 328–329.

Deutsch, M., and J. E. Estes, 1980, Landsat detection of oil from natural seeps: Photogrammetric Engineering and Remote Sensing, v. 46, p. 1313–1322.

Deutsch, M., R. R. Vollmers, and J. P. Deutsch, 1980, Landsat tracking of oil slicks from the 1979 Gulf of Mexico oil well blowout: Proceedings of the 14th International Symposium on Remote Sensing of Environment, p. 1197–1211, Environmental Research Institute of Michigan, Ann Arbor, Mich.

Epstein, E. S., W. M. Callicott, D. J. Cotter, and H. W. Yates, 1984, NOAA satellite programs: IEEE Transactions on Aerospace and Electronic Systems, v. AES 20, p. 325–344.

Estes, J. E., R. E. Crippen, and J. L. Star, 1985, Natural oil seep detection in the Santa Barbara Channel, California, with Shuttle Imaging Radar: Geology, v. 13, p. 282–284.

Fantasia, J. F., and H. C. Ingrao, 1974, Development of an experimental airborne laser remote sensing system for the detection and classification of oil spills: Proceedings of the Ninth Symposium on Remote Sensing of Environment, p. 1711–1745, Environmental Research Institute of Michigan, Ann Arbor, Mich.

Ford, J. P., J. B. Cimino, and C. Elachi, 1983, Space Shuttle Columbia views the world with imaging radar—the SIR–A experiment: Jet Propulsion Laboratory Publication 82-95, Pasadena, Calif.

Fornari, D. J., W. B. F. Ryan, and P. J. Fox, 1984, The evolution of craters and calderas on young seamounts—insights from Sea MARC I and Sea Beam sonar surveys of a small seamount group near the axis of the East Pacific Rise at 10°N: Journal of Geophysical Research, v. 89, p. 11069–11083.

Fu, L. L., and B. Holt, 1982, Seasat views oceans and sea

ice with synthetic aperture radar: Jet Propulsion Laboratory Publication 81-120, Pasadena, Calif.

Gordon, H. R., D. K. Clark, J. L. Mueller, and W. A. Hovis, 1980, Phytoplankton pigments from the Nimbus-7 Coastal Zone Color Scanner—comparison with surface measurements: Science, v. 210, p. 63–66.

Hammack, J. C., 1977, Landsat goes to sea: Photogrammetric Engineering and Remote Sensing, v. 43, p. 683–691.

Hovis, W. A., and others, 1980, Nimbus-7 Coastal Zone Color Scanner—system description and initial imagery: Science, v. 210, p. 60–63.

Hughes, B. A., and J. F. R. Gower, 1983, SAR imagery and surface truth comparisons of internal waves in Georgia Strait, British Columbia, Canada: Journal of Geophysical Research, v. 88, p. 1809–1824.

Justice, C. O., J. R. G. Townshend, B. N. Holben, and C. J. Tucker, 1985, Analysis of the phenology of global vegetation using meteorological satellite data: International Journal of Remote Sensing, v. 6, p. 1271–1318.

Kasischke, E. S., R. A. Shuchman, D. R. Lyzenga, and G. A. Meadows, 1983, Detection of bottom features on Seasat synthetic aperture radar imagery: Photogrammetric Engineering and Remote Sensing, v. 49, p. 1341–1353.

Ketchum, R. D., 1984, Seasat SAR sea ice imagery—summer melt to autumn freeze-up: International Journal of Remote Sensing: v. 5, p. 533–544.

Kirby, M. E., and R. T. Lowry, 1981, Iceberg detectability problems using SAR and SLAR systems in Deutsch, M., D. R. Weisnet, and A. Rango, eds., Satellite hydrology: p. 200–212, American Water Resources Association, Minneapolis, Minn.

Maul, G. A., and H. R. Gordon, 1975, On the use of Earth Resource Technology Satellite (Landsat-1) in optical technology: Remote Sensing of Environment, v. 4, p. 95–128.

Maurer, A., and A. T. Edgerton, 1976, Flight evaluation of U.S. Coast Guard airborne oil surveillance system: Marine Technology Society Journal, v. 10, p. 38–52.

Moore, R. K., 1983, Radar fundamentals and scatterometers in Colwell, R. N., ed., Manual of remote sensing, second edition: ch. 9, p. 369–427, American Society for Photogrammetry and Remote Sensing, Falls Church, Va.

National Research Council, 1985, Oil in the sea—inputs, fates, and effects: National Academy Press, Washington, D.C.

Osborne, A. R., and T. L. Burch, 1980, Internal solitons in the Andaman Sea: Science, v. 208, p. 451–460.

Rouse, L. J., and J. M. Coleman, 1976, Circulation observations in Louisiana bight using Landsat imagery: Remote Sensing of Environment, v. 5, p. 55–66.

Rouse, J. W., 1968, Arctic ice type identification by radar: University of Kansas Center for Research Technical Report 121-1, Lawrence, Kan.

Schuchman, R. A., 1982, Quantification of SAR signatures of shallow water ocean topography: University of Michigan Ph.D. dissertation, Ann Arbor, Mich.

Specht, M. R., D. Needler, and N. L. Fritz, 1973, New color film for water penetration photography: Photogrammetric Engineering, v. 40, p. 359–369.

Tucker, C. J., J. R. G. Townshend, and T. E. Goff, 1985, African land-cover classification using satellite data: Science, v. 227, p. 369–375.

Vizy, K. N., 1974, Detecting and monitoring oil slicks with aerial photos: Photogrammetric Engineering, v. 40, p. 697–708.

ADDITIONAL READING

Cracknell, A. P., ed., 1981, Remote sensing in meteorology, oceanography, and hydrology: Ellis Horwood, West Sussex, England.

Cracknell, A. P., ed., 1983, Remote sensing applications in marine science and technology: D. Reidel Publishing Co., Dordrecht, Holland.

Gower, J. F. R., ed., 1980, Oceanography from space: Plenum Press, N.Y.

Hall, D. K., and J. Martinac, 1985, Remote sensing of ice and snow: Chapman and Hall, N.Y.

Haughton, J. T., A. H. Cook, and H. Charnock, eds., 1985, The study of the ocean and land surface from satellites: Cambridge University Press, N.Y.

Johanssen, C. J., and J. L. Sanders, eds., 1982, Remote sensing for resource management: Soil Conservation Society of America, Ankeny, Iowa.

Johnson, R. W., and others, 1983, The marine environment in Colwell, R. N., ed., Manual of remote sensing, second edition: ch. 28, p. 1371–1495, American Society for Photogrammetry and Remote Sensing, Falls Church, Va.

Maul, G. A., 1985, Introduction to satellite oceanography: Martinus Nijhoff Publishers, Dordrecht, Holland.

NASA, 1984, Passive microwave remote sensing for sea ice research: Report of NASA Science Working Group for the Special Sensor Microwave Imager, NASA Headquarters, Washington, D.C.

Robinson, I. S., 1985, Satellite oceanography: John Wiley & Sons, N.Y.

Weisnet, D. R., and others, 1983, Remote sensing of weather and climate in Colwell, R. N., ed., Manual of remote sensing, second edition: ch. 27, p. 1305–1369, American Society for Photogrammetry and Remote Sensing, Falls Church, Va.

Land-Use and Land-Cover Analysis

Land use describes how a parcel of land is used (such as for agriculture, residences, or industry), whereas *land cover* describes the materials (such as vegetation, rocks, or buildings) that are present on the surface. The land cover of an area may be evergreen forest, but the land use may be lumbering, recreation, oil extraction, or various combinations of activities. Accurate, current information on land use and cover is essential for many planning activities. Remote sensing methods are becoming increasingly important for mapping land use and land cover for the following reasons:

1. Images of large areas can be acquired rapidly.

2. Images can be acquired with a spatial resolution that matches the degree of detail required for the survey.

3. Remote sensing images eliminate the problems of surface access that often hamper ground surveys.

4. Images provide a perspective that is lacking for ground surveys.

5. Image interpretation is faster and less expensive than conducting ground surveys.

6. Images provide an unbiased, permanent data set that may be interpreted for a wide range of specific land use and land cover, such as forestry, agriculture, and urban growth.

There are some disadvantages to remote sensing surveys:

1. Different types of land use may not be distinguishable on images.

2. Most images lack the horizontal perspective that is valuable for identifying many categories of land use.

3. For surveys of small areas, the cost of mobilizing a remote sensing mission may be uneconomical.

Remote sensing interpretations should be supplemented by ground checks of areas that represent various categories of land use and land cover. This chapter describes a system for classifying land use and land cover that is based on the interpretation of remote sensing images. Typical remote sensing images illustrate the system.

CLASSIFICATION PRINCIPLES AND SYSTEMS

Classification systems should recognize both activities (land use) and resources (land cover). Such a classification system that utilizes orbital and aircraft remote sensing data should meet the following criteria (Anderson and others, 1976):

1. The minimum level of accuracy in identifying land-use and land-cover categories from remote sensing data should be at least 85 percent.

2. Accuracy of interpretation for all categories should be approximately equal.

3. Repeatable results should be obtainable from one interpreter to another and from one time of sensing to another.

4. The system should be applicable over extensive areas.

5. The system should be usable for remote sensing data obtained at different times of the year.

6. The system should allow use of subcategories that can be derived from ground surveys or from larger scale remote sensing data.

7. Aggregation of categories should be possible.

8. Comparison with future land-use data should be possible.

9. Multiple land uses should be recognizable.

The U.S. Geological Survey recognized the need for a multilevel classification system that would enable the user to select the type and scale of image that suits the objectives of the survey. Table 10.1 lists the image systems and image scales employed for each of the four classification levels. For example, the level-I classification is suitable for an entire state, whereas level III is suitable for a municipality.

The levels I and II categories of Table 10.2 were slightly modified from those defined by the U.S. Geological Survey (Anderson and others, 1976). The level-III categories were modified from those defined by the Florida Bureau of Comprehensive Planning (1976). The classification system appears straightforward and definitive on paper, but there are some problems and uncertainties in using the system. On the images it may be difficult to recognize definitive examples of the various categories. Distinguishing land from water may seem simple until one considers the problem of classifying seasonally wet areas, tidal flats, or marshes with various kinds of plant cover. Another problem is to define boundaries between different land-use categories that grade into each other. The boundary between the "light industrial" and "heavy industrial" categories (131 and 132 of Table 10.2) may be difficult to draw. In such cases, the boundary is drawn to separate areas where one land use predominates. The categories listed in Table 10.2 are clearly defined, but commonly two or more categories

are intermixed and cannot be mapped separately at the scale of the images. For example, warehouses of the "commercial and services" category (120) are commonly located within predominantly industrial areas (category 130). Where the minor use occupies more than one-third of the area, the "mixed" category (180) is used at level II. An alternative treatment is a compound category (130/120) with the minor category given in the second position. Multiple use of a parcel of land presents another problem to the interpreter. For example, forest lands are commonly used for recreational purposes, such as hunting and camping; a compound category (410/170) may be employed for this situation. The interpreter must explain the "ground rules" and any modifications of the classification system that were used in a project.

The classification may be modified to accommodate specific needs. For example, one can precede irrigated orchards with an "i" (i221) and nonirrigated orchards with an "n" (n211).

It is difficult to map and display any area on an image smaller than 5 mm on a side. As shown in the following examples, however, one can still designate narrow linear features such as freeways and waterways.

USE OF THE MULTILEVEL CLASSIFICATION SYSTEM

The Los Angeles, California, area is used here to demonstrate the use of the multilevel system for classifying land use and land cover. A Landsat MSS image at a scale of 1:1,000,000 (Figure 10.1) was interpreted to produce the level-I land-use map (Figure 10.2). The color composite version of this Landsat image (described in Chapter 4) demonstrates the value of infrared color images for interpreting land use. Figure 10.3 is a 1:120,000-scale aerial photograph, acquired from a NASA high-altitude aircraft, that was interpreted to produce a level-

TABLE 10.1 Images employed in multilevel classification

Level	System	Image scale
I	Landsat MSS images	1:250,000 and smaller
II	Landsat TM images and high-altitude aerial photographs	1:80,000 and smaller
III	Medium-altitude aerial photographs	1:20,000 to 1:80,000
IV	Low-altitude aerial photographs	Larger than 1:20,000

Source: From Anderson and others (1976).

II land-use map (Figure 10.4). The interpretation was done using stereo pairs of the original IR color photographs, which yielded more information than the single black-and-white reproduction shown here. Interpreters should use stereoscopic and color images whenever these are available. Figure 10.5 is a conventional black-and-white aerial photograph at a scale of 1:30,000 that was interpreted stereoscopically with overlapping photographs to produce the level-III map (Figure 10.6).

Definitions and descriptions of the level-I and level-II categories were provided by Anderson and others (1976) and are summarized here, together with comments on some level-III categories.

Urban or Built-up (100)

This category comprises areas of intensive land use where much of the land is covered by structures and streets. The range of uses is shown by the level-II and level-III categories that are assigned to the "urban or built-up" category. As urban development expands, other level-I categories ("agriculture," "forest," and "water") may be enclosed, and small patches will be included with urban areas at the level-I classification. Where these other categories occur on the fringes of urban land, they will be mapped separately, except where they are surrounded and dominated by urban development. Where criteria for more than one category are met, the "urban or built-up" category takes precedence over others. On the Landsat image (Figure 10.1) there are some residential areas with sufficient tree cover to meet the "forest" criteria, but these are classed as "urban or built-up."

On Landsat IR color composite MSS images, the "urban or built-up" category may be recognized by the following criteria:

1. There is a dense network of streets, with only the major arteries recognizable. The streets may fall into a regular grid (Chicago), a radial pattern (New Orleans), or an irregular pattern (Boston).

2. The central business and industrial section has a blue signature caused by the absence of vegetation and by the concentration of roofs and pavement.

3. The central section is surrounded by residential areas with characteristic pink and red signatures caused by landscape vegetation. Interspersed bright red patches are caused by parks, golf courses, and cemeteries.

Finer details are recognizable on Landsat TM images.
   The level-I "urban or built-up" category is subdivided into a number of level II categories, which are described below.

Residential (110) On the high-altitude aerial photograph (Figure 10.3), there are extensive residential areas, but individual structures are difficult or impossible to resolve at this scale. The fine texture, the regular pattern of closely spaced streets, and the distribution around urban core areas are keys for recognizing this category. On color and IR color photographs, lawns and trees provide useful color tones. At level II the residential areas may be separated from the various nonresidential buildings that are larger or less regular in size and shape.

At level III the "residential" category is subdivided into single-family or multiple-family occupancy. As shown in Table 10.2, single-family units are classed as low, medium, or high density based on the number of *dwelling units per acre* (DUPA). DUPA may be estimated by using a 10-acre square, as shown in Figure 10.6. Trace the square onto a transparent sheet and then position it over a residential area on the aerial photograph (Figure 10.5). Count the units within the square and divide by 10 to obtain DUPA. A 5-hectare template is provided for the metric system. For the three categories of residential density, the equivalent *dwelling units per hectare* (DUPH) are: low density (111) has less than 5 DUPH (2 DUPA), medium density (112) has 5 to 15 DUPH (2 to 6 DUPA), and high density (113) has greater than 15 DUPH (6 DUPA).

Areas of sparse residential land use, such as farmsteads and farm-labor housing, are assigned to a level-I category, such as "agriculture." Rural residential and recreational subdivisions, however, are included in the "residential" category because the land is primarily used for residences, even though it may have forest or rangeland types of cover. Housing facilities at military bases, resorts, colleges, and universities are assigned to the appropriate "institutional" (160) level-III category.

Commercial and Services (120) Facilities in this category are used predominantly for the distribution and sale of products and services. Included are facilities that support the basic uses such as office buildings, warehouses, landscaped areas, driveways and parking lots. Commercial areas may include some noncommercial areas that are too small to be mapped separately. Churches, schools, residences, and industry may be enclosed within commercial areas. Where these noncommercial areas exceed one-third of the total commercial area, the "mixed" category (180) is employed at level II. Facilities in the "commercial and services" typically occur in three distinct settings: (1) concentrations in central urban cores, (2) strips along major streets and highways, and (3) shopping centers and malls adjacent to residential areas.

Facilities in the "commercial and services" category are distinguished from residential areas (category 110)

TABLE 10.2 Land-use and land-cover classification system

Level I	Level II	Level III
100 Urban or built-up	110 Residential	111 Single unit, low-density (less than 2 DUPA*) 112 Single unit, medium-density (2 to 6 DUPA) 113 Single unit, high-density (greater than 6 DUPA) 114 Mobile homes 115 Multiple dwelling, low-rise (2 stories or less) 116 Multiple dwelling, high-rise (3 stories or more) 117 Mixed residential
	120 Commercial and services	121 Retail sales and services 122 Wholesale sales and services (including trucking and warehousing) 123 Offices and professional services 124 Hotels and motels 125 Cultural and entertainment 126 Mixed commercial and services
	130 Industrial	131 Light industrial 132 Heavy industrial 133 Extractive 134 Industrial under construction
	140 Transportation	141 Airports, including runways, parking areas, hangars, and terminals 142 Railroads, including yards and terminals 143 Bus and truck terminals 144 Major roads and highways 145 Port facilities 146 Auto parking facilities (where not directly related to another land use)
	150 Communications and utilities	151 Energy facilities (electrical and gas) 152 Water supply plants (including pumping stations) 153 Sewage-treatment facilities 154 Solid-waste disposal sites
	160 Institutional	161 Educational facilities, including colleges, universities, high schools, and elementary schools 162 Religious facilities, excluding schools 163 Medical and health-care facilities 164 Correctional facilities 165 Military facilities 166 Governmental, administrative, and service facilities 167 Cemeteries
	170 Recreational	171 Golf courses 172 Parks and zoos 173 Marinas 174 Stadiums, fairgrounds, and race tracks
	180 Mixed	
	190 Open land and other	191 Undeveloped land within urban ares 192 Land being developed; intended use not known
200 Agriculture	210 Cropland and pasture	211 Row crops 212 Field crops 213 Pasture
	220 Orchards, groves, vineyards, nurseries, and ornamental horticultural areas	221 Citrus orchards 222 Noncitrus orchards 223 Nurseries 224 Ornamental horticultural 225 Vineyards
	230 Confined feeding operations	231 Cattle 232 Poultry 233 Hogs
	240 Other agriculture	241 Inactive agricultural land 242 Other
300 Rangeland	310 Grassland	

TABLE 10.2 *(Continued)*

Level I	Level II	Level III
	320 Shrub and brushland	321 Sagebrush prairies 322 Coastal scrub 323 Chaparral 324 Second-growth brushland
	330 Mixed rangeland	
400 Forest land	410 Evergreen forest	411 Pine 412 Redwood 413 Other
	420 Deciduous forest	421 Oak 422 Other hardwood
	430 Mixed forest	431 Mixed forest
	440 Clearcut areas	
	450 Burned areas	
500 Water	510 Streams and canals	
	520 Lakes and ponds	
	530 Reservoirs	
	540 Bays and estuaries	
	550 Open marine waters	
600 Wetlands	610 Vegetated wetlands, forested	611 Evergreen 612 Deciduous 613 Mangrove
	620 Vegetated wetlands, nonforested	621 Herbaceous vegetation 622 Freshwater marsh 623 Saltwater marsh
	630 Nonvegetated wetlands	631 Tidal flats 632 Other nonvegetated wetlands
700 Barren land	710 Dry lake beds	
	720 Beaches	
	730 Sand and gravel other than beaches	
	740 Exposed rock	
800 Tundra		
900 Perennial snow or ice	910 Perennial snowfields 920 Glaciers	

Source: Modified from Anderson and others (1976) and from Florida Bureau of Comprehensive Planning (1976).

*DUPA = dwelling units per acre.

by the criteria given earlier. Where "commercial and services" facilities consist of multistory buildings or shopping centers, they are distinguishable from the "industrial" category (130). In many areas, however, warehouses and wholesale stores of the "commercial and services" category are intermixed with industrial buildings (category 130) and cannot be distinguished on high-altitude aerial photographs. The area directly southeast of downtown Los Angeles (center right margin of Figures 10.3 and 10.4) includes both commercial and industrial facilities and is classified as "mixed" (180).

At level III many of the "commercial and services" categories cannot be interpreted from images without the aid of ground information. The "mixed" category (126) is employed where more than one-third of the area is occupied by uses other than the predominant category.

Industrial (130) This category designates manufacturing facilities, which are separated at level III into the "light industrial" (131) and "heavy industrial" (132) categories. Light industries are those that design, assemble, finish, process, and package products. Many light in-

FIGURE 10.1 Landsat MSS band-5 image of Los Angeles region, acquired March 7, 1976. Scale is 1:1,000,000.

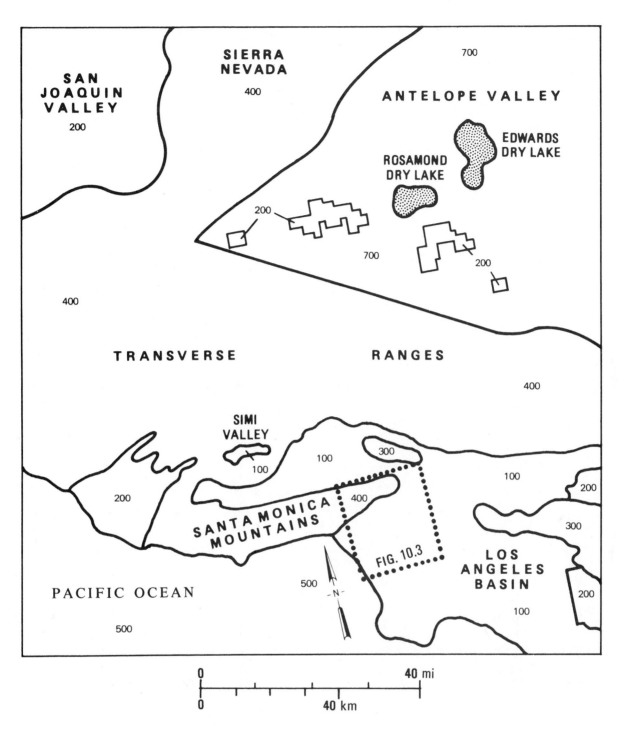

FIGURE 10.2 Level-I land-use classification map for Figure 10.1. Numbers designate categories explained in Table 10.2.

FIGURE 10.3 High-altitude photograph of the western portion of the Los Angeles area, acquired by NASA aircraft, September 2, 1979. Scale is 1:120,000.

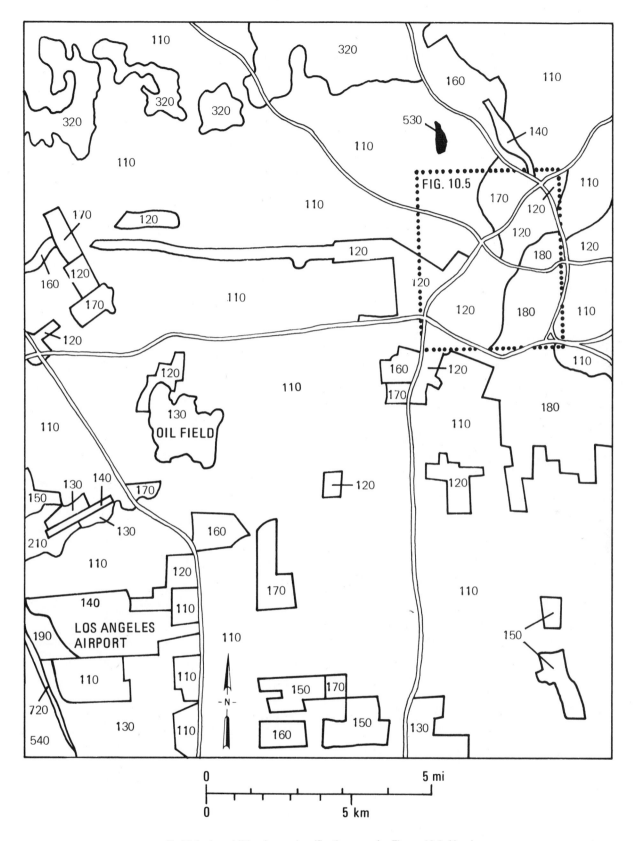

FIGURE 10.4 Level-II land-use classification map for Figure 10.3. Numbers designate categories explained in Table 10.2.

FIGURE 10.5 Medium-altitude photograph of central Los Angeles acquired October 25, 1972. Scale is 1:30,000.

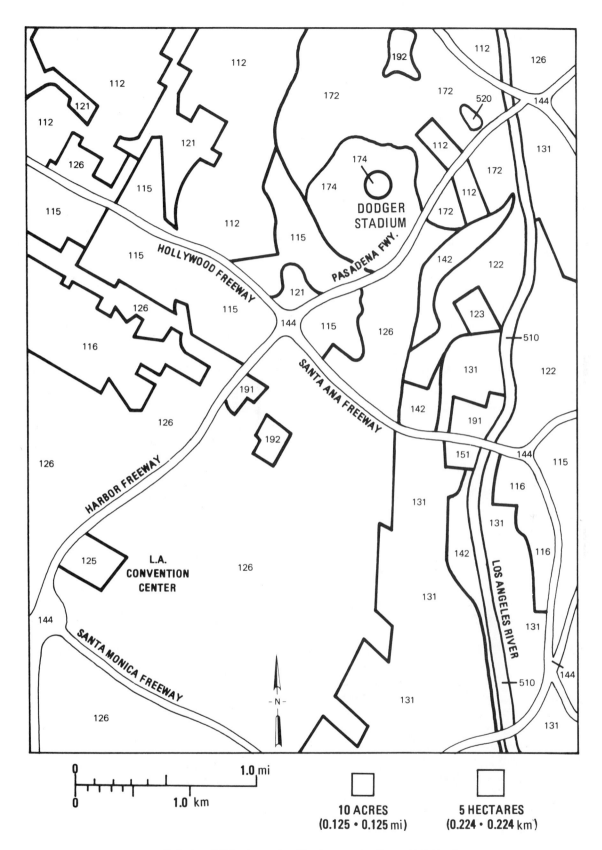

0 ————————————— 1.0 mi
|‒‒‒‒‒‒‒‒‒‒‒‒‒‒‒‒‒‒‒‒‒‒‒|
0 ————————————— 1.0 km

☐ **10 ACRES**
(0.125 • 0.125 mi)

☐ **5 HECTARES**
(0.224 • 0.224 km²)

FIGURE 10.6 Level-III land-use classification map for Figure 10.5. Numbers designate categories explained in Table 10.2.

dustries are concentrated in industrial parks, which may be located adjacent to airports, residential areas, or in open country.

Heavy industries use raw materials such as timber, iron ore, and crude oil. Pulp and lumber mills, steel mills, oil refineries and tank farms, chemical plants, and brickmaking plants are typical facilities. Food-processing plants (canneries, grain storage, and milling) are included in the "heavy industrial" category. Heavy industries are recognized on images by stockpiles of raw materials, waste disposal areas, and transportation facilities for bulk shipments. The oil refinery adjacent to the coast in the southwest corner of Figure 10.3 is recognizable by the numerous round storage tanks.

Extractive facilities such as mines, quarries, oil fields, sand pits, and gravel pits belong to the level-III subdivision (category 133) of the "industrial" category. Some classification systems assign extractive facilities to the "barren land" (700) category, but this assignment ignores the industrial aspect of extractive activities.

Transportation (140)   At level II, only the largest transportation facilities, such as airports, seaport facilities, and major highways are recognizable, as shown in Figures 10.3 and 10.4. At this level, most of the transportation facilities are included with the associated level-II categories, such as 120 or 130.

More detailed interpretation of transportation facilities is possible at level III, but only the larger areas are mapped. On the conventional aerial photograph (Figure 10.5), it is possible to recognize railroad yards and terminals (category 142) that could not be distinguished on the high-altitude photograph.

Communications and Utilities (150)   At level II this category is difficult to recognize. At level III, however, the categories shown in Table 10.2 are generally recognizable. The "energy facilities" category (151) includes electricity generating plants, natural-gas storage tanks, and related facilities. This category does not include coal mines or oil and gas fields, which belong to the "extractive" category. Oil refineries belong to the "heavy industrial" category (132). On Figures 10.5 and 10.6, energy facilities are recognizable from the storage tanks for natural gas. Transmission lines for electricity and telephones as well as pipelines for gas, oil, and water are narrow and rarely constitute a dominant land use; therefore, these are not assigned a separate category.

Institutional (160)   The level-III categories listed in Table 10.2 describe the facilities assigned to the "institutional" subdivision. A few of the categories are recognizable by their characteristics on stereo aerial pho-

tographs. Many educational facilities may be recognized by the associated athletic facilities (ball fields and tracks). Cemeteries are recognizable by the expanses of lawns with driveways and absence of other facilities. Supplemental ground information is generally required to identify the remaining level-III institutional categories.

Recreational (170)   Recreational land uses include the level-III categories listed in Table 10.2. One problem in classification is that the "golf courses" and "parks and zoos" categories (171 and 172) may be confused with the open landscaped areas associated with some institutional facilities (category 160). On the Los Angeles images, Dodger Stadium and the surrounding parking lots are recognizable at both level II (top right of Figure 10.3) and level III (top center of Figure 10.5).

Mixed (180)   This category is used for areas where the preceding level-II categories (110 through 170) cannot be separated at the mapping scale. Where more than one-third of an area has an intermixture of two or more uses, the area is classified as "mixed." Where the intermixed uses occupy less than one-third of the area, the dominant land-use category is used.

Open Land and Other (190)   In many areas belonging to the "urban and built-up" category, there are tracts of vacant land not being used for any of the preceding activities. These tracts are not "rangeland" (category 300); therefore they are classified as "undeveloped land within urban areas" (191) or as "land being developed; intended use not known" (192). An example of category 192 is shown in Figure 10.5 where the bright-toned area 1 km north of Dodger Stadium has been graded and terraced but the ultimate use is not apparent. Elsewhere the ultimate use of land being developed may be inferred from the images. For example, open land adjacent to residential areas that is being graded and laid out with a street grid can be assigned to the "residential" category (110).

Agriculture (200)

In some parts of the world, one can readily recognize land used for crops and orchards on satellite images by rectangular or circular patterns. The cultivation pattern is clearly shown in the San Joaquin Valley, in the northwest portion of the Landsat image (Figure 10.1). Rectangular dark patches within the Antelope Valley indicate areas where the desert is being reclaimed for irrigated farming. Circular areas of centerpoint irrigation are also easily recognized. In much of Canada and the United States, agricultural fields are subdivisions of the basic land survey unit, which is a section 1 mi square containing 640 acres. In Russia, however, wheat fields are typically larger than the 40- and 160-acre fields of the United States and Canada. In other regions, such as India and southeast Asia, individual fields are too small to be recognized on MSS images and the patterns do not form a regular grid. Agricultural land (category 200) is distinguishable from urban land (category 100), which has indicators of population concentrations.

Cropland and Pasture (210) This category includes a wide variety of crops, which are subdivided at level III into three categories. Row crops (category 211) are distinguished by their pattern from field crops (category 212), such as wheat and alfalfa, which uniformly cover the area. "Pasture" (category 213) refers to relatively small areas of grazing land commonly interspersed with croplands.

Ground-survey information is generally needed to identify specific crops. However, experienced interpreters who are familiar with a region and its crop cycles may derive accurate crop estimates from images alone. Crop types and yields have been estimated by digital processing of MSS images acquired at several dates during the growing season. The Large Area Crop Inventory Experiment (LACIE), sponsored by NASA and the U.S. Department of Agriculture, developed this technique to predict wheat production in major growing areas of the world during the late 1970s (Myers and others, 1983, p. 2200).

Orchards, Groves, Vineyards, Nurseries, and Ornamental Horticultural Areas (220) This aggregate category is employed at level II, and the individual components listed in Table 10.2 are used at level III. Knowledge of the region and ground information are generally needed to recognize most of the level-III categories.

Confined Feeding Operations (230) Stockyards, hog and cattle feedlots, confined dairy feeding operations, and large poultry farms constitute this category of land use. In spite of the relatively small areas occupied, waste products from concentrations of these animals cause environmental problems that justify a separate category. On high-altitude images, these facilities are recognized by the built-up appearance, access paths, waste-disposal areas, and lack of vegetation.

Other Agriculture (240) "Inactive agricultural lands" (241) is a major level-III subdivision of "other agriculture." Also at level III, the "other" category (242) is used for farmsteads, corrals, and other relatively small areas associated with the major agricultural activities.

Rangeland (300)

Rangeland is land covered by natural grasses, shrubs, and forbs, which include nonwoody plants such as weeds and flowers. Rangeland is capable of supporting native or domesticated grazing animals. Some rangelands have been modified by eradicating nonproductive plants, such as sagebrush and mesquite, and by planting grasses.

Grassland (310) This category has no level-III subdivisions but includes a wide range of grass types that were summarized by Anderson and others (1976).

Shrub and Brushland (320) A wide range of plant communities make up this category. "Sagebrush prairies" (category 321) includes semiarid lands that also support shadscale, greasewood, and creosote bush. The "coastal scrub" category (322) includes extensive areas in southern Texas and Florida that do not qualify as wetlands. "Chaparral" (category 323) refers to a dense growth of evergreen shrubs that include manzanita, mountain mahogany, and scrub oaks. "Second-growth brushland" (category 324) occurs widely in the eastern United States and consists of former croplands or pastures (cleared from original forest land) that have now grown up in brush.

Mixed Rangeland (330) When more than one-third of an area is an intermixture of grassland with shrub and brushland, the area is classified as "mixed rangeland."

400 Forest Land

Forested lands have a *crown density* (also called the *crown closure percentage*) of 10 percent or more and support trees capable of producing timber or other wood products. The "forest" category also includes lands from which trees have been removed to a crown density of less than 10 percent but have not been developed for other uses. Lumbering, fire, and disease are some of the agents that reduce crown density. On the Landsat image (Figure 10.1), forested lands occur in the mountains of the Transverse Ranges and the southern end of the Sierra Nevada. In this and other regions, there is a transition from forest at higher elevations to brush of the "rangeland" category (300) at lower elevations. Without additional information, however, it is usually difficult to separate these two categories.

Evergreen forests (category 410) are predominantly trees that remain green all year. Deciduous forests (category 420) are predominantly trees that seasonally lose their leaves. Where more than one-third of an area is a mixture of deciduous trees with evergreen trees, the area is classified as "mixed forest" (430). "Clearcut areas" (440) and "burned areas" (450) are level-II categories that are recognizable on high-altitude aerial photographs. In some areas the system may be expanded to recognize planted and cultivated forests.

Water (500)

Few comments are needed for this category. Lakes and ponds (category 520) are natural water bodies, whereas reservoirs (category 530) are artificially impounded.

Wetlands (600)

Most wetlands are located adjacent to water bodies and include marshes, swamps, tidal flats, and many river floodplains. Areas that are only seasonally flooded and do not support typical wetland vegetation are assigned to other categories. Cultivated wetlands such as rice fields and developed cranberry bogs are classified as agricultural land (category 200). Wetlands from which uncultivated products such as timber are harvested or that are grazed by cattle are retained in the "wetlands" category. Shallow water areas covered by floating vegetation (such as water lilies and water hyacinths) are classed as wetlands, but where aquatic vegetation is submerged the area is classed as water (category 500).

Vegetated Wetlands, Forested (610) Management and environmental planning requirements for wetlands are very different from those for dry land. For this reason, forested land occurring in swamps, marshes, and seasonally flooded bottomland is assigned to the "wetlands" category (600) rather than to "forest land" (400). The forested wetlands are divided into the level-III categories of "evergreen" (611), "deciduous" (612), and "mangrove" (613).

Vegetated Wetlands, Nonforested (620) Nonforested wetlands are divided into the level-III categories of "herbaceous vegetation" (621), "freshwater marsh" (622), and "saltwater marsh" (623). The "herbaceous vegetation" category includes nonwoody plants such as grasses, sedges, rushes, mosses, waterlilies, and water hyacinths.

Nonvegetated Wetlands (630) Tidal flats are the major category of nonvegetated wetlands.

Barren Land (700)

Barren land has a limited ability to support life, and less than one-third of the area has vegetation cover. The surface is predominantly thin soil, sand, or rocks. Any vegetation present is more scrubby and widely spaced than that in the "shrub and brushland" category (320). Land that is barren because of human activities (plowed agricultural land, clearcut forested land, and waste and tailings dumps) is assigned to the associated use category rather than the "barren land" category. On the Landsat image (Figure 10.1), the Antelope Valley is typical of barren land in the Mojave Desert. The valley margins are covered by alluvial fans made up of gravel eroded from the adjacent mountains. Sand, thin soil, dry lake beds, and rock exposures cover the central portion of the valley. In this example, level-II categories are readily mapped on an image at the scale of level I.

Dry Lake Beds (710) In many arid regions the floors of closed valleys are covered by salt and silt deposited in lakes that are normally dry (playas or salt flats). These deposits are readily identified on images by their location and distinctive bright tone.

Beaches (720) Beaches are the deposits of sand and gravel along shorelines.

Sand and Gravel Other than Beaches (730) The sand of this category occurs as windblown sheets and dune fields. Mixtures of sand and gravel occur in floodplains and as barren alluvial fans surrounding mountains in arid areas.

Exposed Rock (740) This category includes area of bedrock exposure, volcanic deposits, and talus. When these rock types occur in tundra areas, they are assigned to the "bare-ground tundra" category (830).

Tundra (800)

"Tundra" refers to treeless regions beyond the limit of the boreal forest. Regions above the timberline in high mountain ranges are also classed as tundra. Tundra vegetation is low and dwarfed and it commonly forms a complete mat. The presence of permafrost and the prevalence of subfreezing temperatures most of the year are characteristics of tundra. Where these conditions prevail, areas that would otherwise be assigned to "wetlands" (such as the Arctic Coastal Plain, Alaska) or to "barren land" (such as the Brooks Range, Alaska) are assigned to the "tundra" category (800). Late summer images are best for interpreting tundra categories.

Perennial Snow or Ice (900)

This category is used for areas where snow and ice accumulations persist throughout the year as snowfields or glaciers.

Perennial Snowfields (910) Perennial snowfields are accumulations of snow and firn (coarse, compacted granular snow) that did not entirely melt during previous summers. Snowfields lack flow features, which distinguishes them from glaciers.

Glaciers (920) This category of flowing ice consists of both ice caps and valley glaciers. Flowage is indicated by crevasses and by lateral and medial moraines that form dark streaks against the bright background of the glacial ice. The major problem in recognizing glaciers is the presence of snow that may obscure the moraines and crevasses.

SEASAT IMAGE OF PHOENIX, ARIZONA

The land-use and land-cover classification system was established primarily for use with Landsat images and aerial photographs, but may also be used with other remote sensing images. Figure 10.7A is a Seasat image of the city of Phoenix and vicinity in central Arizona acquired August 2, 1978. Figure 10.7B is a Landsat RBV image acquired eight days later. Figure 10.8 is a map that shows level-II categories of land use and land cover. Phoenix is located in a broad desert valley partly surrounded by mountain ranges. Much of the desert was originally reclaimed for irrigated agriculture, but urban areas are expanding and replacing the croplands. Kozak, Berlin, and Chavez (1981) originally described the Seasat image.

The "barren land" category is represented by sand and gravel (730) and exposed rock (740). In the Seasat image (Figure 10.7A), sand has a dark signature because the fine grain size forms smooth surfaces. Gravel occurs at the foot of the mountains and in the channels of the rivers and washes, all of which were dry when the images were acquired. Gravel has an intermediate to bright Seasat signature because of its coarse grain size and rough surface. Native trees and brush growing along the drainage channels also contribute to their bright radar signatures. The "exposed rock" category (740) is represented by the mountain ranges that are identified in Figure 10.8. The look direction for the Seasat image is toward the northeast; therefore the southwest-facing slopes have bright highlights and the northeast-facing slopes have dark shadows. Radar layover (described in Chapter 6) causes the mountain crests to be displaced toward the west.

The cities and towns belong to the "urban, mixed" category (180) and have intermediate to bright signatures on the Seasat image. One would expect the radar signature of the urban areas to be brighter than it is in Figure 10.7A because of the abundant corner reflectors in urban environments. The reduced brightness is explained by the orientation of the street patterns and buildings relative to the Seasat look direction. Street patterns in this Arizona image are oriented north-south and east-west; therefore the northeast Seasat look direction is oblique to most buildings, which reduces the intensity of radar backscatter. This orientation effect was described in Chapter 6. The retirement community of Sun City (Figure 10.8) northwest of Phoenix is an exception to the orthogonal street pattern, because the streets are laid out in curved and circular patterns. The bright patches within Sun City (Figure 10.7A) are caused by groups of houses oriented with some walls normal to the northeast radar look direction.

The dark linear features in the urban areas are major highways and irrigation canals. Airport runways (Figure 10.8) also have dark signatures. The narrow irregular dark features in Sun City are golf courses and parks with smooth lawns.

The irrigated agricultural areas surrounding Phoenix belong to the major categories of "cropland and pasture" (210); and "orchards, groves, vineyards, nurseries, and ornamental horticulture" (220). In the Seasat image, the fields east of Phoenix have a distinctly darker tone than the fields west of Phoenix. Charles Farr of the University of Arizona Agriculture Extension Service (personal communication) provided the following information on agricultural conditions in the Phoenix area in 1978, when the Seasat and Landsat images were acquired.

Irrigation water is relatively expensive in the areas east of Phoenix, where cotton, wheat, and alfalfa are the major crops. Water is less expensive in the western area, where vegetables, citrus fruits, and grapes are grown along with some cotton and wheat. The relatively bright Seasat signatures of the western area may be caused by the crop types and more intensive irrigation, which results in higher soil moisture and brighter radar signatures.

DIGITAL CLASSIFICATION

In preceding sections of this chapter, the land-use classification system was employed for interpreting images visually. One can also classify land use and land cover by employing the digital methods of image processing described in Chapter 7. The following sections illustrate two applications of image processing:

1. A supervised digital classification of Landsat MSS data of an agricultural area (Oxnard Plain, California)

2. An unsupervised digital classification of Landsat TM data of an urban area (Las Vegas, Nevada)

A. SEASAT IMAGE (23.5 cm) ACQUIRED AUGUST 2, 1978.

B. LANDSAT RBV IMAGE (0.50 TO 0.75 μm) ACQUIRED
AUGUST 10, 1978.

FIGURE 10.7 Seasat and Landsat RBV images of Phoenix and vicinity, Arizona.

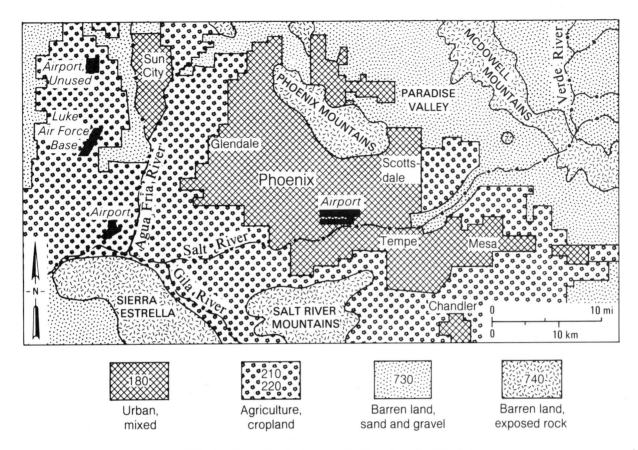

180	210 220	730	740
Urban, mixed	Agriculture, cropland	Barren land, sand and gravel	Barren land, exposed rock

FIGURE 10.8 Land-use classification map for Phoenix and vicinity, Arizona. Compare with images of Figure 10.7.

Agricultural Land Use, Oxnard Plain, California

Some of the best agricultural land in California is located adjacent to cities. Rapid expansion of cities is reducing the state's agricultural base. At one time California considered legislation to protect prime agricultural land from urban expansion. An early step was to evaluate the feasibility of inventorying agricultural land by digital processing of Landsat data. One study area was the Oxnard Plain, a fertile farming area on the California coast 100 km west of Los Angeles. The Landsat MSS image (Plate 15A) shows the vegetated fields in various red tones and the city of Oxnard in blue. The supervised classification map (Plate 15B) was prepared by Estes and others (1979) of the University of California at Santa Barbara using the approach described in Chapter 7. The classification method recognized level-II and level-III categories. Colors and corresponding land-use categories for the classification map are as follows:

Red	Commercial and industrial (categories 120 and 130)
Yellow	Residential (110)
Brown	Shrub and brushland (320)
Dark green	Orchards (220)
Medium green	Field crops (212)
Light blue	Bare fields (241)
Dark blue	Open marine water (550)
White	Sand and gravel (720 and 730)

This example shows the feasibility of classifying land use in a predominantly agricultural area. By repeating the analysis on images acquired at intervals of several years, one can monitor changes in land use.

The Oxnard example is only a small portion of one Landsat scene. In the late 1970s the U.S. Bureau of

Land Management needed to develop a land-use plan for federal lands in the Mojave Desert which amounted to tens of thousands of square kilometers and required a number of Landsat scenes for coverage. The Landsat images were classified and provided a base for land-use planning.

Urban Land Use, Las Vegas, Nevada

Plate 15C is a Landsat TM infrared color image of much of Las Vegas, Nevada, and vicinity. The central part of the city is the dark area in the northwest part of this image. A runway of McCarran Airport is visible in the southwest corner. The east and southeast margins of the image are desert. Much of the image is a mixture of residential areas, commercial developments. and various types of landscape vegetation. The unsupervised classification map (Plate 15D) was prepared from the six TM bands of visible and reflected IR data. The classification produced 16 classes. The analyst combined 7 of these with other classes to produce the nine categories listed below, with their display colors in Plate 15D:

Violet	Residential (category 110)
Orange	Commercial (120)
Black	Streets and parking lots (144 and 146)
Gray	Construction sites (192)
Blue	Open land (191)
Dark green	Irrigated vegetation (170 and 210)
Medium green	Mixed rangeland (330)
Light green	Shrub and brushland (320)
Yellow	Sand and gravel (730)

Agricultural areas, as in the Oxnard Plain, typically consist of homogeneous fields many hectares in area that are well suited for the 79-by-79-m ground resolution cell of Landsat MSS. Urban areas, however, are much more diverse even at the higher resolution of Landsat TM. A typical suburban residential lot is approximately the size of one TM 30-by-30-m ground resolution cell. The lot will include some or all of the following materials: trees and shrubs, lawns, paving and sidewalks, roofs, and water for a swimming pool. For such a cell, the digital numbers of the TM bands for that pixel are a composite of the spectral reflectance of the various materials. Despite these problems the Landsat classification map (Plate 15D) clearly portrays the major categories of land use and land cover.

Business and commercial activities (orange) are concentrated along major streets and include streets and parking lots (black). Residential areas (violet) occur as rectangles adjacent to commercial areas and in outlying areas. Three types of vegetation were classified. Irrigated vegetation (dark green) occurs in golf courses, parks, and irrigated agriculture; in the IR color image (Plate 15C) this category has a bright red signature. Mixed rangeland (medium green) includes areas where houses and vegetation are intermingled in approximately equal proportions. Shrub and brushland (light green) occurs largely around the outskirts of Las Vegas. Sand and gravel (yellow) is unvegetated and has undergone little human disturbance. Open land (blue) has scattered vegetation and isolated buildings. Construction sites (gray) are a minor but distinctive category of open ground that is thoroughly disturbed by building activity. A good example occurs in the northeast corner of the images (Plate 15C,D).

ARCHAEOLOGY

Archaeologic patterns of prehistoric land use may be recognized in remote sensing images. In Europe many towns and roads of Roman and earlier times are now covered by agricultural fields. These sites commonly cause differences in character, moisture content, and vegetation cover of the overlying soils that are recognizable in IR color photographs.

In northern Arizona, cornfields cultivated by prehistoric Indians have been recognized in the pattern of alternating warm and cool strips of land in daytime thermal IR images. Berlin and others (1977) have established through field investigation that the warm strips correlate with bands of volcanic ash used as mulch by the Indian farmers. The low-density ash has lower thermal inertia than adjacent soil and therefore warmer daytime radiant temperatures.

Archaeologists once thought that the Mayan civilization in Yucatan practiced a slash-and-burn system of agriculture, but this system could not have supported the estimated Mayan population. Aerial reconnaissance flights and ground investigations of the region were hampered by swampy terrain with dense vegetation cover and persistent cloud cover. Aircraft radar surveys (Adams, Brown, and Culbert, 1981) revealed an extensive network of canals by which the Maya drained the swamps for cultivation. The region is now overgrown with jungle, but on the radar images the canals are expressed as narrow gray lines surrounded by bright signatures of tropical vegetation. The canals are also covered by vegetation, but the canopy height is lower there than the surrounding area, which accounts for the darker signature.

COMMENTS

The world's population is increasing at a progressively faster rate with attendant changes in land use and land cover. Agricultural areas are being converted to urban uses; forest lands are being stripped for timber or conversion to agriculture. Although many of these changes are detrimental to the overall environment, it is highly unlikely that the process will halt. The alternative is to plan and regulate the changes to minimize negative impacts on the environment. Planning requires an accurate inventory of existing patterns of land use and land cover. Remote sensing images provide a database for such inventories at scales ranging from regional to local.

The classification system of Table 10.2 is designed for use with remote sensing data at different levels of spatial resolution. The system includes all major categories of land use and land cover and can be expanded for special situations. The system is also applicable to digital methods of image processing such as supervised and unsupervised classifications. Digital processing is becoming essential because as spatial resolution and spectral coverage of images increase, the volume of data also increases. For example, a Landsat TM image consists of 244×10^6 pixels, which is almost an order of magnitude greater than in a Landsat MSS image. Images acquired on different dates may be digitally registered and compared to produce change-detection images that emphasize changes in land use and land cover.

QUESTIONS

1. For your area, assemble a set of images for levels I, II, and III (images from Landsat TM and MSS, a photograph from the National High Altitude Photography Program, and a larger-scale aerial photograph). Use Table 10.2 and prepare land-use maps for each of the three levels.

2. The kinds of images mentioned in question 1 are all acquired in the visible and reflected IR bands. Images acquired in other spectral bands may also be employed to analyze land use. Examine the thermal IR image of Ann Arbor, Michigan (Chapter 5), and the radar images of New Orleans and the Imperial Valley, California (Chapter 6). Evaluate the advantages and disadvantages of these images relative to visible-band images for land-use interpretation.

3. For your area, assume that the population will double in the next five years. Use the images and maps of question 1 as a base and develop a plan for adding this population. You must provide for all the facilities required for this growth.

REFERENCES

Adams, R. E. W., W. E. Brown, and T. P. Culbert, 1981, Radar mapping, archaeology, and ancient Maya land use: Science, v. 213, p. 1457–1463.

Anderson, J. R., E. T. Hardy, J. T. Roach, and R. E. Witmer, 1976, A land use and land cover classification system for use with remote sensor data: U.S. Geological Survey Professional Paper 964.

Berlin, G. L., J. R. Ambler, R. H. Hevly, and G. G. Schaber, 1977, Identification of a Sinagua agricultural field by aerial thermography, soil chemistry, pollen/plant analysis, and archaeology: American Antiquity, v. 42, p. 588–600.

Estes, J., L. Tinney, D. Stow, and S. Norman, 1979, Conceptual design and preliminary demonstration of a prime agricultural land component of an integrated remote sensing system: Geography Remote Sensing Unit, University of California, Santa Barbara, Calif.

Florida Bureau of Comprehensive Planning, 1976, The Florida land use and cover classification system: Florida Bureau of Comprehensive Planning Report DSP-BCP-17-76, Tallahassee, Fla.

Kozak, R. C., G. L. Berlin, and P. S. Chavez, 1981, Seasat radar image of Phoenix, Arizona region: International Journal of Remote Sensing, v. 2, p. 295–298.

Myers, V. I., and others, 1983, Remote sensing applications in agriculture *in* Colwell, R. N., ed., Manual of remote sensing, second edition: ch. 33, p. 2111–2228, American Society for Photogrammetry and Remote Sensing, Falls Church, Va.

ADDITIONAL READING

Cooley, M. E., and R. M. Turner, 1982, Application of Landsat products in range- and water-management problems in the Sahelian zone of Mali, Upper Volta, and Niger: U.S. Geological Survey Professional Paper 1058.

Ebert, J. I., and others, 1983, Archaeology, anthropology, and cultural resources management *in* Colwell, R. N., ed., Manual of remote sensing, second edition: ch. 26, p. 1233–1304, American Society for Photogrammetry and Remote Sensing, Falls Church, Va.

Ellis, M. Y., ed., 1978, Coastal mapping handbook: U.S. Government Printing Office, Washington, D.C.

Jensen, J. R., and others, 1983, Urban/suburban land use analysis *in* Colwell, R. N., ed., Manual of remote sensing, second edition: ch. 30, p. 1571–1666, American Society for Photogrammetry and Remote Sensing, Falls Church, Va.

Loelkes, G. L., G. E. Howard, E. L. Schwartz, P. D. Lambert, and S. W. Miller, 1983, Land use/land cover and environmental photointerpretation keys: U.S. Geological Survey Professional Paper 1600.

Pettinger, L. R., 1982, Digital classification of Landsat data for vegetation and land-cover mapping in the Blackfoot River watershed, southeastern Idaho: U.S. Geological Survey Professional Paper 1219.

Reed, W. E., and J. E. Lewis, 1978, Land use and land cover

information and air-quality planning: U.S. Geological Survey Professional Paper 1099–B.

Robinove, C. J., 1979, Integrated terrain mapping with digital Landsat images in Queensland, Australia: U.S. Geological Survey Professional Paper 1102.

Wronski, W., and K. J. Davies, 1972, Photointerpretation for planners: Kodak Publication M-81, Rochester, N.Y.

Natural Hazards

Earthquakes, landslides, volcanic eruptions, and floods are natural hazards that kill thousands of people and destroy billions of dollars of property each year. These losses will increase as world population increases and more people reside in areas that are subject to these hazards. Dams can control flood hazards, and proper engineering design can reduce landslide risks. Aside from steps such as these, there is little that people can do to prevent the occurrence of natural hazards. However, the following actions will minimize their effects:

1. Analyze the risk that natural hazards will occur in a given area. One example is to identify volcanoes that have the potential for eruption. In addition to recognizing hazards, risk analysis should delineate areas on the basis of their relative susceptibility to damage.

2. Provide warning in advance of specific hazardous events. The Chinese, for example, successfully predicted a major earthquake in 1975 and were able to minimize casualties. The United States and Russia are also conducting research programs for earthquake prediction. Volcanic eruptions in Hawaii and elsewhere have been predicted on the basis of ground movements.

3. Assess the damage caused by a hazardous event. An early evaluation of damage caused by floods and

earthquakes is essential for carrying out rescue, relief, and rehabilitation efforts.

Various forms of remote sensing have been used for these activities. Use of remote sensing will undoubtedly increase as we improve our ability to interpret and understand the imagery.

EARTHQUAKES

Earthquakes are caused by the abrupt release of strain that has built up in the earth's crust. Most zones of maximum earthquake intensity and frequency occur at the boundaries between the moving plates that form the crust of the earth. Earthquakes are common along the San Andreas fault that forms part of the boundary between the North America and Pacific plates. Major earthquakes also occur within the interior of crustal plates such as those in the southeast United States, China, and Russia. The science of *earthquake prediction* is just developing, and remote sensing seems to have limited application for predicting specific earthquakes, or *seismic events*. Remote sensing, however, is very useful for *seismic risk analysis*, which estimates the geographic distribution, frequency, and intensity of seismic activity without attempting to pinpoint specific earthquakes. This analysis is essential for locating and designing dams, power plants, and other projects in seismically active areas.

One method of seismic risk analysis is based on the study of instrumentally recorded earthquakes and historic records of earthquakes. The instrumental records and even the historical records, which cover some 2000 years in Japan and 3000 years in China, do not cover enough time for valid extrapolations of future earthquakes.

The second method of seismic risk analysis is based on the recognition of *active faults*, which are defined as breaks along which movement has occurred in Late Quaternary, or Holocene time (past 11,000 years). Remote sensing analyses and field studies of active faults provide a geologic record that extends our instrumental and historic records. Surface faulting during large shallow earthquakes is more universal than has been recognized; analysis of this geomorphic evidence and radiometric age dating of earlier events are two techniques that have not been fully utilized (Allen, 1975, p. 1041). California, Alaska, China, and Pakistan provide the following examples of remote sensing images for seismic risk analysis.

California

The following general relationships between faulting and earthquakes in California were noted by Allen (1975, p. 1043).

1. Virtually all large earthquakes (magnitude greater than 6.0 on the Richter scale) have resulted from ruptures along faults that had been recognized by field geologists prior to the events.

2. All of these faults have a history of earlier displacements in Quaternary and possibly Holocene times.

3. All the earthquakes have been relatively shallow, not exceeding about 20 km in depth. Most earthquakes larger than magnitude 6.0 have been accompanied by surface faulting, as have many of the smaller events.

4. The larger earthquakes have generally occurred on the longer faults, although there has been sufficiently wide variation to indicate caution in blindly applying any single formula for this relationship.

5. Generally only a small segment of the entire length of a fault zone has broken during any single earthquake, although there are some conspicuous and significant exceptions.

Comparing the Landsat MSS image of the Los Angeles region (Plate 4) with the map of faults having Quaternary displacements (Figure 11.1A) shows that most of these faults are evident on the Landsat image. Additional faults are recognizable in TM images because of their higher spatial resolution. The map of major earthquakes in southern California between 1912 and 1974 (Figure 11.1B) shows that, with few exceptions, these have occurred in the areas of Quaternary faulting. Even the smaller earthquakes (not shown on Figure 11.1B) generally follow the same trend (Allen, 1975, p. 1043). The lack of seismic events during most of this century in the Death Valley region and along the Garlock fault and central segment of the San Andreas fault is only a temporary pause, for these faults are marked by surface features characteristic of active faults.

Evidence for Holocene movement includes: (1) historic earthquakes, (2) rock units younger than 11,000 years that are faulted, and (3) certain topographic features caused by faulting. Typical topographic features associated with active strike-slip faults are shown on the diagram of Figure 11.2 and on the Skylab image of Figure 11.3. These features are formed by repeated horizontal and vertical displacements of a few centimeters or meters along the fault. As opposing fault blocks slide laterally along numerous individual breaks, some blocks sink to form depressions that fill with water and are called sag ponds. In arid environments these are dry lakes. Topographic ridges may be shifted laterally to shut off drainage channels and are called shutter ridges. Other narrow fault blocks are called benches and linear ridges. Scarps are the surface expression of fault planes; they may cut a topographic ridge to form a faceted ridge. A fault may block the movement of groundwater, causing it to emerge at the surface as a spring. Linear valleys result from increased erosion of fractured rocks along a fault. Offset drainage channels are especially significant because they also indicate the sense and amount of displacement along the fault. The left-lateral, strike-slip displacement of the Garlock fault is indicated by the offset drainage channels in Figure 11.3. Preservation of these topographic features is evidence for active faulting; had the features formed before the Holocene epoch, most of them would have been obliterated by erosion and deposition. As illustrated in Figure 11.2, a fault zone is comprised of active and inactive faults; the entire belt of faults may be hundreds of kilometers in length and several kilometers in width. A *fault trace* is the surface expression of an individual active fault.

Landsat MSS images are excellent for recognizing the continuity and regional relationships of faults. The higher spatial resolution of Landsat RBV and TM images records many of the topographic features indicative of active faulting. Stereo viewing of aircraft and LFC photographs provides detailed information on geomorphic features formed by faulting. Thermal IR images of arid and semiarid areas may record the presence of active

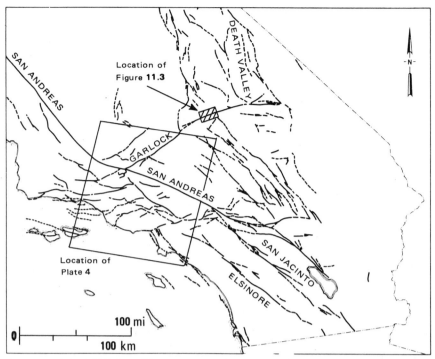

A. FAULTS WITH QUATERNARY DISPLACEMENTS. AREAS SHOWN
BY PLATE 4 AND FIGURE 11.3 ARE OUTLINED.

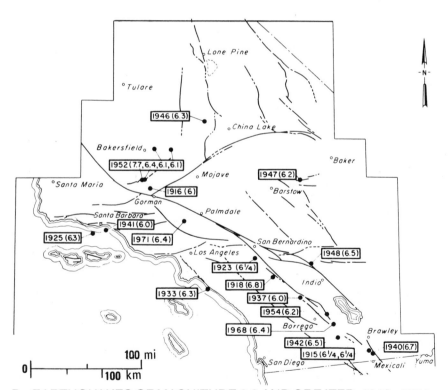

B. EARTHQUAKES OF MAGNITUDE 6.0 AND GREATER, 1912–1974.

FIGURE 11.1 Relationship of Quaternary faulting to earthquakes in southern
California. From Allen (1975, Figures 3 and 4).

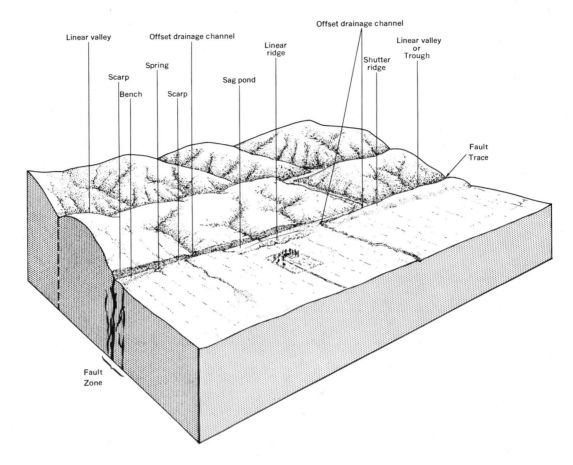

FIGURE 11.2 Generalized diagram of topographic features indicating active faults. From Vedder and Wallace (1970).

faults with little or no surface expression, such as the San Andreas and Superstition Hills faults, which were interpreted in Chapter 5. The highlighting and shadowing effects on low-sun-angle aerial photographs can emphasize topographic scarps associated with active faults, such as in the Carson Range, Nevada (Chapter 2). Radar images also emphasize subtle features along active fault zones, as shown in the SIR-A image of the Sorong fault in Irian Jaya, Indonesia (Chapter 6).

Alaska

Mosaics of Landsat images are used to analyze regional relationships of earthquake *epicenters* (the point on the earth's surface directly above the initial point of rupture) to faults and lineaments. Known faults are well expressed on the Landsat MSS mosaic of the Cook Inlet region in Alaska (Figure 11.4) together with numerous lineaments that had not previously been mapped (Figure 11.5). Gedney and Van Wormer (1974) plotted the epicenters of earthquakes with magnitudes (*M*) of 4 or greater

that occurred in 1972 and noted the following relationships:

1. Epicenters in the southern part of the area bear little relationship to faults and lineaments. These seismic events were not caused by near-surface faulting but were deep-seated earthquakes related to the subduction zone between the Pacific and North America plates.

2. A few earthquakes are associated with the Denali fault, particularly near Mt. McKinley. There is an alignment of epicenters along the Lake Clark fault.

3. Most of the Landsat lineaments are sites of one or more epicenters and probably should be classified as active faults. The epicenter of the 1964 Good Friday earthquake was very close to earthquake epicenters A and B along the lineament in the east-central part of Figure 11.5, although it is not clear whether this lineament relates to the 1964 earthquake. The sharp escarpment along the west flank of the Kenai Moun-

FIGURE 11.3 Skylab earth-terrain-camera photograph of the Garlock fault, California. The following features indicate that the fault is active: D = depression; SR = shutter ridge scarp; OC = offset channel; LR = linear ridge; LV = linear valley; FR = faceted ridge. From Merifield and Lamar (1975, Figure 2). Courtesy P. M. Merifield, California Earth Science Corporation.

tains passes just east of Anchorage and was the site of three 1972 earthquakes.

In the central interior of Alaska, epicenters of the largest earthquakes occur at the intersections of lineaments or faults that are recognizable on Landsat images (Gedney and Van Wormer, 1974). Combining geophysical data and remote sensing information is a promising approach to earthquake risk analysis.

China

Much can be learned about the great earthquakes of China from Landsat images. Figure 11.6 is an MSS image of the southwest end of the Shansi graben in north-ern China, including Sian and the Wei Ho Valley. The 1556 earthquake near Sian caused more than 820,000 deaths and could have been caused by any one of several prominent Quaternary faults visible on the image (Allen, 1975). There is little doubt that the bounding faults of the graben structure are very active features. Chinese geologists note that the most intense earthquake activity has occurred along the bounding faults with the greatest Cenozoic displacements. In addition to southwest-trending boundary faults, there are impressive west-trending faults in the bedrock of the southeast part of Figure 11.6. At the west boundary of the image a major fault cuts both bedrock and alluvium for 48 km and trends eastward toward the area of maximum seismic intensity in the 1556 earthquake.

Landsat images such as Figure 11.6 were compiled into regional mosaics of China that Tapponnier and Molnar (1977, 1979) interpreted and combined with earthquake epicenter data in a manner similar to that used to produce the Alaska map in Figure 11.5. Active faults of great length are recognized in the China mosaics and are the site of most earthquakes. The predominant type of active faulting was also inferred for major regions. In central China the three major west-trending fault systems are left-lateral, strike-slip faults. Normal faulting and right-lateral, strike-slip faulting occur in the Yunan Province, in contrast to thrust faulting in the Szechwan Province. Extensional tectonics and basaltic volcanism dominate northeast China.

Pakistan

Landsat-3 RBV images with 40-m ground resolution cells are well suited for identifying and mapping active faults. Figure 11.7 is a portion of an RBV image that covers 100 km of the Chaman fault in northwest Pakistan. The image has a number of the geomorphic indicators of active faulting shown in the diagram of Figure 11.2. Faceted ridges, dry sag ponds, shutter ridges, and fault scarps in young alluvium all indicate active faulting of a strike-slip nature. Channels of ephemeral streams are offset in a left-lateral sense. In 1892 an earthquake along the Chaman fault located north of Figure 11.7 offset the Quetta-Chaman railroad 75 cm in the left-lateral sense. In 1975 a magnitude-6.6 earthquake caused 4 cm of left-lateral offset along the fault a short distance north of Figure 11.7 (Lawrence and Yeats, 1979).

LANDSLIDES

Landslides occur on the land and on the seafloor in areas underlain by unstable materials. Individual landslide events are difficult or impossible to predict by remote sensing or other methods. Existing slides may be recognized on images, however, which indicates that an area is slide-prone and should be avoided or used with great care.

Stereo pairs of aerial photographs (black-and-white, normal color, and IR color) have long been used to recognize slides and slide-prone terrain. Rib and Liang (1978) published an extensive collection of stereo pairs together with interpretation criteria. High soil moisture lubricates unstable material and is a major factor in landslides. Thermal IR images have been used to recognize damp ground associated with landslides in California (Blanchard, Greeley, and Goettleman, 1974, Figure 1). Evaporative cooling of the damp ground produces a cool signature on aircraft thermal IR images. Satellites, ra-

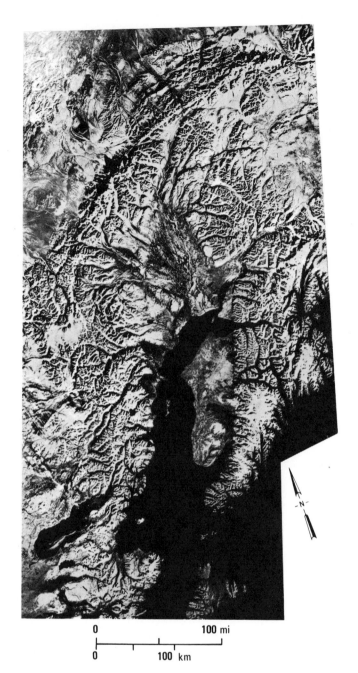

FIGURE 11.4 Landsat mosaic of south-central Alaska compiled from Landsat MSS band-5 images acquired in early November 1972. Mosaic by U.S. Department of Agriculture Soil Conservation Service.

dar, and side-scanning sonar have expanded our capability to recognize unstable terrain where slides occur, as shown by the following examples.

Blackhawk Landslide, Southern California

The prehistoric Blackhawk landslide originated on the north flank of Blackhawk Ridge in the San Bernardino

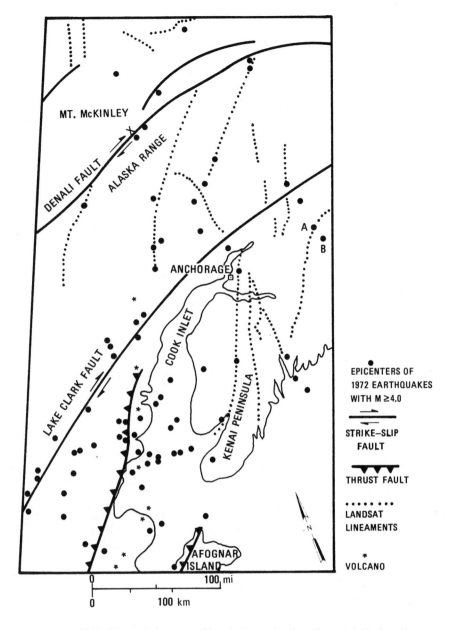

FIGURE 11.5 Interpretation map of Landsat mosaic of south-central Alaska with epicenters plotted for earthquakes in 1972. The epicenter of the major 1964 earthquake was very near epicenters A and B. From Gedney and Van Wormer (1974).

Mountains and moved northward for 9 km into the Mojave Desert. The slide has a maximum width of 3.2 km and includes a volume of 2.7×10^9 m^3 of crushed rock (Shreve, 1968). In form and structure the Blackhawk slide is similar to the smaller and well-known historic slides at Elm in Switzerland, Frank in Alberta, Canada, and on the Sherman Glacier in Alaska, and to the great prehistoric slide at Saidmarreh in Iran. These similarities make the Blackhawk slide a good example for remote sensing analysis.

Figure 11.8 shows two remote sensing images and a map of the Blackhawk slide. Erosion had created an over-steepened slope in the highly fractured bedrock, which collapsed along an arcuate front. The debris trapped a cushion of air during its descent that lubricated the mass, enabling it to flow several kilometers (Shreve, 1968). In the Skylab photograph (Figure 11.8A) the headward scarp of the original debris fall is shadowed, but the characteristic crescent form is apparent. The distal lobe of the slide has a characteristic hummocky ap-

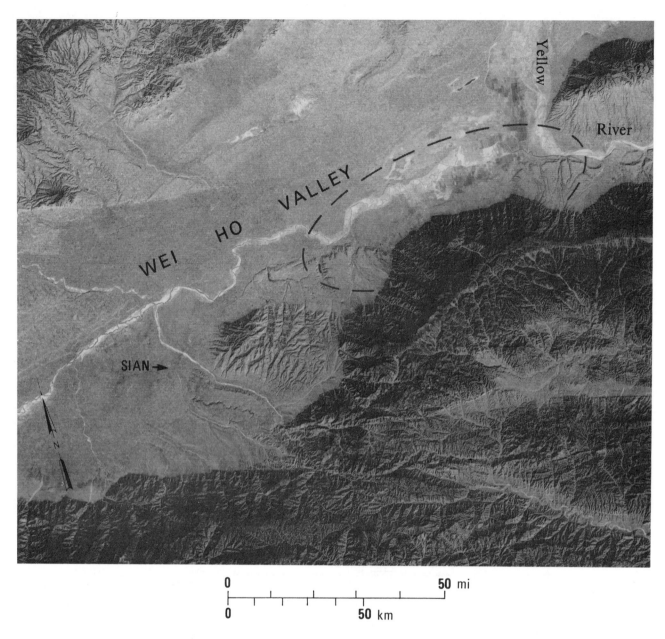

FIGURE 11.6 Landsat MSS band-5 image of the Sian region in Shansi Province, northern China. Dashed line outlines area of shaking greater than intensity XI (Mercalli scale) during the great 1556 earthquake. Annotation from Allen (1975, Figure 28).

pearance. The toe and lateral margins of the slide form a pressure ridge 15 to 30 m high that stands slightly above the surface of the slide. A northwest-flowing ephemeral stream has deposited younger alluvium over the central portion of the slide. In the Skylab photograph, the slide debris is darker than the stream and desert alluvium. The radar image (Figure 11.8B) has poorer resolution than the Skylab photograph but portrays major features of the slide. The coarse debris and hummocky surface are rough and produce a bright signature relative to the darker signature of the finer-grained alluvium. The pressure ridges have narrow, bright signatures.

Submarine Landslides, Mississippi Delta

The Mississippi Delta consists of unconsolidated, water-saturated, fine-grained sediment. The gentle depositional slopes around the margin of the delta are unstable and may collapse to form submarine landslides and related features. Because the delta is an oil-producing region, it is important to map slide-prone areas before installing production platforms and seafloor pipelines. The turbid water prevents remote sensing by visible means. The Coastal Studies Institute of Louisiana State University and the U.S. Geological Survey acquired parallel strips of side-scanning sonar images that were digitally processed and compiled into the mosaic of Figure 11.9A. As described in Chapter 9, bright signatures in sonar images record shadow zones or smooth surfaces. Dark signatures record strong returns from scarps facing the sonar pulse and from rough surfaces.

As shown in the interpretation map (Figure 11.9B), the slides originate at lobate slump areas where the sediment collapses to form irregular blocks. The slump material moves downslope through narrow, steep-sided channels (chutes) that merge at junctions. At the junctions, the slump material may leave the chute and form spillover deposits. Extensive systems of slumps and chutes have been mapped in the delta with sonar.

Piper and others (1985) acquired and interpreted sonar images near the epicenter of the 1929 Grand Banks earthquake on the seafloor south of Newfoundland. The earthquake triggered a major submarine landslide and turbidity currents. The sonar images clearly show the following features: landslide scarps, slumps, debris flows, a lineated seafloor, channels, and gullies. Sonar images can identify unstable areas that should be avoided for offshore engineering projects.

LAND SUBSIDENCE

Land subsidence, which is downward movement, is a different type of hazard from landslides, in which lateral movement predominates. Subsidence results from removal of underlying support and can have serious effects on man-made structures.

Surface subsidence up to several meters in depth can be caused by extraction of subsurface fluids. Where subsidence occurs in populated and industrialized coastal areas, the resulting flooding can be a major problem. A classic example is the subsidence associated with the Long Beach oil field in southern California, which was stopped through injection of water to replace the extracted oil.

In the past 30 years, up to 2.5 m of land subsidence has occurred in metropolitan Houston, Texas, and is related to the following causes (Verbeek and Clanton, 1981):

1. Active geologic processes of faulting and compaction of young sedimentary deposits

FIGURE 11.7 Landsat RBV image of the Chaman fault, northwest Pakistan.

2. Extraction of oil and gas

3. Widespread pumping of groundwater from shallow depths, which is thought to be a major cause of subsidence

A. PHOTOGRAPH FROM NASA SKYLAB EARTH–TERRAIN CAMERA.

B. AIRCRAFT SYNTHETIC–APERTURE, X–BAND RADAR IMAGE.

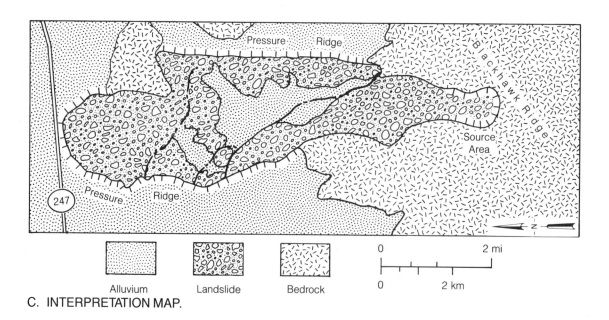

C. INTERPRETATION MAP.

FIGURE 11.8 Blackhawk landslide, San Bernardino Mountains, California.

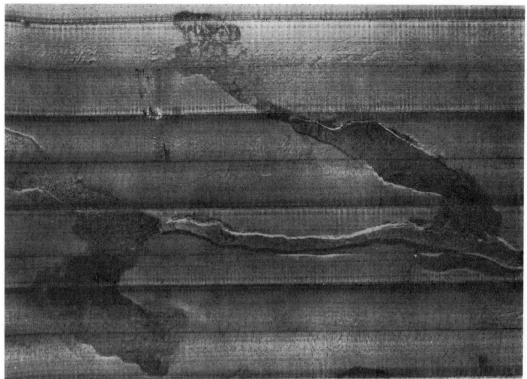

A. MOSAIC OF SIDE–SCANNING SONAR IMAGES. FROM PRIOR, COLEMAN, AND GARRISON (1979, FIGURE 2). COURTESY D. B. PRIOR, COASTAL STUDIES INSTITUTE, LOUISIANA STATE UNIVERSITY.

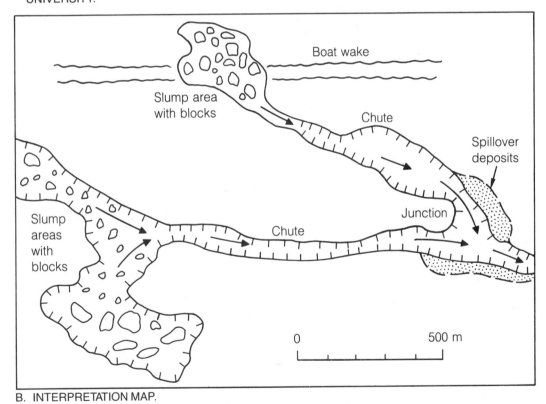

B. INTERPRETATION MAP.

FIGURE 11.9   Submarine landslides, Mississippi Delta, Gulf of Mexico.

The rate of subsidence is increasing and is concentrated along active faults, which have caused extensive damage to buildings, roads, and especially underground utilities. Clanton and Verbeek (1981) acquired oblique aerial photographs with normal color and IR color film and interpreted them using the following criteria to identify traces of active faults:

1. Fault scarps

2. Sag ponds along fault traces

3. Differences in drainage patterns on opposite sides of faults

4. Linear tonal anomalies caused by higher soil moisture on the downthrown sides of faults

5. Vegetation anomalies

These criteria have been used to prepare maps showing the locations of active faults.

In areas of limestone terrain, groundwater can dissolve the bedrock to produce underground caverns that may collapse and cause surface subsidence. In Florida these collapse areas, called *sinkholes*, are up to several hundred meters wide and 10 m deep and may form in a few hours. Areas of incipient sinkholes have anomalously high surface moisture. A few aircraft thermal IR images show cool areas that may be indicators of future sinkholes.

VOLCANOES

On the basis of their activity, volcanoes may be classified as

Active erupted at least once in historic time
Dormant no historic eruptions, but probably capable of erupting
Extinct incapable of further eruptions

Hazards to people and property from active volcanoes include (1) lava flows; (2) hot avalanches, mudflows, and floods; and (3) volcanic ash and gases. Despite these hazards, many areas of known active volcanoes are densely populated, including Japan, Java, and Italy. Fortunately eruptions occur infrequently and cause low annual losses in comparison to earthquakes, floods, and landslides. Nevertheless, eruptions can be costly and dangerous.

Warning signs of increasing volcanic activity and means of detection are summarized as follows (Foxworthy and Hill, 1982, Figure 3):

1. Earthquakes caused by magma intrusion are detected by seismometers.

2. Changes in shape of a volcano caused by magma intrusion are recorded by tiltmeters and by repeated, precise surveys, including aerial photographs.

3. Steam-driven ejection of old rock and ash is monitored by visual observation, radar observation, photography, and studies of ejecta (material erupted by the volcano).

4. Increased emission of heat and gas may be detected from visible fumes, increased stream flow, avalanches from melted snow, and changes in chemistry of the meltwater. Ground temperature may be measured directly with probes or remotely by IR scanners and radiometers.

Mount St. Helens, Washington

Mount St. Helens, in the southwest part of the state of Washington, erupted several times in the 1800s and has long been recognized as an active volcano. The major explosive eruption of May 18, 1980 was thoroughly monitored and studied for two reasons: (1) there were abundant warnings of the impending eruption, which gave scientists time to mobilize, and (2) the volcano was a threat because of its proximity to population centers. For these reasons the 1980 event was probably the most thoroughly monitored and analyzed eruption in history.

The first warning was a strong earthquake at shallow depth beneath the volcano on March 20, 1980. On March 27, explosive hydrothermal activity began at the summit, accompanied by formation of a small crater, ground fracturing, and the beginning of a topographic bulge high on the north flank. Strong seismic activity and relatively mild steam-blast eruptions continued intermittently into mid-May. During that time the new crater gradually enlarged and the bulge became conspicuous. On the morning of May 18, an earthquake, similar to earlier activity, caused multiple fractures of the bulging north flank and initiated great avalanches of rock debris. This unloading led to a northward-directed lateral blast, partly driven by steam explosions, that devastated an area of nearly 600 km². These events in turn triggered an explosive eruption that ejected a vertical column of ash more than 25 km high, causing ash falls more than 1500 km to the east. Pyroclastic flows occurred on the north flank. Catastrophic mudflows and floods were generated, in part by rapid melting of snow and ice. Smaller eruptions have occurred intermittently since the main blast. U.S. Geological Survey reports by Lipman and Mullineaux (1981) and by Foxworthy and Hill (1982) document the events.

Numerous remote sensing images, such as the pair of aerial photographs in Figure 11.10, were acquired before and after the main eruption. The preblast photograph (Figure 11.10A) shows the symmetrical form of the mountain, which lacked a summit crater. The May 18 blast created a large crater and blew out the north rim, as shown in the postblast photograph (Figure 11.10B). The map (Figure 11.11) shows the major postblast destructional features and volcanic deposits. Timber north of the volcano was flattened by the blast and partially covered by deposits from the blast, debris avalanches, and pyroclastic flows. The deposits blocked the outlet of Spirit Lake; as a result the lake is larger in the postblast photograph and has a gray signature, which is caused by floating logs. The east, south, and west flanks of the volcano are modified by mudflows triggered by melted snow and ice. The upper slopes were scoured by the flowing mud, and the lower slopes were covered by mudflow deposits. In the postblast photograph (Figure 11.10B), steam clouds conceal a lava dome that formed on the floor of the crater.

Figure 11.12A is a Seasat radar image acquired 2 years before the blast. The high elevation of the volcano and the steep depression angle of Seasat cause the severe displacement of the peak toward the radar look direction. The postblast X-band aircraft image (Figure 11.12B) was acquired at a low depression angle, which enhances topographic features such as the crater walls, lava dome, and mudflows on the south flank. Although steam clouds were probably present, the radar image clearly shows the lava dome.

Thermal IR images were acquired before and after the eruption. Kieffer, Frank, and Friedman (1981) analyzed the pre-eruption images. The initial explosive event occurred on March 27. A thermal IR image (Figure 11.13A) of the resulting crater was acquired on May 16, 2 days before the main blast. The image is not geometrically corrected for scanner distortion. In the interpretation map (Figure 11.13B) the cross-track scale, in the north-south direction, is compressed relative to the along-track scale. The image was processed to emphasize the hottest temperatures, greater than 12°C, which have very bright signatures in Figure 11.13A and are shown as dark patches in the map. The hot spots coincided with deep pits, many of which were the source of steam plumes. Kieffer, Frank, and Friedman (1981, p. 272) inferred from the image data that a temperature of 400°C occurred at a depth of only 40 m and was caused by intense hydrothermal circulation. The cluster of hot spots on the north flank of the volcano mark the bulge that was subsequently obliterated by the May 18 blast.

Numerous aircraft thermal IR images were acquired and analyzed in the months following the eruption (Friedman and others, 1981). Figure 11.14A is a nighttime image of the crater and vicinity that is also not corrected for scanner distortion; hence the semicircular crater has an oval outline in the image. Despite the distortion, thermal patterns in the image correlate readily with features shown in the geologic map (Figure 11.11). Flanks of the volcano are cool (dark signature). Pyroclastic flow deposits with intermediate temperatures (gray signature) fill the crater and the depression on the north flank that was created by the blast. The highest radiant temperatures (bright signatures) were identified by density-slicing the data and are shown in black in Figure 11.14B. The lava dome formed in August appears as the hot oval on the crater floor. A semicircular pattern of hot spots occurs along the inner wall of the crater. The irregular hot spots in the southeast floor of the crater are fumaroles where steam and gas are vented. A linear hot feature extending south from the rim is an igneous dike 15 m wide. Other hot linear features occur in the new pyroclastic deposits and represent fractures probably caused by compaction. A northwest-trending alignment of hot spots (thermal lineament) is probably related to a major structural feature cutting across the core of the volcano (Friedman and others, 1981, p. 287).

Judged by its volume of ejecta (0.6 km³), Mount St. Helens was a moderate event compared to the 1883 eruption of Krakatoa, Indonesia (26 km³), or the 1815 eruption of Tambora, Indonesia (46 km³). Because of the advance warnings, authorities restricted access to Mount St. Helens, and only about 60 people were killed in the eruption. Property damage is estimated at 2 to 3 billion dollars. The fact that the relatively moderate eruption of Mount St. Helens caused so much damage is reason to continue improving methods of volcano monitoring and prediction.

Thermal IR Surveys of Other Volcanoes

Moxham (1971, p. 103) summarized the history and status through 1970 of thermal surveillance of volcanoes with remote sensing and contact temperature devices. He concluded that "results to date are more tantalizing than elucidating. At some volcanoes thermal forerunners [of eruptions] have been well documented; elsewhere, temperature measurements have been inconclusive or negative." Friedman and Williams (1968, Table 1) listed all the IR surveys of volcanoes reported up to 1968. More than 23 volcanoes throughout the world were surveyed, but generally only with a single flight or, at best, several flights within a few weeks.

Moxham (1967) analyzed three IR surveys of the Taal Volcano, Philippines, during the 9-month quiescent period between the 1965 and 1966 eruptions. The Taal is

0   4 mi

0   4 km

A. PHOTOGRAPH ACQUIRED SEPTEMBER 12, 1975.   B. PHOTOGRAPH ACQUIRED JUNE 19, 1980.

FIGURE 11.10   Aerial photographs of Mount St. Helens acquired before and after the May 18, 1980, eruption.

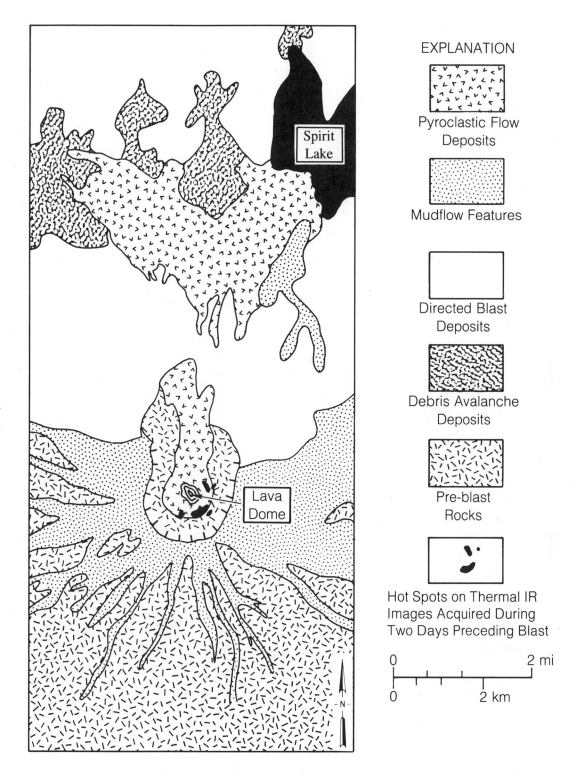

EXPLANATION

Pyroclastic Flow
Deposits

Mudflow Features

Directed Blast
Deposits

Debris Avalanche
Deposits

Pre-blast
Rocks

Hot Spots on Thermal IR
Images Acquired During
Two Days Preceding Blast

FIGURE 11.11 Volcanic deposits and features of the 1980 eruption of Mount St. Helens. From Lipman and Mullineaux (1981, Plate 1).

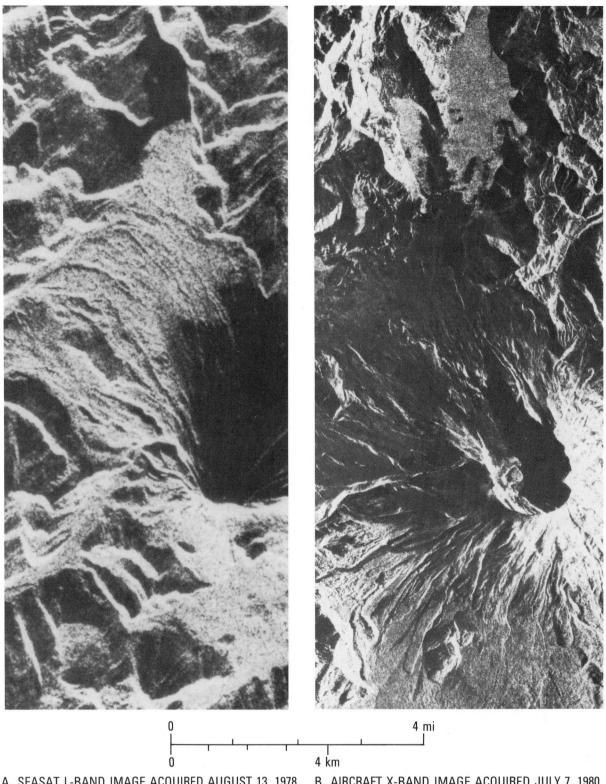

0   4 mi

0   4 km

A. SEASAT L-BAND IMAGE ACQUIRED AUGUST 13, 1978. B. AIRCRAFT X-BAND IMAGE ACQUIRED JULY 7, 1980.
  LOOK DIRECTION TOWARD NORTHEAST.   LOOK DIRECTION TOWARD WEST

FIGURE **11.12** Radar images of Mount St. Helens acquired before and after the May 18, 1980, eruption.

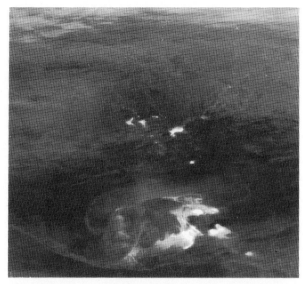

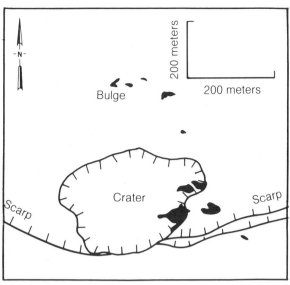

A. THERMAL IR IMAGE (8 TO 14 μm).
   BRIGHT SIGNATURES ARE HOT AREAS.

B. INTERPRETATION MAP. BLACK SPOTS
   INDICATE HOTTEST TEMPERATURES.

FIGURE 11.13 Nighttime thermal IR image of the summit of Mount St. Helens acquired May 16, 1980. From Kieffer, Frank, and Friedman, 1981, Figure 158). Courtesy H. H. Kieffer, U.S. Geological Survey.

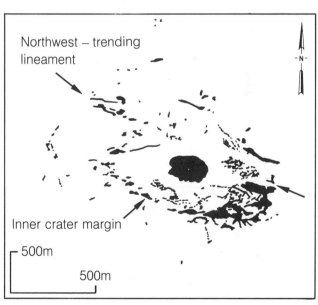

A. THERMAL IMAGE (8 TO 14 μm). BRIGHT
   SIGNATURES ARE HOT AREAS.

B. INTERPRETATION MAP. BLACK SPOTS
   INDICATE HOTTEST TEMPERATURES.

FIGURE 11.14 Nighttime thermal IR image of Mount St. Helens acquired August 20, 1980. Location of the image is shown in Figure 11.11. From Friedman and others (1981, Figure 174). Courtesy J. D. Friedman, U.S. Geological Survey.

well suited for IR surveys because most of the volcano is below the water table and there is both a crater lake and a surrounding water body. Because of the uniform and high emissivity of water, small temperature changes are readily detectable and there are no masking effects of vegetation or moisture differences. The first IR survey showed the two known fumarole fields, plus several previously unknown fumaroles on the north flank of the central cone. Both IR and ground investigations showed new hot springs along the flanks of the 1965 explosion crater, one of which enlarged prior to the 1966 eruption. Hydrothermal activity persisted around the rim of the 1965 cinder cone, which was the site of the 1966 eruption midway between the two areas of maximum thermal water discharge.

IR surveys were made over Italian volcanoes in 1970 and 1973 to monitor any changes and to provide reference data for any future eruptions (Cassinis, Marino, and Tonelli, 1974).

In the early 1960s, seismic and thermal activity increased at the crest of Mount Rainier, Washington, and the U.S. Geological Survey flew IR surveys in 1964, 1966, and 1969. Moxham (1970) interpreted the images and reported that fumaroles lining the rims of the two summit craters were similar in extent and apparent thermal intensity in the 1964 and 1966 images. On the 1969 image, however, several new warm anomalies appeared that coincided with areas of higher snow melt.

Mount Baker, Washington, was the site of volcanic activity from 1843 to 1859. Intermittent fumarole activity has continued since then. The U.S. Geological Survey flew seven thermal IR surveys between 1970 and 1973. Because of increased activity in 1975, the Army National Guard acquired additional images. All of the images and other data were analyzed by Friedman and Frank (1980), who concluded that heat was discharged in 1975 at seven times the 1972 rate.

No major eruptions have been predicted for Mount Rainier or Mount Baker, and the activity appears to have subsided. However, when Mount St. Helens erupted in 1980, it did so catastrophically after a relatively short sequence of warning events. These cases illustrate the need for additional data and research in using remote sensing information to predict eruptions.

Satellite Images

The repetitive coverage, regional scale, and low cost of thermal IR images from satellites make this an attractive method for monitoring volcanoes. However, the 1-km ground resolution cell of the AVHRR system is too large for mapping individual fissures and vents. Landsat TM band 6 acquires thermal IR images with a ground resolution cell of 120 m, which should prove useful for monitoring volcanic activity, especially on nighttime images, when the interfering effects of solar heating are absent.

Although the spatial resolution of NOAA environmental satellite systems is too coarse to record details of surface thermal patterns, the plumes of smoke and ash from volcanoes are detectable. Less than 33 hours after the January 25, 1973, eruption of Heimaey, Iceland, the plume was recorded on a NOAA-2 thermal IR image (Williams and others, 1974, Figure 11). The Tobalchik Volcano on the Kamchatka Peninsula erupted on July 6, 1975, and the plume was first observed on NOAA-4 thermal IR images acquired on July 9. Jayaweera, Seifert, and Wendler (1976) measured the length and azimuth of the plume on 14 NOAA images acquired from July 9 through August 17, 1975. The most spectacular observation was made July 18, when the length of the plume was at least 960 km. From the distance between the plume and its shadow, which was measured from the visible image, and from the known sun elevation, the height of the plume on July 28 was estimated at 6.5 km. The plume from Mount St. Helens was tracked on images from GOES West and from U.S. Air Force satellites (Rice, 1981). On the NOAA thermal IR images, volcanic plumes have a cold (bright) signature because they have cooled to the temperature of the surrounding atmosphere; on visible images the plumes have dark signatures. The plume from the 1973 eruption of the Russian volcano Tiatia on Kunashir Island shows clearly on Landsat MSS images. The ability to monitor size and distribution of volcanic plumes may prove useful in planning the rehabilitation of areas covered by ash.

FLOODS

In the spring of 1973, record rains caused extensive flooding throughout the Mississippi River Valley. The sky over the flooded area was relatively clear on two successive days, and Landsat acquired images of the Mississippi Valley. Figure 11.15 of the St. Louis area compares an MSS image of the river during its normal stage with an MSS image acquired during the flood stage. Band 7 is the optimum MSS band for mapping the boundary of flooded areas for reasons given in bathymetry section of Chapter 9. The map in Figure 11.15C shows the extent of flooding. A more extensive Landsat analysis of flooding in the entire Mississippi Valley was carried out by the U.S. Geological Survey (Deutsch and Ruggles, 1974).

A decade later, Landsat 4 acquired TM images of floods in the lower Mississippi River Valley (Figure 11.16B). TM band 4 (0.76-to-0.90-μm wavelength), which covers the approximate spectral range of MSS band 7,

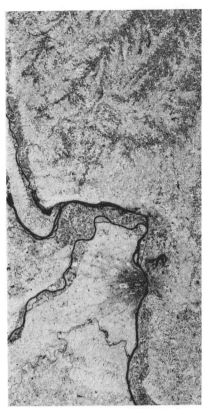

A. NORMAL STAGE, OCTOBER 2, 1972.

B. FLOOD STAGE, MARCH 31, 1973.

C. INTERPRETATION MAP.

FIGURE 11.15 Landsat MSS band-7 images showing flooding in upper Mississippi River Valley.

is the optimum TM band for flood monitoring. Features of the TM images are shown in the map of Figure 11.17. In the dry-season image (Figure 11.16A), water in the lakes and rivers has a dark signature; agricultural fields with standing crops are bright, and those with bare soil are dark; native vegetation along the waterways and in the southeast portion of the image has an intermediate gray signature. In the wet-season image (Figure 11.16B), the high water of the Mississippi has submerged many of the sandbars and the lakes are expanded. Lowland areas in the southeast portion of the image are extensively flooded. Flooded areas are shown in the map of Figure 11.17, which was prepared in 1 hour.

In April 1975, Landsat images were used to delineate flooded areas in Louisiana; these were compared with land-use classification maps to determine the extent of flood damage to urban areas, farmland, and other categories. The state used these data to analyze the damage rapidly and to document the need for federal disaster-relief funds. Landsat images have also been used for flood-damage assessment in South Dakota, Iowa, Arizona, and Pakistan. Additional examples of flood mon-

itoring are given in Deutsch and Ruggles (1978) and in Kruus and others (1981).

Flooded areas are typically cloud covered and are rarely recorded on Landsat images. However, radar, with its cloud-penetration capability, is well suited to monitor floods. It is desirable to have an operational radar satellite system, such as Seasat or SIR, that can provide all-weather images for damage assessment and reclamation work.

SUBSURFACE COAL FIRES

Coal mining produces large volumes of waste rock, which includes combustible material such as contaminated coal, carbonaceous shale, and pyrite (an iron sulfide mineral). Large refuse dumps containing millions of tons of this material are present at coal-mining areas throughout the world. Surface fires or spontaneous combustion may ignite the dumps, and the resulting fire may spread to abandoned coal mines. Such fires in refuse dumps and mines are common, and some have been burning for

A. NORMAL STAGE, SEPTEMBER 23, 1982.

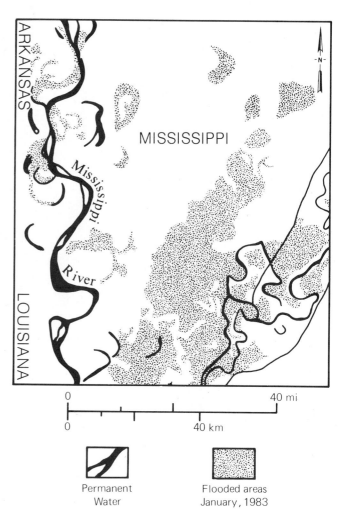

FIGURE 11.17 Flooded areas interpreted from Landsat-4 TM images.

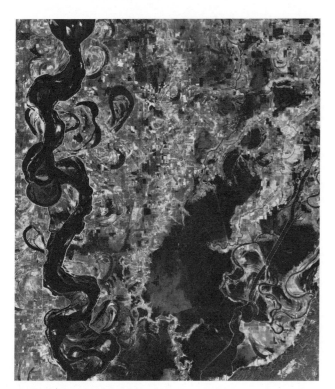

B. FLOOD STAGE, JANUARY 13, 1983.

FIGURE 11.16 Landsat TM band-4 images showing flooding in the lower Mississippi River Valley.

over 50 years. Subsurface fires constitute three major hazards:

1. The large volumes of carbon monoxide and sulfur dioxide gases released at the surface are potentially dangerous.

2. Fires in abandoned mines may cause surface-subsidence problems.

3. The fires may interfere with mining operations.

Coal fires are controlled by shutting off the air supply. This is accomplished either by coating the dumps with clay or by injecting water and slurries through holes drilled into the combustion zone. For such measures to be effective, the underground fire must be accurately located. Thermal IR images have proven useful for this

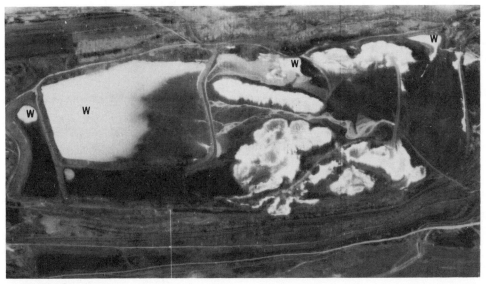

A. ORIGINAL IMAGE.

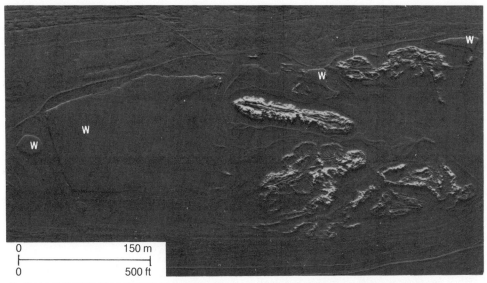

| 0 | | 150 m |
| 0 | | 500 ft |

B. IMAGE AFTER PROCESSING TO ENHANCE AREAS OF MAXIMUM TEMPERATURES (WHITE TONES), WHICH INDICATE FIRES WITHIN THE REFUSE BANK.

FIGURE 11.18 Original and enhanced nighttime thermal IR images (3 to 5 μm) acquired April 28, 1967, of a burning coal refuse bank near Wilkes-Barre, Pennsylvania. Areas of open water are indicated by *W*. From Knuth and Fisher (1967, Figures 4 and 5). Images courtesy HRB-Singer, Incorporated.

purpose. Figure 11.18A is a nonquantitative nighttime image of a burning coal refuse dump near Wilkes-Barre, Pennsylvania. On this unenhanced image the relative temperature scale is such that surface water and subsurface fires have similar bright signatures and are difficult to separate. Processing the magnetic-tape record enhances the areas of maximum surface radiant tem-

perature (Figure 11.18B) so that coal fires are readily distinguishable from water (Knuth and Fisher, 1967). The enhanced image emphasizes the active burning fronts, and it is apparent that the narrow, elongate dump is the site of the most intense burning. IR images are valuable for planning fire-control programs and for evaluating the effectiveness of control measures.

Greene, Moxham, and Harvey (1969) acquired thermal IR images of coal-mine fires in Pennsylvania and correlated these with temperature measurements in boreholes. They reached the following conclusions:

1. Fires at shallow depths (less than 10 m) were readily detectable.

2. Fires at intermediate depths (10 to 30 m) were detectable where heat was carried to the surface by convection in open cracks or where the fire had been burning long enough (several years or more) for heat to reach the surface by conduction.

3. Deep fires (greater than 30 m) were detectable only where heat reached the surface by convection in open cracks.

COMMENTS

Remote sensing methods are useful for recognizing areas subject to natural hazards, such as earthquakes, volcanic eruptions, landslides, and floods. Images are also useful for assessing the damage caused by these hazardous events and for planning relief and rehabilitation efforts. Prediction of specific events is a desirable, but much more difficult, objective.

QUESTIONS

1. Prepare an interpretation map for the Landsat image of the Sian region in China (Figure 11.6). Identify the lineaments that may be the expression of active faults and give reasons for your identification.

2. Describe how you would interpret the following images for evidence of active faults: stereo pairs of aerial photographs, nighttime thermal IR images, and SIR images.

3. Describe how you would interpret the same set of images for evidence of landslides.

4. Plan a remote sensing system using available aircraft and satellite technology to monitor volcanoes such as those of the Cascade Range in Washington and Oregon.

5. Describe an all-weather satellite system for monitoring the extent of flooded terrain.

6. Describe the conditions of an underground coal-mine fire that could be detected on Landsat TM band-6 nighttime images.

REFERENCES

Allen, C. R., 1975, Geological criteria for evaluating seismicity: Geological Society of America Bulletin, v. 86, p. 1041–1057.

Blanchard, M. B., R. Greeley, and R. Goettleman, 1974, Use of visible, near-infrared, and thermal infrared remote sensing to study soil moisture: Proceedings of the Ninth International Symposium of Remote Sensing of Environment, p. 693–700, Environmental Research Institute of Michigan, Ann Arbor, Mich.

Cassinis, R., C. M. Marino, and A. M. Tonelli, 1974, Remote sensing techniques applied to the study of Italian volcanic areas—the results of the repetition of the airborne survey compared to the previous data: Proceedings of the Ninth International Symposium on Remote Sensing of Environment, v. 3, p. 1989–2004, Environmental Research Institute of Michigan, Ann Arbor, Mich.

Clanton, U. S., and E. R. Verbeek, 1981, Photographic portrait of active faults in the Houston metropolitan area, Texas *in* Etter, E. M., ed., Houston area environmental geology—surface faulting, ground subsidence, and hazard liability: p. 70–113, Houston Geological Society, Houston, Texas.

Deutsch, M., and F. H. Ruggles, 1974, Optical data processing and projected applications of the ERTS-1 imagery covering the 1973 Mississippi River floods: Water Resources Bulletin, v. 10, p. 1023–1039.

Deutsch, M., and F. H. Ruggles, 1978, Hydrological applications of Landsat imagery used in the study of the 1973 Indus River flood, Pakistan: Water Resources Bulletin, v. 14, p. 261–274.

Foxworthy, B. L., and M. Hill, 1982, Volcanic eruptions of 1980 at Mount St. Helens, the first 100 days: U.S. Geological Survey Professional Paper 1249.

Friedman, J. D., and D. Frank, 1980, Infrared surveys, radiant flux, and total heat discharge at Mount Baker volcano, Washington, between 1970 and 1975: U.S. Geological Survey Professional Paper 1022-D.

Friedman, J. D., and R. S. Williams, 1968, Infrared sensing of active geologic processes: Proceedings of the Fifth International Symposium on Remote Sensing of Environment, p. 787–815, University of Michigan, Ann Arbor, Mich.

Friedman, J. D., D. Frank, H. H. Kieffer, and D. L. Sawatzky, 1981, Thermal infrared surveys of the May 18 crater, subsequent lava domes, and associated deposits *in* Lipman, P. W. and D. L. Mullineaux, eds., The 1980 eruption of Mount St. Helens, Washington: U.S. Geological Survey Professional Paper 1250, p. 279–293.

Gedney, L., and J. Van Wormer, 1974, Earthquakes and tectonic evolution in Alaska: Third ERTS-1 Symposium, NASA SP-351, v. 1, p. 745–756.

Greene, G. W., R. M. Moxham, and A. H. Harvey, 1969, Aerial infrared surveys and borehole temperature measurements of coal mine fires in Pennsylvania: Proceedings of the Sixth International Symposium on Remote Sensing of Environment, p. 517–525, Environmental Research Institute Michigan, Ann Arbor, Mich.

Jayaweera, K. O. L. F., R. Seifert, and G. Wendler, 1976, Satellite observations of the eruption of Tolbachik Volcano: Transactions of the American Geophysical Union, v. 57, p. 196–200.

Kieffer, H. H., D. Frank, and J. D. Friedman, 1981, Thermal infrared surveys at Mount St. Helens prior to the eruption of May 18 in Lipman, P. W. and D. L. Mullineaux, eds., The 1980 eruption of Mount St. Helens, Washington: U.S. Geological Survey Professional Paper 1250, p. 257–277.

Knuth, W. M., and W. Fisher, 1967, Detection and delineation of subsurface coal fires by aerial infrared scanning (abstract) in Abstracts for 1967 Meetings: Geological Society of America Special Paper 115, p. 67–68.

Kruus, J., M. Deutsch, P. L. Hansen, and H. L. Ferguson, 1981, Flood applications of satellite imagery in Deutsch, M., D. R. Wiesnet, and A. R. Rango, eds., Satellite hydrology: p. 292–301, American Water Resources Association, Minneapolis, Minn.

Lawrence, R. D., and R. S. Yeats, 1979, Geological reconnaissance of the Chaman fault in Pakistan in Farah, A. and K. A. DeJong, eds., Geodynamics of Pakistan: p. 351–357, Geological Survey of Pakistan, Quetta, Pakistan.

Lipman, P. W., and D. L. Mullineaux, eds., 1981, The 1980 eruption of Mount St. Helens, Washington: U.S. Geological Survey Professional Paper 1250.

Merifield, P. M., and D. L. Lamar, 1975, Active and inactive faults in Southern California viewed from Skylab: NASA Earth Resources Survey Symposium, NASA TM X-58168, v. 1, p. 779–797.

Moxham, R. M., 1967, Changes in surface temperature at Taal Volcano, Philippines 1965–1966: Bulletin Volcanologique, v. 31, p. 215–234.

Moxham, R. M., 1970, Thermal features at volcanoes in the Cascade Range, as observed by aerial infrared surveys: Bulletin Volcanologique, v. 34, p. 77–106.

Moxham, R. M., 1971, Thermal surveillance of volcanoes in The surveillance and prediction of volcanic activity: p. 103–124, UNESCO, Paris.

Piper, D. J. W., and others, 1985, Sediment slides and turbidity currents on the Laurentian fan—sidescan sonar investigation near the epicenter of the 1929 Grand Banks earthquake: Geology, v. 13, p. 538–541.

Prior, D. B., J. M. Coleman, and L. E. Garrison, 1979, Digitally acquired undistorted side-scan sonar images of submarine landslides, Mississippi River delta: Geology, v. 7, p. 423–425.

Rib, H. T., and T. Liang, 1978, Recognition and identification in Schuster, R. L., and R. J. Krizek, eds., Landslides—analysis and control: ch. 3, p. 34–69, National Academy of Sciences Special Report 176, Washington, D.C.

Rice, C. J., 1981, Satellite observations of the Mount St. Helens eruption of 18 May, 1980: Aerospace Corporation Report SD-TR-81-70, El Segundo, Calif.

Shreve, R. L., 1968, The Blackhawk landslide: Geological Society of America Special Paper No. 108, Boulder, Colo.

Tapponnier, P., and P. Molnar, 1977, Active faulting and tectonics in China: Journal of Geophysical Research, v. 82, p. 2905–2930.

Tapponnier, P., and P. Molnar, 1979, Active faulting and Cenozoic tectonics of the Tien Shan, Mongolia, and Baykal regions: Journal of Geophysical Research, v. 84, p. 3425–3459.

Vedder, J. G., and R. E. Wallace, 1970, Map showing recently active breaks along the San Andreas and related faults between Cholame Valley and Tejon Pass, California: U.S. Geological Survey, Miscellaneous Geologic Investigations, Map I-574.

Verbeek, E. R., and U. S. Clanton, 1981, Historically active faults in the Houston metropolitan area, Texas in Etter, E. M., ed., Houston area environmental geology—surface faulting, ground subsidence, and hazard liability: p. 28–68, Houston Geology Society, Houston, Texas.

Williams, R. S., and others, 1974, Environmental studies of Iceland with ERTS-1 imagery: Proceedings of the Ninth International Symposium on Remote Sensing of Environment, v. 1, p. 31–81, Environmental Research Institute of Michigan, Ann Arbor, Mich.

ADDITIONAL READING

Hays, W. W., ed., 1981, Facing geologic and hydrologic hazards, earth science considerations: U.S. Geological Survey Professional Paper 1240-B.

Mintzer, O., 1983, Engineering applications in Colwell, R. N., ed., Manual of remote sensing, second edition: ch. 32, p. 1955–2109, American Society for Photogrammetry and Remote Sensing, Falls Church, Va.

Rosenfeld, C. A., 1980, Observations on the Mount St. Helens eruption: American Scientist, v. 68, p. 494–509.

Sabins, F. F., 1973, Engineering geology applications of remote sensing in Moran, D. E., ed., Geology, Seismicity, and Environmental Impact: Association of Engineering Geologists Special Publication, p. 141–155, Los Angeles, Calif.

CHAPTER

12

Comparison of Image Types

Previous chapters described the major remote sensing systems and related image signatures to terrain characteristics for each of the wavelength regions. The present chapter will combine this information across the electromagnetic spectrum by comparing images of the same locality acquired at different wavelengths. Two of the most intensively and repetitively imaged areas in the world are in California: Death Valley and Pisgah Crater. These areas are selected to compare images because of their extensive databases and ready accessibility for field checking.

DEATH VALLEY

Death Valley, in the Mojave Desert of California, has been the site of many remote sensing investigations for the following reasons:

1. A wide range of rocks and recent deposits are well exposed.

2. The geology is well known.

3. The weather is almost always clear for image acquisition.

Earlier in this book, TIMS images (Chapter 5) and radar images (Chapter 6) of Death Valley were interpreted.

Regional Geology and Images

Death Valley is bounded on the west by the Panamint Range and on the east by the Funeral and Black mountains (Figure 12.1), which consist of complexly deformed metamorphic, volcanic, and sedimentary rocks. The floor of the valley includes the lowest elevation in the United States, 86 m below sea level, at Badwater. The high point of the Panamint Range is only 30 km west of Badwater and has an elevation of 3368 m. The floor of Death Valley is a closed salt pan covered by evaporite deposits. Intermittent streams on the valley floor are bordered by floodplain deposits of clay and silt. The valley floor is bordered by alluvial fan deposits of gravel eroded from the adjacent mountains. As shown in Figure 12.1, the fans on the west margin of the valley are broad and extensive, while those on the east margin are small. This asymmetry is caused by continuing subsidence of the valley floor along the zone of active border faults at the foot of the Black Mountains. The fans at the foot of the Panamint Range form broad coalescing depositional surfaces; at the foot of the Black Mountains, older fans are down-dropped along the border faults and are buried by younger gravels. Active movement along the border faults also results in (1) slightly eroded fault scarps formed on the basement rocks; (2) wineglass-shaped canyons along the west face of the Black Mountains; and (3) fault scarps cutting the recent

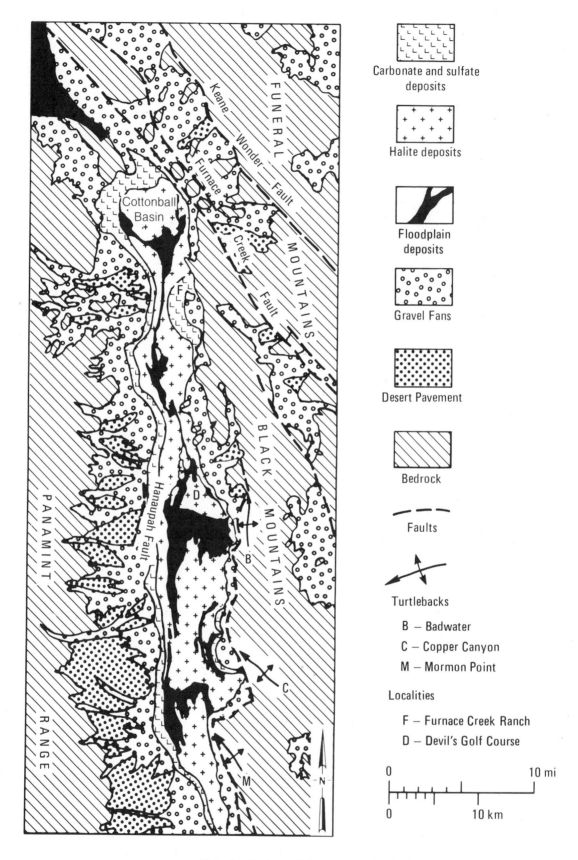

FIGURE 12.1 Regional map of Death Valley, California.

gravel deposits. The geology of the area has been described by Hunt and Mabey (1966), Drewes (1963), and others.

The images selected for this study (Table 12.1) include all bands of the electromagnetic spectrum that are commonly employed in remote sensing. The regional satellite images of Death Valley illustrated in Figure 12.2 were acquired by Seasat, LFC, and Landsat TM bands 5 and 6. These images represent the radar, visible, reflected IR, and thermal IR spectral regions. The regional geologic map (Figure 12.1) matches the scale of the images. Heat Capacity Mapping Mission images and Landsat MSS images were also examined but are not discussed here; their spatial resolution is inferior to that of TM images and aircraft images, which provide comparable spectral information.

Plate 16A is an IR color Landsat TM image that was digitally enhanced using the contrast stretch, IHS, and nondirectional edge-enhancement techniques described in Chapter 7. The general absence of red tones indicates the lack of vegetation. The prominent red patch at Furnace Creek Ranch (Figure 12.1) is caused by irrigated lawns and trees. Linear red streaks on the alluvial fan at Furnace Creek are caused by mesquite trees along drainage channels. Along the foot of alluvial fans in the southwest part of the image, a narrow, faint red band is caused by shrubs and bushes. Comparing Plate 16A with the individual black-and-white TM images (Figure 12.2C,D) illustrates the superiority of color images.

Plate 16B is an image of ratio 5/1 that was density-sliced and displayed in color. A number of different ratio images were examined; the 5/1 image was selected because its colors distinguish most terrain types, as shown by comparison with the geologic map (Figure 12.1).

Materials

Table 12.2 lists typical materials of Death Valley. Their distribution is shown in the geologic map (Figure 12.1). Most of the materials were illustrated in Chapter 6.

Bedrock In the Seasat image (Figure 12.2A), bedrock is characterized by highlights and shadows caused by topographic relief. The look direction for this image was toward the east, resulting in layover of ridges and peaks toward the west. In the LFC photograph (Figure 12.2B) and TM reflected image (Figure 11.2C), bedrock signatures are dominated by highlights and shadows caused by solar illumination from the south. In the daytime thermal IR image of TM band 6 (Figure 12.2D), bedrock is characterized by bright (warm) and dark (cool) signatures. These are caused by high and low thermal radiation from the sunlit and shadowed slopes, respectively.

Coarse Gravel Along the margins of Death Valley, gravel fills the stream channels and forms the alluvial fans. The gravel exceeds the rough criterion (relief greater than 6 cm) and produces bright signatures on the Seasat image. In the LFC photograph and TM reflected IR image, the fans have a medium gray signature and major channels have bright signatures. In the daytime thermal IR image (band 6), fans and channels have an intermediate gray signature indicating an intermediate radiant temperature. Gravel is absent from the channels and floodplain deposits on the floor of Death Valley.

Desert Pavement Desert pavement is produced by mechanical disintegration of gravel and boulders into slabs and flakes that form a smooth, mosaiclike surface.

TABLE 12.1 Characteristics of remote sensing systems used to acquire images of Death Valley

System	Date	Wavelength	Spatial resolution, m	Property recorded
Large-format camera	October 1984	0.5 to 0.7 μm	~5	Albedo and topography
Landsat-4 thematic mapper				
Bands 1–5, 7	November 1982	0.45 to 2.35 μm	30	Reflected solar energy
Band 6	November 1982	10.4 to 12.5 μm	120	Radiant temperature
Thermal IR, aircraft				
Daytime and nighttime	March 1977	7.95 to 13.5 μm	10	Radiant temperature
Radar				
Aircraft X-band	July 1976	3.0 cm	10	Surface roughness and topography
Seasat L-band	September 1978	23.5 cm	25	Surface roughness and topography
SIR–B L-band	October 1984	23.5 cm	40	Surface roughness and topography

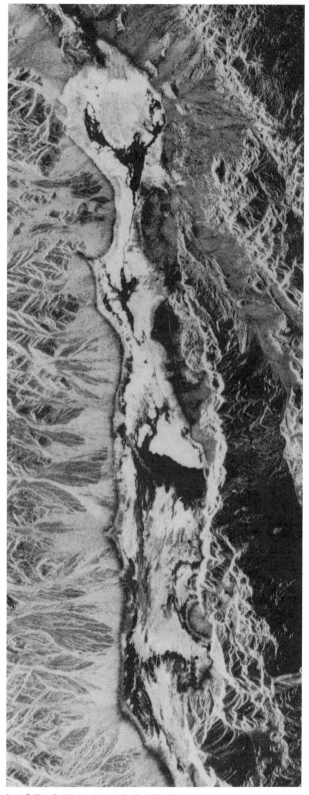 

A. SEASAT L–BAND RADAR (23.5 cm).   B. LARGE–FORMAT CAMERA (0.5 TO 0.7 μm).

FIGURE **12.2**   Regional satellite images of Death Valley.

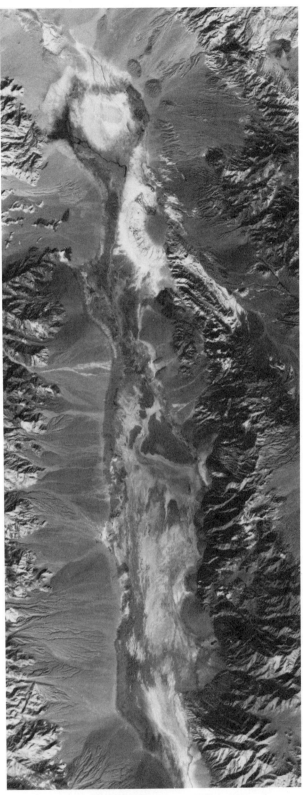

C. LANDSAT TM BAND 5 (1.55 TO 1.75 μm).

D. LANDSAT TM BAND 6 (10.4 TO 12.5 μm).

TABLE 12.2 Materials at Death Valley

Material	Relief, cm	Color	Typical occurrences
Rough halite	29	Brown	Margins of Cottonball Basin and Devil's Golf Course
Intermediate halite	6	White to brown	Central part of Cottonball Basin
Coarse gravel	12	Medium gray	Alluvial fans and channels
Carbonate and sulfate deposits	2.0	Light gray	Margins of valley floor
Sand and fine gravel	1.0	Gray	Margins of alluvial fans
Desert pavement	1.0	Black	West-side fans
Floodplain deposits	0.2	White	Badwater Basin and Cottonball Basin
Vegetation	Variable	Green	Furnace Creek Ranch and washes
Bedrock	High	Varied	Panamint, Funeral, and Black mountains

Extensive areas of desert pavement occur on older gravel deposits along the west side of Death Valley. These older gravel deposits are elevated above the level of adjacent deposits of coarse gravel. Areas of desert pavement are dissected by drainage channels that produce an irregular pattern on images. In the Seasat image (Figure 12.2A) the smooth desert pavement forms a distinctive dark signature that contrasts with the bright signature of the adjacent coarse gravel. The dark expanses of desert pavement are cut by the networks of bright lines formed by gravel in the drainage channels.

The irregular dissected pattern of the desert pavement helps distinguish it from the active fans in the LFC visible image and reflected IR image. Desert pavement is generally coated with desert varnish, causing a dark signature in the LFC image (Figure 12.2B) that contrasts with the bright signature of active coarse gravel. In the TM reflected IR image (Figure 12.2C), however, much of the desert pavement has a brighter signature than the active gravel.

In the daytime thermal IR image (Figure 12.2D), desert pavement has a distinctly warmer radiant temperature than the active gravel. Two factors probably contribute to this warmer temperature:

1. The darker color of the desert pavement may cause a higher emissivity.

2. The smoother surface may be a more efficient radiator of heat.

Sand and Fine Gravel   Deposits of sand and fine gravel form a narrow, nearly continuous belt at the foot of the fans along the western margin of the valley. A similar belt also occurs at the foot of some of the fans on the eastern margin. The sand and gravel, which was deposited along the shore of an ancient lake, is volumetrically insignificant but is described here because it has a conspicuous signature on both the radar and thermal IR images. The belt of sand is too narrow to portray in the regional geologic map (Figure 12.1); it occurs between the fan gravel and the belt of carbonate and sulfate deposits at the margin of the valley floor. Details of the sand deposits are illustrated later in images of Copper Canyon fan.

In the Seasat image (Figure 12.2A) the smooth surface of the sand produces a prominent, narrow dark band between the medium-to-bright signature of the active gravel and the very bright signature of the carbonate and sulfate deposits. In the TM thermal IR image (Figure 12.2D), the sand and fine-gravel deposits form a distinct light gray belt. The sand belt is inconspicuous in the LFC photograph and TM reflected IR image (Figure 12.2B,C), where it has a slightly brighter signature than the darker adjacent units. The prominent warm signature of the sand belt in the TM thermal IR image is probably due to the low thermal inertia caused in part by the low density of this unconsolidated porous material.

Halite   In their geologic map, Hunt and Mabey (1966) divided evaporite deposits on the floor of Death Valley into various units based on field observations of mineralogy, weathering characteristics, and depositional setting. Such detailed distinctions are not possible on the regional images of Figure 12.2. The evaporites are separated into two units: (1) halite deposits and (2) carbonate and sulfate deposits. Halite deposits occupy the central portion of the valley, and the carbonate and sulfate deposits occur around the margins.

The geologic maps in this chapter show the halite deposits as a single unit. On the basis of vertical relief, however, two subdivisions are recognizable: rough halite and intermediate halite. Rough halite, which is well developed at Devil's Golf Course, consists of jumbled slabs of rock salt a meter or more in diameter that are covered with sharp pinnacles. Vertical relief is 29 cm or more. Intermediate halite occurs at Cottonball Basin and in the eastern and southern portions of the valley. Seasonal flooding in these areas results in subdued vertical relief on the order of 6 cm for intermediate halite. Vertical relief of rough halite greatly exceeds the rough criterion for the Seasat radar and produces bright signatures on the image (Figure 12.2A). Relief of intermediate halite corresponds to the transition from intermediate to rough response and produces a light gray signature on the Seasat image. In printed reproductions of Seasat images, the signatures of both halite units are generally bright and may be indistinguishable.

In the TM reflected IR image (Figure 12.2C), halite signatures range from white to light gray because of differences in silt content, moisture, and distribution of surface coatings. For example, Devil's Golf Course is dark because the halite includes clay and silt, whereas the cleaner halite farther south is bright. In the LFC photograph, Devil's Golf Course is very bright and indistinguishable from surrounding floodplain deposits. The LFC photograph was acquired 2 years after the TM image (Table 12.1); apparently a coating of white salt formed at Devil's Golf Course during that time. On the daytime TM thermal IR image, the halite deposits have an intermediate radiant temperature.

Carbonate and Sulfate Deposits

These evaporite deposits, which occur around the margins of Death Valley, contain a wide range of silt and sand. The surface is commonly undulating or puffy with a relief of 6 cm, which is near the boundary between intermediate and rough surfaces for Seasat. As a result, these materials have intermediate to bright signatures on Seasat that are similar to those of intermediate halite.

In the LFC and TM images, carbonate and sulfate deposits range from dark gray to white tones, depending on clastic content, moisture, and surface coatings. In the daytime TM image, these deposits have intermediate to warm radiant temperatures.

Floodplain Deposits

Portions of the floor of Death Valley are seasonally flooded to form flat floodplains that consist of brine-saturated silt and clay. Floodplain deposits are typically covered with a salt crust ranging from a thin coating to a hard crust up to 10 cm thick.

Because of the smooth, uniform surface, the floodplains have a conspicuous dark signature on the Seasat image. In the LFC photograph the floodplains have a uniform bright signature because of the salt crust. In the TM reflected IR image, however, there are irregular patterns of medium gray tones that may be due to variations in moisture content and in thickness or composition of the salt crust. In the TM thermal IR image, most of the floodplains have a dark signature (cool radiant temperature) caused by evaporative cooling of absorbed moisture.

Distribution of flood plains changes markedly with the seasonal rainfall, which explains some of the differences between the Seasat image (1978), the TM image (1982), and the LFC photograph (1984).

Geologic Structure

Major structural features are shown in the geologic map (Figure 12.1). The following discussion of faults and turtlebacks emphasizes their appearance on the images rather than an analysis of the tectonic history of Death Valley.

Faults

The major faults are the Keane Wonder fault, the Furnace Creek fault, and the frontal fault along the western margin of the Black Mountains (Figure 12.1). At the surface these are high-angle faults that bring bedrock of the upthrown blocks against unconsolidated gravel deposits of the downthrown blocks. The north- and northwest-trending fault scarps produce very bright linear patterns on the Seasat image (Figure 12.2A) with its eastward look direction. At the southern end of the Furnace Creek fault, there is a graben in which the east-facing scarp produces a prominent dark linear radar shadow. The LFC photograph was illuminated from the southwest; shadows and highlights are oriented similar to those in the Seasat image but are less pronounced.

In the TM images, solar illumination was from the southeast, essentially parallel with the Keane Wonder and Furnace Creek faults, which therefore are not enhanced by highlights or shadows. Portions of the north-trending frontal fault are enhanced by shadows. On the LFC and TM images, the faults are recognizable by the linear contact between bedrock and alluvial deposits with their contrasting topographic expressions.

On the western margin of Death Valley, the Hanaupah fault brings a desert pavement surface on the west into contact with down-dropped gravel deposits on the east. In the Seasat image, this fault is clearly expressed by the linear contact between the dark desert pavement and gravel deposits with an intermediate gray tone.

Turtlebacks The turtleback structures of the Black Mountains (Figure 12.1) are anticlinal arches of Precambrian metamorphic rocks once covered by a sheet of tectonic breccia of Tertiary age that has been largely removed by erosion. Wright, Otton, and Troxel (1974) review various explanations for the origin of these features.

The turtlebacks are not recognizable in the Seasat image, in part because of the pronounced radar layover (Chapter 6) of these features. In the LFC and TM images, however, the Copper Canyon turtleback forms a prominent northwest-plunging nose, partially rimmed by outcrops of the breccia. Strong erosion of the flanks of the Mormon Point and Badwater turtlebacks makes them less recognizable on these images.

Cottonball Basin

Cottonball Basin, which is located in the northern part of Death Valley (Figure 12.1), includes examples of all the materials described earlier. Figure 12.3A–F compares images taken by Seasat and SIR–B satellite radar, by LFC, by aircraft nighttime thermal IR, and by Landsat TM bands 4 and 7. A map and explanation (Figure 12.3G) show the distribution of materials. Except for the Seasat example, these images are different from the regional images in Figure 12.2. The two radar images (Figure 12.3A,B) were acquired at identical wavelengths and similar spatial resolution. The depression angle for SIR–B was gentler than for Seasat; therefore, smooth and rough criteria are somewhat different. Both images were digitally processed—Seasat at Jet Propulsion Laboratory and SIR–B at Chevron Oil Field Research Company. Differences in gray tones between the images are probably due to different contrast enhancement methods rather than differences in terrain backscatter for the two radar systems. In both radar images the signatures of materials are related to their roughness. Gravel of the active portions of the alluvial fans has bright to intermediate signatures, depending on the size of the gravel fragments. Carbonate and sulfate deposits form a band around the margin of Cottonball Basin, adjacent to the alluvial fans. These deposits have an intermediate gray signature in the radar images that is consistent with their relief. Rough and intermediate halite deposits, although not mapped separately in Figure 12.3G, are distinguishable in the radar images. The intermediate halite occurs in the center of Cottonball Basin and has a light gray signature. Rough halite forms a rim around the basin and extends southward as a band adjacent to the carbonate and sulfate deposits. The bright signature of rough halite distinguishes it from the gray signatures of the other evaporite deposits, which have less surface relief.

In the aircraft nighttime thermal IR image (Figure 12.3D), darker signatures represent cooler radiant temperatures and brighter signatures are warmer. Brightest (warmest) signatures are associated with the rough halite and with native and cultivated vegetation of Furnace Creek Ranch and vicinity. Floodplain deposits and intermediate halite have intermediate temperatures. Carbonate and sulfate deposits have distinct, cool signatures in this image. Coarse gravel has an intermediate temperature, and desert pavement is recognizable by a cooler signature.

Landsat band 4 (Figure 12.3E) represents the short-wavelength portion (0.76 to 0.90 μm) of the reflected IR region, and band 7 (Figure 12.3F) represents the long-wavelength (2.08 to 2.35 μm) portion. Vegetation is bright in band 4 and dark in band 7, which agrees with spectral reflectance data for foliage. Desert pavement is dark in band 4 but has an intermediate tone in band 7.

PISGAH CRATER

Pisgah Crater is an extinct cinder cone with associated basalt lava flows in the Mojave Desert of southeast California (Figure 12.4). The volcanic rocks provide a different geologic setting from the sedimentary deposits at Death Valley. Satellite views of the region are shown in the Landsat RBV and Seasat images of Figure 12.5. The Mojave Desert consists of rugged, eroded, fault-block mountains separated by broad valleys of alluvial deposits and dry lakes. Persistent winds from the west have deposited streaks and patches of sand over local areas. The Pisgah Crater region is drained by the east-flowing Mojave River, which is dry most of the year. The major geologic structures are right-lateral, strike-slip faults that trend west and northwest through the area. Native vegetation is very sparse in this hot, arid region. Some irrigated fields are present in Troy Valley, where they have dark RBV signatures and bright Seasat signatures.

Satellite Images and Regional Structure

The map (Figure 12.4) and images (Figure 12.5) are shown at the same scale and orientation to facilitate comparison. Spatial resolution of the Landsat-3 RBV image (40 m) and Seasat image (25 m) are comparable. The different dates of acquisition are not significant because the Mojave Desert shows little seasonal change, aside from temperature differences.

As explained earlier, signatures in the visible spectral region are determined by albedo, whereas in the radar region they are largely determined by roughness and topography. These properties cause a general reversal of signatures between the RBV and Seasat images of the Pisgah region. Bedrock of the mountains and lava flows has a low albedo and dark RBV signature. Bedrock signatures are bright in the Seasat image, however, because of the rugged topography and rough surfaces. For Seasat the smooth criterion is 1 cm and the rough criterion is 6 cm. Alluvial deposits, dry lakes, and windblown sand have high albedos and bright RBV signatures. These deposits are relatively smooth and have dark Seasat signatures.

Bedrock of the mountains has a uniform bright signature in the Seasat image, so different rock types cannot be distinguished. Albedo differences in the RBV image, however, allow one to distinguish dark basalts from lighter-colored rocks. Detailed interpretation of the RBV image would distinguish several other rock units that are not recognizable in the Seasat image.

Faults are more obvious in the Seasat image, where the eastward look direction causes highlights and shadows from fault scarps. For example, the eastern part of the Lava Bed Mountains (Figure 12.4) consists of basalt flows with a dark RBV signature. These flows are cut and offset by the Pisgah fault, which has prominent scarps. Neither the offsets nor the scarps are visible in the RBV image but are clearly seen in the Seasat image. The Calico fault crosses Troy Valley in the northwest part of the area (Figure 12.4) but is not recognizable in the RBV image. In the Seasat image, however, the portion of Troy Valley west of the Calico fault is distinctly brighter than the area to the east. Field checking showed that the active fault is a barrier to the movement of subsurface ground water. Consequently, the water table is shallower on the west side of the fault and supports a sparse growth of mesquite trees. The trees trap the east-moving windblown sand and form a dune field that ends abruptly at the fault trace. The rough, hummocky dune terrain produces the bright Seasat signature, that differs from the dark signature of the smooth terrain east of the fault.

The enhanced expression of faults in this Seasat image is not a universal phenomenon. In other areas, many older faults lack topographic expression and must be recognized by the juxtaposition of different rock types. Should the rocks on opposite sides of a fault have similar roughness, they would not be distinguishable in a radar image.

Aerial Photograph and Surface Materials

The black-and-white photograph of Pisgah Crater and vicinity (Figure 12.6A) was reproduced from a high-altitude IR color photograph acquired by NASA aircraft. The geologic map (Figure 12.6B) matches the scale and orientation of the photograph. The Pisgah flows consist of fine-grained, porphyritic, vesicular basalt that has two distinct phases, aa and pahoehoe. The basalt flowed eastward onto the surface of Lavic dry lake. Pisgah Crater is actually a cinder cone that rises about 80 m above the lava surface. The cinder cone was undisturbed when the photograph was acquired in 1971, but subsequently much of the cone has been excavated for construction material. Bedrock of the Bullion Mountains crops out north of Lavic dry lake. The dark Sunshine Basalt occurs south of the lake. Materials of the Pisgah area are illustrated in Figure 12.7 and listed in Table 12.3 with their signatures on various images. Characteristics of the materials are summarized below.

Aa (Figure 12.7A)   Black, rough basalt that is difficult to walk across.

Pahoehoe (Figure 12.7B)   Gray, relatively smooth basalt with an undulating ropey surface with flow structures. In the aerial photograph (Figure 12.6A), pahoehoe has a gray signature that distinguishes it from the black aa.

TABLE **12.3**   Properties and image signatures of materials at Pisgah Crater

Material	Albedo	Roughness, cm	L-band radar signature	Radiant temperature	
				Day	Night
Aa basalt	Low	15	Bright	Cool	Warm
Pahoehoe basalt	Low	5	Intermediate	Warm	Medium
Cinders	Low	3	Intermediate	Warm	Cool
Alluvium	Medium	1	Dark	Medium	Cool
Windblown sand	High	0.5	Dark	Medium	Cool
Dry-lake deposits	High	0.5	Dark	Medium	Cool

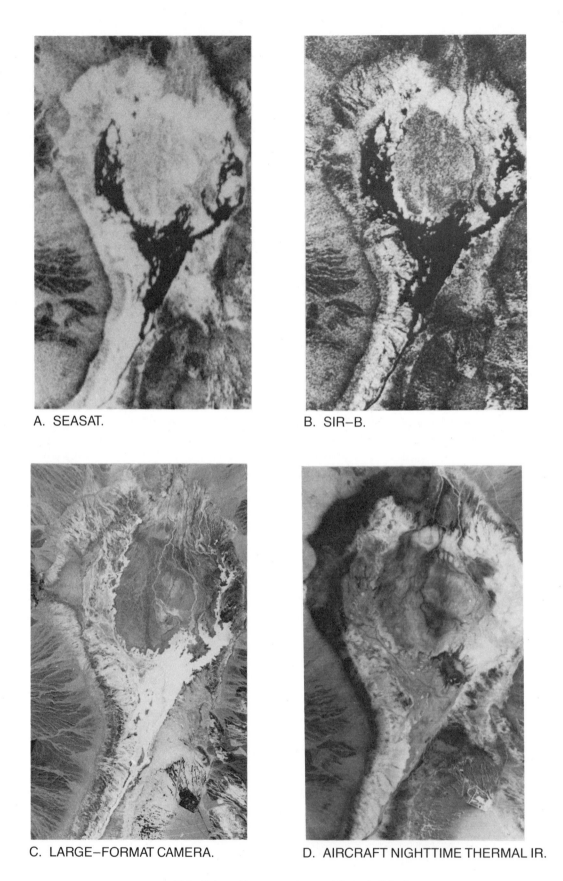

A. SEASAT.

B. SIR-B.

C. LARGE-FORMAT CAMERA.

D. AIRCRAFT NIGHTTIME THERMAL IR.

FIGURE 12.3 Enlarged images and map of Cottonball Basin.

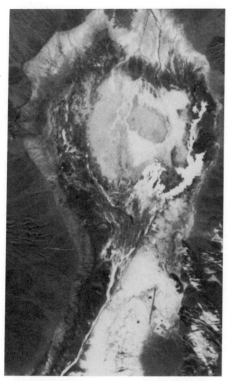

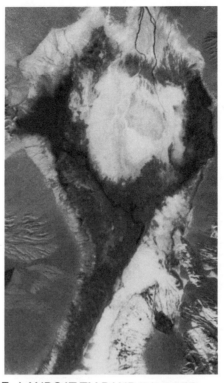

E. LANDSAT TM BAND 4 (0.76 TO
   0.90 µm).

F. LANDSAT TM BAND 7 (2.08 TO
   2.35 µm).

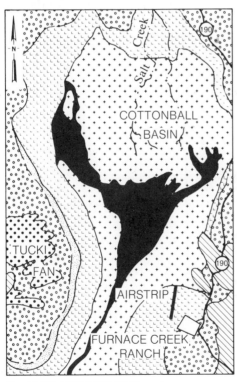

G. MAP AND LEGEND.

SYMBOL	MATERIAL
	Carbonate and sulfate deposits
	Halite deposits
	Floodplain deposits
	Sand and fine gravel
	Alluvial fan gravel
	Desert pavement
	Bedrock

0   4 mi

0   4 km

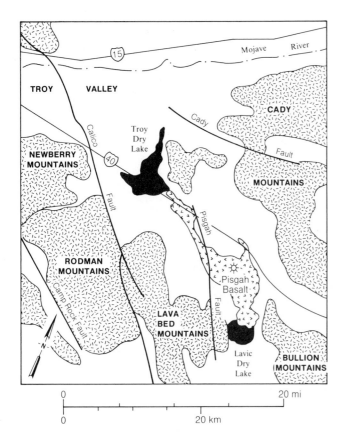

FIGURE 12.4 Regional map of Pisgah Crater and vicinity, San Bernardino County, California.

A. LANDSAT 3 RBV IMAGE ACQUIRED APRIL 30, 1981.

B. SEASAT RADAR IMAGE ACQUIRED AUGUST 27, 1978.

FIGURE 12.5 Regional satellite images of Pisgah Crater and vicinity, California.

Cinders (Figure 12.7C) Gravel-sized fragments of basalt pumice that are dark gray to red, depending on the degree of oxidation of iron minerals. Cinders have a much lower density than basalt.

Alluvium (Figure 12.7D) Gravel eroded from older bedrock, with a light gray signature. Older alluvium is more consolidated than younger alluvium.

Windblown sand (Figure 12.7E) Light-colored sheets and streaks of sand that lap onto and cover the western and southern margins of the basalt to various depths. The sand-covered basalt is indicated on the map (Figure 12.6B).

Dry-lake deposits (Figure 12.7F) Silt and clay deposits of Lavic dry lake with a very bright albedo and smooth surface with numerous mud cracks.

Radar Images

Multiple-polarized L-band aircraft images (Figure 12.8) of the Pisgah flows were acquired at a depression angle of 50° in the center of each image. For these images, the smooth criterion is 1.2 cm and the rough criterion is 7.0 cm. Scale of the radar images matches that of the aerial photograph and map (Figure 12.6). There is no vegetation to cause volume scattering of incident radar energy. As a result, the parallel-polarized and cross-polarized images have similar signatures. The rough aa basalt is distinguished by its very bright signature from the intermediate signature of the pahoehoe. Roughness characteristics of these lava types are shown in the ground photographs (Figure 12.7). The cinder cone causes highlights and shadows relative to the northward look-direction of the radar.

Windblown sand, younger alluvium, and dry-lake deposits all have less surface relief than the smooth criterion and therefore have dark signatures. The rough topography formed by older alluvium and bedrock has bright signatures.

Interstate Highway 40 has a smooth surface and is not distinguished from the dark signatures of adjacent sand and alluvium. Guard rails and large vehicles cause local bright signatures along the highway. The metal rails of the Santa Fe railroad produce a narrow bright signature. The 2.5-km stretch of railroad with the wider, bright signature is probably caused by the additional tracks of a siding.

The sand-covered basalt has a darker radar signature than the bare basalt, which indicates that little, if any, of this sand is effectively penetrated by the radar. There is adequate rainfall to cause the sand to be moist at some depth, which would restrict radar penetration.

Thermal IR Images

On March 30, 1975, a NASA aircraft acquired daytime and nighttime thermal IR images of the Pisgah cinder cone and vicinity, together with a daytime visible image (Figure 12.9). Figure 12.10 shows a geologic map of the area. In the visible-band image, materials have the same signature as in the aerial photograph (Figure 12.6A). The daytime thermal IR image (Figure 12.9B) is dominated by topographic effects. The south-facing flank of the cinder cone is sunlit and has a warm signature, whereas the north slope is shadowed and cool. A large lava dome east of the cone is cut by deep intersecting fissures that are warm on the sunlit side and cool on the shadowed side. The pahoehoe basalt has a warmer daytime radiant temperature than the aa basalt.

Most of the topographic effects dissipated by the time the nighttime thermal IR image (Figure 12.9C) was acquired. Cool nighttime signatures are associated with lower-density materials: sand, cinders, and alluvium. Basalt, with its higher density, is relatively warm at night, and there is a tendency for aa to be warmer than pahoehoe. The density of the two lava types is similar; the highly pitted surface of aa apparently traps and reradiates the thermal energy at night.

A. HIGH–ALTITUDE PHOTOGRAPH ACQUIRED BY NASA AIRCRAFT, APRIL 1971.

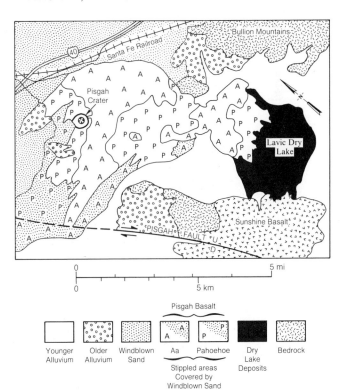

B. GEOLOGIC MAP.

FIGURE 12.6 Aerial photograph and geologic map of Pisgah Crater and vicinity.

Fissures on the lava dome have warm signatures in the nighttime image. Energy radiating from one vertical wall is absorbed and reradiated from the opposite wall, which retains heat in the fissures.

The map of apparent thermal inertia (*ATI*) in Figure 12.10B was prepared by Jet Propulsion Laboratory using

A. AA BASALT.

B. PAHOEHOE BASALT.

C. CINDERS.

D. ALLUVIUM.

E. WINDBLOWN SAND.

F. DRY—LAKE DEPOSITS.

FIGURE 12.7 Ground photographs of materials at Pisgah Crater. Each photograph is 7 cm wide.

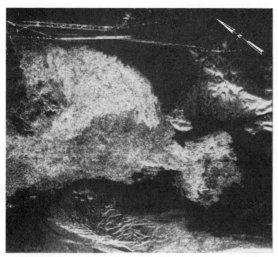

A. HH (PARALLEL–POLARIZED).

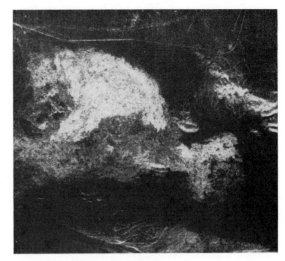

B. HV (CROSS–POLARIZED).

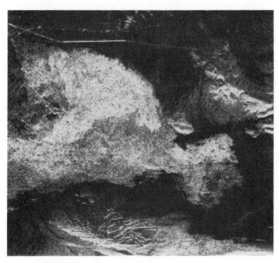

C. VV (PARALLEL–POLARIZED).

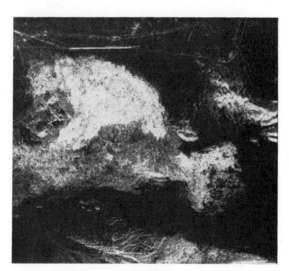

D. VH (CROSS–POLARIZED).

FIGURE 12.8 Multiple-polarized, L-band (25-cm wavelength), synthetic-aperture, aircraft radar images acquired June 3, 1984. Look direction is toward the upper margin of the images. Each image is 12 km wide. Courtesy Jet Propulsion Laboratory.

the method described in Chapter 5. The daytime and nighttime images were digitally registered using ground-control points, as described in Chapter 7. For each pixel, ΔT was calculated by subtracting the nighttime radiant temperature from the daytime temperature. Albedo (A) is determined from the daytime visible-image data.

Figure 12.10B displays the resulting values of *ATI* as a map such that bright tones represent high *ATI* values. This map has not been corrected for topographic effects; shadowed areas, such as the north flank of the cinder cone and some fissure walls, have high *ATI* values that do not correlate with thermal inertia. Elsewhere in the map, the *ATI* values agree with relative density of materials. Aa and pahoehoe have the highest *ATI* values; cinders, sand, and alluvium have the lowest values. Gillespie and Kahle (1977) published additional thermal IR and *ATI* images of Pisgah Crater.

COMMENTS

The wide range of materials and geologic features at Death Valley and the Pisgah Crater area have been imaged by remote sensing systems that extend from the

A. DAYTIME REFLECTANCE (ALBEDO) (0.4 TO 0.7 μm).

B. DAYTIME THERMAL IR (8 TO 14 μm).

C. NIGHTTIME THERMAL IR (8 TO 14 μm).

0   1.0 mi

0   1.0 km

FIGURE 12.9 Visible and thermal IR images acquired March 30, 1975, by NASA and processed by Jet Propulsion Laboratory. From Gillespie and Kahle (1977, Figures 6 and 7). Courtesy A. B. Kahle, Jet Propulsion Laboratory.

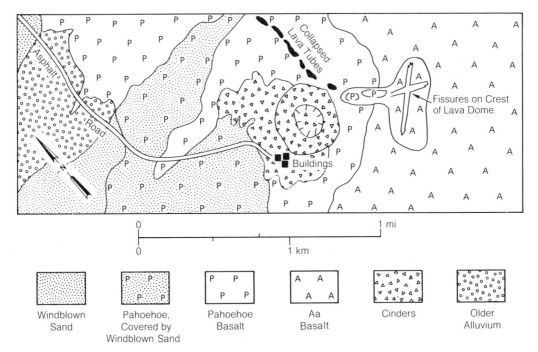

A. GEOLOGIC MAP.

Legend:

Windblown Sand

Pahoehoe, Covered by Windblown Sand

Pahoehoe Basalt

Aa Basalt

Cinders

Older Alluvium

B. MAP OF APPARENT THERMAL INERTIA. BRIGHT TONES ARE HIGH *ATI* VALUES. FROM GILLESPIE AND KAHLE (1977, FIGURE 11). COURTESY A.B. KAHLE, JET PROPULSION LABORATORY.

FIGURE 12.10 Geologic map and apparent thermal inertia map.

visible through the microwave spectral regions. Comparing these images demonstrates that no single system provides all the needed information about an area. In Death Valley, for example, floodplain deposits, halite, and carbonate and sulfate deposits may all have high albedos in the visible region and be indistinguishable on photographs and Landsat images. Because of roughness differences, however, these deposits are readily distinguished on aircraft and satellite radar images.

In highly dissected bedrock areas such as the Mojave Desert, the surface relief of all types of bedrock may exceed the rough criterion for radar. The resulting bright signatures do not differentiate among rock types. Images in the visible band, however, record albedo differences that may distinguish rocks such as basalt, rhyolite, and granite.

Thermal IR images distinguish materials on the basis of their thermal inertia, which is primarily determined by density for rocks and soils. Variations in moisture content also influence thermal IR signatures and have been used to map faults in arid and semiarid regions.

The interpreter should keep in mind the distinction between distinguishing and identifying materials. A material may be *distinguished*, or separated, from the surrounding materials and mapped as a unit on the basis of such physical properties as albedo, thermal inertia, or roughness. These physical properties alone are rarely sufficient to identify the material because many materials may have similar combinations of physical properties. *Identifying* a material is the process of recognizing and classifying it as, for example, sandstone, shale, deciduous forest, or wheat fields. Identification frequently requires information in addition to data on the physical properties determined from remote sensing images. This additional information may be obtained by visiting the area under study. The author visited many of the areas described in this text to identify the materials responsible for signatures in the images. Other identifications were made from published maps and reports.

In practice the difference between distinguishing and identifying materials is not clear-cut. An experienced interpreter working with images of an unfamiliar area can distinguish and map materials based on their physical properties. By analogy to known areas, the interpreter can then identify many of the materials without field visits or reference to published information. For example, given radar and TM images of an arid basin in Mongolia, an interpreter could not only distinguish and map the materials and geologic structures but also identify features such as desert pavement, floodplain deposits, faults, and so forth. However, field checking would be essential to verify these identifications.

New methods such as the airborne imaging spectrometer (Chapter 2) and thermal IR multispectral scanner (Chapter 5) measure detailed spectral characteristics that may enable many materials to be identified remotely.

Other studies have been made to compare and evaluate the usefulness of various remote sensing images. Newton (1981) and Newton, Viljoen, and Longshaw (1981) compared various types of aerial photographs, Landsat images, and other images for geologic mapping in South Africa. Just as in the Death Valley and Pisgah Crater examples, the results of these and other studies are specific for the areas and objectives of the investigation. In humid, cloud-covered, vegetated tropical regions, Landsat and thermal IR images are of limited value whereas radar images are essential. Agricultural and forestry investigations usually find IR color photographs and Landsat images far more useful than radar or thermal IR images. The investigator must select the image types based on the sensor capability, climatic conditions, and objectives of the investigation.

QUESTIONS

1. For your local area, list the major terrain categories (types of vegetation, rocks, land use, land cover, and so forth). For each category, predict the signature in following images: Landsat TM IR color image, Seasat radar image, and daytime and nighttime thermal IR images.

2. Explain the reason for each predicted signature.

3. Acquire and interpret various types of remote sensing images of your area. For each terrain category, compare the actual image signatures with your predictions.

4. Explain any differences between your predictions and the actual signatures. Field checking should help resolve these discrepancies.

5. Bedrock in the Pisgah Crater region consists of different crystalline and volcanic rocks that cannot be distinguished in the Seasat image (Figure 12.5B). In Indonesia, however, several different lithologic terrains were distinguished in SIR–A images (Chapter 6). Explain this regional difference, which is not related to radar system differences, such as depression angle.

REFERENCES

Drewes, H., 1963, Geology of the Funeral Peak Quadrangle, California, on the east flank of Death Valley: U.S. Geological Survey Professional Paper 413.

Gillespie, A. R., and A. B. Kahle, 1977, Construction and interpretation of a digital thermal inertia image: Photogrammetric Engineering and Remote Sensing, v. 43, p. 983–1000.

Hunt, C. B., and D. R. Mabey, 1966, Stratigraphy and structure Death Valley, California: U.S. Geological Survey Professional Paper 494–A.

Newton, A. R., 1981, Evaluation of remote sensing methods in a test area near Krugersdorp, Transvaal, I—Aerial Photography: Transactions of the Geological Society of South Africa, v. 84, p. 207–216.

Newton, A. R., R. P. Viljoen, and T. G. Longshaw, 1981, Evaluation of remote sensing methods in a test area near Krugerdorp, Transvaal, II—Landsat imagery: Transactions of the Geological Society of South Africa, v. 84, p. 217–275.

Wright, L. A., J. K. Otton, and B. W. Troxel, 1974, Turtleback surfaces of Death Valley viewed as a phenomenon of extensional tectonics: Geology, v. 2, p. 53–54.

ADDITIONAL READING

Kahle, A. B., J. P. Schieldge, M. J. Abrams, R. E. Alley, and C. J. LeVine, 1981, Geologic applications of thermal inertial imaging using HCMM data: Jet Propulsion Laboratory Publication 81-55, Pasadena, Calif.

Sabins, F. F., 1984, Geologic mapping of Death Valley from thematic mapper, thermal infrared, and radar images: International Symposium on Remote Sensing of Environment, Proceedings of the Third Thematic Conference, Remote Sensing for Exploration Geology, p. 139–152, Environmental Research Institute of Michigan, Ann Arbor, Mich.

Summary

Remote sensing is a diverse field, in both its technology and applications. Chapter 1 of this book described the properties of electromagnetic radiation and its interaction with matter. The nature of this interaction, which is specific for different wavelengths of radiation and different types of matter, is detected and recorded by remote sensing systems. Table 13.1 summarizes the characteristics of the principal remote sensing systems. This table will aid in selecting the optimum sensing system for a particular application. At one time the concept of multiple sensor imagery was popular; for any interpretation project, it was felt that all possible types of imagery should be acquired and interpreted. This approach is rarely applied today. Few of the remote sensing systems can be used simultaneously to acquire optimum images. For example, the high altitude and side-scanning geometry of aircraft radar acquisition are incompatible with other systems such as thermal IR. The expense of multiple flights with different aircraft and imaging systems is excessive for reconnaissance surveys. Today's strategy is to select the remote sensing system that will provide the maximum amount of relevant information for the project. For example, geologic structure in forested areas with dense cloud cover is best mapped using radar images. Thermal plumes and currents in water bodies, on the other hand, are best mapped using thermal IR images. Imaging systems for other applications can be selected from Table 13.1.

LIMITATIONS AND PRECAUTIONS

Previous chapters have illustrated applications of remote sensing for a variety of activities ranging from oil and mineral exploration to environmental monitoring. These examples have also described some of the limitations and precautions associated with the various remote sensing methods. For example, in multispectral classification for mineral prospects, clay deposits in dry stream channels may have spectral signatures similar to those of hydrothermally altered rocks that contain clay minerals. On thermal IR images, the cool radiant temperatures of oil films may be identical to those of cold-water currents. Experienced interpreters have learned to cope with these problems. The linear branching pattern of stream-deposited clay aids in distinguishing this material from the more localized pattern of hydrothermal alteration zones. The experienced interpreter will refer to simultaneously acquired UV and visible images to distinguish oil films from water currents.

Experienced interpreters are those who understand (1) the capabilities and limitations of the imaging system

TABLE 13.1 Remote sensing systems and image types

Image types	Wavelength region detected	Properties detected	Imaging systems
Ultraviolet images and photographs	0.3 to 0.4 μm	Reflectance and fluorescence from solar radiation.	Aerial cameras and scanners.
Visible and reflected IR photographs	0.4 to 0.9 μm	Spectral reflectance of solar energy in the visible and short-wavelength IR regions. Restricted to spectral sensitivity range of film.	Aerial cameras, hand-held satellite cameras, and Space Shuttle large-format camera.
Visible and multispectral reflected IR images	0.4 to 3.0 μm	Spectral reflectance of solar energy in the visible and reflected IR regions.	Landsat MSS, Landsat TM, aerial scanners, AIS, SPOT, AVHRR, and CZCS.
Thermal IR	3.0 to 14.0 μm	Radiant temperature, which is determined by kinetic temperature, emissivity, and thermal inertia.	Aerial scanners, HCMM, Landsat TM band 6, and TIMS.
Radar	1 to 30 cm	Surface roughness and dielectric properties.	Aerial systems, Seasat, and SIR–A, B, and C.

and (2) the characteristics of the features being interpreted, such as land use, geology, or oceanography. If the image data have been digitally processed, the interpreter must understand the effect that processing has on the image. For this reason, interpreters should participate in the interactive processing of the data that will be used in their projects. This book provides the technical background and procedures for understanding and applying remote sensing technology. Books alone, however, cannot provide the all-important factor of experience, which is gained only by working with images. Students in my remote sensing classes are sometimes discouraged when I point out features they missed while interpreting an image. A common reaction is to say, "The fault is so obvious when you point it out; how did I miss it?" My response is that the instructor sees the image from the perspective of many years of experience, whereas the student has only a few weeks of practice. There is no substitute for experience.

The science of remote sensing has matured perceptibly over the past decade. Only rarely are flamboyant claims made that some new method will "find oil fields and mines." The true capabilities and limitations that have always been understood by remote sensing professionals are becoming more generally understood. In the field of resource exploration, for example, few people expect that remote sensing alone will discover oil fields and mines. However, small-scale satellite images of relatively low spatial resolution, such as Landsat MSS, provide regional data from which one can recognize sedimentary basins and potential metal-bearing trends. Satellite and aircraft images of higher spatial resolution, such as Landsat TM and aircraft radar and thermal IR images, enable the interpreter to recognize local structures. Multispectral image data, when digitally processed, may indicate hydrothermally altered rocks where mineral deposits may occur. In all these cases, however, remote sensing is an initial step that must be followed by field checking, sampling, core drilling, and geophysical surveys before wells are drilled or shafts are sunk. Such caution is required because nonrenewable resources occur at some depth in the earth, and remote sensing images only record surface phenomena from which subsurface exploration targets must be inferred.

The relationship between images and targets is more straightforward for other users of images. Renewable resources such as surface water and vegetation occur at the surface and can be interpreted more directly. This direct relationship, however, does not eliminate the requirement for the field checking of image interpretations.

FUTURE DEVELOPMENTS

Future developments and change will occur in two areas: technical and political.

Over the past decade many technical advances have occurred. High-quality radar images have been acquired from satellites. The spatial and spectral resolution of images throughout the electromagnetic spectrum have improved. Digital processing of image data is more advanced and more widely used. Future technical advances will provide even finer spatial and spectral resolution with the goal of identifying materials rather than simply distinguishing broad categories of materials. Major technical advances are occurring in the image-processing field. Costs of hardware and software are dropping, while capabilities are increasing.

Several of the recent technical advances have demonstrated the feasibility of acquiring new types of data, but the technology has not become operational. Seasat and SIR, for example, have demonstrated the capability of acquiring radar images from space; however, no operational radar satellite system exists. In a similar fashion, the Skylab earth-terrain camera and the Shuttle large-format camera programs have demonstrated the feasibility and value of acquiring high-resolution stereo photographs from space. It would appear that LFC could be routinely deployed on future Space Shuttle missions, but no such plans have been announced.

Political developments influencing the future of remote sensing are occurring within both the United States and the international community. The U.S. government transferred operation of the Landsat system to the commercial sector in 1985. Time will tell whether this was a wise and timely decision. Although the federal government will cease to operate satellites for remote sensing of land areas, it is essential that the government maintain a leading role in research and development of new and improved technology for use in future systems.

Since the inception of remote sensing, the United States has been the leading nation in acquiring and disseminating data. In the past, other nations did not develop their own satellite imaging systems for various reasons, including (1) the cost of such systems; and (2) the liberal U.S. policy on data distribution, which has reduced the need for other satellite systems. In the future, however, other nations will launch their own satellite imaging systems. It remains to be seen whether data from these future satellites will be as universally available and cost effective as those from the Landsat and other U.S. programs.

Appendix

Basic Geology for Remote Sensing

On images of land areas, the terrain is a direct expression of the geology of the area. Many interpreters are concerned with nongeologic subjects (such as land use, environment, or forestry), but an understanding of the geology will contribute to the overall understanding of the image. This brief review emphasizes the major rock types and geologic structures that are expressed on images. Additional information is given in geology texts, such as those by Shelton (1966) and by Press and Siever (1986).

ROCK TYPES

Rocks belong to three major categories: sedimentary, igneous, and metamorphic. These are described in the following sections.

Sedimentary Rocks

Material that has been transported and deposited by water or wind forms sedimentary rocks characterized by layers, called *beds* or *strata*, formed during deposition. The surfaces separating strata are called *bedding planes*. Outcrops of sedimentary strata typically have a banded appearance on images. *Sedimentary rocks* are divided into the broad categories of clastic and chemical rocks.

Clastic Rocks   Erosion of older rocks produces fragments and particles that are transported and deposited to form *clastic rocks*. After deposition, the fragments are compressed and cemented to form rocks. The consolidated rocks are classified on the basis of particle size before consolidation, as follows:

Consolidated rocks	Unconsolidated particles
Conglomerate	Boulders and gravel
Sandstone	Sand
Siltstone	Mud
Shale	Clay

Clastic rocks differ in their resistance to erosion: sandstone and conglomerate typically form ridges, but shale and siltstone form valleys.

Chemical Rocks   Minerals dissolved in water may be removed from solution to form *chemical rocks*, either by chemical precipitation or by uptake into organisms whose shells and skeletons form sediments after death. Algae and shellfish remove calcium carbonate from seawater as the mineral calcite, which accumulates to form the rock called *limestone*. Half of the calcium atoms in calcite may be replaced by magnesium to form the mineral and rock called *dolomite*. Evaporation of seawater produces deposits of gypsum, anhydrite, and salt. Because of its low density, salt may migrate upward into the overlying strata to form cylindrical plugs called *salt domes*.

Open spaces between the grains of sedimentary rocks are called *pores* and contain fresh or salt water or, less commonly, oil and gas.

Igneous Rocks

Igneous rocks are rocks that have cooled from molten material. They are assigned to the classes of intrusive and extrusive rocks.

Intrusive Rocks Molten rock, called *magma*, that invades country rock (areas of older rock) cools to form *intrusive rocks*. Based on the relationship to the country rock, intrusive rocks are classed as batholiths, dikes, and sills (Figure A.1). *Batholiths* are large, irregularly shaped masses of intrusive rock that cut across the structure of the country rock. Erosion of the overlying country rocks exposes outcrops of the batholith rock. The outcrops are commonly cut by intersecting fractures, called *joints*, that give a distinctive appearance to these rocks on images. Typical batholithic rocks crop out in the Peninsular Ranges of southern California and the Adirondack Mountains of New York, which are illustrated on Landsat images in Chapter 4.

As shown in Figure A.1, country rocks are commonly layered or stratified. Tabular bodies of igneous rock that are intruded parallel with the layers are *sills*, such as the Palisades sill that crops out along the west bank of the Hudson River, New York.

Tabular bodies of intrusive rock that cut across the layers of country rock are called *dikes*. Erosion of the overlying country rock may expose dikes in the form of ridges, such as the Chinese Wall at Yellowstone National Park. Some dikes are less resistant to erosion than the surrounding country rock and weather to form depressions. On images, dikes are recognized by their linear shape and cross-cutting relationships to the country rock.

As the intruded magma slowly cools, silicate minerals form relatively coarse crystals that are visible to the unaided eye. There are two major types of intrusive rocks, classed according to silica content:

1. *Granite*, which has a high silica content and is typically light gray to pink.

2. *Gabbro*, which has a relatively low silica content and

high content of iron- and magnesium-bearing minerals. Gabbro is dark and has a higher density than granite.

Extrusive Rocks Magma may reach the surface as a liquid called *lava*, which cools to form lava flows. The lava may also be explosively ejected into the air as cinders and ash that can accumulate as volcanoes, or *cinder cones* (Figure A.1). Because lava cools rapidly at the surface, the resulting *extrusive rocks* are very fine-grained. Three major categories of extrusive rocks are

1. *Rhyolite*: the extrusive equivalent of granites. Rhyolite has a high silica content and is typically pink.

2. *Andesite*: a rock having intermediate silica content. Andesite volcanoes form the "ring of fire" around the Pacific Ocean and make up the Andes of South America and the Cascade Range of northwestern United States. Mount St. Helens (Chapter 11) is an andesite volcano.

3. *Basalt*: the extrusive equivalent of gabbro. Basalt has a low silica content and a high content of iron- and magnesium-bearing minerals. Basalt is dark and has a relatively high density. Flows of basalt with rough surfaces are called *aa*; those with smooth, ropey surfaces are called *pahoehoe*. Both flow types occur in the Pisgah Crater area (Chapter 12) together with a volcanic cone of basalt cinders.

Metamorphic Rocks

Heat and pressure may transform igneous and sedimentary rocks into *metamorphic rocks*. As shown in Figure A.1, the heat and pressure associated with intrusive rocks may convert original country rock into the following kinds metamorphic rock:

Original rock	Metamorphic equivalent
Sandstone	Quartzite
Shale and siltstone	Schist
Limestone	Marble

Intensive regional compression and deep burial can also produce metamorphic rocks.

STRUCTURAL GEOLOGY

Stresses within the earth's crust may deform the rocks to produce geologic structures that may be mapped on

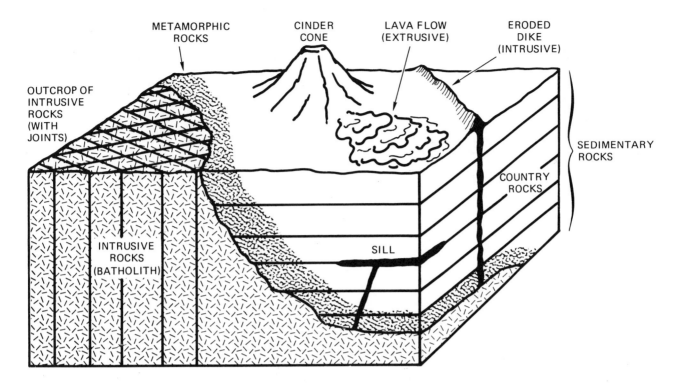

FIGURE A.1. Intrusive and extrusive rocks.

various types of images. Most sedimentary rocks were deposited as horizontal strata, but later uplift and tilting caused the strata to become inclined.

Strike and Dip

Geologists use the terms *strike* and *dip* to describe the orientation and degree of inclination. The line formed by the intersection between an inclined surface and a horizontal plane is the *strike* of the inclined surface (Figure A.2). The orientation of the strike is described by its geographic azimuth, in this case N45°E. *Dip* is measured in the vertical plane oriented normal to the strike and measures the inclination below horizontal of the inclined surface in degrees. In Figure A.2, the dip is 30° toward the southeast. The strike-and-dip symbol is used to record the *attitudes* of structures on geologic maps. Exposed bedding planes of dipping strata are called *dip slopes*. Erosion of dipping strata produces ledges called *antidip scarps* (Figure A.2) that face the opposite direction from dip slopes.

The ability to recognize strike and dip is fundamental for interpreting geologic structure from remote sensing images. Except for highly deformed areas, beds generally dip less than 45°, and the following criteria apply. Dip slopes are relatively broad and are traversed by relatively long streams that flow in the direction of dip.

Antidip scarps are narrow and have a few short drainage channels that flow opposite to the dip direction. The orientation of shadows and highlights is an important key for interpreting images acquired with inclined illumination, such as low-sun-angle aerial photographs and radar images. If the scene in Figure A.2 were photographed in the morning with sun shining from the southeast, the dip slope would form a bright expanse and the shadowed antidip scarp would form a narrow dark band; in the afternoon the highlights and shadows would be reversed.

Folds

Compressive stresses form folds called anticlines and synclines. As shown in Figure A.3, in *anticlines* the beds dip away from the axis; in *synclines* the beds dip toward the axis. Note orientation of the strike-and-dip symbols and the attitude of the dip slopes. Also, note the arcuate outcrop patterns that mark the *plunge* of a fold. Erosion exposes older beds in the center of anticlines and younger beds in the center of synclines. On geologic maps the axes of folds are shown by long lines with short crossing arrows that point away from the crest of anticlines and toward the center of synclines (Figure A.3). Chapter 2 showed aerial photographs of the Alkali anticline and syncline in Wyoming.

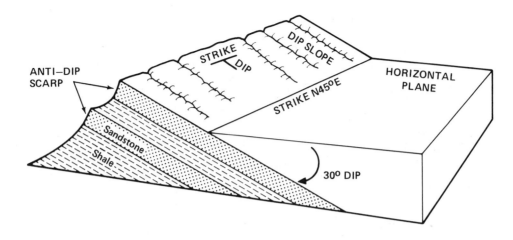

FIGURE A.2. Strike and dip of inclined beds.

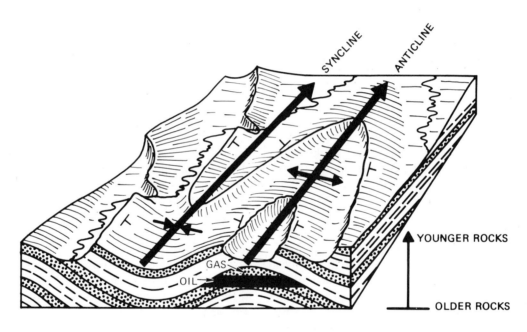

FIGURE A.3. Anticline, syncline, and oil field.

Organic matter contained in shales (*source rocks*) generates oil and gas, which migrate into the pore spaces of sandstones and limestones (*reservoir rocks*). Rocks in the subsurface are also saturated with water. Because hydrocarbons are less dense than water, oil and gas float on the water, becoming concentrated in the high points of structures, such as crests of anticlines. Such concentrations may form oil fields (Figure A.3). Chapter 4 shows a Landsat TM image of anticlinal oil fields in the Thermopolis area of Wyoming. A Landsat MSS image of the giant Rangely field anticline in Colorado is shown in

Chapter 8. Oil fields also occur along faults and in ancient reefs, sandbars, and river channels.

Joints and Faults

Joints, mentioned earlier in connection with batholiths, are intersecting sets of fractures along which no movement has occurred (Figure A.1). Erosion along joints produces linear depressions that are readily interpreted on images. *Faults* are fractures along which appreciable movement has occurred. Strike-and-dip terminology is also used to describe the attitude of fault surfaces. Faults are assigned to the following categories according to the kind of relative movement.

Normal Faults   The relative movement of rocks on either side of a fault is indicated by arrows, as shown in Figure A.4. *Normal faults* dip less than 90° and the rocks on the upper side of the fault have moved downward relative to those on the lower side. Topographic escarpments caused by faults are called *fault scarps*. In the Basin and Range Province of the United States, normal faults separate the uplifted mountain ranges from the intervening down-dropped basins, as shown in the Landsat mosaic of Nevada in Chapter 4. The uplifted blocks are called *horsts* and the down-dropped basins are called *graben.*

Thrust Faults   *Thrust faults* are horizontal or gently dipping faults in which the rocks overlying the fault have moved up and over the rocks below the fault (Figure A.4). Thrust faults are common in the Central Range of Irian Jaya, as seen in the Space Shuttle radar images and maps in Chapter 6.

Strike-Slip Faults   Faults along which the rocks have moved laterally are *strike-slip faults* (Figure A.4). Because of their steep dip, strike-slip faults have linear surface traces, as seen by the traces of the San Andreas and Garlock faults in the Landsat image of the Los Angeles region (Plate 4). The offset of features on opposite sides of a strike-slip fault is used to determine the relative displacement. In Figure A.4 an observer standing on the stream and looking across the fault at the offset channel will note that the lateral displacement of the stream is toward the right, making this a right-lateral, strike-slip fault. In left-lateral faults the features are displaced to the left.

REFERENCES

Press, F., and R. Siever, 1986, Earth, fourth edition: W. H. Freeman and Co., N.Y.

Shelton, J. S., 1966, Geology illustrated: W. H. Freeman and Co., N.Y.

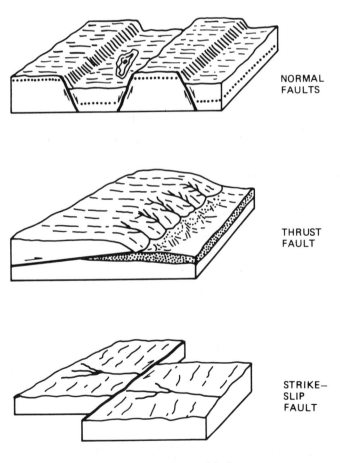

NORMAL FAULTS

THRUST FAULT

STRIKE–SLIP FAULT

FIGURE **A.4.**   Types of faults.

Glossary

The following glossary includes terms, abbreviations, and acronyms commonly employed in remote sensing. The glossary definitions refer to the applications for which the terms are used in this text and omit applications outside the field of remote sensing. Definitions of geologic and geographic terms may be found in any standard text on those subjects or in the *Glossary of Geology*, edited by R. L. Bates and J. A. Jackson and published in 1980 by the American Geological Institute.

absorption band Wavelength interval within which electromagnetic radiation is absorbed by the atmosphere or by other substances.

absorptivity Capacity of a material to absorb incident radiant energy.

active remote sensing Remote sensing methods that provide their own source of electromagnetic radiation to illuminate the terrain. Radar is one example.

additive primary colors Blue, green, and red. Filters of these colors transmit the primary color of the filter and absorb the other two colors.

adiabatic cooling Refers to decrease in temperature with increasing altitude.

advanced very high resolution radiometer (AVHRR) Cross-track multispectral scanner on a NOAA polar-orbiting satellite that acquires five spectral bands of data (0.55 to 12.50 μm) with a ground resolution cell of 1.1 by 1.1 km.

air base Ground distance between optical centers of successive overlapping aerial photographs.

aerial magnetic survey Survey that records variations in the earth's magnetic field.

airborne imaging spectrometer (AIS) Along-track multispectral scanner with spectral bandwidth of 0.01 μm.

airborne visible and infrared imaging spectrometer (AVIRIS) Experimental airborne along-track multispectral scanner under development at JPL to acquire 224 images in the spectral region from 0.4 to 2.4 μm.

AIS Airborne imaging spectrometer.

albedo (A) Ratio of the amount of electromagnetic energy reflected by a surface to the amount of energy incident upon it.

along-track scanner Scanner with a linear array of detectors oriented normal to flight path. The IFOV of each detector sweeps a path parallel with the flight direction.

alteration Changes in color and mineralogy of rocks surrounding a mineral deposit that are caused by the solutions that formed the deposit. Suites of alteration minerals commonly occur in zones.

amplitude For waves, the vertical distance from crest to trough.

analog display A form of data display in which values are shown in graphic form, such as curves. Differs from digital displays, in which values are shown as arrays of numbers.

angular beam width In radar, the angle subtended in the horizontal plane by the radar beam.

angular field of view Angle subtended by lines from a remote sensing system to the outer margins of the strip of terrain that is viewed by the system.

angular resolving power Minimum separation between two resolvable targets, expressed as angular separation.

anomaly An area on an image that differs from the surrounding, normal area. For example, a concentration of vegetation within a desert scene constitutes an anomaly.

antenna Device that transmits and receives microwave and radio energy in radar systems.

aperture Opening in a remote sensing system that admits electromagnetic radiation to the film or detector.

Apollo U.S. lunar exploration program of satellites with crews of three astronauts.

apparent thermal inertia (ATI) An approximation of thermal inertia calculated as one minus albedo divided by the difference between daytime and nighttime radiant temperatures.

ASA index Index of the American Standards Association designating film speed, or sensitivity to light. Higher values indicate higher sensitivity. The ASA index has been replaced by the ISO index.

ATI Apparent thermal inertia.

atmosphere Layer of gases that surrounds some planets.

atmospheric correction Image-processing procedure that compensates for effects of selectively scattered light in multispectral images.

atmospheric window Wavelength interval within which the atmosphere readily transmits electromagnetic radiation.

attitude Angular orientation of a remote sensing system with respect to a geographic reference system.

AVIRIS Airborne visible and infrared imaging spectrometer.

azimuth Geographic orientation of a line given as an angle measured in degrees clockwise from north.

azimuth direction In radar images, the direction in which the aircraft is heading. Also called *flight direction*.

azimuth resolution In radar images, the spatial resolution in the azimuth direction.

background Area on an image or the terrain that surrounds an area of interest, or target.

backscatter In radar, the portion of the microwave energy scattered by the terrain surface directly back toward the antenna.

backscatter coefficient A quantitative measure of the intensity of energy returned to a radar antenna from the terrain.

band A wavelength interval in the electromagnetic spectrum. For example, in Landsat images the bands designate specific wavelength intervals at which images are acquired.

base–height ratio Air base divided by aircraft height. This ratio determines vertical exaggeration on stereo models.

batch processing Method of data processing in which data and programs are entered into a computer that carries out the entire processing operation with no further instructions.

bathymetry Configuration of the seafloor.

beam A focused pulse of energy.

binary Numerical system using the base 2.

bit Contraction of *binary digit*, which in digital computing represents an exponent of the base 2.

blackbody An ideal substance that absorbs all the radiant energy incident on it and emits radiant energy at the maximum possible rate per unit area at each wavelength for any given temperature. No actual substance is a true blackbody, although some substances, such as lampblack, approach its properties.

brightness Magnitude of the response produced in the eye by light.

brute-force radar See real-aperture radar.

byte A group of eight bits of digital data.

calibration Process of comparing an instrument's measurements with a standard.

calorie Amount of heat required to raise the temperature of 1 g of water by 1°C.

camouflage detection photographs Another term for *IR color photograph*.

cardinal point effect In radar, very bright signatures caused by optimally oriented corner reflectors, such as buildings.

cathode ray tube (CRT) A vacuum tube with a phosphorescent screen on which images are displayed by an electron beam.

CCD Charge-coupled detector.

CCT Computer-compatible tape.

centerpoint The optical center of a photograph.

change-detection images A difference image prepared by digitally comparing images acquired at different times. The gray tones or colors of each pixel record the amount of difference between the corresponding pixels of the original images.

charge-coupled detector (CCD) A device in which electrons are stored at the surface of a semiconductor.

chlorosis Yellowing of plant leaves resulting from an imbalance in the iron metabolism caused by excess concentrations of copper, zinc, manganese, or other elements in the plant.

circular scanner Scanner in which a faceted mirror rotates about a vertical axis to sweep the detector IFOV in a series of circular scan lines on the terrain.

classification Process of assigning individual pixels of an image to categories, generally on the basis of spectral reflectance characteristics.

coastal zone color scanner (CZCS) A satellite-carried multispectral scanner designed to measure chlorophyll concentrations in the oceans.

color composite image Color image prepared by projecting individual black-and-white multispectral images, each through a different color filter. When the projected images are superposed, a color composite image results.

color ratio composite image Color composite image prepared by combining individual ratio images for a scene using a different color for each ratio image.

complementary colors Two primary colors of light (one additive and the other subtractive) that produce white light when added together. Red and cyan are complementary colors.

computer-compatible tape (CCT) The magnetic tape on which the digital data for Landsat MSS and TM images are distributed.

conduction Transfer of electromagnetic energy through a solid material by molecular interaction.

contact print A reproduction from a photographic negative in direct contact with photosensitive paper.

contrast enhancement Image-processing procedure that improves the contrast ratio of images. The original narrow range of digital values is expanded to utilize the full range of available digital values.

contrast ratio On an image, the ratio of reflectances between the brightest and darkest parts of the image.

convection Transfer of heat through the physical movement of heated matter.

corner reflector Cavity formed by two or three smooth planar surfaces intersecting at right angles. Electromagnetic waves entering a corner reflector are reflected directly back toward the source.

COSMIC Computer Software Management and Information Center, University of Georgia. This facility distributes computer programs developed by U.S. government-funded projects.

cross-polarized Describes a radar pulse in which the polarization direction of the return is normal to the polarization direction of the transmission. Cross-polarized images may be HV (horizontal transmit, vertical return) or VH (vertical transmit, horizontal return).

cross-track scanner Scanner in which a faceted mirror rotates about a horizontal axis to sweep the detector IFOV in a series of parallel scan lines oriented normal to the flight direction.

CRT Cathode ray tube.

cycle One complete oscillation of a wave.

CZCS Coastal zone color scanner.

data collection system (DCS) On Landsats 1 and 2, the system that acquired information from seismometers, flood gauges, and other measuring devices. These data were relayed to ground receiving stations.

densitometer Optical device for measuring the density of photographic transparencies.

density, of images Measure of the opacity, or darkness, of a negative or positive transparency.

density, of materials (ρ) Ratio of mass to volume of a material, typically expressed as grams per cubic centimeter.

density slicing Process of converting the continuous gray tones of an image into a series of density intervals, or slices, each corresponding to a specific digital range. The density slices are then displayed either as uniform gray tones or as colors.

depolarized Refers to a change in polarization of a transmitted radar pulse as a result of various interactions with the terrain surface.

depression angle (γ) In radar, the angle between the imaginary horizontal plane passing through the antenna and the line connecting the antenna and the target.

detectability Measure of the smallest object that can be discerned on an image.

detector Component of a remote sensing system that converts electromagnetic radiation into a recorded signal.

developing Chemical processing of an exposed photographic emulsion to produce an image.

dielectric constant Electrical property of matter that influences radar returns. Also referred to as *complex dielectric constant*.

difference image Image prepared by subtracting the digital values of pixels in one image from those in a second image to produce a third set of pixels. This third set is used to form the difference image.

diffuse reflector Surface that reflects incident radiation nearly equally in all directions.

digital display A form of data display in which values are shown as arrays of numbers.

digital image processing Computer manipulation of the digital-number values of an image.

digital number (DN) Value assigned to a pixel in a digital image.

digitization Process of converting an analog display into a digital display.

digitizer Device for scanning an image and converting it into numerical format.

directional filter Mathematical filter designed to enhance on an image those linear features oriented in a particular direction.

distortion On an image, changes in shape and position of objects with respect to their true shape and position.

diurnal Daily.

Doppler principle Describes the change in observed frequency that electromagnetic or other waves undergo as a result of the movement of the source of waves relative to the observer.

dwell time Time required for a detector IFOV to sweep across a ground resolution cell.

earth-terrain camera (ETC) A high-resolution camera carried on Skylab.

EDC EROS Data Center.

edge enhancement Image-processing technique that emphasizes the appearance of edges and lines.

Ektachrome A Kodak color positive film.

electromagnetic radiation Energy propagated in the form of an advancing interaction between electric and magnetic fields. All electromagnetic radiation moves at the speed of light.

electromagnetic spectrum Continuous sequence of electromagnetic energy arranged according to wavelength or frequency.

emission Process by which a body radiates electromagnetic energy. Emission is determined by kinetic temperature and emissivity.

emissivity (ϵ) Ratio of radiant flux from a body to that from a blackbody at the same kinetic temperature.

emulsion Suspension of photosensitive silver halide grains in gelatin that constitutes the image-forming layer on photographic film.

energy flux Radiant flux.

enhancement Process of altering the appearance of an image so that the interpreter can extract more information.

EOSAT The commercial company that took over operations of the Landsat system in 1985.

EREP Earth Resources Experiment Package, carried on Skylab and consisting of cameras and a multispectral scanner.

EROS Earth Resource Observation System.

EROS Data Center (EDC) Facility of the U.S. Geological Survey at Sioux Falls, South Dakota, that archives, processes, and distributes images.

ERTS Earth Resource Technology Satellite, now called Landsat.

ETC Earth-terrain camera.

evaporative cooling Temperature drop caused by evaporation of water from a moist surface.

false color photograph Another term for *IR color photograph*.

far range The portion of a radar image farthest from the aircraft or spacecraft flight path.

film Light-sensitive photographic emulsion and its base.

film speed Measure of the sensitivity of photographic film to light. Larger numbers indicate higher sensitivity.

filter, digital Mathematical procedure for modifying values of numerical data.

filter, optical A material that, by absorption or reflection, selectively modifies the radiation transmitted through an optical system.

flight path Line on the ground directly beneath a remote sensing aircraft or spacecraft. Also called *flight line*.

fluorescence Emission of light from a substance following exposure to radiation from an external source.

f-number Representation of the speed of a lens determined by the focal length divided by diameter of the lens. Smaller numbers indicate faster lenses.

focal length In cameras, the distance from the optical center of the lens to the plane at which the image of a very distant object is brought into focus.

format Size of an image.

forward overlap The percent of duplication by successive photographs along a flight line.

frequency (ν) Number of wave oscillations per unit time or the number of wavelengths that pass a point per unit time.

f-stop Focal length of a lens divided by the diameter of the len's adjustable diaphragm. Smaller numbers indicate larger openings, which admit more light to the film.

GCP Ground-control point.

Gemini U.S. program of two-man earth-oribiting spacecraft in 1965 and 1966.

geometric correction Image-processing procedure that corrects spatial distortions in an image.

geostationary Refers to satellites traveling at the angular velocity at which the earth rotates; as a result, they remain above the same point on earth at all times.

Geostationary Operational Environmental Satellite A NOAA satellite that acquires visible and thermal IR images for meteorologic purposes.

geothermal Refers to heat from sources within the earth.

Goddard Space Flight Center The NASA facility at Greenbelt, Maryland, that is also a Landsat ground receiving station.

GMT Greenwich mean time. This international 24-h system is used to designate the time at which Landsat images are acquired.

GOES Geostationary Operational Environmental Satellite.

gossan Surface occurrence of iron oxide formed by the weathering of metallic sulfide ore minerals.

granularity Graininess of developed photographic film that is determined by the texture of the silver grains.

gray scale A sequence of gray tones ranging from black to white.

ground-control point A geographic feature of known location that is recognizable on images and can be used to determine geometric corrections.

ground range On radar images, the distance from the ground track to an object.

ground-range image Radar image in which the scale in the range direction is constant.

ground receiving station Facility that records data transmitted by a satellite, such as Landsat.

ground resolution cell Area on the terrain that is covered by the IFOV of a detector.

ground swath Width of the strip of terrain that is imaged by a scanner system.

GSFC Goddard Space Flight Center.

harmonic Refers to waves in which the component frequencies are whole-number multiples of the fundamental frequency.

heat capacity (c) Ratio of heat absorbed or released by a material to the corresponding temperature rise or fall. Expressed in calories per gram per degree centigrade. Also called *thermal capacity*.

Heat Capacity Mapping Mission (HCMM) NASA satellite orbited in 1978 to record daytime and nighttime visible and thermal IR images of large areas.

highlights Areas of bright tone on an image.

hue In the IHS system, represents the dominant wavelength of a color.

IFOV Instantaneous field of view.

IHS Intensity, hue, and saturation system of colors.

image Pictorial representation of a scene recorded by a remote sensing system. Although *image* is a general term, it is commonly restricted to representations acquired by nonphotographic methods.

image swath See *ground swath*.

incidence angle In radar, the angle formed between an imaginary line normal to the surface and another connecting the antenna and the target.

incident energy Electromagnetic radiation impinging on a surface.

index of refraction (n) Ratio of the wavelength or velocity of electromagnetic radiation in a vacuum to that in a substance.

instantaneous field of view (IFOV) Solid angle through which a detector is sensitive to radiation. In a scanning system, the solid angle subtended by the detector when the scanning motion is stopped.

intensity In the IHS system, brightness ranging from black to white.

interactive processing Method of image processing in which the operator views preliminary results and can alter the instructions to the computer to achieve desired results.

interpretation The process in which a person extracts information from an image.

interpretation key Characteristic or combination of characteristics that enable an interpreter to identify an object on an image.

IR Infrared region of the electromagnetic spectrum that includes wavelengths from 0.7 μm to 1 mm.

IR color photograph Color photograph in which the red-imaging layer is sensitive to photographic IR wavelengths, the green-imaging layer is sensitive to red light, and the blue-imaging layer is sensitive to green light. Also known as *camouflage detection photographs* and *false-color photographs*.

ISO index Index of the International Standards Organization, designating film speed in photography. Higher values indicate higher sensitivity.

isotherm Contour line connecting points of equal temperature. Isotherm maps are used to portray surface-temperature patterns of water bodies.

Johnson Space Flight Center A NASA facility in Houston, Texas.

JPL Jet Propulsion Laboratory, a NASA facility at Pasadena, California, operated under contract by the California Institute of Technology.

Ka band Radar wavelength region from 0.8 to 1.1 cm.

kernel Two-dimensional array of digital numbers used in digital filtering.

kinetic energy The ability of a moving body to do work by virture of its motion. The molecular motion of matter is a form of kinetic energy.

kinetic temperature Internal temperature of an object determined by random molecular motion. Kinetic temperature is measured with a contact thermometer.

Kodachrome A Kodak color positive film.

LACIE Large Area Crop Inventory Experiment.

Landsat A series of unmanned earth-orbiting NASA satellites that acquire multispectral images in various visible and IR bands.

Laplacian filter A form of nondirectional digital filter.

large-format camera (LFC) An experiment first carried on the Space Shuttle in October 1984.

latent image Invisible image produced by the photochemical effect of light on silver halide grains in the emulsion of film. The latent image is not visible until after photographic development.

layover In radar images, the geometric displacement of the top of objects toward the near range relative to their base.

L band Radar wavelength region from 15 to 30 cm.

lens One or more pieces of glass or other transparent material shaped to form an image by refraction of light.

LFC Large-format camera.

light Electromagnetic radiation ranging from 0.4 to 0.7 μm in wavelength that is detectable by the human eye.

light meter Device for measuring the intensity of visible radiation and determining the appropriate exposure of photographic film in a camera.

lineament Linear topographic or tonal feature on the terrain and on images, maps, and photographs that may represent a zone of structural weakness.

linear Adjective that describes the straight line–like nature of features on the terrain or on images and photographs.

line-pair Pair of light and dark bars of equal widths. The number of such line-pairs aligned side by side that can be distinguished per unit distance expresses the resolving power of an imaging system.

lineation The one-dimensional alignment of internal components of a rock that cannot be depicted as an individual feature on a map.

look direction Direction in which pulses of microwave energy are transmitted by a radar system. The look direction is normal to the azimuth direction. Also called *range direction*.

low-sun-angle photograph Aerial photograph acquired in the morning, evening, or winter when the sun is at a low elevation above the horizon.

luminance Quantitative measure of the intensity of light from a source.

Mercury U.S. program of one-man, earth-orbiting spacecraft in 1962 and 1963.

microwave Region of the electromagnetic spectrum in the wavelength range of 0.1 to 30 cm.

minimum ground separation Minimum distance on the ground between two targets at which they can be resolved on an image.

minus-blue photographs Black-and-white photographs acquired using a filter that removes blue wavelengths to produce higher spatial resolution.

modular optoelectric multispectral scanner (MOMS) An along-track scanner carried on the Space Shuttle that recorded two bands of data.

modulate To vary the frequency, phase, or amplitude of electromagnetic waves.

modulation transfer function (MTF) A method of describing spatial resolution.

MOMS Modular optoelectric multispectral scanner.

mosaic Composite image or photograph made by piecing together individual images or photographs covering adjacent areas.

MSS Multispectral scanner system of Landsat that acquires images at four wavelength bands in the visible and reflected IR regions.

multiband camera System that simultaneously acquires photographs of the same scene at different wavelengths.

multispectral classification Identification of terrain categories by digital processing of data acquired by multispectral scanners.

multispectral scanner Scanner system that simultaneously acquires images of the same scene at different wavelengths.

nadir Point on the ground directly in line with the remote sensing system and the center of the earth.

NASA National Aeronautical and Space Administration.

near range Refers to the portion of a radar image closest to the aircraft or satellite flight path.

negative photograph Photograph on film or paper in which the relationship between bright and dark tones is the reverse of that of the features on the terrain.

NHAP National High Altitude Photography program of the U.S. Geological Survey.

NOAA National Oceanic and Atmospheric Administration.

noise Random or repetitive events that obscure or interfere with the desired information.

nondirectional filter Mathematical filter that treats all orientations of linear features equally.

nonsystematic distortion Geometric irregularities on images that are not constant and cannot be predicted from the characteristics of the imaging system.

normal color film Film in which the colors are essentially true representations of the colors of the terrain.

NSSDC National Space Science Data Center.

oblique photograph Photograph acquired with the camera intentionally directed at some angle between horizontal and vertical orientations.

OMS Orbital maneuvering system.

orbit Path of a satellite around a body such as the earth, under the influence of gravity.

overlap Extent to which adjacent images or photographs cover the same terrain, expressed as a percentage.

panchromatic film Black and white film that is sensitive to all visible wavelengths.

parallax Displacement of the position of a target in an image caused by a shift in the observation position.

parallel-polarized Describes a radar pulse in which the polarization of the return is the same as that of the transmission. Parallel-polarized images may be HH (horizontal transmit, horizontal return) or VV (vertical transmit, vertical return).

pass In digital filters, refers to the spatial frequency of data transmitted by the filter. High-pass filters transmit high-frequency data; low-pass filters transmit low-frequency data.

passive remote sensing Remote sensing of energy naturally reflected or radiated from the terrain.

path-and-row index System for locating Landsat MSS and TM images.

pattern Regular repetition of tonal variations on an image or photograph.

periodic line dropout Defect on Landsat MSS or TM images in which no data are recorded for every sixth or sixteenth scan line, causing a black line on the image.

periodic line striping Defect on Landsat MSS or TM images in which every sixth or sixteenth scan line is brighter or darker than the others. Caused by the sensitivity of one detector being higher or lower than the others.

photodetector Device for measuring energy in the visible-light band.

photogeology Mapping and interpretation of geologic features from aerial photographs.

photograph Representation of targets on film that results from the action of light on silver halide grains in the film's emulsion.

photographic IR Short-wavelength portion (0.7 to 0.9 μm) of the IR band that is detectable by IR color film or IR black-and-white film.

photographic UV Long-wavelength portion of the UV band (0.3 to 0.4 μm) that is transmitted through the atmosphere and is detectable by film.

photomosaic Mosaic composed of photographs.

photon Minimum discrete quantity of radiant energy.

picture element In a digitized image, the area on the ground represented by each digital number. Commonly contracted to *pixel*.

pitch Rotation of an aircraft about the horizontal axis normal to its longitudinal axis that causes a nose-up or nose-down attitude.

pixel Contraction of *picture element*.

polarization The direction of orientation in which the electrical field vector of electromagnetic radiation vibrates.

positive photograph Photographic image in which the tones are directly proportional to the terrain brightness.

previsual symptom A vegetation anomaly that is recognizable on IR film before it is visible to the naked eye or on normal color photographs. It results when stressed vegetation loses its ability to reflect photographic IR energy and is recognizable on IR color film by a decrease in brightness of the red hues.

primary colors A set of three colors that in various combinations will produce the full range of colors in the visible spectrum. There are two sets of primary colors, additive and subtractive.

principal-component (PC) image Digitally processed image produced by a transformation that recognizes maximum variance in multispectral images.

principal point Optical center of an aerial photograph.

printout Display of computer data in alphanumeric format.

pulse Short burst of electromagnetic radiation transmitted by a radar antenna.

pulse length Duration of a burst of energy transmitted by a radar antenna, measured in microseconds.

pushbroom scanner An alternate term for an along-track scanner.

radar Acronym for *radio detection and ranging*. Radar is an active form of remote sensing that operates in the microwave and radio wavelength regions.

radar shadow Dark signature on a radar image representing no signal return. A shadow extends in the far-range direction from an object that intercepts the radar beam.

radian Angle subtended by an arc of a circle equal in length to the radius of the circle 1 rad = 57.3°.

radiant energy peak (λ_{max}) Wavelength at which the maximum electromagnetic energy is radiated at a particular temperature.

radiant flux Rate of flow of electromagnetic radiation measured in watts per square centimeter.

radiant temperature Concentration of the radiant flux from a material. Radiant temperature is the kinetic temperature multiplied by the emissivity to the one-fourth power.

radiation Propagation of energy in the form of electromagnetic waves.

radiometer Device for quantitatively measuring radiant energy, especially thermal radiation.

random line dropout In scanner images, the loss of data from individual scan lines in a nonsystematic fashion.

range direction See *look direction*.

range resolution In radar images, the spatial resolution in the range direction, which is determined by the pulse length of the transmitted microwave energy.

raster pattern Pattern of horizontal lines swept by an electron beam across the face of a CRT that constitute the image display.

ratio image An image prepared by processing digital multispectral data as follows: for each pixel, the value for one band is divided by that of another. The resulting digital values are displayed as an image.

Rayleigh criterion In radar, the relationship between surface roughness, depression angle, and wavelength that determines whether a surface will respond in a rough or smooth fashion to the radar pulse.

RBV Return-beam vidicon.

real-aperture radar Radar system in which azimuth resolution is determined by the transmitted beam width, which is in turn determined by the physical length of the antenna and by the wavelength.

real time Refers to images or data made available for inspection simultaneously with their acquisition.

recognizability Ability to identify an object on an image.

rectilinear Refers to images with no geometric distortion in which the scales in the horizontal and vertical directions are identical.

reflectance Ratio of the radiant energy reflected by a body to the energy incident on it. Spectral reflectance is the reflectance measured within a specific wavelength interval.

reflected energy peak Wavelength (0.5 μm) at which maximum amount of energy is reflected from the earth's surface.

reflected IR Electromagnetic energy of wavelengths from 0.7 μm to about 3 μm that consists primarily of reflected solar radiation.

reflectivity Ability of a surface to reflect incident energy.

refraction Bending of electromagnetic rays as they pass from one medium into another when each medium has a different index of refraction.

registration Process of superposing two or more images or photographs so that equivalent geographic points coincide.

relief Vertical irregularities of a surface.

relief displacement Geometric distortion on vertical aerial photographs. The tops of objects appear in the photograph to be radially displaced from their bases outward from the photograph's centerpoint.

remote sensing Collection and interpretation of information about an object without being in physical contact with the object.

reseau marks Pattern of small crosses added to photographs.

resolution Ability to separate closely spaced objects on an image or photograph. Resolution is commonly expressed as the most closely spaced line-pairs per unit distance that can be distinguished. Also called spatial resolution.

resolution target Series of regularly spaced alternating light and dark bars used to evaluate the resolution of images or photographs.

resolving power A measure of the ability of individual components, and of remote sensing systems, to separate closely spaced targets.

reststrahlen band In the thermal IR region, refers to absorption of energy as a function of silica content.

return In radar, a pulse of microwave energy reflected by the terrain and received at the radar antenna. The strength of a return is referred to as *return intensity*.

return-beam vidicon (RBV) A system in which images are formed on the photosensitive surface of a vacuum tube; the image is scanned with an electron beam and transmitted or recorded. Landsat 3 used a pair of RBVs to acquire images.

roll Rotation of an aircraft that causes a wing-up or wing-down attitude.

roll compensation system Component of an airborne scanner system that measures and records the roll of the aircraft. This information is used to correct the imagery for distortion due to roll.

rough criterion In radar, the relationship between surface roughness, depression angle, and wavelength that determines whether a surface will scatter the incident radar pulse in a rough or intermediate fashion.

roughness In radar, the average vertical relief of small-scale irregularities of the terrain surface. Also called *surface roughness*.

satellite An object in orbit around a celestial body.

saturation In the IHS system, represents the purity of color. Saturation is also the condition where energy flux exceeds the sensitivity range of a detector.

scale Ratio of distance on an image to the equivalent distance on the ground.

scan line Narrow strip on the ground that is swept by IFOV of a detector in a scanning system.

scanner An imaging system in which the IFOV of one or more detectors is swept across the terrain.

scanner distortion Geometric distortion that is characteristic of cross-track scanner images.

scan skew Distortion of scanner images caused by forward motion of the aircraft or satellite during the time required to complete a scan.

scattering Multiple reflections of electromagnetic waves by particles or surfaces.

scattering coefficient curves Display of scatterometer data in which relative backscatter is shown as a function of incidence angle.

scatterometer Nonimaging radar device that quantitatively records backscatter of terrain as a function of incidence angle.

scene Area on the ground that is covered by an image or photograph.

Seasat NASA unmanned satellite that acquired L-band radar images in 1978.

sensitivity Degree to which a detector responds to electromagnetic energy incident on it.

sensor Device that receives electromagnetic radiation and converts it into a signal that can be recorded and displayed as either numerical data or an image.

Shuttle imaging radar (SIR) L-band radar system deployed on the Space Shuttle.

sidelap Extent of lateral overlap between images acquired on adjacent flight lines.

side-looking airborne radar (SLAR) An airborne side scanning system for acquiring radar images.

side-scanning sonar Active system for acquiring images of the seafloor using pulsed sound waves.

side scanning system A system that acquires images of a strip of terrain parallel with the flight or orbit path but offset to one side.

signal Information recorded by a remote sensing system.

signature Set of characteristics by which a material or an object may be identified on an image or photograph.

silver halide Silver salts that are especially sensitive to visible light and convert to metallic silver when developed.

Skylab U.S. earth-orbiting workshop that housed three crews of three astronauts in 1973 and 1974.

skylight Component of light that is strongly scattered by the atmosphere and consists predominantly of shorter wavelengths.

slant range In radar, an imaginary line running between the antenna and the target.

slant-range distance Distance measured along the slant range.

slant-range distortion Geometric distortion of a slant-range image.

slant-range image In radar, an image in which objects are located at positions corresponding to their slant-range distances from the aircraft flight path. On slant-range images, the scale in the range direction is compressed in the near-range region.

SLAR Side-looking airborne radar.

smooth criterion In radar, the relationship between surface roughness, depression angle, and wavelength that determines whether a surface will scatter the incident radar pulse in a smooth or intermediate fashion.

software Programs that control computer operations.

sonar Acronym for *sound navigation ranging*. Sonar is an active form of remote sensing that employs sonic energy to image the seafloor.

Space Shuttle U.S. manned satellite program in the 1980s, officially called the Space Transportation System (STS).

spectral reflectance Reflectance of electromagnetic energy at specified wavelength intervals.

spectral sensitivity Response, or sensitivity, of a film or detector to radiation in different spectral regions.

spectral vegetation index An index of relative amount and vigor of vegetation. The index is calculated from two spectral bands of AVHRR imagery.

spectrometer Device for measuring intensity of radiation absorbed or reflected by a material as a function of wavelength.

spectrum Continuous sequence of electromagnetic energy arranged according to wavelength or frequency.

specular Refers to a surface that is smooth with respect to the wavelength of incident energy.

SPOT Système Probatoire d'Observation de la Terre. Unmanned French remote sensing satellite orbiting in the late 1980s.

Stefan-Boltzmann constant 5.68×10^{-12} W $\cdot$ cm$^{-2} \cdot$ °K^{-4}.

Stefan-Boltzmann law States that radiant flux of a blackbody is equal to the temperature to the fourth power times the Stefan-Boltzmann constant.

stereo base Distance between a pair of correlative points on a stereo pair that are oriented for stereo viewing.

stereo model Three-dimensional visual impression produced by viewing a pair of overlapping images through a stereoscope.

stereo pair Two overlapping images or photographs that may be viewed stereoscopically.

stereoscope Binocular optical device for viewing overlapping images or diagrams. The left eye sees only the left image, and the right eye sees only the right image.

subscene A portion of an image that is used for detailed analysis.

subtractive primary colors Yellow, magenta, and cyan. When used as filters for white light, these colors remove blue, green, and red light, respectively.

sunglint Bright reflectance of sunlight caused by ripples on water.

sun-synchronous Earth satellite orbit in which the orbit plane is nearly polar and the altitude is such that the satellite passes over all places on earth having the same latitude twice daily at the same local sun time.

supervised classification Digital-information extraction technique in which the operator provides training-site information that the computer uses to assign pixels to categories.

surface phenomenon Interaction between electromagnetic radiation and the surface of a material.

surface roughness See *roughness*.

synthetic-aperture radar (SAR) Radar system in which high azimuth resolution is achieved by storing and processing data on the Doppler shift of multiple return pulses in such a way as to give the effect of a much longer antenna.

synthetic stereo images Stereo images constructed through digital processing of a single image. Topographic data are used to calculate parallax.

system Combination of components that constitute an imaging device.

systematic distortion Geometric irregularities on images that are caused by known and predictable characteristics.

target Object on the terrain of specific interest in a remote sensing investigation.

TDRS Tracking and Data Relay Satellite.

telemeter To transmit data by radio or microwave links.

terrain Surface of the earth.

texture Frequency of change and arrangement of tones on an image.

thematic mapper (TM) A cross-track scanner deployed on Landsat that records seven bands of data from the visible through the thermal IR regions.

thermal capacity (c) See *heat capacity*.

thermal conductivity (K) Measure of the rate at which heat will pass through a material, expressed in calories per centimeter per second per degree Centigrade.

thermal crossover On a plot of radiant temperature versus time, the point at which temperature curves for two different materials intersect.

thermal diffusivity (k) Governs the rate at which temperature changes within a substance, expressed in centimeters squared per second.

thermal inertia (P) Measure of the response of a material to temperature changes, expressed in calories per square centimeter per square root of second.

thermal IR IR region from 3 to 14 μm that is employed in remote sensing. This spectral region spans the radiant power peak of the earth.

thermal IR image Image acquired by a scanner that records radiation within the thermal IR band.

thermal IR multispectral scanner (TIMS) Airborne scanner that acquires multispectral images within the 8-to-14-μm band of the thermal IR region.

thermal model Mathematical expression that relates thermal and other physical properties of a material to its temperature. Models may be used to predict temperature for given properties and conditions.

thermography Medical applications of thermal IR images. Images of the body, called *thermograms,* have been used to detect tumors and monitor blood circulation.

TIMS Thermal IR multispectral scanner.

TM Thematic mapper.

tone Each distinguishable shade of gray from white to black on an image.

topographic inversion An optical illusion that may occur on images with extensive shadows. Ridges appear to be valleys, and valleys appear to be ridges. The illusion is corrected by orienting the image so that the shadows trend from the top margin of the image to the bottom.

topographic reversal A geomorphic phenomenon in which topographic lows coincide with structural highs and vice versa. Valleys are eroded on crests of anticlines to cause topographic lows, and synclines form ridges, or topographic highs.

Tracking and Data Relay Satellite (TDRS) Geostationary satellite used to communicate between ground receiving stations and satellites such as Landsat.

trade-off As a result of changing one factor in a remote sensing system, there are compensating changes elsewhere in the system; such a compensating change is known as a trade-off.

training site Area of terrain with known properties or characteristics that is used in supervised classification.

transmissivity Property of a material that determines the amount of energy that can pass through the material.

transparency Image on a transparent photographic material, normally a positive image.

transpiration Expulsion of water vapor and oxygen by vegetation.

travel time In radar, the time interval between the generation of a pulse of microwave energy and its return from the terrain.

unsupervised classification Digital information extraction technique in which the computer assigns pixels to categories with no instructions from the operator.

UV Ultraviolet region of the electromagnetic spectrum ranging in wavelengths from 0.01 to 0.4 μm.

vegetation anomaly Deviation from the normal distribution or properties of vegetation. Vegetation anomalies may be caused by faults, trace elements in soil, or other factors.

vertical exaggeration In a stereo model, the extent to which the vertical scale appears larger than the horizontal scale.

visible radiation Energy at wavelengths from 0.4 to 0.7 μm that is detectable by the human eye.

volume scattering In radar, interaction between electromagnetic radiation and the interior of a material.

watt (W) Unit of electrical power equal to rate of work done by one ampere under a potential of one volt.

wavelength (λ) Distance between successive wave crests or other equivalent points in a harmonic wave.

Wien's displacement law Describes the shift of the radiant power peak to shorter wavelengths as temperature increases.

X band Radar wavelength region from 2.4 to 3.8 cm.

yaw Rotation of an aircraft about its vertical axis so that the longitudinal axis deviates left or right from the flight line.

Locality Index

Hillside fault, 59
Houston, 385
  northwest, 224
  Texas lineament, 61
  Van Horn, 59, 61
Transverse Range province, 81

Utah, 64
  Buckhorn Plateau, 165
  Kaibab Plateau, 62
  Kaiparowits Plateau, 62
  Lisbon Valley uranium district,
    295
  San Rafael Desert, 165

San Rafael swell, 163
Wasatch Plateau, 165

Venezuela, 211, 214

Washington
  Mount Baker, 394
  Mount Rainier, 394
  Mount St. Helens, 388–393,
    426
Washington, D.C., 49, 50
West Virginia, 168
  Lost River gas field, 309
Wyoming, 90, 259, 260

Alkali anticline, 35, 36, 37, 427
Big Horn Basin, 35, 37, 94, 95
Copper Mountain uranium district,
  295
Owl Creek, 94
Owl Creek Mountains, 91
Patrick Draw oil field, 309
Thermopolis, 94, 96, 238, 252, 262,
  263, 269, 428
Wind River, 91, 238
Wind River Mountains, 91
Yellowstone Park, 313, 426

Zaire, 298

Subject Index
(Includes facilities)

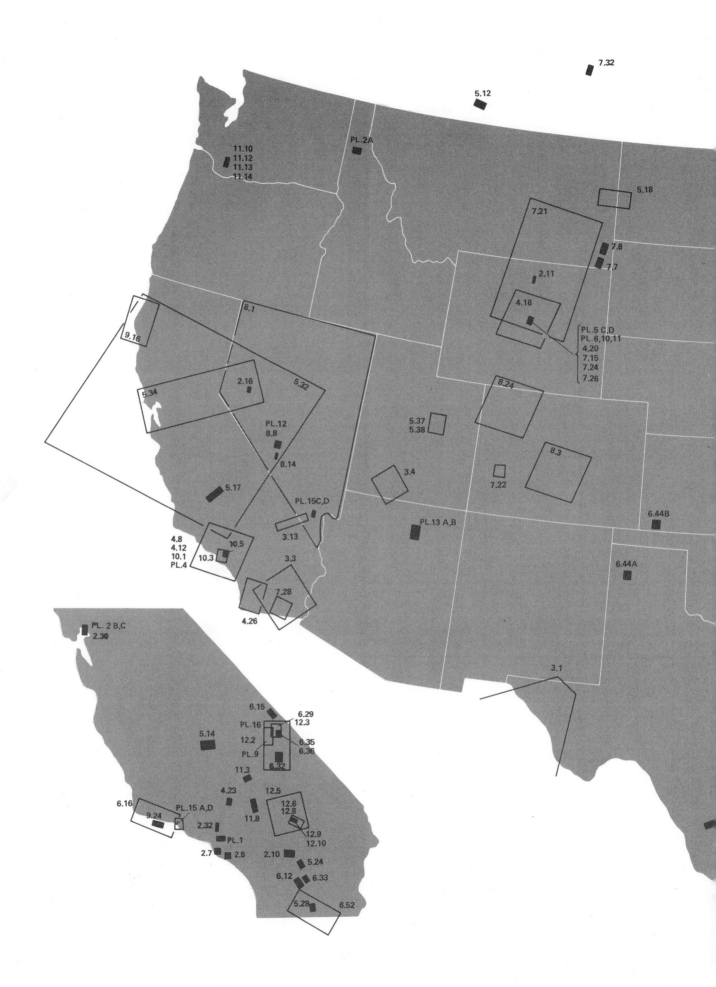